T0281013

# Lecture Notes in Computer Science 14508

The series Lecture Notes in Computer Science (LNCS), including its subseries Lecture Notes in Artificial Intelligence (LNAI) and Lecture Notes in Bioinformatics (LNBI), has established itself as a medium for the publication of new developments in computer science and information technology research, teaching, and education.

LNCS enjoys close cooperation with the computer science R & D community, the series counts many renowned academics among its volume editors and paper authors, and collaborates with prestigious societies. Its mission is to serve this international community by providing an invaluable service, mainly focused on the publication of conference and workshop proceedings and postproceedings. LNCS commenced publication in 1973.

Subrahmanyam Kalyanasundaram ·
Anil Maheshwari
Editors

# Algorithms and Discrete Applied Mathematics

10th International Conference, CALDAM 2024
Bhilai, India, February 15–17, 2024
Proceedings

 Springer

*Editors*
Subrahmanyam Kalyanasundaram ⓘ
Indian Institute of Technology Hyderabad
Sangareddy, India

Anil Maheshwari ⓘ
Carleton University
Ottawa, ON, Canada

ISSN 0302-9743      ISSN 1611-3349 (electronic)
Lecture Notes in Computer Science
ISBN 978-3-031-52212-3      ISBN 978-3-031-52213-0 (eBook)
https://doi.org/10.1007/978-3-031-52213-0

This Springer imprint is published by the registered company Springer Nature Switzerland AG
The registered company address is: Gewerbestrasse 11, 6330 Cham, Switzerland

Paper in this product is recyclable.

# Preface

This volume contains the papers presented at CALDAM 2024 (the 10th International Conference on Algorithms and Discrete Applied Mathematics) held on February 15–17, 2024, at the Indian Institute of Technology Bhilai (IIT Bhilai), Chhattisgarh, India. CALDAM 2024 was organised by IIT Bhilai and the Association for Computer Science and Discrete Mathematics (ACSDM), India. The program committee consisted of 32 highly experienced and active researchers from various countries.

The conference had papers in the areas of algorithms and complexity, discrete applied mathematics, computational geometry, graph theory, graph coloring, graph partitioning, and domination in graphs. We received 57 submissions from authors from all over the world. Program committee members and other expert reviewers extensively reviewed each paper. Single-blind review was used, where each paper received 3 reviews. The committee decided to accept 22 papers for presentation. The program included three invited talks by Bhawani Sankar Panda (Indian Institute of Technology Delhi), Iztok Peterin (University of Maribor), and Saket Saurabh (Institute of Mathematical Sciences, Chennai).

As volume editors, we thank the authors of all submissions for considering CALDAM 2024 for the potential presentation of their works. We are very much indebted to the program committee members and the external reviewers for providing serious reviews within a very short period of time. Our sincerest thanks to the invited speakers, Bhawani Sankar Panda, Iztok Peterin, and Saket Saurabh, for accepting our invitation to give a talk. We thank the organizing committee chaired by Rishi Ranjan Singh of Indian Institute of Technology Bhilai for the smooth conduct of CALDAM 2024 and the Indian Institute of Technology Bhilai for providing the necessary facilities. We are very grateful to the chair of the steering committee, Subir Ghosh, for his active help, support, and guidance. We thank the PC chairs of the previous edition, Amitabha Bagchi and Rahul Muthu, for assisting with various process details. We thank Springer for publishing the proceedings in the Lecture Notes in Computer Science series and for supporting the best paper presentation awards. We thank the EasyChair conference management system, which effectively handled the entire process.

February 2024

Subrahmanyam Kalyanasundaram
Anil Maheshwari

# Organization

## Steering Committee

| | |
|---|---|
| Subir Kumar Ghosh (Chair) | Ramakrishna Mission Vivekananda Educational and Research Institute, India |
| Gyula O. H. Katona | Hungarian Academy of Sciences, Hungary |
| János Pach | École Polytechnique Fédérale De Lausanne, Switzerland |
| Nicola Santoro | Carleton University, Canada |
| Swami Sarvottamananda | Ramakrishna Mission Vivekananda Educational and Research Institute, India |
| Chee Yap | New York University, USA |

## Program Committee

| | |
|---|---|
| Amitabha Bagchi | IIT Delhi, India |
| Niranjan Balachandran | IIT Bombay, India |
| Sayan Bandyapadhyay | Portland State University, USA |
| Aritra Banik | NISER Bhubaneswar, India |
| Ahmad Biniaz | University of Windsor, Canada |
| Sergio Cabello | University of Ljubljana, Slovenia |
| Manoj Changat | University of Kerala, India |
| Keerti Choudhary | IIT Delhi, India |
| Sandip Das | ISI Kolkata, India |
| Mark de Berg | Eindhoven University of Technology, The Netherlands |
| Jean-Lou De Carufel | University of Ottawa, Canada |
| Florent Foucaud | Université Clermont Auvergne, France |
| Sumit Ganguly | IIT Kanpur, India |
| Daya Gaur | University of Lethbridge, Canada |
| Cyril Gavoille | University of Bordeaux, France |
| Sathish Govindarajan | IISc Bengaluru, India |
| Subrahmanyam Kalyanasundaram (Co-chair) | IIT Hyderabad, India |
| Sushanta Karmakar | IIT Guwahati, India |
| Matthew Katz | Ben-Gurion University of the Negev, Israel |
| Anil Maheshwari (Co-chair) | Carleton University, Canada |

| | |
|---|---|
| Rogers Mathew | IIT Hyderabad, India |
| Neeldhara Misra | IIT Gandhinagar, India |
| Apurva Mudgal | IIT Ropar, India |
| Wolfgang Mulzer | Freie Universität Berlin, Germany |
| Rahul Muthu | DA-IICT, India |
| Arti Pandey | IIT Ropar, India |
| Deepak Rajendraprasad | IIT Palakkad, India |
| Vlady Ravelomanana | Université Paris Cité, France |
| Bodhayan Roy | IIT Kharagpur, India |
| Sagnik Sen | IIT Dharwad, India |
| Uéverton Souza | Fluminense Federal University, Brazil |
| C. R. Subramanian | IMSc Chennai, India |

## Organizing Committee

| | |
|---|---|
| Amit Kumar Dhar | IIT Bhilai, India |
| Barun Gorain | IIT Bhilai, India |
| Vinod Reddy | IIT Bhilai, India |
| Anurag Singh | IIT Bhilai, India |
| Rishi Ranjan Singh (Chair) | IIT Bhilai, India |

## Additional Reviewers

Arun Anil
Dhanyamol Antony
Sina Bagheri Nezhad
Susobhan Bandopadhyay
Suman Banerjee
Sriram Bhyravarapu
Jan Bok
Dibyayan Chakraborty
Dipayan Chakraborty
Supraja D. K.
Hill Darryl
Tapas Das
Hiranya Dey
J. Geetha
Shiwali Gupta
Mohammad Hashemi
Mahya Jamshidian
Sangram Kishor Jena

Lijo Jose
Anjeneya Swami Kare
Atílio G. Luiz
Raghunath Reddy Madireddy
Kutty Malu V. K.
Bobby Miraftab
Suchismita Mishra
Tapas Kumar Mishra
Rajat Mittal
Ramin Mousavi
Karthik Murali
Aravind N. R.
Satyadev Nandakumar
Meghana Nasre
Mauro Nigro
Saeed Odak
Nacim Oijid
Fabiano Oliveira

Subhabrata Paul
Marcille Pierre-Marie
Veena Prabhakaran
Nidhi Purohit
Shruti Rastogi
Arpan Sadhukhan
Lakshay Saggi
Brahadeesh Sankarnarayanan

Caroline A. P. Silva
Anuj Tawari
Raghunath Tewari
Vikash Tripathi
Seshikanth Varma
S. Venkitesh
Shaily Verma
Vijaykumar Visavaliya

# Abstracts of Invited Talks

# Total Coloring Conjecture and Classification Problem

B. S. Panda

Department of Mathematics, Indian Institute of Technology Delhi,
New Delhi 110016, India
bspanda@maths.iitd.ac.in

A graph $G = (V, E)$ is said to have a total coloring if the elements of $V(G) \cup E(G)$ are colored so that: adjacent vertices receive different colors; adjacent edges receive different colors; and if edge $e$ is incident to vertex $v$ then $e$ and $v$ receive different colors. The total chromatic number $\chi''(G)$ is the least number of colors needed to totally color $G$. The Total Coloring Conjecture (see [1,4]) which claims that for any graph $G$, $\Delta(G) + 1 \leq \chi''(G) \leq \Delta(G) + 2$, where $\Delta(G)$ is the maximum degree of $G$ is one of the important open problems that has driven the research in total coloring of graphs. This conjecture has been proved to be true for various graph classes but it is still open. A natural and well-studied problem for the graphs for which the total coloring conjecture hold is the classification of graphs according to their total chromatic numbers. If $\chi''(G) = \Delta(G) + 1$, then $G$ is said to be of Type 1 and if $\chi''(G) = \Delta(G) + 2$, $G$ is said to be of Type 2. The classification problem is open for various classes of graphs for which the total coloring conjecture has been verified to hold true. One such class is complete multipartite graphs. The complete $p$-partite graph $K = K[V_1, \ldots, V_p]$ is the simple graph with vertex set $V(K) = \cup_{i=1}^{p} V_i$ (each set $V_i$ is called a part), where $V_i \cap V_j = \emptyset$ for $i \neq j$, in which two vertices are adjacent if and only if they belong to different parts of $K$. If the names of the vertex sets are unimportant then $K$ is simply referred to as $K(r_1, \ldots, r_p)$, where $|V_i| = r_i$ for $1 \leq i \leq p$.

The graph $K$ is of sufficient complexity that settling the values of its graph parameters is often a challenge. Finding the chromatic index $\chi'(K)$ is a typical example. Of course, the classic result of Vizing [4] shows that $\chi'(G)$ is $\Delta(G)$ or $\Delta(G) + 1$, thereby giving rise to the classification of whether a graph is Class 1 or Class 2, respectively. It was finally shown in 1992 that $K$ is a Class 2 graph if and only if it is overfull [5] (A graph $G = (V, E)$ is overfull if $|E| > \Delta(G)\lfloor \frac{n}{2} \rfloor$). Bermond settled the type of $K$ when it is regular [2]. Yap [7] proved that $\chi''(K) \leq \Delta + 2$, and Chew and Yap [3] showed that if $K$ has an odd number of vertices then it is of Type 1. It has been conjectured [6] that the total chromatic number $\chi''(K)$ of the complete $p$-partite graph $K = K(r_1, \ldots, r_p)$ is $\Delta(K) + 1$ if and only if $K \neq K_{r,r}$ and if $K$ has an even number of vertices then $def(K) = \Sigma_{v \in V(K)}(\Delta(K) - d_K(v))$ is at least the number of parts of odd size. This conjecture is still open.

In this talk, we will discuss the progress made towards resolving the total coloring conjecture and the classification problem for some graph classes for which the total coloring conjecture holds true.

# References

1. Behzad, M.: Graphs and their chromatic numbers, Doctoral Thesis, Michigan State University (1965)
2. Bermond, J.-C.: Nombre chromatique total du graphe r-parti complete (French). J. London Math. Soc. **9**, 279–285 (1974)
3. Chew, K.H., Yap, H.P.: Total chromatic number of complete r-partite graphs. J. Graph Theory **16**, 629–634 (1992)
4. Vizing, V.G.: Some unsolved problems in graph theory. (Russian) Uspehi Mat. Nauk **23**, 117–134 (1968)
5. Hoffman, D.G., Rodger, C.A.: The chromatic index of complete multipartite graphs. J. Graph Theory **16**, 159–163 (1992)
6. Hoffman, D.G., Rodger, C.A.: The total chromatic number of complete multipartite graphs. Festschrift for C. St. J. A. Nash-Williams. Congr. Numer. **113**, 205–220 (1996)
7. Yap, H.P.: Total colourings of graphs. Bull. London Math. Soc. **21**, 159–163 (1989)

# Two Heuristic Approaches for Some Special Colorings of Graphs

Iztok Peterin

Faculty of Electrical Engineering and Computer Science,
University of Maribor, Maribor, Slovenia
iztok.peterin@um.si

Let $G$ be a graph. A mapping $c : V(G) \to \{1, \ldots, k\}$ is a (proper) coloring if $c(u) \neq c(v)$ for every $uv \in E(G)$. A vertex with all the colors in its closed neighborhood is called a b-vertex. Set $V_i = \{v \in V(G) : c(v) = i\}$ is the $i$-th color class for every $i \in \{1, \ldots, k\}$. The chromatic number $\chi(G)$ of $G$ is the minimum number of colors in a proper coloring. It is well known that determining $\chi(G)$ is an NP-hard problem which yields a space for heuristic approach to get an approximate value for $\chi(G)$. Probably two most known approaches are via greedy coloring and b-coloring. For a greedy coloring, one assigns to an uncolored vertex the minimum color not present in its open neighborhood. On the other hand we try to reduce the number of colors in a given coloring by local re-colorings of all vertices of one color to get a b-coloring. This is possible whenever there exists a color class without a b-vertex. By the use of heuristics is desired to know the worst case scenario that can happen. Both mentioned methods are well studied from this perspective: the Grundy chromatic number $\Gamma(G)$ for the greedy approach and b-chromatic number $\chi_b(G)$ for b-colorings.

A special coloring is a coloring with some additional conditions. There are numerous special colorings and let us mentioned three of them. A coloring is acyclic if any two color classes $V_i \cup V_j$ induces a forest, i.e. there are no cycles with two colors. The minimum number of colors in an acyclic coloring is acyclic chromatic number $A(G)$ of $G$. A star coloring is a coloring where the union of any two color classes $V_i \cup V_j$ induces a family of stars. The minimum number of colors in a star coloring is the star chromatic number $S(G)$ of $G$. A packing coloring is a $k$-coloring where $d(u, v) > i$ for any two different vertices $u, v \in V_i$ for any $i \in \{1, \ldots, k\}$. The packing chromatic number $\chi_\rho(G)$ is the minimum number of colors in a packing coloring of $G$.

It seems that not many heuristic approaches for special colorings are known in the literature. Therefore we started a wide project to introduce some of them based on greedy and b-coloring approach. We introduce acyclic b-chromatic number $A_b(G)$ [1,2], Grundy acyclic chromatic number $\Gamma_a(G)$ [5], star b-chromatic number $S_b(G)$ [3] and Grundy packing chromatic number $\Gamma_p(G)$ [4]. We present several properties for them, derive diverse bounds, give some exact results and pose several questions about them.

# References

1. Anholcer, M., Cichacz, S., Peterin, I.: On b-acyclic chromatic number of a graph, Comput. Appl. Math. **42**(21), 20 (2023)
2. Anholcer, M., Cichacz, S., Peterin, I.: On acyclic b-chromatic number of cubic graphs, in preparation
3. Božović, D., Peterin, I., Štesl, D.: On star b-chromatic number of a graph, in preparation
4. Gözüpek, D., Peterin, I.: Grundy packing chromatic number of a graph, in preparation
5. Pawlik, B., Peterin, I.: On Grundy acyclic chromatic number of a graph, in preparation

# Random Deselection

Saket Saurabh

Institute of Mathematical Sciences, HBNI, Chennai 600113, India
saket@imsc.res.in

In most randomized algorithms, we select an object that belongs to the solution. In this talk, we will survey some recent algorithms in which we will select an object that does not belong to the solution, which we will call the *method of deselection*, and exploit it to design good algorithms. In particular, we will also discuss a recent 2-approximation algorithm for Feedback Vertex Set in Tournaments that is based on picking a vertex at random and declaring it to not be part of the solution. In the second part, using the deselection methodology, we will give a framework to design exponential-time approximation schemes for basic graph partitioning problems such as $k$-WAY CUT, MULTIWAY CUT, STEINER $k$-CUT and MULTICUT, where the goal is to minimize the number of *edges* going across the parts.

# Contents

## Algorithms and Complexity

Consecutive Occurrences with Distance Constraints ........................ 3
  *Waseem Akram and Sanjeev Saxena*

Parameterized Aspects of Distinct Kemeny Rank Aggregation ............... 14
  *Koustav De, Harshil Mittal, Palash Dey, and Neeldhara Misra*

Monitoring Edge-Geodetic Sets in Graphs: Extremal Graphs, Bounds,
Complexity ......................................................... 29
  *Florent Foucaud, Pierre-Marie Marcille, Zin Mar Myint,*
  *R. B. Sandeep, Sagnik Sen, and S. Taruni*

Distance-2-Dispersion with Termination by a Strong Team .................. 44
  *Barun Gorain, Tanvir Kaur, and Kaushik Mondal*

On Query Complexity Measures and Their Relations for Symmetric
Functions .......................................................... 59
  *Rajat Mittal, Sanjay S. Nair, and Sunayana Patro*

## Computational Geometry

Growth Rate of the Number of Empty Triangles in the Plane ................. 77
  *Bhaswar B. Bhattacharya, Sandip Das, Sk. Samim Islam, and Saumya Sen*

Geometric Covering Number: Covering Points with Curves .................. 88
  *Arijit Bishnu, Mathew Francis, and Pritam Majumder*

Improved Algorithms for Minimum-Membership Geometric Set Cover ........ 103
  *Sathish Govindarajan and Siddhartha Sarkar*

Semi-total Domination in Unit Disk Graphs ............................. 117
  *Sasmita Rout and Gautam Kumar Das*

## Discrete Applied Mathematics

An Efficient Interior Point Method for Linear Optimization Using
Modified Newton Method ............................................ 133
  *Sajad Fathi Hafshejani, Daya Gaur, and Robert Benkoczi*

Unique Least Common Ancestors and Clusters in Directed Acyclic Graphs . . . .    148
   *Ameera Vaheeda Shanavas, Manoj Changat, Marc Hellmuth,*
   *and Peter F. Stadler*

The Frobenius Problem for the Proth Numbers . . . . . . . . . . . . . . . . . . . . . . . . . . . .    162
   *Pranjal Srivastava and Dhara Thakkar*

## Graph Algorithms

Eternal Connected Vertex Cover Problem in Graphs: Complexity
and Algorithms . . . . . . . . . . . . . . . . . . . . . . . . . . . . . . . . . . . . . . . . . . . . . . . . . . . . .    179
   *Kaustav Paul and Arti Pandey*

Impact of Diameter and Convex Ordering for Hamiltonicity
and Domination . . . . . . . . . . . . . . . . . . . . . . . . . . . . . . . . . . . . . . . . . . . . . . . . . . . . .    194
   *R. Mahendra Kumar and N. Sadagopan*

On Star Partition of Split Graphs . . . . . . . . . . . . . . . . . . . . . . . . . . . . . . . . . . . . . . .    209
   *D. Divya and S. Vijayakumar*

Star Covers and Star Partitions of Cographs and Butterfly-free Graphs . . . . . . . .    224
   *Joyashree Mondal and S. Vijayakumar*

Open Packing in *H*-free Graphs and Subclasses of Split Graphs . . . . . . . . . . . . . .    239
   *M. A. Shalu and V. K. Kirubakaran*

## Graph Theory

Location-Domination Type Problems Under the Mycielski Construction . . . . . . .    255
   *Silvia M. Bianchi, Dipayan Chakraborty, Yanina Lucarini,*
   *and Annegret K. Wagler*

On Total Chromatic Number of Complete Multipartite Graphs . . . . . . . . . . . . . . .    270
   *Aseem Dalal and B. S. Panda*

The Weak-Toll Function of a Graph: Axiomatic Characterizations
and First-Order Non-definability . . . . . . . . . . . . . . . . . . . . . . . . . . . . . . . . . . . . . . .    286
   *Lekshmi Kamal K. Sheela, Manoj Changat, and Jeny Jacob*

Total Coloring of Some Graph Operations . . . . . . . . . . . . . . . . . . . . . . . . . . . . . . . .    302
   *T. Kavaskar and Sreelakshmi Sukumaran*

Star Colouring of Regular Graphs Meets Weaving and Line Graphs ........... 313
     *M. A. Shalu and Cyriac Antony*

**Author Index** ...................................................... 329

# Algorithms and Complexity

# Consecutive Occurrences with Distance Constraints

Waseem Akram[✉] and Sanjeev Saxena

Department of Computer Science and Engineering, Indian Institute of Technology,
Kanpur, Kanpur 208 016, India
{akram,ssax}@iitk.ac.in

**Abstract.** A consecutive occurrence of a pattern $P[1:m]$ in a text $T[1:n]$ is a pair $(i,j)$, with $i < j$, of indices in $T$ such that $P$ occurs at $i$ and $j$, but not at any index between them. We give deterministic solutions to the following two problems using simple and classical data structures.

The first problem is to preprocess the text $T$ so that one can efficiently answer *bounded gap queries*: "given a pattern $P$ and a range $[\alpha, \beta]$ such that $1 \leq \alpha \leq \beta \leq n$, report all the consecutive occurrences $(i,j)$ of the pattern $P$ in the text $T$ with distance $j - i \in [\alpha, \beta]$". We present an $O(n \log n)$-space data structure that supports bounded gap queries in $O(m + \log \alpha + \#output)$-time. The time needed to build the data structure is $O(n^2)$. Moreover, the query time can be improved to $O(m + \#output)$ if $\alpha$ or $\beta$ is known at the time of preprocessing.

The second problem is the string indexing for top-$k$ close consecutive occurrences problem, which asks to preprocess the input text $T[1:n]$ so that one can quickly answer *top-k queries*: "given an integer $k > 0$ and a pattern $P$, report the $k$ closest consecutive occurrences of $P$ in $T$". Using the same data structure mentioned above, we can answer a top-$k$ query in $O(m + \#output)$-time.

**Keywords:** Pattern Matching · Segment Intersection · Data Structures · Algorithms

## 1 Introduction

The pattern matching problem is a fundamental problem in text processing [1]. The problem asks to preprocess a given text $T[1:n]$ so that all the occurrences of a pattern $P[1:m]$ in the text $T$ can be reported efficiently [1,2]. An index $i$ of the string $T$ is said to be an occurrence of a string $P[1:m]$ if $P[j] = T[i+j-1]$ for all $1 \leq j \leq m$. A pair $(i,j), i < j$, of indices is said to be a consecutive occurrence of $P$ in $T$ if $i$ and $j$ are occurrences of $P$, and $P$ has no occurrence between $i$ and $j$. The distance of a consecutive occurrence $(i,j)$ is defined as $j - i$.

In this paper, we study two natural variants of the pattern matching problem. The first problem we study is the string indexing for top-$k$ close consecutive

S. Kalyanasundaram and A. Maheshwari (Eds.): CALDAM 2024, LNCS 14508, pp. 3–13, 2024.
https://doi.org/10.1007/978-3-031-52213-0_1

occurrence (called SITCCO in [3]). In this problem, we want to preprocess the text $T$ so that subsequent queries of the following type can answered efficiently.

**top-$k$ query:** given a pattern $P$ and an integer $k > 0$, report the $k$ consecutive occurrences of the pattern $P$ in the text $T$ that are closest.

Bille et al. [3] introduced the problem and solved it using a partially persistent data structure in $O(n \log n)$ space and $O(\log n + k)$ query time. We present a solution that uses simpler and classical data structures and achieves the same space and time bounds.

The other problem we study is to preprocess the text $T$ into an efficient data structure so that subsequent queries of the following type can be answered quickly.

**bounded-gap query:** given a pattern $P$ and two real numbers $\alpha$ and $\beta$ with $0 \leq \alpha \leq \beta$, report all the consecutive occurrences $(i, j)$ of pattern $P$ in text $T$ with distance $j - i \in [\alpha, \beta]$.

Navarro and Thankachan [4] introduced the problem and gave a solution with $O(n \log n)$ space and $O(\log n + \#output)$ query time. We give a solution that takes $O(n \log n)$ space and $O(m + \log \alpha + \#output)$ time. Our solution is fully deterministic and uses simpler data structures compared to Navarro and Thankachan's solution [4]. Moreover, if one of the endpoints ($\alpha$ or $\beta$) of query ranges is known at the time of indexing, then we can improve the query time bound to $O(m + \#output)$.

## 1.1   Previous and Related Works

Bille et al. [3] investigated the top-$k$ query problem and gave an $O(n \log n)$ space solution with optimal $O(m + \#output)$ query time. They transformed the problem into a geometric problem, namely, the orthogonal segment intersection problem, which they solved using a persistent linked list data structure. They also presented an $O(\frac{n}{\epsilon})$ space data structure that supports queries in $O(m + k^{1+\epsilon})$ time, where $\epsilon$ is a constant in $(0, 1]$.

Navarro and Thankachan [4] solved the bounded gap query problem and gave an $O(n \log n)$ space solution with optimal $O(m + \#output)$ query time. To deal with query patterns of length more than $\log \log n$, they transformed the problem into the orthogonal segment intersection problem, for which they employed a van Emde Boas tree-based data structure. For query patterns with length at most $\log \log n$, a 1-d range reporting data structure based on hash functions is used. Recently, Bille et al. [5] studied a more general problem where the goal is to preprocess $T$ so that, given query patterns $P_1$ and $P_2$ and a range $[\alpha, \beta]$, all the consecutive occurrences of pairs of $P_1$ and $P_2$ with distance in the range $[\alpha, \beta]$ can be quickly reported. They gave an $\tilde{O}(n)$ space solution with $\tilde{O}(|P_1| + |P_2| + n^{2/3}occ^{1/3})$ query time, where $occ$ is the output size. The $\tilde{O}$ notation hides the logarithmic factors.

The non-overlapping indexing is a related problem where the goal is to preprocess the text $T$ so that all non-overlapping occurrences of a query pattern $P$

can be found efficiently. Two occurrences of $P$ in the text $T$ are said to be non-overlapping if they are separated by at least $|P|$ characters. Cohen and Porat [6] gave an optimal $O(m + \#output)$ time solution that uses $O(n)$ word space. In their solution, the suffix tree for $T$ is augmented with $O(n)$ space structure. Ganguly et al. [7] provided a simpler, more space-efficient solution. They gave an algorithm that takes $O(m + \#output)$ time to answer a query using a suffix tree only; no augmentation of the suffix tree is required.

## 1.2 Preliminaries

The *suffix tree* is a classical data structure in string processing [1,8,9]. The suffix tree built for a given string stores all the suffixes of the string as their keys and positions in the string as their values. Each node, except the leaves, has at least two children. Each edge is labeled with a non-empty substring; no two edges starting out of a node can have labels beginning with the same character. The number of leaves in the tree equals the string length; each leaf corresponds to a unique suffix. Concatenating edge labels on a root-to-leaf path gives the corresponding suffix of the string. The suffix tree can be built in $O(n)$ time and space [1], where $n$ is the length (the number of characters) of the string. Many string operations can efficiently be implemented using the suffix tree data structure, e.g., substring check, finding the longest repeated substring, building suffix array, and computing the longest common substring [9].

The *suffix array* for a text $T[1 : n]$ is an array, denoted by $SA[1 : n]$, that stores the indices of lexicographically ordered suffixes of $T$. In particular, $SA[i]$ stores the index of the $i^{th}$ lexicographically smallest suffix of $T$.

A *heavy path* of a tree is a root-to-leaf path in which each node has a size no smaller than the size of any of its siblings (the size of a node is the number of nodes in the subtree rooted at that node). *Heavy path decomposition* is a process of decomposing a tree into heavy paths. First, we find a heavy path by starting from the tree's root and choosing a child with the maximum size at each level. We follow the same procedure recursively for each subtree rooted at a node that is not on the heavy path, but its parent node is (on the heavy path). As a result, a collection of (disjoint) heavy paths is obtained [10]. Some of its properties are:

1. each node $v$ belongs to exactly one heavy path.
2. any path from the root to a leaf can pass through at most $\log n$ heavy paths.
3. The number of heavy paths equals the number of leaves in the tree.

The second property follows from the following result due to Sleator and Tarjan [10].

**Lemma 1.** *The number of heavy paths intersected by any root-to-leaf path is at most $\log n$, where $n$ is the number of leaves in the tree.*

The *orthogonal segment intersection problem* asks to preprocess a given set of horizontal segments such that whenever a vertical segment comes as a query, we can efficiently report all the horizontal segments intersected by the vertical segment. Chazelle [11] gave a linear space data structure that can answer a query in $O(\log N + \#output)$, where $N$ is the number of horizontal segments. In the preprocessing phase, an orthogonal subdivision is built using the endpoints of the horizontal segments, and then it is processed for point location. Given a vertical segment, the face (or region) containing the lower endpoint of the query segment is computed using a point location query. Then, the output segments are reported one by one in the non-decreasing order of their $y$-coordinates while moving toward the upper endpoint along the vertical segment. Thus, the data structure can be used to report the $k$ horizontal segments with the smallest $y$-coordinates that are intersected by the query segment in time $O(\log N + k)$ (for details, see Sect. 4 in [11]).

### 1.3 Notations

Let $T[1:n]$ and $P[1:m]$ be the text and pattern strings respectively. We denote by $T[i]$ the $i^{th}$ character from the start of $T[1:n]$. For any two indices $i$ and $j$ with $i \leq j$, the contiguous subsequence of characters $T[i], T[i+1], ..., T[j]$ is called a substring, denoted by $T[i:j]$. A substring of the form $T[1:i]$ is called a prefix, and a substring of the form $T[i:n]$ is called a suffix of the string $T$. We denote by $|S|$ the length (the number of characters) of a string $S$.

Let $\mathcal{T}$ denote the suffix tree built for the text $T[1:n]$ with the property that the children of each non-leaf node are sorted by the first characters of their edge labels. Note that $i^{th}$ leftmost leaf in $\mathcal{T}$, denoted by $l_i$, corresponds to the $i^{th}$ lexicographically smallest suffix of text $T$. We denote by $path(u)$ the string obtained by concatenating the edge labels on the path from the root to node $u$. We use $locus(P)$ to denote the highest node $u \in \mathcal{T}$ such that $P$ is a prefix of $path(u)$. The lowest common ancestor of any two nodes $u$ and $v$ is denoted by $lca(u, v)$.

Recall that in a heavy path decomposition, each heavy path contains exactly one leaf node. The heavy path containing the leaf $l_i$ is called $i^{th}$ heavy path and is denoted by the number $i$. We use $apex(h)$ to represent the highest node in the heavy path $h$. For a node $v \in \mathcal{T}$, $hp(v)$ denotes the heavy path containing $v$. For each heavy path $i$ in the decomposition, $H(i)$ is the set of all heavy paths $h$ such that the parent of $apex(h)$ is on the $i^{th}$ heavy path. Formally, as defined in [4],

$$H(i) = \{j : hp(lca(l_i, l_j)) = i\}$$

By Lemma 1, a heavy path can contribute to at most $\log n$ such sets, which implies that $\sum_{i=1}^{n} |H(i)| = O(n \log n)$.

## 2      Algorithms

Navarro et al. [4] and Bille et al. [3] solved the bounded gap query problem and the top-$k$ query problem, respectively, by transforming them into the orthogonal segment intersection problem. Using similar techniques, we build a data structure consisting of simple textbook data structures. The data structure supports both (top-$k$ and bounded gap) types of queries.

## 2.1   Data Structure

We build a suffix tree $\mathcal{T}$ for the given text $T[1:n]$ [1]. Using the method of Sleator and Tarjan [10], we decompose the tree $\mathcal{T}$ into a collection of $n$ heavy paths. As in [4], we define a set of horizontal segments for each heavy path $h$ as follows.

For each $j \in H(h)$, let $P_j = path(lca(l_j, l_h))$ i.e. the string corresponding to the lowest common ancestor of $j^{th}$ and $h^{th}$ leaves from the left. Let $a$ be the occurrence of $P_j$ just before $SA[j]$ and $b$ be the occurrence of $P_j$ just after $SA[j]$. Two horizontal segments corresponding to $j$ are created, provided $a$ and $b$ exist.

1. Let $P'_j$ be the smallest non-empty prefix of $P_j$ that does not have any occurrence in the index range $[a+1, SA[j]-1]$. We create a horizontal segment whose $y$-coordinate is $SA[j] - a$ and $x$-coordinates are in the range $[|P'_j|, |P_j|]$. Associate the pair $(a, SA[j])$ with the segment.

2. Let $P''_j$ be the smallest non-empty prefix of $P_j$ that does not have any occurrence in the index range $[SA[j]+1, b-1]$. We create a horizontal segment whose $y$-coordinate is $b - SA[j]$ and $x$-coordinates are in the range $[|P''_j|, |P_j|]$. Associate the pair $(SA[j], b)$ with the segment.

The set of horizontal segments corresponding to heavy path $h$ is represented by $I_h$ (as in [4]). An algorithm for computing the horizontal segments for every heavy path is described in Sect. 3.

For each heavy path $h$ in the decomposition, we preprocess the set $I_h$ of horizontal segments for the orthogonal segment intersection queries using Chazelle's method [11]. We denote the data structure for $I_h$ by $D_h$. For each node $u \in \mathcal{T}$ with $|path(u)| \leq \log n$, we store the consecutive occurrences of $path(u)$ in a linear list $D(u)$ in the non-decreasing order of their distances. Our final data structure consists of the suffix tree $\mathcal{T}$ augmented with lists $D(u)$ and the structure $D_h$ for each heavy path $h$ in the decomposition.

**Lemma 2.** *The data structure uses $O(n \log n)$ space.*

*Proof.* The space used by suffix tree $\mathcal{T}$ is $O(n)$. For each $j \in H(h)$, at most, two horizontal line segments are being created. As $\sum_{i=1}^{n} |H(i)| \leq n \log n$, the total number of horizontal segments created for all heavy paths will be $O(n \log n)$.

Each edge label in the tree $\mathcal{T}$ consists of at least one character, so the level of a node $u$ with $|path(u)| \leq \log n$ can be at most $\log n$. The subtrees rooted at any two nodes of the same level in $\mathcal{T}$ have no leaves (occurrences) in common, so the space used to store occurrences at a particular level is $O(n)$. The total space requirement for all $\log n$ levels will be $O(n \log n)$. □

Navarro and Thankachan [4] have shown the following (See Lemma 3 in [4]).

**Lemma 3.** *Let $P[1:m]$ and $[\alpha, \beta]$ be the input parameters of a query and let $h = hp(locus(P))$. Then, the set of pairs associated with all those horizontal* segments in $I_h$, which are stabbed by the vertical segment $m \times [\alpha, \beta]$ (i.e., the segment connecting the points $(m, \alpha)$ and $(m, \beta)$) forms the output to the bounded gap query problem.

We have the following Lemma for the top-$k$ query problem.

**Lemma 4.** *Given a pattern $P[1 : m]$ and an integer $k > 0$ as query parameters, the set of pairs associated with the $k$ horizontal segments in $I_h$ with smallest $y$-coordinates that are intersected by the vertical segment $m \times (-\infty, \infty)$ forms the output to the top-k query problem. Here, $h = hp(locus(P))$.*

*Proof.* Let $(i, j)$ be one of the $k$ closest consecutive occurrences of pattern $P$ in text $T$, and $(x_1, x_2, y)$ be the horizontal segment corresponding to $(i, j)$. Let $u = locus(P)$. Since the pair $(i, j)$ is a consecutive occurrence of pattern $P$, so $m = |P| \in [x_1, x_2]$. The $y$-coordinate of a horizontal segment represents the distance of the corresponding consecutive occurrence. The segment $(x_1, x_2, y)$ corresponds to the pair $(i, j)$, which is among the $k$ closest consecutive occurrences of $P$, so it would be among the first $k$ horizontal segments that are intersected by the vertical ray emanating from point $(m, -\infty)$.

Let $(x'_1, x'_2, y')$ be one of the first $k$ horizontal segments intersected by the vertical ray emanating from the point $(m, -\infty)$. We need to show that pair $(i', j')$ associated with the segment $(x'_1, x'_2, y')$ is among the $k$ closest consecutive occurrences of $P$. Since the ray intersects the segment $m \in [x'_1, x'_2]$, in other words, the $(i', j')$ belongs to the set of consecutive occurrences of $P$ in $T$. Since the segment is one of the first $k$ segments intersected by the ray and the $y$-coordinate of a segment represents the distance of the corresponding consecutive occurrence, the pair $(i', j')$ is one of the $k$ closest consecutive occurrences of pattern $P$ in text $T$.                                            $\square$

## 2.2    Answering Queries

*Top-k Query*: Let pattern $P[1 : m]$ and integer $k > 0$ be the query parameters. We first compute $locus(P)$ in suffix tree $T$. Let $u = locus(P)$ and $h = hp(u)$. If the pattern length $m$ is not more than $\log n$, we simply report the first $k$ consecutive occurrences from the sorted list $D(u)$. Otherwise (i.e., $m > \log n$), we query the structure $D_h$ with the vertical segment $s_v : m \times (-\infty, \infty)$ and return only the first $k$ horizontal segments, if they exist, intersected by the segment $s_v$ as we move from the lower endpoint $(m, -\infty)$ along the segment $s_v$. We report the consecutive occurrence associated with each returned segment.

Computing $locus(P)$ takes $O(m)$ time. For the case, $m \leq \log n$, the additional time spent reporting the output is $O(k)$. So, the query time, in this case, is $O(m + k)$. When $m > \log n$, we spend additional $O(\log n + k)$ time to compute the output using the structure $D_h$. In this case, the time needed to answer the query is also $O(m+k)$; the log factor is absorbed by the pattern length $m$. Thus, a top-$k$ query takes $O(m + k)$ time.

*Bounded-Gap Query*: Let pattern $P[1 : m]$ and real numbers $\alpha$ and $\beta$ with $0 \leq \alpha \leq \beta$ be the query parameters. Again, we first compute $locus(P)$, say it is the node $u$. If the pattern length $m$ is no more than $\log n$, we compute the position of value $\alpha$ in the sorted list $D(u)$ using an exponential search [12]. From the computed position in $D(u)$, we start reporting the consecutive occurrences while moving down in the list and stop as soon as a consecutive occurrence with a distance greater than $\beta$ is found. When $m > \log n$, we query the structure $D_h$ with the vertical segment $m \times [\alpha, \beta]$. We report the consecutive occurrences associated with the returned horizontal segments.

We spend $O(m)$ time to compute $locus(P)$. For the case $m > \log n$, additional $O(\log n + k)$ time is used for a segment intersection query to the structure $D_h$. As $m > \log n$, the query time, in this case, will be $O(m + k)$. For the other case $m \leq \log n$, we spend additional $O(\log \alpha + k)$ time; $O(\log \alpha)$ time for an exponential search and $O(k)$ time for reporting the consecutive occurrences. Finding the position of $\alpha$ in the list $D(u)$ using an exponential search will take $O(\log r)$ time, where $r$ is the rank of $\alpha$ in the list [12]. Since distances of consecutive occurrences are integer values from the range $[1, n - 1]$, so $r \leq \alpha$. Therefore, the exponential search takes $O(\log \alpha)$ time. Thus, the time spent to answer the query when $m \leq \log n$ is $O(m + \log \alpha + k)$.

The total query time is dominated by the query time for the case $m \leq \log n$, which is $O(m + \log \alpha + k)$. We have the following theorem.

**Theorem 1.** *We can preprocess a string $T$ of length $n$ into an $O(n \log n)$ space data structure so that, given string $P$ of length $m$ and a positive integer $k$, we can report the $k$ consecutive occurrences of $P$ in $T$ that are closest in $O(m + k)$ time. If two real numbers $\alpha$ and $\beta$ with $0 \leq \alpha \leq \beta$ come as query parameters along with $P$, the data structure can be used to report all the consecutive occurrences of $P$ in $T$ with distance in the range $[\alpha, \beta]$ in $O(m + \log \alpha + \#output)$ time.*

If one of the endpoints of query ranges ($\alpha$ or $\beta$) is known at the time of indexing, the solution can be modified so that queries can be answered in optimal $(m + k)$ time. Suppose $\alpha$ is given along with the text $T[1 : n]$ at the time of indexing. In each node $u$ in the suffix tree $\mathcal{T}$ with $|path(u)| \leq \log n$, we compute the position of $\alpha$ in the structure $D(u)$ in the preprocessing phase to save $\log \alpha$ factor in the query time. Thus, the query time will be $O(m + k)$.

**Corollary 1.** *We can preprocess a text $T[1 : n]$ and an integer $\alpha \geq 0$ so that, given a query pattern $P[1 : m]$ and an integer $\beta \geq \alpha$, all the consecutive occurrences of $P$ in $T$ can be reported in $(m + \#output)$ time. The space used is $O(n \log n)$.*

Bille et al. [3] considered this special case and solved it using $O(\frac{n}{\epsilon})$ space and $O(m + k^{1+\epsilon})$ time.

## 3   Computation of Horizontal Segments

In this section, we describe an algorithm to compute the horizontal segments corresponding to every heavy path. The algorithm takes $O(n^2)$ time and $O(n \log n)$

space. The algorithm follows an approach that is very similar to the plane sweep technique [13]. We use notations of Sect. 2.

The *level* of a node $v \in T$ is defined as the number of edges on the path from root to node $v$. Equivalently, $level(root) = 0$ and $level(v) = level(u) + 1$ for every other node $v$, where $u$ is the parent of $v$. We compute the level of each node in the suffix tree $T$ using the level order traversal. We decompose the tree $T$ using the heavy path decomposition [10]. The $apex(h)$ of every heavy path $h$ stores a pointer to its parent in $T$.

We process the nodes of the suffix tree $T$ in a bottom-up fashion where nodes at higher levels are processed first. In other words, the nodes at level $i$ will be processed before any node of level $i-1$. Without loss of generality, we process the nodes at a particular level in the left-to-right order. After processing a node on a heavy path $h$, some (possibly zero) new horizontal segment(s) corresponding to $h$ gets created.

While processing the nodes of a particular level, we maintain a doubly linked list for each heavy path $h$ that has a node $v$ on the current level. The list corresponding to $h$ after processing a node $v$ on $h$, denoted by $L_h(v)$, stores the occurrences of $path(v)$ in the text order (i.e., the $i^{th}$ node from the start of the list stores the $i^{th}$ occurrence in the text order). Note that the occurrences corresponding to two consecutive nodes in $L_h(v)$ form a consecutive occurrence of $path(v)$. We maintain two global lookup tables $LT[1 : n]$ and $LT^{-1}[1 : n]$ that maintain information about consecutive occurrences: Let $v \in h$ be the node being processed, and $w$ be the child of $v$ on $h$. If $(i, j)$ is a consecutive occurrence of $path(v)$ but not of $path(w)$, then $LT[i]$ stores the tuple $(j, h, v)$ and $LT^{-1}[j]$ stores the value $i$. We initialize $LT[i] \leftarrow \phi$ and $LT^{-1}[i] \leftarrow -1$, for every $i \in [1, n]$, to indicate that the lists are initially empty.

Let $l$ denote the number of levels in the tree $T$. For each level $i \in \{l, l - 1, ..., 2, 1\}$, in order, we consider the nodes of level $i$ in the left-to-right order. Let $u$ be the current node being processed, and let $h' = hp(u)$. For each child $v$ (of node $u$) which is not on the path $h'$, we create a horizontal segment for every consecutive occurrence of $path(v)$ using the list $L_h(v)$, where $h = hp(v)$. See Fig. 1. Note that we have $L_h(v)$ as the node $v$ has already been processed. Let $i$ and $j$ correspond to two consecutive nodes in the list $L_h(v)$. If $LT[i] = (j, h, w)$, we add the horizontal segment $[|path(v)|, |path(w)|] \times (j-i)$ to set $I_h$. In addition, we assign $LT[i] \leftarrow \phi$ and $LT^{-1}[j] \leftarrow -1$ to indicate that these entries are now empty. Finally, we merge the linked lists $L_h(v)$ along with $L'_h(w)$, where $w$ is the child of $u$ on $h'$, and update the lookup tables accordingly. The resultant list is $L'_h(u)$, and the old (merged) lists become obsolete. While updating the lookup tables, we create horizontal segments as follows.

We scan the merged linked list in left-to-right order. While scanning the list, we update the lookup tables and create new horizontal segments. Let $i$ and $j$ be the occurrences corresponding to the current consecutive pair of nodes in the list $L'_h(u)$.

- **case 1:** if both the occurrences $i$ and $j$ are not present in the list $L(u)$, we set $T[i] \leftarrow (j, h', u)$ and $T^{-1}[j] \leftarrow i$.

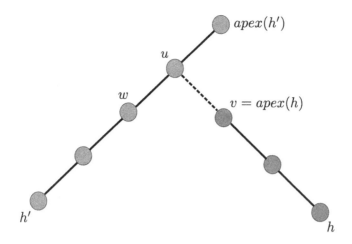

**Fig. 1.** Two heavy paths $h$ and $h'$ such that $apex(h)$ has its parent in path $h'$.

- **case 2:** if occurrence $j$ is present in $L(u)$ but not $i$, then two scenarios are possible. If $LT^{-1}[j]$ is empty, then set $LT[i] \leftarrow (j, h', u)$ and $LT^{-1}[j] \leftarrow i$. Otherwise, let $LT^{-1}[j] = k$ and $LT[k] = (j, h', w)$. We first create the horizontal segment $[|path(u)|, |path(w)|] \times (j - k)$ and add it to the set $I_{h'}$. In addition, we make the following assignments.
  - $LT[k] \leftarrow (i, h', u)$ and $LT^{-1}[i] \leftarrow k$
  - $LT[i] \leftarrow (j, h', u)$ and $LT^{-1}[j] \leftarrow i$
- **case 3:** if occurrence $i$ is present in $L(u)$ but not $j$, two scenarios are possible. If $LT[i]$ is empty, then set $LT[i] \leftarrow (j, h', u)$ and $LT^{-1}[j] \leftarrow i$. Otherwise, let $LT[i] = (k, h', w)$. We create the segment $[|path(u)|, |path(w)|] \times (k - i)$ and add it to the set $I_{h'}$. In addition, we make the following assignments.
  - $LT[i] \leftarrow (j, h', u)$ and $LT^{-1}[j] \leftarrow i$
  - $LT[j] \leftarrow (k, h', u)$ and $LT^{-1}[k] \leftarrow j$
- **case 4:** When both the occurrences $i$ and $j$ are present in $L(u)$, no action will be taken.

**Lemma 5.** *We can compute the horizontal segments corresponding to every heavy path in $O(n^2)$ time and $O(n \log n)$ space.*

*Proof.* As the number of leaves (i.e., occurrences) in the subtree rooted at a node can not be more than $n$. The linked list corresponding to a heavy path consists of at most $n$ nodes (each corresponds to an occurrence). Recall that there are $n$ heavy paths resulting from the heavy path decomposition. Two sorted doubly linked lists can be merged in time linear in the sum of their sizes, so merging the linked lists of all the heavy paths will take $O(n^2)$ time in total. We scan the merged list once to update the lookup tables and create horizontal segments due to merging. Therefore, this task will also take total $O(n^2)$ time. So, the total time to compute the horizontal segments for all the heavy paths is $O(n^2)$.

We are maintaining two tables $LT[\,]$ and $LT^{-1}[\,]$ over the algorithm, each of size $n$. Observe that the sum of the number of leaves in the subtrees rooted at the nodes of a particular level is precisely $n$. At any point during the algorithm, we maintain linked lists only for those heavy paths with a node at the current level. Thus, the space used by the algorithm to maintain the linked lists is $O(n)$. Total $O(n \log n)$ horizontal segments are created for all the heavy paths, so the space complexity is dominated by the $O(n \log n)$ factor.

## 4   Conclusion

We study two variants of computing consecutive occurrences under different constraints: $(i)$ finding consecutive occurrences of a query pattern with distance in a range and $(ii)$ computing $k$ consecutive occurrences of a query pattern that are closest. We propose deterministic algorithms for solving these problems using simple and classical data structures. After preprocessing the given text, both queries can be answered efficiently.

We show that the required preprocessing could be done in $O(n^2)$ time and $O(n \log n)$ space. We believe that the preprocessing time could be reduced.

## References

1. Weiner, P.: Linear pattern matching algorithms. In: 14th Annual Symposium on Switching and Automata Theory (SWAT 1973), pp. 1–11 (1973). https://doi.org/10.1109/SWAT.1973.13
2. Knuth, D.E., Morris Jr., J.H., Pratt, V.R.: Fast pattern matching in strings SIAM J. Comput. **6**(2), 323–350 (1977). https://doi.org/10.1137/0206024
3. Bille, P., Gortz, I.L., Pedersen, M.R., Rotenberg, E., Steiner, T.A.: String indexing for top-k close consecutive occurrences. Theor. Comput. Sci. 133–147 (2022). https://doi.org/10.1016/j.tcs.2022.06.004
4. Navarro, G., Thankachan, S.V.: Reporting consecutive substring occurrences under bounded gap constraints. Theor. Comput. Sci. **638**, 108–111 (2016). https://doi.org/10.1016/j.tcs.2016.02.005
5. Bille, P., Gørtz, I.L., Pedersen, M.R., Steiner, T.A.: Gapped indexing for consecutive occurrences. Algorithmica 1432–0541 (2022). https://doi.org/10.1007/s00453-022-01051-6
6. Cohen, H., Porat, E.: Range non-overlapping indexing. In: Dong, Y., Du, D.-Z., Ibarra, O. (eds.) ISAAC 2009. LNCS, vol. 5878, pp. 1044–1053. Springer, Heidelberg (2009). https://doi.org/10.1007/978-3-642-10631-6_105
7. Ganguly, A., Shah, R., Thankachan, S.V.: Succinct non-overlapping indexing. Algorithmica **82**(1), 107–117 (2015). https://doi.org/10.1007/s00453-019-00605-5
8. Ukkonen, E.: On-line construction of suffix trees. Algorithmica **14**(3), 249–260 (1995). https://doi.org/10.1007/BF01206331
9. Gusfield, D.: Algorithms on Strings, Trees, and Sequences: Computer Science and Computational Biology. Cambridge University Press, Cambridge (1997). https://doi.org/10.1017/CBO9780511574931
10. Sleator, D., Tarjan, R.E.: A data structure for dynamic trees. J. Comput. Syst. Sci. **26**(3), 362–391 (1983). https://doi.org/10.1016/0022-0000(83)90006-5

11. Chazelle, B.: Filtering search: a new approach to query-answering. SIAM J. Comput. **15**(3), 703–724 (1986). https://doi.org/10.1137/0215051

12. Bentley, J.L., Yao, A.C.: An almost optimal algorithm for unbounded searching. Inf. Process. Lett. **5**(3), 82–87 (1976). https://doi.org/10.1016/0020-0190(76)90071-5

13. Berg, M., Cheong, O., Krevel, M., Overmars, M.: Line Segment Intersection In Computational Geometry: Algorithms and Applications, 3rd edn., pp. 19–43 (2008)

# Parameterized Aspects of Distinct Kemeny Rank Aggregation

Koustav De[1]([⊠]) [iD], Harshil Mittal[2], Palash Dey[1] [iD], and Neeldhara Misra[2] [iD]

[1] Indian Institute of Technology Kharagpur, Kharagpur 721302, West Bengal, India
koustavde7@kgpian.iitkgp.ac.in, palash.dey@cse.iitkgp.ac.in
[2] Indian Institute of Technology Gandhinagar, Gandhinagar 382424, Gujarat, India
{mittal_harshil,neeldhara.m}@iitgn.ac.in

**Abstract.** The Kemeny method is one of the popular tools for rank aggregation. However, computing an optimal Kemeny ranking is NP-hard. Consequently, the computational task of finding a Kemeny ranking has been studied under the lens of parameterized complexity with respect to many parameters. We study the parameterized complexity of the problem of computing all distinct Kemeny rankings. We consider the target Kemeny score, number of candidates, average distance of input rankings, maximum range of any candidate, and unanimity width as our parameters. For all these parameters, we already have FPT algorithms. We find that any desirable number of Kemeny rankings can also be found without substantial increase in running time. We also present FPT approximation algorithms for Kemeny rank aggregation with respect to these parameters.

**Keywords:** Diversity · Voting · Kemeny · Kendall-Tau

## 1 Introduction

Aggregating individual ranking over a set of alternatives into one societal ranking is a fundamental problem in social choice theory in particular and artificial intelligence in general. Immediate examples of such applications include aggregating the output of various search engines [14], recommender systems [23], etc. The Kemeny rank aggregation method is often the method of choice in such applications due to its many desirable properties like Condorcet consistency that is electing the Condorcet winner (if it exists), etc. A Condorcet winner is a candidate who defeats every other candidate in pairwise election. The Kemeny method outputs a ranking $R$ with minimum sum of dissatisfaction of individual voters known as *Kemeny score* of $R$; the dissatisfaction of a voter with ranking $Q$ with respect to $R$ is quantified as the number of pairs of candidates that $Q$ and $R$ order differently [20]. This quantity is also called the Kendall-Tau distance between $Q$ and $R$. A ranking with minimum Kemeny score is called the Kemeny ranking.

The computational question of finding optimal Kemeny rankings is intractable in very restricted settings (for instance, even with a constant number of voters). Therefore, it has been well-studied from both approximation and parameterized perspectives. A problem is said to be *fixed-parameter tractable* or FPT with respect to a parameter $k$ if

S. Kalyanasundaram and A. Maheshwari (Eds.): CALDAM 2024, LNCS 14508, pp. 14–28, 2024.
https://doi.org/10.1007/978-3-031-52213-0_2

it admits an algorithm whose running time can be described as $f(k) \cdot n^{O(1)}$ where the input size is $n$, implying that the algorithm is efficient for instances where the parameter is "small" [11]. For the Kemeny rank aggregation problem, the following parameters (among others) have enjoyed attention in the literature:

- *Range.* The range of a candidate in a profile is the difference between its positions in the votes which rank him/her the lowest and the highest [7]. The maximum and average range of a profile is defined as, respectively, the maximum and average ranges of individual candidates. Profiles which are "homogeneous", *i.e.* where most candidates are viewed somewhat similarly by the voters, are likely to have low values for range, while a single polarizing candidate can skew the max range parameter considerably.
- *KT-distance.* The average (respectively, maximum) KT distance is the average (respectively, maximum) of the Kendall-Tau distances between all pairs of votes [7]. Recall that the KT distance between a pair of rankings is the number of pairs that are ordered *differently* by the two rankings under consideration.

A pair of candidates are said to be *unanimous* with respect to a voting profile if all votes rank them in the same relative order. Consider the following "unanimity graph" associated with a profile $P$ and defined as follows: every candidate is represented by a vertex, and there is an edge between a pair of candidates if and only if they are unanimous with respect to the profile. We use $G_P$ to denote this graph. Note that the structure of the *complement* of this graph, denoted $\overline{G_P}$, carries information about candidates about whom the voters are not unanimous in their opinion. In particular, for every pair of candidates $a$ and $b$ that have an edge between them in the complement of the unanimity graph, there is at least one voter who prefers $a$ over $b$ and at least one who prefers $b$ over $a$. Thus every edge signals a lack of consensus, and one could think of the number of edges in this graph as a measure of the distance of the profile from an "automatic consensus", which is one that can be derived from the information about unanimous pairs alone. Motivated by this view, we consider also the following structural parameter:

*Unanimity Width.* For an input voting profile $P$, we define a graph, called "unanimity graph" or the comparability graph, on the set of candidates where we have an edge from $i$ to $j$ if and only if every ranking in $P$ puts $i$ before $j$; we denote its complement by $\overline{G_P}$. We call the pathwidth of $\overline{G_P}$ the unanimity width of $P$ [4] (refer to Sect. 2 for the formal definition of pathwidth).

Our contribution concerns enumerating optimal Kemeny rankings. In recent times, there is considerable research interest in finding a set of diverse optimal or near-optimal solutions of an optimization problem. Indeed, it is often difficult to encode all aspects of a complex system into a neat computational problem. In such scenarios, having a diverse set of optimal solutions for a problem $\Gamma$ allows the user to pick a solution which meets other aspects which are not captured in $\Gamma$. In the context of rank aggregation, such other external constraints may include gender fairness, demographic balance, etc. For the Kemeny rank aggregation method, Arrighi et al. [5] present a parameterized algorithm to output a set of diverse Kemeny rankings with respect to unanimity width as the parameter.

However, note that external requirements are often independent of the constraints in the optimization problem, and consequently they may not be correlated with diversity based on distance parameters. In particular, for useful externalities like gender fairness or geographic balance—these features of the candidates may not have any relation with their position in the voters' rankings, and therefore, diversity *between* solutions may not imply diversity *within* any of the solutions. This becomes particularly stark when most near-optimal rankings do not meet the external requirements. Indeed, there is a substantial literature [1, 25] that considers the problem of accounting for these requirements explicitly, and studies trade-offs between optimality of solutions and the degree to which demands of diversity can be met.

In this paper, we shift our focus from finding diverse solutions to finding as many *distinct* solutions as possible. Enumerating solutions is a fundamental goal for any optimization problem. The literature on counting optimal Kemeny rankings is arguably limited considering that even finding one is hard in very restricted settings, and that instances could have exponentially many rankings—which would be too expensive to enumerate. Indeed, consider a profile that consists of two votes over $m$ candidates, where one vote ranks the candidates in lexicographic order and the other ranks the candidates in reverse lexicographic order. For this instance, every ranking is an optimal ranking. However, note that real world preferences often have additional structure: for example, profiles with an odd number of votes that are single-peaked [10] or single-crossing [10] have unique optimal solutions. To address scenarios where the number of optimal solutions is large, we allow the user to specify the number $r$ of optimal solutions that she wants the algorithm to output. In our problem called DISTINCT OPT KEMENY RANKING AGGREGATION, the input is a set of rankings over a set of candidates and an integer $r$, and we need to output $\max\{r, \text{number of optimal solutions}\}$ Kemeny rankings.

## 1.1    Our Contributions

*Algorithms for Distinct Kemeny Rank Aggregation.*    The first parameter that we consider is the optimal Kemeny score $k$, also called the *standard parameter*. Many applications of rank aggregation, for example, faculty hiring, etc. exhibit correlation among the individual rankings—everyone in the committee may tend to prefer some candidate with strong academic background than some other candidate with weak track record. In such applications, the optimal Kemeny score $k$, average Kendall-Tau distance $d$ (a.k.a. Bubble sort distance) among input rankings, maximum range of the positions of any candidate $r_{\text{max}}$, and unanimity width $w$ will be small, and an FPT algorithm becomes useful. We show that there is an algorithm for DISTINCT OPT KEMENY RANKING AGGREGATION running in time $\mathcal{O}^*\left(2^k\right)$ [Theorem 1]. We next consider the number of candidates, $m$ as the parameter and present an algorithm running in time $\mathcal{O}^*\left(2^m r^{\mathcal{O}(1)}\right)$ [Theorem 2] where $r$ is the required number of solutions. For $d$ and $r_{\text{max}}$, we present algorithms with running time $\mathcal{O}^*(16^d)$ and $\mathcal{O}^*\left(32^{r_{\text{max}}}\right)$ [Theorems 3 and 4] respectively. Our last parameter is the unanimity width $w$ which is the pathwidth of the co-comparability graph of the unanimity order and we present an algorithm running in time $\mathcal{O}^*\left(2^{\mathcal{O}(w)} \cdot r\right)$ [Theorem 5].

Some instances may have a few optimal solutions, but have many close-to-optimal solutions. To address such cases, we study the DISTINCT APPROXIMATE KEMENY RANKING AGGREGATION problem where the user gives a real number $\lambda \geq 1$ as input and looks for max$\{r$, number of optimal solutions$\}$ rankings with Kemeny score at most $\lambda$ times the optimal Kemeny score. For this problem, we design algorithms with running time $\mathcal{O}^* \left(2^{\lambda k}\right)$ [Corollary 1], $\mathcal{O}^* \left(2^m r^{\mathcal{O}(1)}\right)$ [Corollary 2] and $\mathcal{O}^* \left(16^{\lambda d}\right)$ [Theorem 6].

We observe that the running time of all our algorithms are comparable with the respective parameterized algorithms for the problem of finding one Kemeny ranking. We note that this phenomenon is in sharp contrast with the diverse version of Kemeny rank aggregation where we have an FPT algorithm only for unanimity width as the parameter. To begin with, the algorithm, as presented in [4], cannot be used to find only optimal solutions; it can find only approximately optimal solutions. However, one can set the parameters of the algorithm in [4] to find all $\lambda$ approximate rankings in time $\mathcal{O}^*((w!(\lambda - 1)\text{OPT})^{\mathcal{O}(m!)})$ where OPT is the optimal Kemeny score of the input rankings.

## 1.2  Related Work

Kemeny rule [20] shows us its most significant and popular mechanism for ranking aggregation. However, Bartholdi et al. [6] have established that KEMENY SCORE is NP-complete even if we apply the restriction of having only four input rankings [15]. Fixed-parameter algorithms for Kemeny voting rule have been proved to be an effective and important area for research by Betzler et al. [7] considering structural parameterizations such as "number of candidates", "solution size *i.e.* Kemeny Score", "average pairwise distance", "maximum range", "average range" of candidates in an election. A multi-parametric algorithm for DIVERSE KEMENY RANK AGGREGATION over partially ordered votes has been studied in [4]. A small error in the construction proof from [15] has been rectified by Biedl et al. [8] and they have established the approximation factor of $2 - 2/k$, improving from the previous approximation factor of 2.

Further classification in more exact manner of the classical computational complexity of Kemeny elections has been provided by Hemaspaandra et al. [19]. With respect to the practical relevance of the computational hardness of the KEMENY SCORE, polynomial-time approximation algorithms have been developed where a factor of $8/5$ is seen in [26] and a factor of $11/7$ is proved in [2]. Kenyon-Mathieu and Schudy [21] proposed a polynomial-time approximation scheme (PTAS) for finding a Kemeny ranking. However, their algorithm is not very useful in practice. There are quite a few works which develop practical heuristics for this problem [9,12,24].

Polynomial time algorithms producing good solutions for rank aggregation rule is a consequence of thorough computational studies [3,27]. Cornaz et al. [10] have established polynomial time computability of the single-peaked and single-crossing widths and have proposed new fixed-parameter tractability results for the computation of an optimal ranking according to the Kemeny rule by following the results of Guo te al. [16]. In social choice theory [6,22], the ideas related to diverse sets of solutions have found tremendous applicability. The study in [28] introduced the $(j, k)$-Kemeny rule which is a generalization of Kemeny's voting rule that aggregates ballots containing weak orders with $j$ indifference classes into a weak order with $k$ indifference classes. In social choice theory, different values of $j$ and $k$ yield various rules of the interest of the community turning up as special cases. The minimum Kendall-Tau distance between pairs of solutions has a nice analogy with $min$ Hamming distance over all pairs of solutions as shown in [17, 18].

## 2   Preliminaries

For an integer $\ell$, we denote the set $\{1, \ldots, \ell\}$ by $[\ell]$. For two integers $a, b$, we denote the set $\{i \in \mathbb{N} : a \le i \le b\}$ by $[a, b]$. Given two integer tuples $(x_1, \ldots, x_\ell), (y_1, \ldots, y_\ell) \in \mathbb{N}^\ell$, we say $(x_1, \ldots, x_\ell) >_{\text{lex}} (y_1, \ldots, y_\ell)$ if there exists an integer $i \in [\ell]$ such that we have (i) $x_j = y_j$ for every $j \in [i-1]$, and (ii) $x_i > y_i$.

Let $\mathcal{C}$ be a set of candidates and $\Pi = \{\pi_1, \ldots, \pi_n\}$ a multi-set of $n$ rankings (complete orders) on $\mathcal{C}$. For a ranking $\pi$ and a candidate $c$ let us define $\text{pos}_\pi(c)$ to be $|\{c' \in \mathcal{C} : c' \succ_\pi c\}|$. We precisely define the *range* $r(c)$ in a set of rankings $\Pi$ to be $\max\limits_{\pi_i, \pi_j \in \Pi} \left\{ |\text{pos}_{\pi_i}(c) - \text{pos}_{\pi_j}(c)| \right\} + 1$. We denote the set of all complete orders over $\mathcal{C}$ by $\mathcal{L}(\mathcal{C})$. The Kemeny score of a ranking $Q \in \mathcal{L}(\mathcal{C})$ with respect to $\Pi$ is

$$\text{Kemeny}_\Pi(Q) = \sum_{i=1}^{n} d_{\text{KT}}(Q, \pi_i)$$

where $d_{\text{KT}}(\cdot, \cdot)$ is the Kendall-Tau distance – the number of pairs of candidates whom the linear orders order differently – between two linear orders, and $N_\Pi(x \succ y)$ is the number of linear orders in $\Pi$ where $x$ is preferred over $y$. A Kemeny ranking of $\Pi$ is a ranking $Q$ which has the minimum $\text{Kemeny}_\Pi(Q)$; the score $\text{Kemeny}_\Pi(Q)$ is called the optimal Kemeny score of $\Pi$.

We now define our problems formally. For a set of rankings $\Pi$, we denote the set of (optimal) Kemeny rankings and rankings with Kemeny score at most some integer $k$ for $\Pi$ respectively by $K(\Pi)$ and $K(\Pi, k)$, and the minimum Kemeny score by $k_{\text{OPT}}(\Pi)$.

**Definition 1** (DISTINCT OPT KEMENY RANKING AGGREGATION). *Given a set of rankings (complete orders) $\Pi$ over a set of candidates $\mathcal{C}$ and integer $r$, compute $\ell = \min\{r, |K(\Pi)|\}$ distinct Kemeny rankings $\pi_1, \ldots, \pi_\ell$. We denote an arbitrary instance of it by $(\mathcal{C}, \Pi, r)$.*

For a set of rankings $\Pi$ over a set of candidates $\mathcal{C}$, we say that a complete order $\pi$ respects unanimity order if we have $x \succ_\pi y$ whenever $x \succ y$ for all $\succ \in \Pi$.

**Definition 2** (DISTINCT APPROXIMATE KEMENY RANKING AGGREGATION). *Given a set of ranking (complete order) $\Pi$ over a set of candidates $\mathcal{C}$, an approximation factor $\lambda \ge 1$, and integer $r$, compute $\ell = \min\{r, |K(\Pi, \lambda \cdot k_{OPT}(\Pi))|\}$ distinct rankings $\pi_1, \ldots, \pi_\ell$ such that each ranking $\pi_i, i \in [\ell]$ respects unanimity order with respect to $\Pi$ and the Kemeny score of each ranking $\pi_i, i \in [\ell]$ is at most $\lambda \cdot k_{OPT}(\Pi)$. We denote an arbitrary instance of it by $(\mathcal{C}, \Pi, \lambda, r)$.*

**Definition 3** (DISTINCT KEMENY RANKING AGGREGATION). *Given a list of partial votes $\Pi$ over a set of candidates $\mathcal{C}$, and integers $k$ and $r$, compute $\ell = \min\{r, |K(\Pi, k)|\}$ distinct rankings $\pi_1, \ldots, \pi_\ell$ such that the Kemeny score for each ranking $\pi_i$ is at most $k$ and each $\pi_i, i \in [\ell]$ respects unanimity order. We denote an arbitrary instance of it by $(\mathcal{C}, \Pi, k, r)$.*

We use $\mathcal{O}^*(\cdot)$ to hide polynomial factors. That is, we denote $\mathcal{O}(f(k)\text{poly}(n))$ as $\mathcal{O}^*(f(k))$ where $n$ is the input size.

We define a path decomposition of a graph $G = (V, E)$ by a tuple $\mathcal{P} = (\mathcal{B}_i)_{i \in [t]}$ where each bag $\mathcal{B}_i \subseteq V$, $t$ is the number of bags in $\mathcal{P}$ and $\mathcal{P}$ satisfies the following additional constraints : (1) $\bigcup_{i \in [t]} \mathcal{B}_i = V$, (2) $\exists i \in [t]$ such that $u, v \in \mathcal{B}_i$ for each $(u, v) \in E$ and (3) $\mathcal{B}_i \cap \mathcal{B}_k \subseteq \mathcal{B}_j$ for each $i, j, k \in [t]$ satisfying $i < j < k$. The width of $\mathcal{P}$ denoted by $w(\mathcal{P})$ is defined as $\max_{i \in [t]} |\mathcal{B}_i| - 1$. The *pathwidth* of $G$ is denoted by $pw(G)$ which is defined as the minimum width of a path decomposition of $G$.

## 3   Algorithms for DISTINCT KEMENY RANKING AGGREGATION

We start with an easy Turing reduction from DISTINCT OPT KEMENY RANKING AGGREGATION to DISTINCT KEMENY RANKING AGGREGATION. In the interest of space, we defer the proof of some of our results to [13]. They are marked $\star$.

**Observation 1.** *Suppose there exists an algorithm for* DISTINCT KEMENY RANKING AGGREGATION *running in time $\mathcal{O}(f(m, n))$ where $m$ is the number of candidates and $n$ is the number of input votes. Then there exists an algorithm for* DISTINCT OPT KEMENY RANKING AGGREGATION *running in time $\mathcal{O}(f(m, n) \log(mn))$.*

*Proof.* We note that the optimal Kemeny score belongs to the set $\{0, 1, \ldots, n\binom{m}{2}\}$. To solve DISTINCT OPT KEMENY RANKING AGGREGATION, we perform a binary search in the range from 0 to $n\binom{m}{2}$ to find the smallest $k$ such that the algorithm for DISTINCT KEMENY RANKING AGGREGATION returns at least one ranking.

We now present a bounded search based FPT algorithm for DISTINCT KEMENY RANKING AGGREGATION parameterized by the optimal Kemeny score. Hence, we also have an FPT algorithm for DISTINCT OPT KEMENY RANKING AGGREGATION parameterized by the optimal Kemeny score.

**Theorem 1.** *Let $k$ be the Kemeny score of a Kemeny ranking. There is an FPT algorithm for* DISTINCT KEMENY RANKING AGGREGATION *parameterized by $k$ which runs in time $\mathcal{O}^*(2^k)$. Hence, we have an FPT algorithm for* DISTINCT OPT KEMENY RANKING AGGREGATION *parameterized by $k_{OPT}$ which runs in time $\mathcal{O}^*(2^{k_{OPT}})$.*

*Proof.* Due to Observation 1, it is enough to present an algorithm for DISTINCT KEMENY RANKING AGGREGATION. We design an algorithm for a more general problem DISTINCT KEMENY RANKING AGGREGATION$'$ where every output ranking needs to respect the relative order of some set of pair of candidates given as input. If the set of pairs of candidates is empty, then the new problem is the same as DISTINCT KEMENY RANKING AGGREGATION.

Let $(\mathcal{C}, \Pi, k, r)$ be an arbitrary instance of DISTINCT KEMENY RANKING AGGRE-GATION. We define $\mathcal{X} = \{a \succ b : a, b \in \mathcal{C},$ every ranking in $\Pi$ prefers $a$ over $b\}$ to be the unanimity order of $\Pi$. We find a solution of DISTINCT KEMENY RANKING AGGREGATION' instance $(\mathcal{C}, \Pi, k, r, \mathcal{X})$. We now design a bounded search based algorithm. We maintain a set $\mathcal{S}$ of solutions, which is initialized to the empty set. If every pair of candidates belong to $\mathcal{X}$ and $k \geq 0$, then we put the ranking induced by $\mathcal{X}$ in $\mathcal{S}$. If $k < 0$, then we discard this branch. Otherwise, we pick a pair $(a, b)$ of candidates not present in $\mathcal{X}$, solve $(\mathcal{C}, \Pi, k - |\{\pi \in \Pi : b \succ a$ in $\pi\}|, r$, transitive closure of $\mathcal{X} \cup \{a \succ b\})$ and $(\mathcal{C}, \Pi, k - |\{\pi \in \Pi : a \succ b$ in $\pi\}|, r$, transitive closure of $\mathcal{X} \cup \{b \succ a\})$ recursively, and put solutions found in $\mathcal{S}$. We note that, since $(a, b)$ is not a unanimous order of $\Pi$, the target Kemeny score $k$ decreases by at least one on both the branches of the search tree. Hence, the height of the search tree is at most $k$. Thus, the number of leaves and nodes in the search tree are at most respectively $2^k$ and $2 \cdot 2^k$. After the search terminates, we output $\min\{r, |\mathcal{S}|\}$ rankings from $\mathcal{S}$. If $\mathcal{S}$ remains empty set, report that there is no ranking whose Kemeny score is at most $k$. The computation at each node of the search tree (except the recursive calls) clearly takes a polynomial amount of time. Hence, the runtime of our algorithm is $\mathcal{O}^* \left(2^k\right)$. The correctness of our algorithm follows from the observation that every ranking $R$ whose Kemeny score is at most $k$, appears in a leaf node of the search tree of our algorithm.

Running the algorithm in Theorem 1 with target Kemeny score $\lambda k$ where $k$ is the optimal Kemeny score gives us the following result.

**Corollary 1.** *There is an algorithm for* DISTINCT APPROXIMATE KEMENY RANKING AGGREGATION *running in time* $\mathcal{O}^* \left(2^{\lambda k}\right)$ *parameterized by both $\lambda$ and $k$.*

We now consider the number of candidates as parameter and present a dynamic programming based FPT algorithm for DISTINCT KEMENY RANKING AGGREGATION.

**Theorem 2.** *There is an algorithm for* DISTINCT KEMENY RANKING AGGREGATION *which runs in time* $\mathcal{O}^* \left(2^m r^{\mathcal{O}(1)}\right)$. *In particular,* DISTINCT KEMENY RANKING AGGREGATION *and* DISTINCT OPT KEMENY RANKING AGGREGATION *are* FPT *parameterized by the number of candidates since the number $r$ of output rankings can be at most $m!$.*

*Proof.* Let $(\mathcal{C}, \Pi, k, r)$ be an arbitrary instance of DISTINCT KEMENY RANKING AGGREGATION. We maintain a dynamic programming table $\mathcal{T}$ indexed by the set of all possible non-empty subsets of $\mathcal{C}$. For a subset $\mathcal{S} \subseteq \mathcal{C}, \mathcal{S} \neq \emptyset$, the table entry $\mathcal{T}[\mathcal{S}]$ stores at most $\min\{r, |\mathcal{S}|!\}$ distinct rankings on $\mathcal{S}$ which have the least Kemeny score when the votes are restricted to $\mathcal{S}$. Let us define $\kappa = \min\{r, |\mathcal{S}|!\}$. We initialize table $\mathcal{T}$ for the trivial cases like $\mathcal{T}[\mathcal{S}] = ()$ when $|\mathcal{S}| = 0$, $\mathcal{T}[\mathcal{S}] =$ (the element from $\mathcal{S}$) when $|\mathcal{S}| = 1$ and $\mathcal{T}[\mathcal{S}] = (x \succ y)$ when $\mathcal{S} = \{x, y\}$ and $x \succ y$ has the least Kemeny score when $\Pi$ is restricted to $\{x, y\}$ or $\mathcal{T}[\mathcal{S}] = (x \succ y, y \succ x)$ when $\mathcal{S} = \{x, y\}$ and both $x \succ y$ and $y \succ x$ have the least Kemeny score when $\Pi$ is restricted to $\{x, y\}$. To update the table entry $\mathcal{T}[\mathcal{S}]$ for $|\mathcal{S}| \geq 3$, we include to that entry $\min\{r, |\mathcal{S}|!\}$ rankings that have the least Kemeny score (when the votes are restricted to $\mathcal{S}$) among all rankings of the form $c > \pi$, where $c$ is a candidate in $\mathcal{S}$ and $\pi$ is a ranking stored in $\mathcal{T}[\mathcal{S} \setminus \{c\}]$. Updating each table entry takes at most $\mathcal{O}^\star(r^{\mathcal{O}(1)})$ time. As there are $2^m - 1$ table entries, the running time of our algorithm is at most $\mathcal{O}^\star\left(2^m r^{\mathcal{O}(1)}\right)$.

We now present the proof of correctness of our algorithm. Suppose we have $\mathcal{S} = \{c_1, ..., c_\ell\}$ and $c_1 \succ ... \succ c_\ell$ be a ranking in $\mathcal{T}[\mathcal{S}]$. Then $c_1 \succ ... \succ c_\ell$ is a Kemeny ranking if the votes in $\Pi$ are restricted to $\mathcal{S}$. But then $c_2 \succ ... \succ c_\ell$ is a Kemeny ranking if votes are restricted to $\mathcal{S} \setminus \{c_1\}$. If not, then suppose $c'_2 \succ ... \succ c'_\ell$ be a ranking with Kemeny score less than $c_2 \succ ... \succ c_\ell$. Then the Kemeny score of $c_1 \succ c'_2 \succ ... \succ c'_\ell$ is less than the Kemeny score of $c_1 \succ c_2 \succ ... \succ c_\ell$ contradicting our assumption that $c_1 \succ ... \succ c_\ell$ is a Kemeny ranking when votes restricted to $\mathcal{S}$. Hence, the update procedure of our dynamic programming algorithm is correct. □

Corollary 2 follows from the algorithm presented in the proof of Theorem 2.

**Corollary 2** (⋆). DISTINCT APPROXIMATE KEMENY RANKING AGGREGATION *is FPT parameterized by the number of candidates* $m$.

Our next parameter is the "average pairwise distance (Kendall-Tau distance)" $d$ of the input rankings. We present a dynamic programming based FPT algorithm parameterized by $d$.

**Theorem 3.** *Let* $d$ *be the average KT-distance of an election* $(\Pi, \mathcal{C})$. *There is an* FPT *for* DISTINCT OPT KEMENY RANKING AGGREGATION *parameterized by* $d$ *which runs in time* $\mathcal{O}^\star\left(16^d\right)$.

*Proof.* Let $|\mathcal{C}| = m$, $|\Pi| = n$ and average position of a candidate $c \in \mathcal{C}$ in $\Pi$ is defined as $p_{avg}(c) := \frac{1}{n} \cdot \sum_{v \in \Pi} v(c)$ where $v(c) := |\{c' \in \mathcal{C} : c' \succ c \text{ in } v \in \Pi\}|$.

Formally for an election $(\Pi, \mathcal{C})$, $d := \dfrac{\sum_{v \in \Pi} \sum_{w \in \Pi} d_{\mathrm{KT}}(v, w)}{n \cdot (n-1)}$. Following the proof of both Lemma 6 and Lemma 7 from Betzler et al. [7], we have a set of candidates say $P_i := \{c \in \mathcal{C} \mid p_{avg}(c) - d < i < p_{avg}(c) + d\}$ for each position $i \in [m-1]_0$ in an optimal Kemeny Consensus and we know that $|P_i| \leq 4d \ \forall i \in [m-1]_0$. Our FPT dynamic programming algorithm is an extension of the algorithm presented in Fig. 4. of Sect. 6.4 of [7].

Let the subset of candidates that are forgotten at latest at position $i$, be denoted by $F(i) := P_{i-1} \setminus P_i$ and the subset of candidates that are introduced for the first time at position $i$ be denoted by $I(i) := P_i \setminus P_{i-1}$. We maintain a three dimensional dynamic programming table $\mathcal{T}$ indexed by $\forall i \in [m-1]_0, \forall c \in P_i$ and $\forall P'_i \subseteq P_i \setminus \{c\}$ of size at most $\mathcal{O}\left(16^d \cdot d \cdot m\right)$. We define the partial Kemeny Score pK-score$(c, \mathcal{R}) := \sum_{c' \in \mathcal{R}} \sum_{v \in \Pi} d^{\mathcal{R}}_v(c, c')$ where $d^{\mathcal{R}}_v(c, c') := 0$ if $c \succ_v c'$ and $d^{\mathcal{R}}_v(c, c') := 1$ otherwise and

$\mathcal{R} \subseteq \mathcal{C}$. At each table entry $\mathcal{T}(i, c, P'_i)$, we store a sequence of at most $\min(r, 4d)$ number of partial Kemeny Scores sorted in non-decreasing order by considering and iterating over the entries in $\mathcal{T}(i-1, c', (P'_i \cup F(i)) \setminus \{c'\}) \ \forall c' \in P'_i \cup F(i)$ and we store the tuple

$$\left( \mathcal{T}(i-1, c', (P'_i \cup F(i)) \setminus \{c'\}) + \text{pK-score}(c, (P_i \cup \bigcup_{i < j < m} I(j)) \setminus (P'_i \cup \{c\})) \right)_{c' \in P'_i \cup F(i)}$$

in that table entry unlike storing only the minimum partial Kemeny Score at each table entry. K-score of an election is the Kemeny Score of an optimal Kemeny ranking.

$$\text{K-score}(\Pi, \mathcal{C}) = \sum_{i=0}^{m-2} \text{pK-score}(c_i, \mathcal{R}_i).$$

At each entry of the table candidate $c$ takes position $i$ and all of $P_i'$ take position smaller than $i$. The initialization step is same as the algorithm presented in Fig. 4. of Sect. 6.4 of [7] but the difference lies in the update step of that algorithm. Though we are storing Kemeny score in each table entry, we can enumerate Kemeny ranking(s) from them within asymptotic bound of our current run time by iteratively ordering the candidate(s) for which we get minimum partial Kemeny Score in a particular table entry. We Output first $r$ number of optimal Kemeny rankings whose K-scores are stored in the entry $T(m - 1, c, P_{m-1} \setminus \{c\})$ where $r \leqslant 4d \leqslant 4m^2 << m!$. Correctness of Lemma 8 of [7] ensures the correctness of our algorithm for generating at most $min\,(r, 4d)$ number of optimal Kemeny Rankings.

Updating each table entry takes time at most $\min(r, 4d) \cdot (4d + nm \log m)$ time. Hence, the overall runtime is bounded above by $\mathcal{O}^* \left( 16^d \right)$.

We next consider the "maximum range" $r_{max}$ of candidate positions in the input rankings, as our parameter. We again present a dynamic programming based FPT algorithm parameterized by $r_{max}$.

**Theorem 4** (⋆). *Let $r_{max}$ be the maximum candidate position range of an election $(\Pi, \mathcal{C})$. There exists an* FPT *dynamic programming algorithm for* DISTINCT OPT KEMENY RANKING AGGREGATION *parameterized by $r_{max}$ which runs in time* $\mathcal{O}^* \left( 32^{r_{max}} \right)$.

Our final parameter is the unanimity width of the input rankings. We present a dynamic programming based FPT algorithm.

**Theorem 5.** DISTINCT OPT KEMENY RANK AGGREGATION *admits an* FPT *algorithm in the combined parameter unanimity width $w$ and number of rankings $r$, which runs in time* $\mathcal{O}^* \left( 2^{\mathcal{O}(w)} \cdot r \right)$.

*Proof.* The problem of finding a Kemeny consensus is known to admit an FPT algorithm in the parameter $w$ (Sect. 3, [5]). We adapt this algorithm to prove Theorem 5. Consider an instance $(\mathcal{C}, \Pi, r)$ of DISTINCT OPT KEMENY RANKING AGGREGATION. Let $m$ denote the number of candidates in $\mathcal{C}$, and let $n$ denote the number of voters in $\Pi$. For any candidates $a, b \in \mathcal{C}$, let $cost(a, b)$ denote the number of voters in $\Pi$ who prefer $b$ over $a$. Note that for any linear ordering $\pi$ of candidates, $\text{Kemeny}_\Pi(\pi) = \sum_{a,b \in \mathcal{C}:a \succ b \text{ in } \pi} cost(a, b)$. Let $\rho$ denote the unanimity order of $\Pi$. Let $G_\rho$ denote the cocomparability graph of $\rho$. Using Lemma 3 of [4], let's construct a nice $\rho$-consistent path decomposition, say $\mathcal{P} = (B_1, \ldots, B_{2m})$, of $G_\rho$ of width $w' \leq 5w + 4$ in time $\mathcal{O}\left(2^{\mathcal{O}(w)} \cdot m\right)$. For each $1 \leq i \leq 2m$,

- Let $forg(i)$ denote the set of candidates that have been forgotten up to $i^{th}$ bag. That is, $forg(i) = \left(B_1 \cup \ldots \cup B_{i-1}\right) \setminus B_i$.
- For each candidate $v \in B_i$, let $\mathcal{A}(i, v)$ denote the cost incurred by the virtue of placing all candidates of $forg(i)$ before $v$. That is, $\mathcal{A}(i, v) = \sum_{u \in forg(i)} cost(u, v)$.

- For each candidate $v \in B_i$ and each $T \subseteq B_i \setminus \{v\}$, let $\mathcal{B}(i,v,T)$ denote the cost incurred by the virtue of placing all candidates of $T$ before $v$. That is, $\mathcal{B}(i,v,T) = \sum_{u \in T} cost(u,v)$.

- For each $T \subseteq B_i$, let $C(i,T)$ be a set that consists of first $min(r, |forg(i) \uplus T|!)$ orderings, along with their Kemeny scores, if all linear extensions of $\rho$ on $forg(i) \uplus T$ were to be sorted in ascending order of their Kemeny scores. That is, $C(i,T)$ consists of the tuples $(\pi_1, k_1), (\pi_2, k_2), \ldots$, where $\pi_1, \pi_2, \ldots$ are the first $min(r, |forg(i) \uplus T|!)$ orderings in the sorted order, and $k_1, k_2, \ldots$ are their respective Kemeny scores.

Recall that every Kemeny consensus extends $\rho$ (Lemma 1, [7]). So, if all linear extensions of $\rho$ on $\mathcal{C}$ were to be sorted in ascending order of their Kemeny scores, then all Kemeny consensuses would appear in the beginning. Thus, $(\mathcal{C}, \Pi, r)$ is a YES instance if and only if $C(2m, \phi)$ contains $r$ orderings of the same Kemeny score. Let's use DP to find all $\mathcal{A}(\cdot, \cdot)$'s, $\mathcal{B}(\cdot, \cdot, \cdot)$'s and $C(\cdot, \cdot)$'s as follows:

- First, let's compute and store $\mathcal{A}(i, \cdot)$'s in a table for $i = 1, \ldots, 2m$ (in that order) in time $\mathcal{O}(w' \cdot m \cdot \log(m \cdot n))$ as follows: We set $\mathcal{A}(1, u) = 0$, where $u$ denotes the candidate introduced by $B_1$. Now, consider $i \geq 2$ and a candidate $v \in B_i$. Let's describe how to find $\mathcal{A}(i, v)$.
  **Introduce Node:** Suppose that $B_i$ introduces a candidate, say $x$. Note that $forg(i) = forg(i-1)$. So, if $v \neq x$, we set $\mathcal{A}(i, v) = \mathcal{A}(i-1, v)$. Now, suppose that $v = x$. Let's show that $cost(u, x) = 0$ for all $u \in forg(i)$. Consider $u \in forg(i)$. In $\mathcal{P}$, $u$ is forgotten before $x$ is introduced. So, $\{u, x\} \notin E(G_\rho)$. That is, $u$ and $x$ are comparable in $\rho$. Also, due to $\rho$-consistency of $\mathcal{P}$, we have $(x, u) \notin \rho$. Therefore, $(u, x) \in \rho$. That is, all voters in $\Pi$ prefer $u$ over $x$. So, $cost(u, x) = 0$. Thus, we set $\mathcal{A}(i, x) = 0$.
  **Forget Node:** Suppose that $B_i$ forgets a candidate, say $x$. Note that $forg(i) = forg(i-1) \uplus \{x\}$. So, we set $\mathcal{A}(i, v) = \mathcal{A}(i-1, v) + cost(x, v)$.

- Next, let's compute and store all $\mathcal{B}(\cdot, \cdot, \cdot)$'s in a table in time $\mathcal{O}(w' \cdot 2^{w'} \cdot m \cdot \log(m \cdot n))$ as follows: Consider $1 \leq i \leq 2m$ and $v \in B_i$. We have $\mathcal{B}(i, v, \phi) = 0$. Let's set $\mathcal{B}(i, v, T)$ for non-empty subsets $T \subseteq B_i \setminus \{v\}$ (in ascending order of their sizes) as $\mathcal{B}(i, v, T \setminus \{u\}) + cost(u, v)$, where $u$ is an arbitrary candidate in $T$.

- Next, let's compute and store $C(i, \cdot)$'s in a table in time $\mathcal{O}(w' \cdot 2^{w'} \cdot m^2 \cdot r \cdot \log(m \cdot n \cdot r))$ for $i = 1, \ldots, 2m$ (in that order) as follows: We set $C(1, \phi) = \{(, 0)\}$ and $C(1, \{u\}) = \{(u, 0)\}$, where $u$ denotes the candidate introduced by $B_1$. Now, consider $i \geq 2$. Let's describe how to find $C(i, \cdot)$'s.
  **Introduce node:** Suppose that $B_i$ introduces a candidate, say $x$. For each $T \subseteq B_i$ that does not contain $x$, we set $C(i, T) = C(i-1, T)$.
  Now, let's find $C(i, T)$ for all subsets $T \subseteq B_i$ that contain $x$ (in ascending order of their sizes) as follows: First, let's consider $T = \{x\}$. Recall that $(u, x) \in \rho$ for all $u \in forg(i)$. So, $x$ is the last candidate in all linear extensions of $\rho$ on $forg(i) \uplus \{x\}$. Also, in any such ordering, the pairs of the form $(u, x)$, where $u \in forg(i)$, contribute 0 to Kemeny score. Thus, we put the tuples $(\pi_1 > x, s_1), (\pi_2 >$

$x, s_2), \ldots$ in $C(i, \{x\})$, where $(\pi_1, s_1), (\pi_2, s_2), \ldots$ denote the tuples of $C(i-1, \phi)$, and $\pi_1 > x, \pi_2 > x, \ldots$ denote the orderings obtained by appending $x$ to $\pi_1, \pi_2, \ldots$ respectively.

Now, let's consider a subset $T \subseteq B_i$ of size $\geq 2$ that contains $x$. Let's describe how to find $C(i, T)$. Let $\Delta(i, T)$ denote the set of all candidates $c \in T$ such that $c$ is not unanimously preferred over any other candidate of $forg(i) \uplus T$. That is, there is no other candidate $u \in forg(i) \uplus T$ such that $(c, u) \in \rho$. Recall that $x$ appears after all candidates of $forg(i)$ in any linear extension of $\rho$ on $forg(i) \uplus T$. So, it is clear that in any such ordering, the last candidate (say $y$) belongs to $\Delta(i, T)$. Moreover,

- The pairs of the form $(u, y)$, where $u \in forg(i)$, together contribute $\mathcal{A}(i, y)$ to Kemeny score.
- The pairs of the form $(u, y)$, where $u \in T \setminus \{y\}$, together contribute $\mathcal{B}(i, y, T \setminus \{y\})$ to Kemeny score.

So, to find $C(i, T)$, let's proceed as follows: We compute $\Delta(i, T)$. For each possible choice $y \in \Delta(i, T)$ of the last candidate, let's form a set, say $\Gamma(y)$, that consists of the following tuples:

- $\left( \pi_1^y > y, s_1^y + \mathcal{A}(i, y) + \mathcal{B}(i, y, T \setminus \{y\}) \right)$
- $\left( \pi_2^y > y, s_2^y + \mathcal{A}(i, y) + \mathcal{B}(i, y, T \setminus \{y\}) \right)$ and so on

where $(\pi_1^y, s_1^y), (\pi_2^y, s_2^y), \ldots$ denote the tuples of $C(i, T \setminus \{y\})$, and $\pi_1^y > y, \pi_2^y > y, \ldots$ denote the orderings obtained by appending $y$ to $\pi_1^y, \pi_2^y, \ldots$ respectively. Finally, let's sort all tuples of $\biguplus_{y \in \Delta(i,T)} \Gamma(y)$ in ascending order of their Kemeny scores, and put the first $min(r, |forg(i) \uplus T|!)$ of them in $C(i, T)$.

**Forget node:** Suppose that $B_i$ forgets a candidate, say $x$. For each $T \subseteq B_i$, as $forg(i) \uplus T = forg(i-1) \uplus (T \uplus \{x\})$, we set $C(i, T) = C(i-1, T \uplus \{x\})$.

This concludes the proof of Theorem 5.

**Corollary 3.** DISTINCT APPROXIMATE KEMENY RANKING AGGREGATION *is FPT in the combined parameter unanimity width $w$ and number of rankings $r$.*

*Proof.* Consider an instance DISTINCT APPROXIMATE KEMENY RANKING AGGREGATION. As in the algorithm described in the proof of Theorem 5, we find all $\mathcal{A}(\cdot, \cdot)$'s, $\mathcal{B}(\cdot, \cdot, \cdot)$'s and $C(\cdot, \cdot)$'s. Note that $(\mathcal{C}, \Pi, \lambda, r)$ is a YES instance if and only if $C(2m, \phi)$ contains $r$ orderings, and the Kemeny score of the $r^{th}$ ordering is at most $\lambda$ times the Kemeny score of the first ordering. The overall running time of the algorithm is at most $\mathcal{O}^*\left(2^{\mathcal{O}(w)} \cdot r\right)$. This proves Corollary 3.

Our last result is an FPT algorithm for DISTINCT APPROXIMATE KEMENY RANKING AGGREGATION parameterized by the average Kendall-Tau distance $d$ and the approximation parameter $\lambda$. Here we aim to relate the position of a candidate $c$ in a $\lambda$-approximate ranking $\pi$, *i.e.* a ranking whose Kemeny Score denoted by K-score $(\pi)$ has value at most $\lambda \cdot k_{OPT}$ where $k_{OPT}$ denotes the optimal Kemeny Score, to its average position in the set of votes $\Pi$ denoted by $p_{avg}(c)$.

**Lemma 1.** $p_{avg}(c) - \lambda \cdot d \leq \pi(c) \leq p_{avg}(c) + \lambda \cdot d$ *where $\pi(c)$ denotes position of $c$ in $\pi$ and $d$ is average KT-distance.*

*Proof.* There can be two cases for a vote $v \in \Pi$.

**Case 1.** $v(c) \leq \pi(c)$

In Case 1 there are $\pi(c) - 1$ candidates that appear before $c$ in $\pi$. Note that at most $v(c) - 1$ of them can appear before $c$ in $v$. Hence, at least $\pi(c) - v(c)$ of them must appear after $c$ in $v$. Thus, $d_{KT}(v, \pi) \geq \pi(c) - v(c)$.

**Case 2.** $v(c) > \pi(c)$

Here in Case 2, we come up with $d_{KT}(v, \pi) \geq v(c) - \pi(c)$ arguing similarly to Case 1.

$$
\begin{aligned}
\text{K-score}(\pi) &= \sum_{v \in \Pi} d_{KT}(v, \pi) \\
&= \sum_{v \in \Pi : v(c) \leq \pi(c)} d_{KT}(v, \pi) + \sum_{v \in \Pi : v(c) > \pi(c)} d_{KT}(v, \pi) \\
&\geq \sum_{\substack{v \in \Pi: \\ v(c) \leq \pi(c)}} (\pi(c) - v(c)) + \sum_{\substack{v \in \Pi: \\ v(c) > \pi(c)}} (v(c) - \pi(c)) \quad [\text{using } Case\,1 \text{ and } Case\,2]
\end{aligned}
$$

$$(1)$$

Note that

$$
\begin{aligned}
&\sum_{v \in \Pi : v(c) \leq \pi(c)} (\pi(c) - v(c)) + \sum_{v \in \Pi : v(c) > \pi(c)} (v(c) - \pi(c)) \\
&= \sum_{v \in \Pi} v(c) - 2 \sum_{\substack{v \in \Pi: \\ v(c) \leq \pi(c)}} v(c) + \pi(c) \cdot (2 \cdot |\{v \in \Pi : v(c) \leq \pi(c)\}| - n) \\
&= n \cdot p_{avg}(c) - n\pi(c) - 2 \sum_{\substack{v \in \Pi: \\ v(c) \leq \pi(c)}} v(c) + \pi(c) \cdot (2 \cdot |\{v \in \Pi : v(c) \leq \pi(c)\}|) \\
&\geq n \left( p_{avg}(c) - \pi(c) \right)
\end{aligned}
$$

$$(2)$$

Similarly,

$$
\begin{aligned}
&\sum_{v \in \Pi : v(c) \leq \pi(c)} (\pi(c) - v(c)) + \sum_{v \in \Pi : v(c) > \pi(c)} (v(c) - \pi(c)) \\
&= -\sum_{v \in \Pi} v(c) + 2 \sum_{\substack{v \in \Pi: \\ v(c) > \pi(c)}} v(c) + \pi(c) \cdot (-2 \cdot |\{v \in \Pi : v(c) > \pi(c)\}| + n) \\
&= -n \cdot p_{avg}(c) + n\pi(c) + 2 \sum_{\substack{v \in \Pi: \\ v(c) > \pi(c)}} v(c) - \pi(c) \cdot (2 \cdot |\{v \in \Pi : v(c) > \pi(c)\}|) \\
&\geq -n \left( p_{avg}(c) - \pi(c) \right)
\end{aligned}
$$

$$(3)$$

Now let's show that

$$
\text{K-score}(\pi) \leq \lambda \cdot n \cdot d
$$

$$(4)$$

We have

$$d = \frac{\sum\limits_{v \in \Pi} \sum\limits_{w \in \Pi} d_{\mathrm{KT}}(v, w)}{n \cdot (n-1)} \geq \frac{n \cdot \sum\limits_{w \in \Pi, w \neq v^\star} d_{\mathrm{KT}}(v^\star, w)}{n \cdot (n-1)} > \frac{\sum\limits_{w \in \Pi, w \neq v^\star} d_{\mathrm{KT}}(v^\star, w)}{n}$$

$$\left[ \exists v^\star \in \Pi \text{ for which } \sum_{w \in \Pi, w \neq v^\star} d_{\mathrm{KT}}(v^\star, w) \text{ is minimum} \right]$$

$$\implies \text{K-score}(v^\star) < n \cdot d$$

So, $k_{OPT} \leq \text{K-score}(v^\star) < n \cdot d$ \hfill (5)

K-score$(\pi) \leq \lambda \cdot k_{OPT} < \lambda \cdot n \cdot d$ [Using Eq. (5) and proving Eq. (4)] \hfill (6)

Now $\lambda \cdot n \cdot d \geq$ K-score$(\pi) \geq n \cdot (p_{avg}(c) - \pi(c))$ [Using Eq. (1), (2) & (4)]

$$\implies p_{avg}(c) - \lambda \cdot d \leq \pi(c) \tag{7}$$

Again $\lambda \cdot n \cdot d \geq -n \cdot (p_{avg}(c) - \pi(c))$ [Using Eq. (1), (3) & (4)]

$$\implies \pi(c) \leq p_{avg}(c) + \lambda \cdot d \tag{8}$$

Hence $p_{avg}(c) - \lambda \cdot d \leq \pi(c) \leq p_{avg}(c) + \lambda \cdot d$ [Using Eq. (7) & (8)] \hfill (9)

Equation (9) concludes the proof of Lemma 1.

The following Lemma 2 depends on the Lemma 7 from [7].

**Lemma 2** *(⋆).* $|P_i| \leq 4\lambda d - 1 \; \forall i \in [m-1]_0$

We now use the dynamic programming algorithm of Theorem 3 to claim the following Theorem 6. Its proof of correctness follows from Lemma 2.

**Theorem 6.** *There exists an FPT dynamic programming algorithm for* DISTINCT APPROXIMATE KEMENY RANKING AGGREGATION *parameterized by both $\lambda$ and d which runs in time* $\mathcal{O}^*(16^{\lambda d})$.

## 4    Concluding Remarks and Future Work

We consider the problem of finding distinct rankings that have a good Kemeny score in either exact or approximate terms, and propose algorithms that are tractable for various natural parameterizations of the problem. We show that many optimal or close to optimal solutions can be computed without significant increase in the running time compared with the algorithms to output a single solution, which is in sharp contrast with the diverse version of the problem. We also establish a complete comparison between the five natural parameters associated with the problem, and demonstrate these relationships through experiments.

We propose three main themes for future work. The first would be to extend these studies to other voting rules, and possibly identify meta theorems that apply to classes of voting rules. The second would be to understand if the structural parameters that we studied are correlated with some natural distance notion on the solution space: in other words, for a given distance notion, do all similar-looking instances have similar parameter values? Finally, we would also like to establish algorithmic lower bounds for the question of finding a set of diverse solutions that match the best known algorithms in the current literature.

# References

1. Agrawal, R.: The continuum-armed bandit problem. SIAM J. Control Optim. **33**(6), 1926–1951 (1995). https://doi.org/10.1137/S0363012992237273
2. Ailon, N., Charikar, M., Newman, A.: Aggregating inconsistent information: ranking and clustering. J. ACM (JACM) **55**(5), 1–27 (2008)
3. Ailon, N., Charikar, M., Newman, A.: Aggregating inconsistent information: ranking and clustering. J. ACM **55**(5), 23:1–23:27 (2008). https://doi.org/10.1145/1411509.1411513
4. Arrighi, E., Fernau, H., Lokshtanov, D., de Oliveira Oliveira, M., Wolf, P.: Diversity in Kemeny rank aggregation: a parameterized approach. In: Zhou, Z. (ed.) Proceedings of the Thirtieth International Joint Conference on Artificial Intelligence, IJCAI 2021, Virtual Event/Montreal, Canada, 19–27 August 2021, pp. 10–16. ijcai.org (2021). https://doi.org/10.24963/ijcai.2021/2
5. Arrighi, E., Fernau, H., de Oliveira Oliveira, M., Wolf, P.: Width notions for ordering-related problems. In: Saxena, N., Simon, S. (eds.) 40th IARCS Annual Conference on Foundations of Software Technology and Theoretical Computer Science (FSTTCS 2020). Leibniz International Proceedings in Informatics (LIPIcs), vol. 182, pp. 9:1–9:18. Schloss Dagstuhl - Leibniz-Zentrum für Informatik, Dagstuhl (2020). https://doi.org/10.4230/LIPIcs.FSTTCS.2020.9, https://drops.dagstuhl.de/entities/document/10.4230/LIPIcs.FSTTCS.2020.9
6. Bartholdi, J., Tovey, C.A., Trick, M.A.: Voting schemes for which it can be difficult to tell who won the election. Soc. Choice Welfare **6**(2), 157–165 (1989)
7. Betzler, N., Fellows, M.R., Guo, J., Niedermeier, R., Rosamond, F.A.: Fixed-parameter algorithms for Kemeny rankings. Theor. Comput. Sci. **410**(45), 4554–4570 (2009). https://doi.org/10.1016/j.tcs.2009.08.033
8. Biedl, T., Brandenburg, F.J., Deng, X.: On the complexity of crossings in permutations. Discret. Math. **309**(7), 1813–1823 (2009)
9. Conitzer, V., Davenport, A., Kalagnanam, J.: Improved bounds for computing Kemeny rankings. In: AAAI, vol. 6, pp. 620–626 (2006)
10. Cornaz, D., Galand, L., Spanjaard, O.: Kemeny elections with bounded single-peaked or single-crossing width. In: Rossi, F. (ed.) IJCAI 2013, Proceedings of the 23rd International Joint Conference on Artificial Intelligence, Beijing, China, 3–9 August 2013, pp. 76–82. IJCAI/AAAI (2013). http://www.aaai.org/ocs/index.php/IJCAI/IJCAI13/paper/view/6944
11. Cygan, M., et al.: Parameterized Algorithms. Springer, Cham (2015). https://doi.org/10.1007/978-3-319-21275-3
12. Davenport, A., Kalagnanam, J.: A computational study of the Kemeny rule for preference aggregation. In: AAAI, vol. 4, pp. 697–702 (2004)
13. De, K., Mittal, H., Dey, P., Misra, N.: Parameterized aspects of distinct Kemeny rank aggregation (2023). https://doi.org/10.48550/ARXIV.2309.03517
14. Dwork, C., Kumar, R., Naor, M., Sivakumar, D.: Rank aggregation methods for the web. In: Shen, V.Y., Saito, N., Lyu, M.R., Zurko, M.E. (eds.) Proceedings of the Tenth International World Wide Web Conference, WWW 10, Hong Kong, China, 1–5 May 2001, pp. 613–622. ACM (2001). https://doi.org/10.1145/371920.372165
15. Dwork, C., Kumar, R., Naor, M., Sivakumar, D.: Rank aggregation methods for the web. In: Proceedings of the 10th International Conference on World Wide Web, pp. 613–622 (2001)
16. Guo, J., Hüffner, F., Niedermeier, R.: A structural view on parameterizing problems: distance from triviality. In: Downey, R., Fellows, M., Dehne, F. (eds.) IWPEC 2004. LNCS, vol. 3162, pp. 162–173. Springer, Heidelberg (2004). https://doi.org/10.1007/978-3-540-28639-4_15
17. Hebrard, E., Hnich, B., O'Sullivan, B., Walsh, T.: Finding diverse and similar solutions in constraint programming. In: AAAI, vol. 5, pp. 372–377 (2005)

18. Hebrard, E., O'Sullivan, B., Walsh, T.: Distance constraints in constraint satisfaction. In: IJCAI, vol. 2007, pp. 106–111 (2007)
19. Hemaspaandra, E., Spakowski, H., Vogel, J.: The complexity of Kemeny elections. Theoret. Comput. Sci. **349**(3), 382–391 (2005)
20. Kemeny, J.G.: Mathematics without numbers. Daedalus **88**(4), 577–591 (1959). http://www.jstor.org/stable/20026529
21. Kenyon-Mathieu, C., Schudy, W.: How to rank with few errors. In: Proceedings of the Thirty-Ninth Annual ACM Symposium on Theory of Computing, pp. 95–103 (2007)
22. Klamler, C.: The Dodgson ranking and its relation to Kemeny's method and Slater's rule. Soc. Choice Welfare **23**(1), 91–102 (2004)
23. Pennock, D.M., Horvitz, E., Giles, C.L.: Social choice theory and recommender systems: analysis of the axiomatic foundations of collaborative filtering. In: Kautz, H.A., Porter, B.W. (eds.) Proceedings of the Seventeenth National Conference on Artificial Intelligence and Twelfth Conference on on Innovative Applications of Artificial Intelligence, 30 July–3 August 2000, Austin, Texas, USA, pp. 729–734. AAAI Press/The MIT Press (2000). http://www.aaai.org/Library/AAAI/2000/aaai00-112.php
24. Schalekamp, F., Zuylen, A.V.: Rank aggregation: together we're strong. In: 2009 Proceedings of the Eleventh Workshop on Algorithm Engineering and Experiments (ALENEX), pp. 38–51. SIAM (2009)
25. Slivkins, A., Radlinski, F., Gollapudi, S.: Learning optimally diverse rankings over large document collections. In: Proceedings of the 27th International Conference on International Conference on Machine Learning, ICML 2010, pp. 983–990. Omnipress, Madison (2010)
26. Van Zuylen, A., Williamson, D.P.: Deterministic pivoting algorithms for constrained ranking and clustering problems. Math. Oper. Res. **34**(3), 594–620 (2009)
27. van Zuylen, A., Williamson, D.P.: Deterministic algorithms for rank aggregation and other ranking and clustering problems. In: Kaklamanis, C., Skutella, M. (eds.) WAOA 2007. LNCS, vol. 4927, pp. 260–273. Springer, Heidelberg (2008). https://doi.org/10.1007/978-3-540-77918-6_21
28. Zwicker, W.S.: Cycles and intractability in a large class of aggregation rules. J. Artif. Intell. Res. **61**, 407–431 (2018)

# Monitoring Edge-Geodetic Sets in Graphs: Extremal Graphs, Bounds, Complexity

Florent Foucaud[1], Pierre-Marie Marcille[2], Zin Mar Myint[3(✉)], R. B. Sandeep[3], Sagnik Sen[3], and S. Taruni[3]

[1] Université Clermont Auvergne, CNRS, Clermont Auvergne INP, Mines Saint-Étienne LIMOS, 63000 Clermont-Ferrand, France
`florent.foucaud@uca.fr`

[2] Univ. Bordeaux, CNRS, Bordeaux INP, LaBRI, UMR 5800, F-33400 Talence, France
`pierre-marie.marcille@u-bordeaux.fr`

[3] Indian Institute of Technology Dharwad, Dharwad, India
`zinmarmyint.zmm86@gmail.com`, `sandeeprb@iitdh.ac.in`, `sen007isi@gmail.com`, `taruni.sridhar@gmail.com`

**Abstract.** A monitoring edge-geodetic set, or simply an MEG-set, of a graph $G$ is a vertex subset $M \subseteq V(G)$ such that given any edge $e$ of $G$, $e$ lies on every shortest $u$-$v$ path of $G$, for some $u, v \in M$. The monitoring edge-geodetic number of $G$, denoted by $meg(G)$, is the minimum cardinality of such an MEG-set. This notion provides a graph theoretic model of the network monitoring problem.

In this article, we compare $meg(G)$ with some other graph theoretic parameters stemming from the network monitoring problem and provide examples of graphs having prescribed values for each of these parameters. We also characterize graphs $G$ that have $V(G)$ as their minimum MEG-set, which settles an open problem due to Foucaud *et al.* (CALDAM 2023). We also provide a general upper bound for $meg(G)$ for sparse graphs in terms of their girth, and later refine the upper bound using the chromatic number of $G$. We examine the change in $meg(G)$ with respect to two fundamental graph operations: clique-sum and subdivisions. In both cases, we provide a lower and an upper bound of the possible amount of changes and provide (almost) tight examples. Finally, we prove that the decision version of the problem of finding $meg(G)$ is NP-complete even for the family of 3-degenerate, 2 apex graphs, improving the existing result by Haslegrave (Discrete Applied Mathematics 2023).

**Keywords:** Geodetic set · Monitoring edge geodetic set · $k$-clique sum · Subdivisions · Chromatic number · Girth · Computational complexity

© The Author(s), under exclusive license to Springer Nature Switzerland AG 2024
S. Kalyanasundaram and A. Maheshwari (Eds.): CALDAM 2024, LNCS 14508, pp. 29–43, 2024.
https://doi.org/10.1007/978-3-031-52213-0_3

# 1  Introduction

In the field of network monitoring, the networking components are monitored for faults and evaluated to maintain and optimize their availability. In order to detect failures, one of the popular methods for such monitoring processes involves setting up distance probes [2–4,13]. At any given time, a distance probe can measure the distance to any other probe in the network. If there is any failure in the connection, then the probes should be able to detect it as there would be a change in the distances between the components. Such networks can be modeled by graphs whose vertices represent the components and the edges represent the connections between them. We select a subset of vertices of the graph and call them probes. This concept of probes that can measure distances in graphs has many real-life applications, for example, it is useful in the fundamental task of routing [10,15], or using path-oriented tools to monitor IP networks [4], or problems concerning network verification [2,3,5]. Based on the requirements of the networks, there have been various related parameters that were defined on graphs in order to study the problem and come up with an effective solution. To name a few, we may mention the geodetic number [8,9,12,16], the edge-geodetic number [1,25], the strong edge-geodetic number [18,22], the distance-edge monitoring number [13], and the monitoring edge-geodetic number [14,17]. The focus of this article is on studying the *monitoring edge-geodetic number* of a graph. We deal with simple graphs, unless otherwise stated.

*Note:* We have omitted some proofs due to space constraints.

## 1.1  Preliminaries

Given a graph $G$, a *monitoring edge-geodetic set* of $G$, or simply, an *MEG-set* of a graph $G$ is a vertex subset $M \subseteq V(G)$ that satisfies the following: given any edge $e$ in $G$, there exists $u, v \in M$ such that $e$ lies on all shortest paths between $u$ and $v$. In such a scenario, we say that the vertices $u, v$ monitor the edge $e$. The monitoring edge-geodetic number, denoted by $meg(G)$, is the smallest size of an MEG-set of $G$.

There are some other related parameters whose definitions are relevant in our context. For convenience, we list them below.

– A *geodetic set* of a graph $G$ is a vertex subset $S \subseteq V(G)$ such that every vertex of $G$ lies on some shortest path between two vertices $u, v \in S$. The *geodetic number*, denoted by $g(G)$, is the minimum $|S|$, where $S$ is a geodetic set of $G$. The concept was introduced by Harary et al. in 1993 [16] and received considerable attention since then, both from the structural side [8,9,12] and from the algorithmic side [6,20].

– An *edge-geodetic set* of a graph $G$ is a vertex subset $S \subseteq V(G)$ such that every edge of $G$ lies on some shortest path between two vertices $u, v \in S$. The *edge-geodetic number*, denoted by $eg(G)$, is the minimum $|S|$, where $S$ is an edge-geodetic set of $G$. This was introduced in 2003 by Atici et al. [1] and further studied from the structural angle [25] as well as algorithmic angle [7,11].

– A *strong edge-geodetic set* of a graph $G$ is a vertex subset $S \subseteq V(G)$ and an assignment of a particular shortest $u$-$v$ path $P_{uv}$ to each pair of distinct vertices $u, v \in S$ such that every edge of $G$ lies on $P_{uv}$ for some $u, v \in S$. The *strong edge-geodetic number*, denoted by $seg(G)$, is the minimum $|S|$, where $S$ is a geodetic set of $G$. This concept was introduced in 2017 by Manuel et al. [22]. See [18] for some structural studies, and [11] for some algorithmic results.

## 1.2 Motivation of Our Results and Organization of the Paper

– In Sect. 2, we explore the relation between the parameters geodetic number, edge-geodetic number, strong edge-geodetic number, and monitoring edge-geodetic number. As one may have noticed, the above-mentioned graph parameters are closely related and have a natural relation of inclusion. In this section, we construct examples of graphs having prescribed values of the above-mentioned parameters.
– In Sect. 3 we answer an open question posed by Foucaud, Krishna and Ramasubramony Sulochana [14] on characterizing graphs whose minimum MEG-set is the entire vertex set. We provide such a characterization by proving a necessary and sufficient condition of when a vertex $v$ is part of every MEG-set of graph $G$. Additionally, we also prove a sufficient condition of when a vertex is never part of any minimum MEG-set of the graph.
– In Sect. 4, we provide upper bounds on $meg(G)$, where $G$ is a sparse graph. Our upper bound is a function of the order of $G$, and its girth. A refinement of the upper bound is provided using the chromatic number of $G$.
– In Sect. 5, we explore the effect of two fundamental graph operations, namely, the clique-sum and the subdivision on $mcg(G)$. We show that $meg(G)$ is both lower and upper bounded by functions related to the operations and that the bounds are (almost) tight.
– In Sect. 6, we answer another open question that was posed in [14] regarding the computational complexity of finding $meg(G)$. The general question was recently settled by Haslegrave [17], and we give an alternative proof. In fact, our result is stronger, as we show that the decision version of the problem of finding $meg(G)$ is NP-complete even for the restricted class of 3-degenerate, 2-apex graphs.
– In Sect. 7, we share our concluding remarks which also contain suggestions for future works in this direction.

## 2 Relation Between Network Monitoring Parameters

From the definitions, notice that any strong edge-geodetic set is also an edge-geodetic set, and any edge-geodetic set is also a geodetic set (if the graph has no isolated vertices). Moreover, every MEG-set $M$ of a graph is indeed a strong edge-geodetic set, as every edge of $G$ is contained in all the shortest paths between some pair of vertices in $M$. Thus, one can observe the following relations [14],

$$g(G) \leq eg(G) \leq seg(G) \leq meg(G).$$

*Example 1.* Notice that, for any complete graph $K_n$ on $n \geq 2$ vertices, the values of all the parameters are equal to $n$. That is, equality holds in all the inequalities of the above chain of inequalities.

On the other hand, Fig. 1 gives an example of a graph where all the inequalities of the above chain of inequalities are strict. To be specific, in this particular example, the values of the parameters increase exactly by one in each step.    □

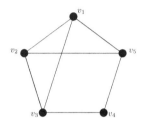

**Fig. 1.** A graph $G$ with $2 = g(G) < 3 = eg(G) < 4 = seg(G) < 5 = meg(G)$. Note that, a minimum geodetic set of $G$ is $\{v_3, v_5\}$, a minimum edge-geodetic set of $G$ is $\{v_1, v_2, v_4\}$, a minimum strong edge-geodetic set of $G$ is $\{v_1, v_2, v_3, v_4\}$, and a minimum MEG-set of $G$ is $\{v_1, v_2, v_3, v_4, v_5\}$.

It is natural to ask the question, given four positive integers $a, b, c, d$ satisfying $2 \leq a \leq b \leq c \leq d$, is there a graph $G_{a,b,c,d}$ such that we have $g(G_{a,b,c,d}) = a$, $eg(G_{a,b,c,d}) = b$, $seg(G_{a,b,c,d}) = c$, and $meg(G_{a,b,c,d}) = d$? We provide a positive answer to this question except for some specific cases. The rest of the section deals with the construction of such graphs.

**Theorem 1.** *For any positive integers $4 \leq a \leq b \leq c \leq d$ satisfying $d \neq c + 1$, there exists a connected graph $G_{a,b,c,d}$ with $g(G) = a, eg(G) = b, seg(G) = c$ and $meg(G) = d$.*

*Proof.* We begin the proof by describing the construction of $G_{a,b,c,d}$.

*Construction of $G_{a,b,c,d}$:* In the first phase of the construction, we start with a $K_{2,2+b-a}$, where the partite set of size two has the vertices $x_1$ and $y$, and the partite set of size $(2 + b - a)$ has the vertices $z_1, z_2$ and $w_1, w_2, \cdots, w_{b-a}$. Moreover, we add some edges in such a way that the set

$$W = \{w_1, w_2, \cdots, w_{b-a}\}$$

becomes a clique. We also add the edge $z_2 w_1$.

In the second phase of the construction, we add $(c - b + 1)$ parallel edges between the vertices $z_1$ and $z_2$. After that we subdivide (once) each of the above-mentioned parallel edges and name the degree two vertices created due to the subdivisions as $v_1, v_2, \cdots, v_{c-b+1}$. The set of these vertices created by the subdivisions is given by

$$V = \{v_1, v_2, \cdots, v_{c-b+1}\}.$$

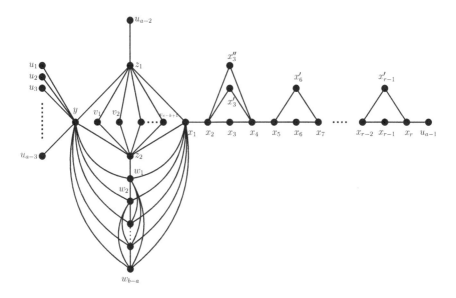

**Fig. 2.** The structure of $G_{a,b,c,d}$.

In the third phase of the construction, we add $(a - 3)$ pendant neighbors $u_1, u_2, \cdots, u_{a-3}$ to $y$, and one pendant neighbor $u_{a-2}$ to $z_1$. Moreover, we attach a long path $x_1 x_2 \cdots x_r u_{a-1}$ with the vertex $x_1$, where $u_{a-1}$ is a pendant vertex and $r = 3\lfloor \frac{d-c}{2} \rfloor + 1$. Next we will add a false twin $x'_{3i}$ to the vertices of the form $x_{3i}$ for $i \in \{1, 2, \cdots, \lfloor \frac{d-c}{2} \rfloor\}$. Additionally, if $(d - c)$ is odd, then we will add another twin $x''_3$ to the vertex $x_3$. For convenience,

$$U = \{u_1, u_2, \cdots, u_{a-1}\}$$

will denote the set of all pendents and

$$X = \{x_{3i}, x'_{3i} | i = 1, 2, \cdots, \lfloor \frac{d-c}{2} \rfloor\} \cup \{x''_3\}.$$

Note that, $x''_3$ exists in $X$ if and only if $(d - c)$ is odd. This completes the description of the construction of the graph $G_{a,b,c,d}$ (see Fig. 2 for a pictorial reference).

As $U$ is the set of all pendants, we know that it will be part of any geodetic set, edge-geodetic set, strong edge-geodetic set, and monitoring edge-geodetic set [14]. However, the vertices of $U$ cannot cover the vertices of $V$ using shortest paths between the vertices of $U$. Therefore, we need at least one more vertex to form a geodetic set. As $U \cup \{z_2\}$ is a geodetic set of $G$, we can infer that

$$g(G_{a,b,c,d}) = (a - 1) + 1 = a.$$

Next, observe that the vertices of $U$ are not able to cover any edge of the clique $W$ using shortest paths between the vertices of $U$. Moreover, the only way to monitor those edges is by taking $W$ in our edge-geodetic set. Observe that $U \cup W$ is still not an edge-geodetic set as they are not able to cover the edge $z_2 w_1$ by any shortest path between the vertices of $U \cup W$. On the other hand, $U \cup W \cup \{z_2\}$ is an edge-geodetic set. Therefore,

$$eg(G_{a,b,c,d}) = (a - 1) + 1 + (b - a) = b.$$

We know that the vertices of $U$ is in any strong edge-geodetic set. Moreover, note that, the vertices of $W$ must be in any strong edge-geodetic set to cover the edges of the clique $W$. Now, let us see how we can cover the edges of the $(c - b + 1)$ 2-paths between $z_1, z_2$, having the vertices of $V$ as their internal vertex. First of all, if we do not take $z_2$ in our strong edge-geodetic set, we have to take all vertices of $V$. Second of all, if we take $z_2$ in our strong edge-geodetic set, then we have to take either (at least) all but one vertices of $V$, or all but two vertices of $V$ along with $z_1$ in the strong edge-geodetic set. That means, we need to take at least $(c - b + 1)$ additional vertices in the strong edge-geodetic set. Moreover, the set $U \cup W \cup V$ is indeed a strong edge-geodetic set. Thus,

$$seg(G_{a,b,c,d}) = (a - 1) + (b - a) + (c - b + 1) = c.$$

Finally for any MEG-set, we have to take the vertices of $U$ (as they are pendants), the vertices of $W$ (to monitor the edges of the clique $W$), the vertices of $X$ (as they are twins [14]). However, even with these vertices, we cannot monitor the edges of the $(c - b + 1)$ 2-paths between $z_1, z_2$, having the vertices of $V$ as their internal vertex. To do so, we have to take all vertices of $V$ in our MEG-set. However, the set $U \cup W \cup V \cup X$ is an MEG-set. Hence,

$$meg(G_{a,b,c,d}) = (a - 1) + (b - a) + (c - b + 1) + 2\lfloor \frac{d - c}{2} \rfloor + \epsilon = d,$$

where $\epsilon = 0$ (resp., 1) if $(d - c)$ is (even (resp., odd), and $d \neq c + 1$. This completes the proof.  □

## 3  Conditions for a Vertex Being in All or No Optimal MEG-sets

In their introductory paper on monitoring edge-geodetic sets, Foucaud, Krishna and Ramasubramony Sulochana [14] asked to characterize the graphs $G$ having $meg(G) = |V(G)|$. We provide a definitive answer to their question, and to this end, we give a necessary and sufficient condition for a vertex to be in any MEG-set of a graph.

**Theorem 2.** *Let $G$ be a graph. A vertex $v \in V(G)$ is in every MEG-set of $G$ if and only if there exists $u \in N(v)$ such that any induced 2-path $uvx$ is part of a 4-cycle.*

*Proof.* For the necessary condition, let us assume that a vertex $v \in V(G)$ is in every MEG-set of $G$. We have to prove that there exists $u \in N(v)$, such that any induced 2-path $uvx$ is part of a 4-cycle. We prove it by contradiction.

Suppose for every $u \in N(v)$, there exists an induced 2-path $uvx$ such that $uvx$ is not part of a 4-cycle. As $uvx$ is an induced 2-path, observe that $u$ and $x$ are not adjacent, also as $uvx$ is not part of a 4-cycle, we get, $d(u,x) = 2$ and the only shortest path between $u$ and $x$ is via $v$. This implies if we take $u$ and $x$ in our MEG-set $S$, then $xv$ and $uv$ are monitored. Hence, all the neighbors of $v$ can monitor all the edges incident to $v$. Therefore, in particular, $V(G) \setminus \{v\}$ is a MEG-set of $G$. This is a contradiction to the fact that $v$ is in every MEG-set of $G$. Thus, the necessary condition for a vertex $v$ to be part of every MEG-set of $G$ is proved according to the statement.

For the sufficient condition, let us assume that for some vertex $v$ of $G$, there exists $u \in N(v)$ such that any induced 2-path $uvx$ is part of a 4-cycle. We need to prove that $v$ is in every MEG-set. Thus, it is enough to show that $S = V(G) \setminus \{v\}$ is not an MEG-set of $G$. Therefore, we would like to find an edge which is not monitored by the vertices of $S$.

We first observe that if there does not exist any induced 2-path of the form $uvx$, then $v$ must be a simplicial vertex, and thus we know that [14] $v$ belongs to every MEG-set of $G$. On the other hand, if there exists an induced 2-path of the form $uvx$, then there are at least two shortest paths between $u$ and $x$. In particular, $u$ and $x$ cannot monitor the edge $uv$. As $x$ is arbitrary, the vertices of $S$ are not able to monitor the edge $uv$ which implies that $v$ has to be part of every MEG-set. This concludes the proof.    □

**Corollary 1.** *Let $G$ be a graph. $meg(G) = n$ if and only if for every $v \in V(G)$, there exists $u \in N(v)$ such that any induced 2-path $uvx$ is part of a 4-cycle.*

*Proof.* The proof directly follows from Theorem 2.    □

Observe that the condition for $meg(G) = n$ can be verified in polynomial time due to Corollary 1.

**Corollary 2.** *If $G \neq K_2$ is a connected graph of order $n$ and girth $g \geq 5$, then $meg(G) \leq n - 1$.*

*Proof.* Notice that, if a vertex $v$ of $G$ satisfies the condition of Theorem 2, then either $v$ has to be part of a 4-cycle, or $v$ has to be simplicial. As $G$ has girth $g \geq 5$, the only way for $v$ to satisfy the condition is by being a pendant vertex. If all vertices of $G$ are pendant vertices, then $G = K_2$, which is not possible. Thus, not every vertex of $G$ can satisfy the condition of Theorem 2. Hence, $meg(G) \leq n - 1$.    □

We also give a sufficient condition for when a vertex is never part of any minimum MEG-set of a graph. This is a useful tool to eliminate such vertices while finding a minimum MEG-set of a given graph.

**Proposition 1.** *Let $G$ be a graph with a path $v_0 v_1 \cdots v_{k-1} v_k$ whose internal vertices have degree two, $v_0$ has degree at least two, and $v_k$ has degree one. Then the vertices $v_0, v_1, \cdots, v_{k-1}$ are never part of any minimum MEG-set.*

*Proof.* Firstly, $v_k$ is part of any MEG-set of $G$ as it is a simplicial vertex [14]. Let $P$ be a shortest path with one of its end points being $v_i$, for some $i \in \{0, 1, \cdots, k-1\}$, while the other end point is $w$ (say). Moreover, assume that $v_j$ is not a vertex of the path $P$, for any value of $j \in \{i+1, i+2, \cdots, k-1\}$. Observe that, the path $P'$ obtained by augmenting the path $v_{i+1} v_{i+2} \cdots v_k$ to $P$ is also a shortest path between $w$ and $v_k$. Thus, in a minimum MEG-set of $G$, the inclusion of any of the vertices $v_0, v_1, \cdots, v_{k-1}$ is redundant. □

**Corollary 3.** *Let $G \neq K_2$ be connected graph. Let $v$ be a vertex of $G$ having a pendant neighbor $u$. Then $v$ is never part of any minimum MEG-set of $G$.*

*Proof.* The proof follows directly from Proposition 1 as a special case where $v = v_0$ and $u = v_k$. □

*Remark 1.* Due to Proposition 1, for a connected graph $G \neq K_2$, every vertex of degree one must be part of any MEG-set, and its neighbor must not be a part of any MEG-set. Therefore, if $meg(G) = |V(G)|$, then $G$ must have minimum degree at least 2.

## 4    Sparse Graphs

Due to Proposition 1, it makes sense to study connected graphs with minimum degree 2 only. In Corollary 2 we noted that $meg(G)$, if $G$ has girth 5 or more, cannot be equal to the order of $G$. Therefore, it is natural to wonder whether $meg(G)$ will become even smaller (with respect to the order of $G$) if $G$ becomes sparser. One way to consider sparse graphs is to study graphs having high girth.

**Theorem 3.** *Let $G$ be a connected graph having minimum degree at least 2. If $G$ has $n$ vertices and girth $g$, then $meg(G) \leq \frac{4n}{g+1}$.*

*Proof.* Let $G$ be a connected graph having minimum degree at least 2, having $n$ vertices, and girth $g$. We construct a vertex subset $M$ of $G$ recursively, and claim that $M$ is an MEG-set of $G$. To begin with, we initialize $M$ by picking an arbitrary vertex of $G$. Next, we add an arbitrary vertex to $M$ that is at a distance at least $\frac{g-3}{4}$ from each vertex of $M$. We repeat this process until every vertex of $V(G) \setminus M$ are at a distance strictly less than $\frac{g-3}{4}$ from some vertex of $M$.

Next, we will show that $M$ is indeed an MEG-set of $G$. Let $uv$ be an arbitrary edge of $G$. Assume without loss of generality that there exists $u' \in M$ such that $d(u, u') \leq \frac{g-3}{4}$ and $d(v, u') > d(u, u')$. Denote $v'$ the vertex of $M$ closest to $v$ that is not $u'$. Let $P_1$ be a shortest path connecting $u'$ and $u$, and $P_2$ be a shortest path connecting $v$ and $v'$. Let $P$ be the path obtained by augmenting the path $P_1$, the edge $uv$, and the path $P_2$. Note that the length of $P$ is at most $\frac{g-1}{2}$, otherwise, there would be a vertex of $P$ at distance more than $\frac{g-3}{4}$ of any vertex of $M$. Moreover, if there exists any other (vertex disjoint) path $P'$ connecting $u$ and $v$ of length $\frac{g-1}{2}$ or less, then it will contradict the fact that $G$ has girth $g$. That means, $P$ is a shortest path between $u'$ and $v'$, and $u', v'$ monitors the edge $uv$. This implies that $M$ is an MEG-set of $G$.

Now we are left with counting the cardinality of $M$. As $G$ is a connected graph with minimum degree at least 2, each vertex $v$ of $M$ is part of a cycle. Thus, $v$ has at least $2 \times \frac{g-3}{8} = \frac{g-3}{4}$ vertices at distance at most $\frac{g-3}{8}$. Let us denote the set of these vertices by $S_v$. Observe that the vertices of $S_v$ do not belong to $M$ as they are too close (within a distance of $\frac{g-3}{8}$ to $v \in M$. Furthermore, notice that, for two vertices $u, v \in M$, we must have $S_u \cap S_v = \emptyset$, as $u$ and $v$ are at distance at least $\frac{g-3}{4}$. Therefore, we must have

$$n = |V(G)| \geq |M| + |M|\left(\frac{g-3}{4}\right) \implies |M| \leq \frac{4n}{g+1}.$$

This completes the proof.    □

*Remark 2.* As girth can be considered as a measure of the sparseness of a graph, the above result shows that $meg(G)$ has a stricter upper bound as the sparseness (in terms of the girth) of $G$ increases. However, the idea used in the proof is quite general and it may be possible to provide a better bound using the same idea for specific families of graphs having more structural information.

The following theorem proves that $meg(G)$ of a sparse graph $G$ is upper bounded by a function of its chromatic number $\chi(G)$.

**Theorem 4.** *Let $G$ be a connected graph with girth at least 5 having minimum degree at least 2. If $G$ has $n$ vertices, and chromatic number $\chi(G)$, then $meg(G) \leq n\left(\frac{\chi(G)-1}{\chi(G)}\right)$.*

Using Theorem 4, we provide a corollary for graphs that have pendant vertices.

**Corollary 4.** *Let $G$ be a graph with girth at least 5, and $\ell$ pendant vertices. If $G$ has $n$ vertices, and chromatic number $\chi(G)$, then $meg(G) \leq n\left(\frac{\chi(G)-1}{\chi(G)}\right) + \frac{\ell}{\chi(G)}$.*

## 5    Effects of Clique-Sum and Subdivisions

Let $G_1$ and $G_2$ be two graphs having cliques $C_1$ and $C_2$ of size $k$, respectively. A *k-clique-sum* of $G_1$ and $G_2$, denoted by $G_1 \oplus_k G_2$, is a graph obtained by gluing the vertices of $C_1$ to the vertices $C_2$ (each vertex of $C_1$ is glued to exactly one vertex of $C_2$), and then deleting one or more edges of the glued clique.

This particular operation between two graphs is a fundamental operation in graph theory and is an important notion in the context of the illustrious graph structure theorem [19,23]. We investigate the changes in $meg(G)$ with respect to the clique-sum operation.

**Theorem 5.** *Let $G_1 \oplus_k G_2$ be a $k$-clique-sum of the graphs $G_1$ and $G_2$ for some $k \geq 2$. Then we have,*

$$meg(G_1) + meg(G_2) - 2k \leq meg(G_1 \oplus_k G_2) \leq meg(G_1) + meg(G_2).$$

*Moreover, both the lower and the upper bounds are tight.*

Let $G$ be a graph. We obtain the graph $S_G^\ell$ by subdividing each edge of $G$ exactly $\ell$ times. The graph operation subdivision is also a fundamental graph operation, integral in the theory of topological minors, which can be used for the specification of a graph. Moreover, subdivision can be considered as the inverse operation to edge contraction, which is another fundamental notion that plays an instrumental role in the famous graph minor theorem [21,24]. The following result proves a relation between $meg(G)$ and $meg(S_G^\ell)$.

**Theorem 6.** *For any graph $G$ and for all $\ell \geq 2$, we have*

$$1 \leq \frac{meg(G)}{meg(S_G^\ell)} \leq 2.$$

*Moreover, the lower bound is tight, and the upper bound is asymptotically tight.*

*Remark 3.* We have observed that $meg(S_G^\ell) = k + 1$ when $G = P_k \square P_2$.

## 6   Computational Complexity

It is known that, for a fixed integer $k$, checking whether a graph has an MEG-set of size at most $k$ can be done in polynomial time [17]. In the same paper, it is shown that the decision version of the problem is NP-complete, which settled an open question asked in [14].

**Theorem 7** ([17]). *The decision problem of determining for a graph $G$ and a natural number $k$ whether $meg(G) \leq k$ is NP-complete.*

In this section, we will improve this result by showing NP-completeness for the restricted family of 3-degenerate, 2-apex graphs. Recall that, a graph is $k$-*degenerate* if every subgraph of it has a vertex of degree at most $k$ and a graph is $k$-*apex* if it contains a set of at most $k$ vertices whose removal yields a planar graph.

In Theorem 7, the reduction was from the Boolean satisfiability problem, whereas, in our proof, we reduce from the VERTEX COVER problem (formally defined below). A *vertex cover* of a graph $G$ is a vertex subset $S \subseteq V(G)$ such that every edge $e$ in $G$ has at least one end point in $S$. We formally state the decision version of the problem of finding the vertex cover of a graph in the following.

VERTEX COVER
**Instance:** A graph $G$, an integer $k$.
**Question:** Does $G$ have a vertex cover of size at most $k$?

It is well known that the decision version of the problem is NP-complete. This problem remains NP-complete even when restricted to the family of sub-cubic planar graphs with degeneracy 2 [26]. We state this result as a theorem below.

**Theorem 8** ([26]). *The* VERTEX COVER *problem is NP-complete even for sub-cubic planar graphs having degeneracy* 2.

Next, let us formally present the decision version of the problem of finding $meg(G) \leq k$ of a graph $G$, where both $G$ and $k$ are given as inputs.

MEG-SET
**Instance:** A graph $G$, an integer $k$.
**Question:** Does $G$ have $meg(G) \leq k$?

We are now ready to state the main result of this section.

**Theorem 9.** *The* MEG-SET *problem is NP-complete even for* 3-*degenerate,* 2-*apex graphs.*

We prove Theorem 9 by giving a reduction from the VERTEX COVER problem, that is, by constructing a graph $\widehat{G}$ and there by proving the equivalence of finding $meg(\widehat{G})$ and finding $\tau(G)$, where $\tau(G)$ is the cardinality of a minimum vertex cover of $G$. Note that, in particular, this improves on the result by Haslegrave [17] which was for general graphs.

**Construction of $\widehat{G}$:** Given a graph $G$, we describe the construction of a new graph $\widehat{G}$ based on the structure of $G$ in the following. For each edge $e = uv \in E(G)$, we add a 5-path $uu'e'v'v$ starting from $u$ and ending at $v$ and add three pendant vertices $u'', e'', v''$ for each of the intermediate vertices $u', e', v'$ having degree 2, respectively. Now, add a common neighbor $y$ for all these $u', v'$'s in the newly added paths. After that, we add a pendant vertex $y^*$ to the vertex $y$. Finally, we add a universal vertex $x$ to all the (original) vertices of the graph $G$, and add a pendant vertex $x^*$ to $x$. We denote the so-obtained graph by $\widehat{G}$. We refer to Fig. 3 for a pictorial reference demonstrating how one edge of $G$ transforms in $\widehat{G}$.

We now prove some key lemmas that will help us prove Theorem 9.

**Lemma 1.** *The new edges of $\widehat{G}$, that is, the edges from the set $E(\widehat{G}) \setminus E(G)$, are monitored by the set of all pendant vertices of $\widehat{G}$.*

*Proof.* Let $e = uv$ be an edge of $G$. Then notice that the pendant vertices of the type $u''$ and $v''$ have a unique shortest path of length 4 connecting them, namely, $u''u'e'v'v''$. This path monitors all the edges on this path, that is, edges of the form $u''u', u'e', e'v'$ and $v'v''$.

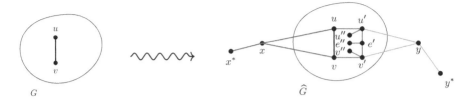

**Fig. 3.** A demonstration of the construction of $\widehat{G}$.

Moreover, between the pendant vertices of the type $u''$ and $x^*$ (resp., $y^*$), there exists a unique shortest path of length 4 (resp., 3), namely, $u''u'uxx^*$ (resp., $u''u'yy^*$), monitoring the edges of the type $u'u, ux$ (resp., $u'y$), and the edge $xx^*$ (resp., $yy^*$).

Finally, observe that the edges of the form $e''e'$ are monitored by the vertices $u'', e''$ via the unique shortest path $u''u'e'e''$ connecting them.    □

**Lemma 2.** *Let $G$ be a graph with $n$ vertices and $m$ edges. If $\tau(G) \leq k$, then $meg(\widehat{G}) \leq k + 3m + 2$.*

*Proof.* Let $S$ be a vertex cover of $G$ having $k$ vertices. Let $Z$ be the set of all pendants of $\widehat{G}$. We want to show that $|Z| = 3m + 2$, and $S \cup Z$ is an MEG-set of $\widehat{G}$.

First observe that, for each edge $e = uv$ of $G$, there are three pendant vertices $u'', e'', v''$. This amounts to $3m$ pendants in $\widehat{G}$. Moreover, counting the pendant neighbors $x^*, y^*$ of $x, y$, respectively, we have $3m + 2$ pendant vertices in total, that is, $|Z| = 3m + 2$.

By Lemma 1, the pendant vertices are enough to monitor all the edges of $\widehat{G}$, except, maybe, the (original) edges of $G$. Let us suppose that the vertex $u$ is in our MEG-set. An immediate consequence of having $u$ and the set $Z$ of all pendants in an MEG-set is that, between $u$ and the pendant vertex $v''$, there is a unique shortest path $uvv'v''$, which monitors the edge $uv$. That means, a vertex $u$ and the vertices of $Z$, monitors all the edges incident to $v$. Thus, $S \cup Z$ monitors every edge of $\widehat{G}$, and hence, $meg(\widehat{G}) \leq |S \cup Z| \leq |S| + |Z| \leq 3m + k + 2$.    □

**Lemma 3.** *Let $G$ be a graph and let $Z$ be the set of pendant vertices in $\widehat{G}$. If $M$ is a minimum MEG-set of $\widehat{G}$, then $S = M \setminus Z \subseteq V(G)$, and $S$ is a vertex cover of $G$. Moreover, $|S| = meg(\widehat{G}) - (3m + 2)$.*

*Proof.* Given a minimum MEG-set $M$ of $\widehat{G}$, we know that all the pendant vertices must belong to $M$, that is, $Z \subseteq M$. As, from the proof of Lemma 2, we know that $|Z| = 3m + 2$, the moreover part of the proof follows assuming the main portion of the statement is true.

Thus, now it remains to prove that $S = M \setminus Z \subseteq V(G)$, and $S$ is a vertex cover of $G$. By Corollary 3 we know that the neighbors of the pendent vertices cannot be part of any minimum MEG-set. Therefore, $S \subseteq V(G)$ as any vertex which does not belong to $V(G)$ is either a pendant or adjacent to a pendant in $\widehat{G}$.

Next, we are going to prove that $S$ must be a vertex cover. Observe that, two pendant vertices of $\widehat{G}$ cannot monitor any edge of $G$. Moreover, as any two vertices of $G$ are connected by a 2-path through the vertex $x$, to monitor an edge $e = uv \in E(G)$ in $\widehat{G}$ using two vertices from $V(G)$, the only option is to consider the vertices $u$ and $v$. On the other hand, from what we had noticed in the proof of Lemma 2, we know that, any vertex $u \in V(G)$, along with the set $Z$ of pendant vertices can monitor every edge incident to $v$ in $G$. Hence, $S$ must contain at least one end point of any edge of $G$. Thus, $S$ is a vertex cover of $G$. $\square$

**Lemma 4.** *Let $G$ be a 2-degenerate sub-cubic planar graph. Then $\widehat{G}$ is a 3-degenerate, 2-apex graph.*

Finally, we are ready to prove the main result of this section.

*Proof of the Theorem 9:* It is easy to verify (in polynomial time), whether a given set $M$ is an MEG-set. This can be done by checking the distances between pairs of vertices in the set, and then checking (and comparing) the distances after removing an edge; and repeating the same for every edge. Therefore, the problem is in NP.

However, we have shown using Lemma 2 and Lemma 3 that finding a minimum MEG-set for $\widehat{G}$ is equivalent (up to polynomial reduction) to finding a minimum vertex cover for $G$. As, it is well known that the minimum vertex cover problem is NP-Hard [26] even when restricted to the family of 2-degenerate sub-cubic planar graphs, using Lemma 4, it follows that the MEG-SET problem is NP-complete even for the class of 3-degenerate, 2-apex graphs. $\square$

## 7  Concluding Remarks

(1) In Sect. 2, we gave examples of graphs $G_{a,b,c,d}$ which attains $g(G_{a,b,c,d}) = a$, $eg(G_{a,b,c,d}) = b$, $seg(G_{a,b,c,d}) = c$, and $meg(G_{a,b,c,d}) = d$ for "almost" all $2 \leq a \leq b \leq c \leq d$. However, for some of the combinations of $a, b, c, d$ we still do not know if an example exists or not. One problem to consider is to decide exactly for which prescribed values of $a, b, c, d$, such a graph $G_{a,b,c,d}$ exists, along with finding an explicit example.

(2) We have proved a sufficient condition for a vertex not to be part of any minimum MEG-set. An open question is to find a necessary and sufficient condition for a vertex not to be in any minimum MEG-set.

(3) In Sect. 5, we deal with the effects on $meg(G)$ with respect to some fundamental graph operations like clique-sums, and subdivisions. It will be interesting to perform similar studies with respect to other fundamental graph operations such as vertex deletion, edge deletion, edge contraction, etc.

(4) We have proved that the problem of finding a minimum MEG-set even for 2-apex graphs is NP-complete. Hence a natural question is to find the computational complexity of MEG-SET for planar graphs. One can ask this question for other graph families as well.

**Acknowledgements.** This work is partially supported by the following projects: IFCAM (MA/IFCAM/18/39), SERB-MATRICS (MTR/2021/000858), French government IDEX-ISITE initiative 16-IDEX-0001(CAP 20-25), ANR project GRALMECO (ANR-21-CE48-0004).

# References

1. Atici, M.: On the edge geodetic number of a graph. Int. J. Comput. Math. **80**(7), 853–861 (2003)
2. Bampas, E., Bilò, D., Drovandi, G., Gualà, L., Klasing, R., Proietti, G.: Network verification via routing table queries. J. Comput. Syst. Sci. **81**(1), 234–248 (2015)
3. Beerliova, Z., et al.: Network discovery and verification. IEEE J. Sel. Areas Commun. **24**(12), 2168–2181 (2006)
4. Bejerano, Y., Rastogi, R.: Robust monitoring of link delays and faults in IP networks. In: IEEE INFOCOM 2003. Twenty-second Annual Joint Conference of the IEEE Computer and Communications Societies (IEEE Cat. No. 03CH37428), vol. 1, pp. 134–144. IEEE (2003)
5. Bilò, D., Erlebach, T., Mihalák, M., Widmayer, P.: Discovery of network properties with all-shortest-paths queries. Theor. Comput. Sci. **411**(14–15), 1626–1637 (2010)
6. Chakraborty, D., Das, S., Foucaud, F., Gahlawat, H., Lajou, D., Roy, B.: Algorithms and complexity for geodetic sets on planar and chordal graphs. In: Cao, Y., Cheng, S., Li, M. (eds.), 31st International Symposium on Algorithms and Computation, ISAAC 2020, 14–18 December 2020, Hong Kong, China (Virtual Conference), volume 181 of LIPIcs, pp. 7:1–7:15. Schloss Dagstuhl - Leibniz-Zentrum für Informatik (2020)
7. Chakraborty, D., Gahlawat, H., Roy, B.: Algorithms and complexity for geodetic sets on partial grids. Theor. Comput. Sci. **979**, 114217 (2023)
8. Chartrand, G., Harary, F., Zhang, P.: On the geodetic number of a graph. Netw. Int. J. **39**(1), 1–6 (2002)
9. Chartrand, G., Palmer, E.M., Zhang, P.: The geodetic number of a graph: a survey. Congressus numerantium **156**, 37–58 (2002)
10. Dall'Asta, L., Alvarez-Hamelin, I., Barrat, A., Vázquez, A., Vespignani, A.: Exploring networks with traceroute-like probes: theory and simulations. Theor. Comput. Sci. **355**(1), 6–24 (2006)
11. Davot, T., Isenmann, L., Thiebaut, J.: On the approximation hardness of geodetic set and its variants. In: Chen, C.-Y., Hon, W.-K., Hung, L.-J., Lee, C.-W. (eds.) COCOON 2021. LNCS, vol. 13025, pp. 76–88. Springer, Cham (2021). https://doi.org/10.1007/978-3-030-89543-3_7
12. Dourado, M.C., Protti, F., Rautenbach, D., Szwarcfiter, J.L.: Some remarks on the geodetic number of a graph. Discret. Math. **310**(4), 832–837 (2010)
13. Foucaud, F., Kao, S.-S., Klasing, R., Miller, M., Ryan, J.: Monitoring the edges of a graph using distances. Discret. Appl. Math. **319**, 424–438 (2022)
14. Foucaud, F., Narayanan, K., Ramasubramony Sulochana, L.: Monitoring Edge-Geodetic Sets in Graphs. In: Bagchi, A., Muthu, R. (eds.) Algorithms and Discrete Applied Mathematics. CALDAM 2023. LNCS, vol. 13947, pp. 245–256. Springer, Cham (2023). https://doi.org/10.1007/978-3-031-25211-2_19
15. Govindan, R., Tangmunarunkit, H.: Heuristics for internet map discovery. In: Proceedings IEEE INFOCOM 2000. Conference on Computer Communications. Nineteenth Annual Joint Conference of the IEEE Computer and Communications Societies (Cat. No. 00CH37064), vol. 3, pp. 1371–1380. IEEE (2000)

16. Harary, F., Loukakis, E., Tsouros, C.: The geodetic number of a graph. Math. Comput. Model. **17**(11), 89–95 (1993)
17. Haslegrave, J.: Monitoring edge-geodetic sets: hardness and graph products. Discret. Appl. Math. **340**, 79–84 (2023)
18. Iršič, V.: Strong geodetic number of complete bipartite graphs and of graphs with specified diameter. Graphs Comb. **34**(3), 443–456 (2018)
19. Joret, G., Wood, D.R.: Complete graph minors and the graph minor structure theorem. J. Comb. Theory Ser. B **103**(1), 61–74 (2013)
20. Kellerhals, L., Koana, T.: Parameterized complexity of geodetic set. J. Graph Algorithms Appl. **26**(4), 401–419 (2022)
21. Lovász, L.: Graph minor theory. Bull. Am. Math. Soc. **43**(1), 75–86 (2006)
22. Manuel, P., Klavžar, S., Xavier, A., Arokiaraj, A., Thomas, E.: Strong edge geodetic problem in networks. Open Math. **15**(1), 1225–1235 (2017)
23. Robertson, N., Seymour, P.D.: Graph minors: Xvii. taming a vortex. J. Comb. Theory Ser. B **77**(1), 162–210 (1999)
24. Robertson, N., Seymour, P.D.: Graph minors. xx. wagner's conjecture. J. Comb. Theory Ser. B **92**(2), 325–357 (2004)
25. Santhakumaran, A., John, J.: Edge geodetic number of a graph. J. Discret. Math. Sci. Cryptogr. **10**(3), 415–432 (2007)
26. Yannakakis, M.: Edge-deletion problems. SIAM J. Comput. **10**(2), 297–309 (1981)

# Distance-2-Dispersion with Termination by a Strong Team

Barun Gorain[1], Tanvir Kaur[2], and Kaushik Mondal[2]([✉])

[1] Indian Institute of Technology Bhilai, Bhilai, India
barun@iitbhilai.ac.in
[2] Indian Institute of Technology Ropar, Rupnagar, India
{tanvir.20maz0001,kaushik.mondal}@iitrpr.ac.in

**Abstract.** Distance-2-Dispersion (D-2-D) problem aims to disperse $k$ mobile robots starting from an arbitrary initial configuration on an anonymous port-labeled graph $G$ with $n$ nodes such that no two robots occupy adjacent nodes in the final configuration, though multiple robots may occupy a single node if there is no other empty node whose all adjacent nodes are also empty. In the existing literature, this problem is solved starting from a rooted configuration for $k(\geq 1)$ robots using $O(m\Delta)$ synchronous rounds with a total of $O(\log n)$ memory per robot, where $m$ is the number of edges and $\Delta$ is the maximum degree of the graph. In this work, we start with $k > n$ mobile robots and improve the run time to $O(m)$ starting from a rooted configuration using the same amount of memory per robot. Further, we achieve D-2-D for an arbitrary initial configuration in $O(pm)$ rounds using $O(\log n)$ memory per robot, where $p$ is the number of nodes containing robots in the initial configuration. Both the algorithms terminate without any global knowledge of $m, n, \Delta, k, p$. As we start with $k > n$ robots, the nodes occupied by robots in the final configuration form a maximal independent set of the graph.

**Keywords:** Mobile robots · Dispersion · Distance-2-Dispersion · Deterministic algorithm · Distributed algorithm

## 1 Introduction

In a recent paper, Kaur et al. [1] propose the problem of Distance-2-Dispersion (D-2-D). The problem is closely linked to the extensively researched dispersion problem [2], which seeks to assign a set of $k$ ($\leq n$) mobile robots to the nodes of an anonymous port-labeled graph with $n$ nodes and $m$ edges, so that, starting from any arbitrary initial configuration, each node contains a maximum of one robot in the final configuration. The authors in [1] consider a constraint on the dispersion problem in the following form: no two adjacent nodes contain robots in the final configuration. As some robots can run out of place in D-2-D, authors allowed the following: an unsettled robot can settle at a node that

S. Kalyanasundaram and A. Maheshwari (Eds.): CALDAM 2024, LNCS 14508, pp. 44–58, 2024.
https://doi.org/10.1007/978-3-031-52213-0_4

already contains a settled robot, only if the unsettled robot has no other node to settle at, maintaining the added constraint. Hence, in the final configuration of D-2-D, many nodes (non-adjacent) can contain multiple robots. Hence, unlike the dispersion problem where there are $k$ ($\leq n$) mobile robots, in D-2-D, the number of robots, $k$, can be any. As a byproduct, authors in [1] demonstrate that nodes with settled robots constitute a maximal independent set of the graph if there are enough robots. The algorithm in [1] solves the D-2-D problem for $k \geq 1$ robots starting from a rooted[1] initial configuration in $O(m\Delta)$ rounds using $O(\log \Delta)$ bits of additional memory per robot, where $\Delta$ is the maximum degree of the graph. However, as robots need to remember their own id, the overall memory requirement is $O(\log n)$ bits per robot.

In this work, we study the D-2-D problem considering there is a strong team of robots, i.e., $k > n$. The power of a strong team is already studied in cases of fundamental problems like gathering [3]. Here, we exploit the same in the case of D-2-D. The purpose is two-fold. We aim to provide a solution quicker w.r.t. [1] for the rooted case exploiting the power of a strong team, and such an assumption guarantees that our algorithm forms a maximal independent set of the graph. Also, we provide a solution to the D-2-D problem, considering the arbitrary initial configurations where robots can start from multiple nodes, say $p$. In both cases, our algorithm terminates without using any global knowledge of $k$, $p$, or any other graph parameters like $n$, $m$, and $\Delta$. Furthermore, as there are more than $n$ robots to start with, the solution of D-2-D ensures the formation of a maximal independent set by the nodes with settled robots.

**The Model:** We consider $G$ to be an arbitrary anonymous undirected connected zero-storage port-labeled graph. $G$ has $n$ nodes, $m$ edges, and maximum degree $\Delta$. Anonymous means the nodes have no id. Port-labeled graph implies each edge associated with any node $v$ has a distinct numbering from the range $[0, \delta(v) - 1]$ where $\delta(v)$ is the degree of node $v$. These port numbers associated with both ends of an edge are independent of each other. By $v(z)$, we mean the node that is reachable from node $v$ through the port number $z$ at $v$. Zero-storage graph means the nodes have no memory.

We assume the presence of $k > n$ robots that are capable of moving along the edges of the graph. Each robot has a unique id in the range $[1, n^c]$, for some constant $c$, and has memory. We consider the face-to-face communication model, where the robots present at a single node can communicate. A robot knows the degree and can see the port numbers of the respective edges associated with a node where it is currently present. A robot knows the port number used by it to enter the current node. All the robots start the algorithm at the same time.

We assume synchronous rounds where, in each round, a robot communicates (with co-located robots), computes, and moves (or does not move). In one round, a robot can traverse only one edge from its current position during the move. The number of synchronous rounds required from the start till the termination of all the robots is the time complexity of the algorithm. The memory requirement per robot to run the algorithm is the memory complexity of the algorithm.

---

[1] all robots start from same node.

**The Problem:** Given a set of $k > n$ robots placed arbitrarily in a port-labeled graph $G$ with $n$ nodes and $m$ edges, the robots need to achieve a configuration by the end of the algorithm where each robot needs to settle at some node satisfying the following two conditions: (i) no two adjacent nodes can be occupied by settled robots, and (ii) a robot can settle in a node where there is already a settled robot only if no other empty node is available satisfying condition (i).

**Our Contribution:** We present an algorithm for the D-2-D problem with rooted initial configuration on arbitrary graphs in $O(m)$ rounds using $O(\log n)$ memory per robot in Sect. 3. We consider $k > n$ robots, and this strong team helps achieve an improved time complexity compared to [1] keeping the memory requirement the same. Then we provide an algorithm to solve the D-2-D problem from any arbitrary initial configuration in Sect. 4 with $k > n$ robots. Our algorithm requires $O(pm)$ rounds, where $p$ is the number of nodes containing robots in the initial configuration. The memory requirement is $O(\log n)$ per robot.

Due to the presence of $k > n$ robots and by virtue of the problem definition, the nodes with settled robots form a maximal independent set of the graph. In both of our algorithms, all the settled robots terminate without any global knowledge regarding any of the parameters $m$, $n$, $\Delta$, $k$, and $p$.

## 2   Related Work

Kaur et al. [1] provide an algorithm that solves D-2-D starting from rooted configuration and terminates after $2\Delta(8m - 3n + 3)$ rounds using $O(\log \Delta)$ additional memory per robot besides the memory required to store their id without using any prior knowledge of the global parameters. They also provide $\Omega(m\Delta)$ lower bound on the number of rounds required by the robots to solve the D-2-D problem. In our work, using a strong team of robots, we provide an improved solution in terms of time complexity as well as a solution for the arbitrary initial configuration.

As our algorithm forms a maximal independent set of the underlying graph, here we discuss a couple of related works that achieve similar objectives. We refrain from going into the vast literature on distributed graph algorithms on the classic maximal independent set finding problem. We discuss only those works that find maximal independent sets using mobile robots. Pramanick et al. propose an algorithm to find a maximal independent set using myopic luminous robots [4] of an arbitrary connected graph under the asynchronous scheduler. However, according to their model, the robots have prior knowledge of $\Delta$ and robots have at least 3 hops visibility. Robots use colors to represent different states as a medium of communication. In a recent preprint, Chand et al. [5] aims to find a minimal dominating set of the underlying graph starting from an arbitrary initial configuration of the robots. They propose an algorithm that runs in $O(l\Delta \log \lambda + nl + m)$ synchronous rounds, where $l$ is the number of nodes with multiple robots in the initial configuration and $\lambda$ is the maximum id length of the robots. However, their algorithm requires prior knowledge of $m$, $\Delta$, and $\lambda$. Our algorithm works without any global knowledge.

The dispersion problem introduced in [2] is a related problem to our work. Till date, there are several works [6–17] on the dispersion problem under different model assumptions. The most efficient algorithm for solving dispersion from arbitrary initial configuration in our model is from [9] that runs in $O(min(m, k\Delta))$ rounds using $\Theta(\log(k + \Delta))$ bits additional memory per robot, ignoring the $O(\log n)$ memory that robots anyhow require to store their id. The drawback of this work is that the algorithm does not terminate. Here in this work, we do dispersion as a part of our algorithm, but we employ a dispersion algorithm starting from an arbitrary initial configuration, that terminates without using any global knowledge, though takes more time which helps the robots to start the next part of our algorithm.

The problem of scattering or uniform distribution is also related to the D-2-D problem. The scattering problem is studied on grids [18] and on rings [19,20]. Both problems are studied with anonymous robots.

## 3   D-2-D from Rooted Initial Configuration

In this section, we present an algorithm based on the Depth-First-Search (DFS) traversal technique to solve D-2-D problem on an arbitrary graph for rooted configuration. The robots achieve D-2-D with termination in $O(m)$ rounds with $O(\log n)$ memory per robot and they do not require any knowledge of the global parameters. We start with a high level idea of our algorithm. The algorithm is divided into two phases. The phase 1 achieves dispersion on the graph using the existing DFS traversal technique similar to [2]. Since $k > n$, we ensure that the robots traverse the whole graph and that at least one robot does not find any vacant position to settle. After the unsettled robots reach the root node with no further ports to explore, they understand that the dispersion is done, i.e., phase 1 is completed for them. The robot settled at the root node, say $r_{min}$, then initiates the phase 2 of the algorithm and as it reaches to other settled robots, they also get to know that phase 2 has started. The robot $r_{min}$ initiates a traversal along the tree edges and reaches the settled robots sequentially, which further decides whether to stay in its original position or not. In other words, phase 2 of the algorithm corresponds to the conversion of the dispersion configuration into a D-2-D configuration. Since $r_{min}$ is only required to traverse the tree edges in phase 2, identifying the *child* robots of a settled robot in the DFS tree poses a challenge, especially considering the limited memory available, which is bounded by $O(\log n)$. To address this challenge, we employ dedicated rounds. In these rounds, robots visit their parents so that parents can locally compute which are the remaining tree edges that $r_{min}$ needs to traverse next. With the assumption of $k > n$, it is ensured that the positions of the settled robots on the graph form a maximal independent set of the graph by virtue of the D-2-D problem definition. The algorithm proceeds in several iterations. Each iteration of the algorithm consists of six rounds, each of which is meant for a specific task to be done by the robots. In each iteration, the first round is for phase 1, rounds two to five are for the settled robots to move to their parents, and the last round is for phase 2. We elaborate on all these later.

### 3.1   The Algorithm

We use the rooted dispersion that happens in phase 1, as a subroutine for the D-2-D problem starting from arbitrary initial configuration as well. The algorithm is similar to the algorithm of [2] except we use some more variables. For the sake of completeness, we provide the algorithm for rooted DFS in short here. Below, we list down the variables used by the robots to store information during the run of the algorithm. Each robot $r_i$ maintains a list of variables defined as follows:

- $r_i.settled$: the value is 1 if robot $r_i$ is settled, otherwise 0.
- $r_i.state$: it indicates if $r_i$ is in *explore* state or *backtrack* state, *state* is initially set to *explore*.
- $r_i.parent$: the port number that is used by $r_i$ in the explore state to enter the node where it is settled. It is initially set to $-1$ for each robot.
- $r_i.recent$: Each time the group of unsettled robots exits a node, say $u$, where robot $r_u$ is settled, the robot $r_u$ updates the value of its *recent* to the port $p_i$ used by the group of unsettled robots to exit from $u$ in the same round. The value of *recent* is initially set to $\perp$.
- $r_i.port\_entered$: the port number through which the unsettled robots enter into a node at the end of a round. It is set to $-1$ initially at the root.
- $r_i.dist$: if a robot $r_i$ (settled or unsettled) is at an even distance from the root node (along the DFS traversal path) during the DFS traversal of phase 1, then it sets $r_i.dist = 0$, else $r_i.dist = 1$. Initially, it is set to 0.
- $r_i.crnt\_port$: If $r_{min}$ is present at a node $u$ with a settled robot, say $r_u$, and $r_{min}.phase = 2$, then $r_u$ sets $r_u.crnt\_port$ to the smallest child port that is not yet taken by the robot $r_{min}$ to move through in the *explore* state.
- $r_i.phase$: initially, all the robots have $r_i.phase = 1$. Whenever a settled robot $r_i$ enters into phase 2, it updates $r_i.phase = 2$.
- $r_i.final\_set$: The robots that finally settle at a node after the execution of phase 2 of the algorithm, set their *final_set* $= 1$, else it is set to 0.

Now we describe dispersion that happens in phase 1 of our algorithm. The robot $r_{min}$ settles at the root node $v_{root}$ and sets $r_{min}.parent = -1$, and $r_{min}.dist = 0$. The unsettled robots exit $v_{root}$ via the port $(r_i.port\_entered + 1)mod(\delta(v_{root}))$. While the robot $r_{min}$ updates $r_{min}.recent = 0$. Each time the unsettled robot $r_i$ visits a new node, the decision is made based on these cases:

- If **$r_i$.state = explore**: The robots set $r_i.dist = r_i.dist+1$. The value of $r_i.dist$ is updated to 0 if it is an even number else it is updated to 1. Further,
  - if the visited node, say $u$, is empty then the minimum id robot from the group of unsettled robots, say $r_j$, settles at $u$ and sets $r_j.parent = r_j.port\_entered$. The remaining unsettled robots update $r_i.port\_entered = (r_i.port\_entered + 1)mod(\delta(u))$, where $\delta(u)$ is the degree of the node $u$. The settled robot $r_j$ sets $r_j.recent = r_i.port\_entered$. If $r_i.port\_entered = r_j.parent$ then the unsettled robots set their $r_i.state = backtrack$, else they exit the node $u$ via $r_i.port\_entered$.

- if the node $u$ that is entered by the unsettled robots is non-empty then they modify $r_i.state = backtrack$ and proceed through $r_i.port\_entered$.
- If $\mathbf{r_j.state = backtrack}$: The robots set $r_i.dist = r_i.dist-1$. If $r_i.dist == -1$, then it is updated to 1. Let $r_j$ be the settled robot at the current node $u$. The unsettled robots $r_i$ after reaching the node $u$, update $r_i.port\_entered = (r_i.port\_entered + 1)mod(\delta(u))$.
  - If $r_i.port\_entered = r_j.parent$ then the unsettled robots exit the node via $r_i.port\_entered$. The settled robot $r_j$ updates $r_j.recent = r_i.port\_entered$.
  - If $r_i.port\_entered \neq r_j.parent$ then the unsettled robots update $r_i.state = explore$ and exit the node through $r_i.port\_entered$. The settled robot $r_j$ updates $r_j.recent = r_i.port$.

Now we explain phase 2 of our algorithm in detail. Since we assume that $k > n$, there is at least one robot, say $r_v$, that does not find any vacant node to settle at. The robot $r_v$ with the other unsettled robots, if any, after exploring the whole graph reaches the root node, and when $r_v.port\_entered = 0$, this implies no unexplored port is left at the root node. Thus, all the remaining unsettled robots settle at the root node and set $r_i.parent = r_{min}.parent$ while $r_{min}$ updates $r_{min}.phase = 2$, and $r_{min}.state = explore$. Thus, phase 2 is initiated by $r_{min}$. The other settled robots at the root also update their *phase* variable to 2. This phase of the algorithm involves transitioning from the dispersion configuration into the D-2-D configuration. The robot $r_{min}$ begins the traversal of the graph via the tree edges. For the robot $r_{min}$ to travel only via the tree edges, it needs the information of the port numbers that lead to the *child* robots of each settled robot, say $r_u$, in the DFS tree. For now, we assume $r_{min}$ somehow gets that information from $r_u.crnt\_port$. We resolve this with the dedicated rounds that we will discuss separately later. Let's proceed with the description of phase 2.

Besides the variables described initially, the robots maintain another two variables which are required explicitly in the phase 2 of the algorithm. These help a robot to decide when to terminate.

- $r_i.final\_port$: The smallest port $i$ of a settled robot $r_i$ such that the robot settled at the node via port $i$, say $r_j$ has $r_j.final\_set = 1$.
- $r_i.count$: Let $r_j$ has $r_j.decision = 1$. The variable *count* represents the number of settled robots in the neighborhood of $r_j$ which have $final\_set = 0$ i.e., their decision is yet to be taken.

The robot $r_{min}$, from node $u$, moves through the $r_u.crnt\_port$ information, where $r_u$ is the settled robot at $u$, to reach node $r_k$. If $r_k.crnt\_port \neq \perp$, then $r_{min}$ further moves through $r_k.crnt\_port$. It moves through the $crnt\_port$ value of the settled robots unless it reaches a settled robot, say $r_l$, that has no child port left to traverse, i.e., $r_l.crnt\_port = \perp$. The robot $r_p$ sets $r_p.decision = 1$ and $r_{min}$ waits for $2\delta(v)$ rounds, where $\delta(v)$ is the degree of the node where $r_l$ is settled at (after waiting for $2\delta(v)$ rounds, $r_{min}$ backtracks through the parent port). The robot $r_l$ moves through all its one-hop neighbors to check the status of the variable $final\_set$ of the settled robots in its one-hop. The decision is made by $r_l$ based on the following:

- If there is at least one neighbor, say $r_j$, of $r_l$ with $r_j.final\_set = 1$, then $r_l$ settles at the node where $r_j$ is settled at and sets $r_l.final\_set = 1$.
- If there are no neighbors of $r_l$ with $final\_set = 1$, then $r_l$ settles at its original position and sets $r_l.final\_set = 1$.

When the robot $r_l$ has $r_l.decision = 1$ and it moves through its ports one by one, if it meets with a settled robot say $r_j$ with $r_j.final\_set = 0$, then $r_l$ increments its value of $count$ by 1. Whereas the smallest port value that is taken by the robot $r_l$ to meet with a settled robot that has $final\_set = 1$ is stored in the variable $r_l.final\_port$. Note that, in case $r_l$ finds a vacant neighboring node, it understands that the settled robot corresponding to that node has already taken its decision and settled elsewhere, and accordingly $r_l$ does not increment its $count$. After visiting all its ports, if $r_l.final\_port$ is $\bot$ then it settles at its original position. Each time a robot visits $r_l$, it decrements its value of $count$ and finally terminates when $count = 0$. However, if $r_l.final\_port \neq \bot$, the robot $r_l$ moves through this $final\_port$ after visiting through all its ports, and settles there, after setting $r_l.final\_set = 1$ and $r_l$ terminates.

After the wait of $2\delta(v)$ rounds, the robot $r_{min}$ backtracks to reach the parent node of $r_p$, where a robot, say, $r_n$ is settled, and checks the value of $r_n.crnt\_port$. If $r_n.crnt\_port \neq \bot$ then the robot $r_{min}$ changes its state to $explore$ and moves through the value of $r_n.crnt\_port$. However, if $r_n.crnt\_port = \bot$, then $r_{min}$ waits for $2\delta(r_n)$ rounds while $r_n$ sets $r_n.decision = 1$. In this way, the decision for each node is made one by one by traversing through the tree edges of the DFS tree constructed in phase 1. A robot $r_i$ for which $r_i.crnt\_port = \bot$, and $r_{min}$ is present with $r_i$ implies that the decision for all its children has been taken, and thus $r_i$ sets $r_i.decision = 1$. Once the robot $r_{min}$ reaches the root node, and $r_j.crnt\_port = \bot$, where $r_j$ are the remaining settled robots at the root node, the settled robots at the root node as well as $r_{min}$ set their value of $decision = 1$. Finally, checking of one-hop of the root node is done by all these robots together and the decision is made collectively. In this way, the robots which are dispersed on the graph in phase 1 of the algorithm, attain a configuration such that no two settled robots are adjacent to each other. An example illustrating the execution of phase 2 of the algorithm is presented in the Fig. 1.

Now we describe how $r_{min}$ gets the information regarding the tree edges during phase 2. To be more specific, let $r_{min}$ be at node $u$. Let $z_1, z_2, ..., z_w$ be the child ports corresponding to $u$ where $w \leq \delta(u)$. To visit via tree edges, while at $u$, $r_{min}$ needs to know the child ports so that it can visit only those from $u$. The problem is the settled robot at $u$ may need $O(\Delta \log \Delta)$ memory to remember all the child ports. In the worst case that can be $O(n \log n)$ whereas the settled robot at $u$ has only $O(\log n)$ memory.

Our algorithm runs in iterations where each iteration contains six rounds; The $i$-th round in each iteration is called the $i$-dedicated round where $i \in [1, 6]$. In the 1-dedicated round of each iteration, each robot $r_i$ with $r_i.phase = 1$ runs phase 1 of the algorithm, and in the 6-dedicated round, each robot $r_i$ with $r_i.phase = 2$ runs the phase 2 of our algorithm. In the 2-dedicated round, each settled robot $r_j$ with $r_j.dist = 0$ move through $r_j.parent$ to reach its $parent$ node, say $v_p$ and

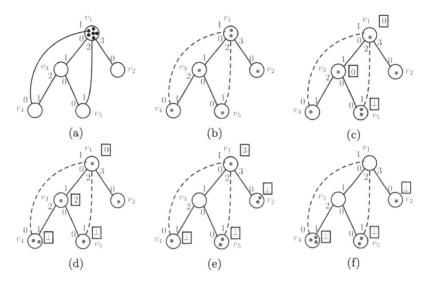

**Fig. 1.** (a) Robots are initially positioned at node $v_1$ (b) In Phase 1, the algorithm constructs a DFS tree with tree edges (shown as dark lines) and non-tree edges (dotted lines). The minimum id robot i.e., $r_{min}$ settles at the root node (marked with a green dot), while other robots settle at their respective nodes (marked with red dots) (c) $r_{min}$ moves through $crnt\_port$ values of the robots settled at $v_1$ and $v_3$ eventually reaching node $v_5$ which has no child ports. The robot that is settled at $v_5$ sets $decision = 1$ and finally settles at its node by setting $final\_set = 1$ which is shown by a blue dot (d) $r_{min}$ backtracks to $v_3$ and further moves through $crnt\_port = 2$ to reach $v_5$. Following the decision of the robot settled at $v_4$, $r_{min}$ backtracks to node $v_3$ (where $crnt\_port = \bot$), thus decision for this node is now taken (e) $r_{min}$ arrives at $v_1$ which now has $crnt_port = 3$ and moves through this port to reach $v_2$. Post the decision for this node, finally $r_{min}$ backtracks to $v_1$ (where $crnt\_port = \bot$). All the robots settled at the root node move through $final\_port = 0$ and settle there (f) The D-2-D configuration. (Color figure online)

meets the settled robot there, say $r_p$. If the robot $r_{min}$ is present with $r_p$ (this happens if in the 6-dedicated round of the last iteration $r_{min}$ moved to $v_p$), then the robot $r_p$ updates $r_p.crnt\_port = min\{r_j.port\_entered \mid r_j.port\_entered > r_p.crnt\_port\}$. In this manner, the value of $r_p.crnt\_port$ is updated each time the robot $r_{min}$ is present with it, and further $r_{min}$ moves through that child port in the 6-dedicated round of the current iteration. In the 3-dedicated round, the robot that moved in the 2-dedicated round returns to its original position through its value of $port\_entered$. In the 4-dedicated round and the 5-dedicated round, a similar procedure is followed by each settled robot $r_j$ with $r_j.dist = 1$. This technique facilitates the traversal of the DFS tree by $r_{min}$ exclusively through the tree edges. Note that, the sequence of edge traversal by $r_j$ might be different than the sequence of tree edges that formed during the DFS traversal in phase 1, and this is not an issue with us. Also, during each of the second,

---

**Algorithm 1:** $D\_2\_D\_Rooted$

---

```
/* 1-dedicated round                                                    */
if rᵢ is an unsettled robot then
|   call algorithm for unsettled robots in phase 1
else if rᵢ is a settled robot then
|   call algorithm for settled robots in phase 1
/* 2-dedicated round                                                    */
if rᵢ is a settled robot and rᵢ.dist = 0 then
|   move through rᵢ.parent
/* 3-dedicated round                                                    */
if rⱼ is a settled robot with rᵢ.dist = 1 and rₘᵢₙ is present with it then
|   rⱼ updates rⱼ.crnt_port using the incoming ports of the robots who entered at its node
|   in the 2-dedicated round
if rᵢ moved in the 2-dedicated round then
|   move through rᵢ.port_entered
/* 4-dedicated round                                                    */
if rᵢ is a settled robot and rᵢ.dist = 1 then
|   move through rᵢ.parent
/* 5-dedicated round                                                    */
if rⱼ is a settled robot with rᵢ.dist = 0 and rₘᵢₙ is present with it then
|   rⱼ updates rⱼ.crnt_port using the incoming ports of the robots who entered at its node
|   in the 4-dedicated round
if rᵢ moved in the 4-dedicated round then
|   move through rᵢ.port_entered
/* 6-dedicated round                                                    */
if rᵢ is the minimum id robot at the root node then
|   call algorithm for rₘᵢₙ in phase 2
else if rᵢ is a settled robot then
|   call algorithm for settled robots in phase 2
```

---

third, fourth, and fifth dedicated rounds, if the parent of the settled robot $r_j$ has a *phase* value of 2, then $r_j$ updates their $r_j.phase = 2$.

Here we explicitly mention what happens in an iteration. The pseudo-code for an iteration is provided in Algorithm 1.

– 1-dedicated round: Robots run phase 1 for one round in this round.
– 2-dedicated round: Each robot $r_j$ that is settled at an even distance from the root node (w.r.t. distance in the DFS traversal), moves through $r_j.parent$.
– 3-dedicated round: The settled robot at odd distant node $u$ updates its $crnt\_port$ if $r_{min}$ is present with it, by seeing the robots who moved in the 2-dedicated round to enter $u$. The robots $r_j$ that moved in the 2-dedicated round return to their original position via the port they entered through.
– 4-dedicated round: Each robot $r_j$ that is settled at an odd distance from the root node (w.r.t. distance in the DFS traversal), moves through $r_j.parent$.
– 5-dedicated round: The settled robot at even distant node $u$ update its $crnt\_port$ if $r_{min}$ is present with it, by seeing the robots who moved in the 4-dedicated round to enter $u$. The robots $r_j$ that moved in the 4-dedicated round return to their original position via the port they entered.
– 6-dedicated round: Robots run phase 2 for one round in this round.

### 3.2   Correctness and Analysis

The proofs of the lemmas are omitted due to page restrictions.

**Lemma 1.** *Phase 1 correctly disperses the robots on the graph. A group of at least two robots understands the termination of phase 1 after phase 1 is done.*

**Lemma 2.** *Phase 2 terminates correctly.*

**Theorem 1.** *Let $G$ be a port-labeled, undirected, and connected graph with $n$ nodes, $m$ edges, and maximum degree $\Delta$. Let $k(> n)$ robots be initially placed at a single node of the graph $G$ and robots have no prior knowledge of any of the global parameters. The robots achieve D-2-D on the graph with termination in $O(m)$ rounds with $O(\log n)$ memory per robot.*

*Proof.* From Lemma 1, after phase 1 of our algorithm, each node of the graph has at least one robot settled on it. In phase 2 of the algorithm, a settled robot $r_i$ vacates its position only if there is at least one settled robot, say $r_j$, with $r_j.final\_set = 1$. As a result, when phase 2 is complete, every node in the graph has at least one neighbor with a robot $r_j$ with $r_j.final\_set = 1$. Hence, robots achieve D-2-D configuration by the end of phase 2. Hence the D-2-D configuration achieved by robots constitutes a maximal independent set of the graph.

k1 involves the DFS traversal of the graph. According to the correctness proof provided in [2], achieving dispersion through DFS traversal necessitates a time complexity of $4m - 2n + 2$ rounds. This can be inferred as each tree edge is traversed twice, while each non-tree edge is traversed at most four times during the process. Our algorithm proceeds in iterations where each iteration consists of 6 rounds and phase 1 of our algorithm runs only in the first round of each iteration. Thus the total number of rounds required for our algorithm to accomplish dispersion on the graph is $6(4m - 2n + 2) = O(m)$.

Following the completion of phase 1, phase 2 is started by $r_{min}$. Robot $r_{min}$ traverses only along the tree edges of the tree constructed during the DFS traversal of phase 1. Consequently, each edge is traversed twice, this takes $2n$ rounds. For a node $v$, the settled robot $r_v$ at $v$, whose decision is to be made, a total of $2\delta(v)$ rounds are required. Thus, at most $2n + 2\Delta n$ rounds are required to execute phase 2 of our algorithm. Again, since phase 2 of our algorithm runs only in the sixth round of each iteration, the total number of rounds required for phase 2 of our algorithm is $6(2n + 2\Delta n) = O(m)$. Hence our Algorithm 1 terminates after $O(m)$ rounds.

The variables $r_i.settled$, $r_i.state$, $r_i.phase$, $r_i.decision$, $r_i.dist$, and $r_i.final\_set$ require 2 bits of memory. The variables $r_i.recent$, $r_i.crnt\_port$, $r_i.parent$, and $r_i.port\_entered$ require $\log \Delta$ memory by the robots. As $\Delta < n$ and robots store their id, the algorithm can be executed with $O(\log n)$ bits of memory per robot.

## 4   D-2-D from Arbitrary Initial Configuration

In this section, we present an algorithm designed for a configuration where $k(> n)$ robots are placed arbitrarily on the graph, occupying $p$ distinct nodes where $1 < p \le n$. We say each of those $p$ nodes contains a group of robots[2] and are known as multiplicity nodes. We start with a high-level description. The

---

[2] Even if a node contains only a single robot, we consider that single robot as a group.

algorithm runs in two phases. The first phase of the algorithm allows all the robots to disperse on the graph. Each group of robots begins their DFS traversal with the aim of dispersing the robots in the group. Each DFS traversal is associated with a *label* that corresponds to the minimum ID robot of the respective group. As each group does its own DFS traversal, the DFSs corresponding to different groups need to merge to continue the process. The merging is done in such a way that at the end there exists only one DFS tree in the graph, and a group of robots can understand the termination of phase 1 so that they can initiate phase 2. Then robots run the algorithm of phase 2 described in Sect. 3 to reach a D-2-D configuration. By the end of phase 1, the idea is go get a single DFS tree associated with the smallest labeled DFS traversal in the graph. When two DFSs meet, we allow the DFS with the smaller *label* to extend its DFS by supplying all the unsettled robots of the larger labeled DFS to it. We say that the DFS corresponding to the larger *label* is merged into the DFS corresponding to the smaller *label*. One of the benefits of our merging technique is that after the merging of two or more DFSs, the settled robots of the collapsed DFS do not vacate their positions. Such settled robots are allowed only to change their parent pointer and the *label* of their associated DFS when they are visited by a smaller labeled DFS. Since we assume that $k > n$, all the surplus unsettled robots eventually become a part of the smallest labeled DFS by changing their *label* and thus complete the DFS traversal of the graph providing a DFS tree with respect to the smallest labeled DFS. Finally, these surplus robots settle at the root node of the smallest labeled DFS. With this, the phase 1 of the algorithm terminates and phase 2 is initiated by the minimum id robot settled at the root node. This phase is the same as the phase 2 of the algorithm discussed for the rooted configuration in Sect. 3. Thus we achieve D-2-D on the graph with the help of the minimum id robot settled at the root node in the phase 2. Note that, similar to the algorithm provided for rooted configuration, the current algorithm proceeds in iterations. Each iteration consists of 6 dedicated rounds, where the 1-dedicated round is for the phase 1 of the algorithm that involves DFS traversal of the graph with merging. The functionalities of the $2 - 6$ dedicated rounds remain the same as discussed in the Algorithm 1 for rooted configuration. We are now ready to discuss our algorithm, specifically, the phase 1 of our algorithm describing the merging technique in detail.

### 4.1   The Algorithm

The phase 1 of the algorithm begins with settling the minimum id robot from each group of unsettled robots at $p$ multiplicity nodes. The remaining unsettled robots set their *label* same as the id of the robot settled at the root node. The robots begin performing dispersion using the method described in Sect. 3 with an additional checking which is performed by the group of unsettled robots after reaching a new node. Since there is more than one DFS running simultaneously, on reaching a new node, the unsettled robots may encounter a settled robot from a different labeled DFS. As a result, the merging of two DFSs takes place.

The idea of the merging of two or more DFSs is that the smaller labeled DFS is allowed to continue its DFS without halting due to the presence of an already settled robot from a different DFS. Moreover, the unsettled robots of the larger labeled DFS are transferred to the smaller labeled DFS. The various scenarios that can occur during this process are described as follows:

– When groups of unsettled robots from two or more DFSs meet at a common node: In this case, the minimum id robot from the minimum labeled DFS (say $l$), is allowed to settle at the current node and the unsettled robots from the other DFSs now become a part of the minimum labeled DFS $l$[3]. Thus they change $r_i.label = l$ and merge with the DFS $l$. In accordance with the DFS $l$, they resume their traversal.

– When the unsettled robots of DFS $l$ meets with a settled robot of DFS $m$:
  - If $l > m$: Let $r_j$ be the settled robot of DFS $m$ and the group of unsettled robots of DFS $l$ meets with $r_j$ when they move in *explore* state to reach a new node. In this case, all the unsettled robots of DFS $l$ change their *label* to $m$ and follow the *recent* ports of each settled robot of the DFS $m$ unless it merges with its group of unsettled robots. Now the following cases may arise:
    * While following the *recent* pointers, the unsettled robots may meet a settled robot with a *label*, say $j$, such that $j < m$. In this case, the group changes its *label* and follows the *recent* ports of the settled robot unless it meets with unsettled group.
    * While following the *recent* pointers, the unsettled robots reach a settled robot, say $r_p$, that has set *recent* $= \perp$, which may occur as $r_p$ was the last robot to settle, and there were no unsettled robots of the DFS $m$ left. The unsettled group then restarts the DFS traversal of the DFS $m$ from the current node by moving through $r_i.port\_entered = r_i.port\_entered + 1$. The settled robot $r_p$ updates $r_p.recent$ same as the port used by the unsettled robots to exit the node i.e. $r_i.port\_entered$.
    * When the group of unsettled robots following the *recent* port of the settled robots meets with the group of unsettled robots of DFS $m$, they merge with the DFS $m$ and continue the traversal of DFS $m$.
  - If $l < m$: Let the unsettled robots of DFS $l$ meet with a settled robot $r_j$ of DFS $m$ while doing its traversal. Since $l < m$, the robot $r_j$ changes $r_j.parent = r_i.port\_entered$ and $r_j.label = r_i.label$, where $r_i$ are the unsettled robots of the DFS $l$. Thus, $r_j$ becomes a part of the DFS $l$. The group of unsettled robots leaves the node by updating $r_i.port\_entered = r_i.port\_entered+1$, while the settled robot $r_j$ updates value of $r_j.recent = r_i.port\_entered$.

An illustrative example demonstrating the execution of phase 1 of the algorithm, which incorporates the DFS with the merging technique, is provided in Fig. 2.

---

[3] We use DFS $w$ to denote the DFS with *label* $w$.

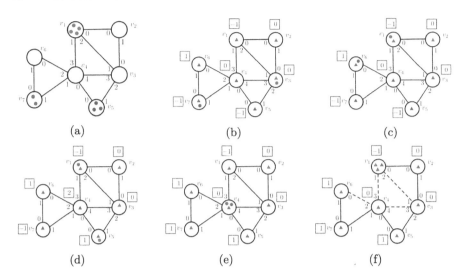

**Fig. 2.** (a) Robots are placed on graph nodes, grouped as red, blue, and green (labeled $r$, $b$, and $g$ respectively), with triangular shapes for settled robots and circles for unsettled ones. Let $r < g < b$ (b) The three groups start their DFS traversal. In round three, an unsettled $b$ robot encounters a settled $g$ robot with a smaller label, so it switches its label to $g$ and follows recent pointers of the settled robots (c) The *recent* pointer of the robot settled at the node $v_6$ is $\perp$. Thus, the unsettled robot continues the DFS $g$ (d) The unsettled $g$ robot reaches the node $v_1$ and encounters a settled robot with label $r$. It changes its label to $r$ and follows *recent* pointers. The unsettled $r$ robot reaches $v_5$ and the robot settled at $v_5$ changes its label to $r$ and *parent* pointer to 1 (e) The unsettled robot of the DFS $r$ moves through nodes $v_4$, $v_7$, and $v_6$, where all the settled robots change their *label* and *parent* pointers accordingly. While another robot follows the *recent* pointers to meet with the unsettled robot of DFS $r$. Finally, they meet at the node $v_4$ and continue the traversal of DFS labeled $r$ (f) Final robot configuration after phase 1, with dark lines indicating tree edges and dotted lines representing non-tree edges. Surplus robots settle at the root node. (Color figure online)

Since $k > n$, we ensure that at least one robot, say $r_u$, that eventually becomes a part of the minimum labeled DFS, completes the DFS traversal of the graph resulting in the construction of a DFS tree with respect to the minimum labeled DFS. Since the phase 1 of the algorithm is concluded, no node is left unexplored. Thus, each settled robot has the *label* set as $r_{min}$, where $r_{min}$ is the robot settled at the root node of the minimum labeled DFS. The robot $r_u$ reaches the root node of this minimum labeled DFS with no further ports to explore. Thus, when $r_u$ updates the value of $r_u.port\_entered = r_u.port\_entered+1$, and it equals 0, the phase 1 of the algorithm concludes. The phase 2 of the algorithm is then initiated by the robot $r_{min}$ in a similar manner as discussed in the algorithm for rooted configuration in Sect. 3. Finally, D-2-D is achieved by the robots after the conclusion of the phase 2 of our algorithm.

## 4.2    Correctness and Analysis

The proofs of the lemmas are omitted due to page restrictions.

**Lemma 3.** *The phase 1 correctly disperses the robots on the entire graph and no node is left vacant.*

**Lemma 4.** *The phase 1 of the algorithm terminates and phase 2 of the algorithm is initiated correctly.*

**Lemma 5.** *At the conclusion of phase 1 of the algorithm, all the settled robots $r_j$ have labels set as the minimum id robot, i.e., $r_j.label = r_{min}$.*

**Theorem 2.** *Let $G$ be a port-labeled, undirected, and connected graph with $n$ nodes, $m$ edges, and maximum degree $\Delta$. Let $k(k > n)$ robots be arbitrarily placed on it at $p$ multiplicity nodes and the robots have no prior knowledge of any of the global parameters $n$, $m$, $k$, $p$, and $\Delta$. Then the algorithm described in the Sect. 4 solves D-2-D with termination in $O(pm)$ rounds, and $O(\log n)$ memory is required by each robot to run the algorithm.*

*Proof.* We assume that in the initial configuration, there are $p$ multiplicity nodes. The settled robots of a DFS can change their *label* at most $p$ times when the unsettled robots from a lower labeled DFS meet them. Thus, an already traversed DFS may need to be traversed again for at most $p$ times. Since rooted dispersion requires $O(m)$ time complexity, the phase 1 of our algorithm requires $O(pm)$ time to complete. The time complexity of phase 2 of our algorithm follows from the proof of Theorem 1. The time complexity of our algorithm is thus $O(pm)$.

The variables maintained by the robots in this algorithm are the same as the variables maintained by the robots in Algorithm 1. Thus, $O(\log \Delta)$ memory is required by the robots. As $\Delta$ can be at most $n - 1$ and the robots store their own id that takes $O(\log n)$ memory, the algorithm is run with $O(\log n)$ memory.

# 5    Conclusion

In this work we provide an improved solution for rooted D-2-D and a solution for D-2-D starting from any arbitrary initial configuration, using a strong team of mobile robots. An $\Omega(m\Delta)$ lower bound for D-2-D problem with $k > 1$ robots is proved in [1]. The proof uses a two robot scenario. It would be interesting to do a lower-bound study of D-2-D starting with $k(> n)$ robots. Also, solving D-2-D in the presence of faults can be another direction of further study.

**Acknowledgements.** Tanvir Kaur acknowledges the support from the CSIR, India (Grant No. 09/1005 (0048)/2020-EMR-I). Barun Gorain and Kaushik Mondal acknowledge the support by the Core Research Grant (no.: CRG/2020/005964) of the SERB, DST, Govt. of India.

# References

1. Kaur, T., Mondal, K.: Distance-2-dispersion: dispersion with further constraints. In: NETYS, pp. 157–173 (2023)
2. Augustine, J., Moses Jr, W.K.: Dispersion of mobile robots: a study of memory-time trade-offs. In: ICDCN, pp. 1:1–1:10 (2018)
3. Molla, A.R., Mondal, K., Moses Jr., W.K.: Fast deterministic gathering with detection on arbitrary graphs: the power of many robots. In: IPDPS, pp. 47–57 (2023)
4. Pramanick, S., Samala, S.V., Pattanayak, D., Mandal, P.S.: Distributed algorithms for filling MIS vertices of an arbitrary graph by myopic luminous robots. Theor. Comput. Sci. **978**, 114187 (2023)
5. Chand, P.K., Molla, A.R., Sivasubramaniam, S.: Run for cover: dominating set via mobile agents. CoRR, abs/2309.02200 (2023)
6. Kshemkalyani, A.D., Ali, F.: Efficient dispersion of mobile robots on graphs. In: ICDCN, pp. 218–227 (2019)
7. Kshemkalyani, A.D., Molla, A.R., Sharma, G.: Fast dispersion of mobile robots on arbitrary graphs. In: ALGOSENSORS, pp. 23–40 (2019)
8. Shintaku, T., Sudo, Y., Kakugawa, H., Masuzawa, T.: Efficient dispersion of mobile agents without global knowledge. In: SSS, pp. 280–294 (2020)
9. Kshemkalyani, A.D., Sharma, G.: Near-optimal dispersion on arbitrary anonymous graphs. In: OPODIS, pp. 8:1–8:19 (2021)
10. Agarwalla, A., Augustine, J., Moses Jr, W.K., Madhav, S.K., Sridhar, A.K.: Deterministic dispersion of mobile robots in dynamic rings. In: ICDCN, pp. 19:1–19:4. ACM (2018)
11. Kshemkalyani, A.D., Molla, A.R., Sharma, G.: Dispersion of mobile robots in the global communication model. In: ICDCN (2020)
12. Kshemkalyani, A.D., Molla, A.R., Sharma, G.: Dispersion of mobile robots on grids. In: WALCOM (2020)
13. Molla, A.R., Moses Jr, W.K.: Dispersion of mobile robots: the power of randomness. In: TAMC, pp. 481–500 (2019)
14. Das, A., Bose, K., Sau, B.: Memory optimal dispersion by anonymous mobile robots. In: CALDAM, vol. 12601, pp. 426–439 (2021)
15. Molla, A., Mondal, K., Moses Jr, W.K.: Efficient dispersion on an anonymous ring in presence of weak byzantine robots. In: ALGOSENSORS, pp. 154–169 (2020)
16. Molla, A.R., Mondal, K., Moses, W.K., Jr.: Optimal dispersion on an anonymous ring in presence of weak byzantine robots. TCS **887**, 111–121 (2021)
17. Pattanayak, D., Sharma, G., Mandal, P.S.: Dispersion of mobile robots tolerating faults. In: ICDCN, pp. 133–138 (2021)
18. Barrière, L., Flocchini, P., Barrameda, E.M., Santoro, N.: Uniform scattering of autonomous mobile robots in a grid. In: IPDPS, pp. 1–8 (2009)
19. Elor, Y., Bruckstein, A.M.: Uniform multi-agent deployment on a ring. TCS **412**(8–10), 783–795 (2011)
20. Shibata, M., Mega, T., Ooshita, F., Kakugawa, H., Masuzawa, T.: Uniform deployment of mobile agents in asynchronous rings. JPDC **119**, 92–106 (2018)

# On Query Complexity Measures and Their Relations for Symmetric Functions

Rajat Mittal[1], Sanjay S. Nair[1(✉)]ⓘ, and Sunayana Patro[2]

[1] Indian Institute of Technology, Kanpur, Kanpur, India
sanjaysnair1995@gmail.com
[2] International Institute of Information Technology, Hyderabad, Hyderabad, India

**Abstract.** The main reason for query model's prominence in complexity theory and quantum computing is the presence of concrete lower bounding techniques: polynomial and adversary method. There have been considerable efforts to give lower bounds using these methods, and to compare/relate them with other measures based on the decision tree.

We explore the value of these lower bounds on quantum query complexity and their relation with other decision tree based complexity measures for the class of symmetric functions, arguably one of the most natural and basic sets of Boolean functions. We show an explicit construction for the dual of the positive adversary method and also of the square root of private coin certificate game complexity for any total symmetric function. This shows that the two values cannot be distinguished for any symmetric function. Additionally, we show that the recently introduced measure of spectral sensitivity gives the same value as both positive adversary and approximate degree for every total symmetric Boolean function.

Further, we look at the quantum query complexity of Gap Majority, a partial symmetric function. It has gained importance recently in regard to understanding the composition of randomized query complexity. We characterize the quantum query complexity of Gap Majority and show a lower bound on noisy randomized query complexity (Ben-David and Blais, FOCS 2020) in terms of quantum query complexity.

**Keywords:** Computational Complexity · Quantum Physics · Query Complexity

## 1 Introduction

The model of query complexity has been essential in the development of quantum algorithms and their complexity: many of the famous quantum algorithms can be best described in this model [15,26] and most of the lower bounds known on complexity of algorithms are obtained through this model [5,8].

The power of this model for analysing complexity of quantum algorithms arises from the fact that there are concrete mathematical techniques to give

---

The full version of this article can be found at the arxiv [22].

S. Kalyanasundaram and A. Maheshwari (Eds.): CALDAM 2024, LNCS 14508, pp. 59–73, 2024.
https://doi.org/10.1007/978-3-031-52213-0_5

lower bounds in this framework. There are two main ways to give lower bounds in this framework.

- Approximate degree and its variants: techniques motivated from capturing the success probability of the algorithm as polynomials [8].
- Adversary bound: techniques motivated from the adversary method and its semi-definite programming characterization [3, 5, 16, 21, 27].

These lower bounds have been motivated by complexity measures of Boolean functions introduced to study deterministic query (decision tree) and randomized query complexity. For deterministic query complexity, measures like Fourier degree, sensitivity, block sensitivity and certificate complexity have been studied extensively [11, 24] (all four are known to lower bound deterministic tree complexity). Similarly randomized certificate complexity is known to be a lower bound randomized query complexity. In last few years many new measures have been introduced to understand these query complexity measures [9, 12]. For example, noisyR was introduced to understand the composition of randomized query complexity [9], and recently Chakraborty et al. introduced the notion of certificate games whose public coin version is a lower bound on certificate as well as randomized query complexity [12]. Huang's landmark result [18] shows that all these measures are polynomially related to sensitivity.

How do these measures relate to each other? Huang [18] showed that these measures are polynomially related. Can we figure out what exponent is needed to bound one complexity measure by another (the exponent will depend on these complexity measures)? A lot of work has been done on these relations [3, 12]. (A very nice table with possible separations is given in Aaronson et al. [3].)

Let us ask a different question, can we compare these quantities for special class of functions? One of the simplest (and well studied) type of functions are the class of symmetric functions; the output of a symmetric function only depends on the Hamming weight of the input. This class contains many of the natural functions (OR, AND, MAJORITY, PARITY) and has been studied extensively in theoretical computer science. More specifically, related to quantum query complexity, Paturi [25] characterized the bounded error approximate degree for any symmetric function. de Wolf [29] showed a tight bound for approximate degree with small error by constructing optimal quantum query algorithms.

The main focus of this work is to examine different lower bound techniques known for quantum query complexity and their relation with other complexity measures for symmetric functions. See the survey by Buhrman and de Wolf [11] for a list of complexity measures based on the query model (we look at these measures when the function is symmetric). For the class of transitive symmetric functions, a study has been initiated by Chakraborty et al. [13].

## 1.1 Our Results

For all results in this paper, assume $\epsilon$ to be a constant less than $1/2$.

We discussed two different lower bounds on bounded error quantum query complexity of a Boolean function: approximate degree and positive adversary.

Our first result shows that for any *total* symmetric function, the positive adversary bound is asymptotically identical to square root of the certificate game complexity [12]. We show it by constructing an explicit solution of the dual of adversary semidefinite program (minimization version, shown in Definition 7) which works for the square root of certificate game complexity too.

**Theorem 1.** *Let $f : \{0,1\}^n \to \{0,1\}$ be a* total *symmetric Boolean function.*

$$\mathrm{Adv}^+(f) = \Theta\left(\sqrt{\mathrm{CG}(f)}\right) = \Theta(\sqrt{t_f \cdot n}). \tag{1}$$

*Here $t_f$ is the minimum $t$ such that $f$ is constant for Hamming weights between $t$ and $n - t$.*

The article [14] introduced the measure expectational certificate complexity to upper bound Las-Vegas randomized query complexity. It had a very similar optimization program to square root of certificate game complexity (only difference being not having the constraint that weights are less than 1). This construction implies that bound on weights in the expectational certificate complexity makes it different from square root of certificate game complexity.

Even though previous results show the value of $\mathrm{Adv}^+(f)$ using the upper bound on $Q_\epsilon(f)$, we give an explicit upper bound using the min-max formulation of Adversary bound, which is found to be $\sqrt{t_f \cdot n}$, where $t_f$ is the minimum $t$ such that $f$ is constant for Hamming weights between $t$ and $n-t$. de Wolf [29] has already shown quantum algorithms with the same quantum query complexity. This shows that $Q_\epsilon(f)$ (bounded error quantum query complexity) and $\mathrm{Adv}^+(f)$ are also asymptotically identical to square root of certificate game complexity for *total* symmetric functions. The bound on adversary method was also shown by [2].

Recently, a lower bound on adversary, called spectral sensitivity, has gained lot of attention and is shown to be a lower bound for approximate degree too [3]. For any *total* symmetric functions, spectral sensitivity gives the same bound as other two techniques.

**Theorem 2.** *Let $f : \{0,1\}^n \to \{0,1\}$ be a* total *symmetric Boolean function. Let $\lambda(f)$ denote the spectral sensitivity of $f$ respectively, then $\lambda(f) = \Theta\left(\sqrt{n \cdot t_f}\right)$.*

This shows that $\lambda(f)$ is identical to the known values of $Q_\epsilon(f)$ and $\mathrm{Adv}^+(f)$ for *total* symmetric functions. The approximate degree is also shown to be $\Theta(\sqrt{t_f \cdot n})$ by Paturi [25]. This shows that almost all lower bounds on quantum query complexity in case of symmetric functions give the same value, $\Theta(\sqrt{t_f \cdot n})$. As far as we know, this only leaves one lower bound for quantum query complexity, known as quantum certificate complexity [1,19]. It is known that this lower bound is equal to $\Theta(\sqrt{n})$ for any symmetric function.

Continuing, we examine the quantum query complexity of Gap Majority problem, a partial symmetric Boolean function. This problem gained a lot of attention recently due to the work of Ben-David and Blais [9] for proving results about composition of randomized query complexity. We prove the following theorem about Gap Majority.

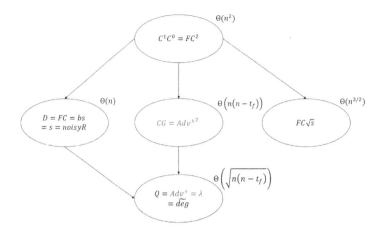

**Fig. 1.** The asymptotic relations on complexity measures of *total* symmetric functions. We show, (1) $\sqrt{\mathrm{CG}(f)}$ and $\lambda(f)$ are asymptotically identical to $\mathrm{Adv}^+(f)$; and (2) $Q_\epsilon(f) = O(\mathrm{noisyR}_\epsilon(f) \cdot \sqrt{n})$;

**Theorem 3.** *Let* $\mathrm{GapMaj}_n$ *denote the Gap Majority function on* $n$ *variables and* $Q_\epsilon(f)$ *denote the quantum query complexity with error* $\epsilon$*. Then,*

$$Q_\epsilon(\mathrm{GapMaj}_n) = \Theta(\sqrt{n}),$$

We prove the result by giving a quantum algorithm for $\mathrm{GapMaj}_n$ based on quantum counting and proving the tight lower bound using adversary method (lower bound also follows from [23]).

Ben-David and Blais [9] recently introduced noisy randomized query complexity (denoted by $\mathrm{noisyR}_\epsilon(f)$); they showed it to be a lower bound on the bounded error randomized query complexity (for definition, see [9]). They also proved that separating noisy randomized query complexity with randomized query complexity is equivalent to giving counterexample for composition of randomized query complexity.

Theorem 3 allows us to prove a lower bound on the noisy randomized query complexity in terms of quantum query complexity.

**Corollary 1.** *Let* $\mathrm{noisyR}_\epsilon(f)$ *($Q_\epsilon(f)$) denote the noisy randomized query complexity (quantum query complexity) with error* $\epsilon$ *respectively. For any total Boolean function* $f : \{0,1\}^n \to \{0,1\}$*,*

$$Q_\epsilon(f) = O(\mathrm{noisyR}_\epsilon(f) \cdot \sqrt{n}).$$

From the definition of noisy randomized query complexity, it is a lower bound on randomized query complexity. Since quantum query complexity is a lower bound on randomized query complexity, Corollary 1 provides a relation between these two lower bounds. All other known lower bounds on randomized query complexity are known to be lower bounds on noisyR.

The known relations on complexity measures for symmetric functions are illustrated in Fig. 1. The preliminaries required for our results are given in Sect. 2. Proof of Theorem 1 is detailed in Sect. 3. The results about Gap Majority and its consequence are given in Sect. 4. The bound on spectral sensitivity is proven in Sect. 5.

## 2    Preliminaries

*Norm:* $\|v\|$ denotes 2-norm of a vector $v$. The spectral norm of a square matrix $\Gamma$ is defined as $\|\Gamma\| = \max_{v:\|v\|=1}\|\Gamma.v\| = \max_{\|u\|=\|v\|=1} u^T \Gamma v$.

**Lemma 1** ([22]). *For any two non-negative $n \times m$ matrices $A$ and $B$, $\|A+B\| \geq \max\{\|A\|, \|B\|\}$*

**Definition 1.** *The Gap Majority function on $n$ variables, called $\mathrm{GapMaj}_n$, is the partial symmetric Boolean function*

$$\mathrm{GapMaj}_n(x) = \begin{cases} 0, & \text{if } |x| = n/2 - \sqrt{n} \\ 1, & \text{if } |x| = n/2 + \sqrt{n} \\ \text{not defined}, & \text{otherwise.} \end{cases}$$

*Quantum Query Complexity:* The bounded error quantum query complexity of a Boolean function $f$ $(Q_\epsilon(f))$ is the minimum number of queries needed to compute $f$ with error $\epsilon$. By repeating the algorithm constant times, success probability can be made $1 - \epsilon$ for any constant $0 < \epsilon < 1/2$. We introduce a few lower bounds on quantum query complexity in the following subsections. (For exact definition and more details about quantum query complexity, see [17].)

*Positive Adversary:* Ambainis [5] introduced the first positive adversary bound, denoted by $\mathrm{Adv}^+(f)$. Later, many modification of it were used to give lower bounds on different problems [6,7,20,30]; all of those were shown to be equivalent [27]. (These methods do not include the generalized (negative) adversary method [16].) The following version is from the original article by Ambainis [5, Theorem2].

**Theorem 4.** *Let $f(x_1, ..., x_n)$ be a function of $n$ $\{0,1\}$-valued variables and $X, Y$ be two sets of inputs such that $f(x) \neq f(y)$ if $x \in X$ and $y \in Y$. Let $R \subset X \times Y$ be such that*

1. *For every $x \in X$, there exist at least $m$ different $y \in Y$ such that $(x, y) \in R$.*
2. *For every $y \in Y$, there exist at least $m'$ different $x \in X$ such that $(x, y) \in R$.*
3. *For every $x \in X$ and $i \in \{1, ..., n\}$, there are at most $l$ different $y \in Y$ such that $(x, y) \in R$ and $x_i \neq y_i$.*
4. *For every $y \in Y$ and $i \in \{1, ..., n\}$, there are at most $l'$ different $x \in X$ such that $(x, y) \in R$ and $x_i \neq y_i$.*

*Then, any quantum algorithm computing $f$ uses $\Omega(\sqrt{\frac{mm'}{ll'}})$ queries.*

Another version used in this paper is called spectral adversary [7].

**Definition 2.** *Let* $f : \{0,1\}^n \to \{0,1\}$ *be a Boolean function. Let* $D_i$, *for all* $i \in [n]$, *be* $2^n \times 2^n$ *Boolean matrices, where indexes for rows and columns are from inputs* $\{0,1\}^n$. *The* $x, y$ *entry of matrix* $D_i$ *is* 1 *iff* $x_i \neq y_i$. *Similarly,* $F$ *is a* $2^n \times 2^n$ *Boolean matrix such that* $F[x, y] = 1 \Leftrightarrow f(x) \neq f(y)$. *Let* $\Gamma$ *be a* $2^n \times 2^n$ *non-negative symmetric matrix, then*

$$\mathrm{SA}(f) = \max_{\Gamma:\Gamma \circ F = \Gamma} \frac{\|\Gamma\|}{\max_{i \in [n]}\|\Gamma \circ D_i\|}. \tag{2}$$

Another version is called the minimax adversary method $\mathrm{MM}(f)$ [20], and is a minimization problem.

**Definition 3.** *Let* $f : S \to \{0,1\}$ *where* $S \subseteq \{0,1\}^n$ *be a Boolean function. Let* $w$ *be a weight function, then*

$$\mathrm{MM}(f) = \min_{w} \max_{x \in Dom(f)} \Sigma_{i \in [n]} w(x, i)$$

$$s.t \sum_{i:x_i \neq y_i} \sqrt{w(x, i)w(y, i)} \geq 1, \forall x, y : f(x) \neq f(y) \tag{3}$$

$$w(x, i) \geq 0, \forall x \in Dom(f), i \in [n].$$

We know that $\mathrm{Adv}^+(f) = \mathrm{MM}(f) = \mathrm{SA}(f) = O(Q_\epsilon(f))$ [5,7,20,27].

*Spectral Sensitivity* $\lambda(f)$: In 2020, Aaronson et al. introduced a new measure based on the sensitivity graph of a Boolean function, which can be used to estimate complexity of the function [3].

**Definition 4.** *For a total Boolean function* $f : \{0,1\}^n \to \{0,1\}$, *the spectral sensitivity is defined as the spectral norm of its adjacency matrix* $A_f$.

$$\lambda(f) = \|A_f\| = \max_{v:\|v\|=1} \|A_f.v\| \tag{4}$$

This spectral relaxation of sensitivity was found to be a lower bound for spectral adversary method by Aaronson et al. [3]. It was also observed that since $G_f$ is symmetric and bipartite, the spectral norm of $A_f$ is simply the largest eigenvalue of $A_f$.

*Expectational Certificate Complexity:* A new complexity measure was introduced by Gavinsky et al. [14] called expectational certificate complexity, defined as follows.

**Definition 5.** *Let* $f : S \to \{0,1\}$ *where* $S \subseteq \{0,1\}^n$ *be a Boolean function. Let* $w$ *be a weight function, then*

$$\mathrm{EC}(f) = \min_{w} \max_{x \in Dom(f)} \Sigma_{i \in [n]} w(x, i)$$

$$s.t \sum_{i:x_i \neq y_i} w(x, i)w(y, i) \geq 1, \forall x, y : f(x) \neq f(y) \tag{5}$$

$$0 \leq w(x, i) \leq 1, \forall x \in Dom(f), i \in [n].$$

It was shown that $EC(f) \geq FC(f)$ ([14][Lemma 7]) and $EC(f) = \Omega(n)$ for all symmetric functions $f$.

# 3    Lower Bounds on Quantum Query Complexity for *Total* Symmetric Functions

In this section we first construct an optimal solution for the min-max formulation of the adversary bound. It turns out that a similar construction gives an optimal bound on the private coin version of certificate game complexity.

$Adv^+(f)$ *for* Total *Symmetric Boolean Functions*

For a Boolean function $f$, we define $t_f$ to be the minimum $t$ such that $f$ is constant between $t$ and $n - t$. We use the min-max formulation of Adversary bound $(MM(f))$ and explicitly show that $Adv^+(f) = O(\sqrt{t_f \cdot n})$.

**Theorem 5.** *For any total symmetric Boolean function $f$, $Adv^+(f) = O(\sqrt{t_f \cdot n})$ (Fig. 2).*

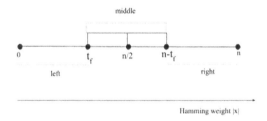

**Fig. 2.** A general *total* symmetric Boolean function viewed on Hamming weights, it is constant in the range $[t_f, n - t_f]$ by definition.

*Proof.* Using $t_f$, $|x|$ can fall in 3 regions; left $(L)$, right $(R)$ and middle $(M)$. From the definition of $MM(f)$ given in Sect. 2, we need to define a weight function $w$.

**Definition 6.** *We define a weight function $w(x, i)$:*

– *For L $(|x| < t_f)$:*

$$w(x, i) = \begin{cases} \sqrt{n/t_f}, & \text{if } x_i = 1 \\ \sqrt{t_f/n}, & \text{if } x_i = 0. \end{cases}$$

– *For R $(|x| > n - t_f)$:*

$$w(x, i) = \begin{cases} \sqrt{n/t_f}, & \text{if } x_i = 0 \\ \sqrt{t_f/n}, & \text{if } x_i = 1. \end{cases}$$

– *For M* $(t_f \leq |x| \leq n - t_f)$*:*

$$\text{For any } t_f \; i's \text{ such that } x_i = 1, \; w(x, i) = \sqrt{n/t_f}.$$

$$\text{For any } t_f \; i's \text{ such that } x_i = 0, \; w(x, i) = \sqrt{n/t_f}.$$

For the constraint $\sum_{i:x_i \neq y_i} \sqrt{w(x,i) \cdot w(y,i)} \geq 1$ for all $x, y : f(x) \neq f(y)$, following cases arise:

– $x \in L$ and $y \in R$: For any $x \in L$ and $y \in R$, there are at least $(n - 2t_f + 2)$ indices such that $x_i = 0$ and $y_i = 1$.

$$\sum_{i:x_i \neq y_i} \sqrt{w(x,i) \cdot w(y,i)} \geq (n - 2t_f + 2)\sqrt{t_f/n}$$

$$\geq 1 \qquad\qquad \text{(true for any } 0 < t_f \leq n/2 \text{ )}.$$

– $x \in L$ and $y \in M$; $x \in R$ and $y \in M$; $x, y \in L$; $x, y \in R$: We know that for any $y \in M$, there are at least $t_f$ indices such that $y_i = 1$. We also know that for any $x \in L$, the maximum number of indices such that $x_i = 1$ is $< t_f$. Then $\exists$ at least one $i$ such that $x_i = 0$ and $y_i = 1$. For this index $i$, we know that $w(x,i) = \sqrt{t_f/n}$ and $w(y,i) = \sqrt{n/t_f}$. Thus $\sum_{i:x_i \neq y_i} \sqrt{w(x,i) \cdot w(y,i)} \geq 1$ for all $x \in L$ and $y \in M$. Similar argument holds for $x \in R$ and $y \in M$ or $x, y \in L$ or $x, y \in R$.
– $x, y \in M$: There are no $x, y \in M$ such that $f(x) \neq f(y)$.

With $w$ as the weight scheme, we can see that the value of $\max_x \Sigma_i w(x, i)$ becomes $\Theta(\sqrt{t_f \cdot n})$. Thus, $\text{Adv}(f) = \text{MM}(f) = O(\sqrt{t_f \cdot n})$. Since we already from Theorem 2 know that $\text{Adv}(f) = \Theta(\sqrt{t_f \cdot n})$, this weight scheme $w$ is an explicit solution for the same.

*Certificate Game Complexity for Symmetric Functions*
The article [12] defined certificate game complexity in various settings. The definition in case of private coin setting is very similar to the min-max definition of the adversary method (Definition 3). In particular, the optimization program for the square root of certificate game complexity does not have a square root in the constraints.

**Definition 7.** *Let $f : S \rightarrow \{0, 1\}$ where $S \subseteq \{0, 1\}^n$ be a Boolean function. Let $w$ be a weight function, then*

$$\sqrt{\text{CG}(f)} = \min_w \max_{x \in Dom(f)} \Sigma_{i \in [n]} w(x, i)$$

$$s.t \sum_{i:x_i \neq y_i} w(x,i)w(y,i) \geq 1, \forall x, y : f(x) \neq f(y) \qquad (6)$$

$$w(x, i) \geq 0, \forall x \in Dom(f), i \in [n].$$

It turns out that the same bound can be obtained for $\sqrt{CG(f)}$ (with the same explicit solution as the one for the adversary bound). Together with the result of previous section, we get Theorem 1.

**Lemma 2.** *For any total symmetric Boolean function f,* $\sqrt{CG(f)} = O(\sqrt{t_f \cdot n})$.

*Proof.* We consider the same weighing scheme $w$ used for $MM(f)$. For the *total* symmetric function f where $t_f$ is the minimum value such that the function value is constant for $(t_f, n - t_f)$, we use a weight function $w(x)$ as defined in Definition 6

First, we verify that $w$ satisfies $\sum_{i:x_i \neq y_i} w(x,i)w(y,i) \geq 1$ for all $x, y : f(x) \neq f(y)$. A case analysis, similar to proof of Theorem 5, verifies the constraint (notice that the square root does not matter because the proof only requires at least one index such that $w(x,i)w(y,i) \geq 1$ ).

The objective value $\max_x \Sigma_i w(x,i)$ remains $\Theta(\sqrt{t_f \cdot n})$. Thus, $\sqrt{CG(f)} = O(\sqrt{t_f \cdot n})$.

*Proof (Proof of Theorem 1).* Suppose $f$ is symmetric. Since it is known that $\sqrt{CG(f)}$ is lower bounded by $Adv^+(f)$, Lemma 2 shows that $\sqrt{CG(f)} = \Theta(Adv^+(f))$. The lower bound on $Adv^+(f)$ follows from [2].

It is easy to see that the other certificate game complexity measures, $CG^{pub}(f) = CG^*(f) = CG^{ns}(f)$ are $\Theta(n)$ for any total symmetric function $f$ [12].

Expectational certificate complexity $(EC(f))$ [14], looks misleadingly similar to $MM(f)$.

Since $EC(f) \geq FC(f)$ ([14][Lemma 7]), $EC(f) = \Omega(n)$ for all symmetric functions $f$. Also Theorem 5 shows that $MM(f) = O(\sqrt{t_f \cdot n})$. There are two differences between $EC(f)$ and $MM(f)$: objective function and the restriction on the weight scheme. Removing the square root in the objective function doesn't change the upper bound (Lemma 2). Though, restricting $w(x,i)$ to be in between 0 and 1 changes the bounds drastically. In particular, we know that there are lot of symmetric functions for which EC is asymptotically bigger than $\sqrt{CG}$.

## 4   Quantum Query Complexity for Gap Majority Function

The Gap Majority function (Definition 1) has been recently used by Ben-David and Blais [9] to understand the composition of randomized query complexity. They used it to characterize the noisy randomized query complexity, a lower bound on randomized query complexity, and were able to show multiple results on composition using this noisy version. As a first step towards understanding the quantum query complexity of partial symmetric functions, we compute the quantum query complexity of $GapMaj_n$ and use it show new lower bounds on noisy randomized query complexity.

The main result of this section, Theorem 3, shows that $Q_\epsilon(\mathrm{GapMaj}_n) = \Theta(\sqrt{n})$. As a corollary, we obtain $Q_\epsilon(f) = O(\mathrm{noisyR}_\epsilon(f) \cdot \sqrt{n})$ for any total Boolean function. It is not known to be true for partial Boolean functions.

*Proof (Proof of Corollary 1).* Ben-David and Blais [9, Theorem 4] showed that for any Boolean function $f$,

$$R_\epsilon(f \circ \mathrm{GapMaj}_n) = \Theta(\mathrm{noisyR}_\epsilon(f) \cdot n)$$
$$\Rightarrow Q_\epsilon(f \circ \mathrm{GapMaj}_n) = O(\mathrm{noisyR}_\epsilon(f) \cdot n). \tag{7}$$

Lee et al. [21, Theorem 1.1, Lemma 5.2] showed that for any total Boolean function $f$,

$$Q_\epsilon(f) \cdot Q_\epsilon(\mathrm{GapMaj}_n) = \Theta(Q_\epsilon(f \circ \mathrm{GapMaj}_n)). \tag{8}$$

We prove in Theorem 3 that

$$Q_\epsilon(\mathrm{GapMaj}_n) = \Theta(\sqrt{n}). \tag{9}$$

Combining the result from Theorem 3 and (Eq. 7) and (Eq. 8), we get the required result for any total Boolean function $f$,

$$Q_\epsilon(f) = O(\mathrm{noisyR}_\epsilon(f) \cdot \sqrt{n}). \tag{10}$$

Comparing our bound with previously known results, Ben-David et al. [9, Lemma 38] show that $\mathrm{noisyR}_\epsilon(f) = \Omega(\mathrm{RC}(f))$ (notice $\mathrm{fbs}(f) = \Theta(\mathrm{RC}(f))$ [28]). Looking at previously known bounds on RC using quantum query complexity [3, Table 1], we know that $Q_\epsilon(f) = O(\mathrm{RC}(f)^3)$. The best possible bound on noisyR in terms of $Q_\epsilon$ becomes

$$Q_\epsilon(f) = O(\mathrm{noisyR}_\epsilon(f)^3). \tag{11}$$

Corollary 1 gives a better bound than the existing bound when $\mathrm{noisyR}_\epsilon(f) = \Omega(n^{\frac{1}{4}})$.

### 4.1   Proof of Theorem 3

We start by showing that there exists a quantum algorithm that can compute the quantum query complexity for Gap Majority in $\Theta(\sqrt{n})$ steps, thus giving us the upper bound for $Q_\epsilon(\mathrm{GapMaj}_n)$.

The main tool is the following lemma from Aaronson and Rall [4, Theorem 1] to estimate the Hamming weight of an input (a modification of quantum approximate counting by Brassard et al. [10, Theorem 15]).

**Lemma 3 (Restatement of Theorem 1 from [4]).** *Let $\epsilon > 0$ and $x \in \{0,1\}^n$ be the input whose Hamming weight we want to estimate, and $t$ be the actual Hamming weight of $x$. Given query access to an input oracle for $x$ and an allowed error rate $\delta > 0$, there exists a quantum algorithm that outputs an estimate $t'$ with probability at least $1 - \epsilon$ satisfying*

$$(1 - \delta)t \le t' \le (1 + \delta)t.$$

*The above algorithm uses $O(\frac{1}{\delta}\sqrt{n/t})$ queries where the constant depends on $\epsilon$.*

The upper bound is a straightforward implication of the previous lemma.

**Lemma 4.** $Q_\epsilon(\text{GapMaj}_n) = O(\sqrt{n})$.

*Proof.* We use approximate counting (Lemma 3) in the following way. For $\text{GapMaj}_n$, we know that $t = n/2 \pm \sqrt{n}$, thus we need to choose a $\delta$ such that the minimum estimate of $n/2 + \sqrt{n}$ is greater than the maximum estimate of $n/2 - \sqrt{n}$. When $\delta = 1/\sqrt{n}$, the minimum estimate for $t = n/2 + \sqrt{n}$ and maximum estimate for $t = n/2 - \sqrt{n}$ is $n/2 \pm \sqrt{n}/2 - 1$ respectively and hence there is no overlap.

Given $t = n/2 \pm \sqrt{n}$ and choosing $\delta = 1/\sqrt{n}$, the quantum algorithm from Lemma 3 can estimate a non-overlapping $t'$ using $O(\sqrt{n})$ queries with probability at least $1 - \epsilon$. Thus $Q_\epsilon(\text{GapMaj}_n) = O(\sqrt{n})$.

The matching lower bound was given by [2] using the positive adversary method. We give a complete proof for the sake of completeness.

**Lemma 5.** $Q_\epsilon(\text{GapMaj}_n) = \Omega(\sqrt{n})$.

*Proof.* There are multiple ways of obtaining this lower bound namely using lemma 29 of [2] or by applying theorem 1.1 from [23] on the function $\text{GapMaj}(f)$. However, we use the positive adversary bound from the original Ambainis article [5]. Using the notation of Theorem 4, let $X$ be the set of all inputs such that $\text{GapMaj}_n(x) = 0$ and $Y$ be the set of all inputs such that $\text{GapMaj}_n(x) = 1$. We take $R$ to be the set of all pairs $(x, y)$ such that the bits which are set to 1 in $x$ are a subset of the bits which are set to 1 in $y$.

For any $x \in X$, the number of $y \in Y$ such that $(x, y) \in R$ are $\binom{\frac{n}{2} + \sqrt{n}}{2\sqrt{n}}$. To enumerate y, we have to look at the number of ways in which we can fill $\frac{n}{2} + \sqrt{n}$ places with $2\sqrt{n}$ ones and $\frac{n}{2} - \sqrt{n}$ zeroes. This is because the rest of the $\frac{n}{2} - \sqrt{n}$ places are the ones which correspond to the ones in $x$. Since this is true for any $x \in X$, the value of $m$ is $\binom{\frac{n}{2} + \sqrt{n}}{2\sqrt{n}}$. Similar argument holds true for $m'$ as well, whose value turns out to be $\binom{\frac{n}{2} + \sqrt{n}}{\frac{n}{2} - \sqrt{n}}$.

For a particular $i$, if $x_i = 0$, the number of $y \in Y$ such that $(x, y) \in R$ and $y_i = 1$ is $\binom{\frac{n}{2} + \sqrt{n} - 1}{2\sqrt{n} - 1}$. This is because we are fixing the value of $y_i$ to be one, and we already have $\frac{n}{2} - \sqrt{n}$ ones filled out from the set bits of x, so we are left with $2\sqrt{n} - 1$ ones to be filled in $\frac{n}{2} + \sqrt{n} - 1$ places. For a particular $i$, if $x_i = 1$, the number of $y \in Y$ such that $(x, y) \in R$ and $y_i = 0$ is 0 because otherwise this pair $(x, y) \notin R$. So, for an $x \in X$, the maximum number of $y \in Y$ such that $(x, y) \in R$ and $x_i \neq y_i$ is $\binom{\frac{n}{2} + \sqrt{n} - 1}{2\sqrt{n} - 1}$. Since this is true for any $x \in X$, the value of $l$ is also the same. Similar argument holds true for $l'$ as well, which equals $\binom{\frac{n}{2} + \sqrt{n} - 1}{\frac{n}{2} - \sqrt{n}}$.

Substituting the values of $m, m', l, l'$, we get $Q_\epsilon(\text{GapMaj}_n) = \Omega(\sqrt{n})$.

*Proof (Proof of Theorem 3).*

From Lemma 4 and Lemma 5, we can conclude that $Q_\epsilon(\text{GapMaj}_n) = \Theta(\sqrt{n})$.

## 5   Spectral Sensitivity of Symmetric Functions

Here, we show that the spectral sensitivity of a *total* symmetric function $f : \{0,1\}^n \to \{0,1\}$ is $\Theta(\sqrt{t_f \cdot n})$ (Theorem 2). Remember that for any symmetric Boolean function $f$ on $n$ variables, we define $t_f$ to be the minimum value such that the function $f$ is constant for Hamming weights between $t_f$ and $n - t_f$.

The spectral sensitivity of a function is given by the spectral norm of the sensitivity graph of the function. First, we will find the spectral sensitivity of threshold functions. Then, we will express the sensitivity graph of a general symmetric functions in terms of the sensitivity graph of threshold functions and obtain tight lower bound on spectral sensitivity.

**Threshold Functions.** A *threshold function* with threshold $k$, $T_k : \{0,1\}^n \to \{0,1\}$, is a symmetric Boolean function defined to be 1 when $|x| \geq k$, and 0 otherwise.

**Theorem 6.** *For threshold function $T_k$, $\lambda(T_k) = \sqrt{k \cdot (n+1-k)}$.*

*Proof.* The adjacency matrix of the sensitivity graph of $T_k$ is denoted by $A_{T_k}$. Remember that $\lambda(T_k) = \|A_{T_k}\| = \max_{v:\|v\|=1} \|A_{T_k} \cdot v\|$

For any $l$, define $v_l$ with indices in $\{0,1\}^n$ to be,

$$v_l(x) = \begin{cases} 1 & if \ |x| = l \\ 0 & otherwise. \end{cases}$$

The length of $v_l$ is $\|v_l\| = \sqrt{\sum_{x:|x|=l} 1^2 + \sum_{x:|x|\neq l} 0^2} = \sqrt{\binom{n}{l}}$.

To prove the lower bound, we will show that $v_k$ achieves a stretch of $\sqrt{k(n+k-1)}$. Expanding the value at any index $x$:

$$(A_{T_k} \cdot v_k)[x] = \sum_{0 \leq |y| \leq n} A_{T_k}[x,y] \cdot v_k[y].$$

Since $v_k[y] = 1 \Leftrightarrow |j| = k$, $(A_{T_k} \cdot v_k)[x] = \sum_{|y|=k} A_{T_k}[x,y] \cdot v_k[y]$.

Notice that $A_{T_k}[x,y] = 1$ and $|y| = k$ then $|x| = (k-1)$. This implies that $(A_{T_k} \cdot v_k)[x] = 0$ if $|x| \neq k-1$. Also, for any $x : |x| = (k-1)$, there are $(n+1-k)$ possible $y$'s such that Hamming distance between $x$ and $y$ is 1 and $|y| = k$. So,

$$(A_{T_k} \cdot v_k)[x] = (n+1-k) \quad if \ |x| = k-1. \tag{12}$$

In other words, $A_{T_k} \cdot v_k = (n+1-k) \cdot v_{k-1}$. Hence, the stretch in the length of vector $v_k$ when multiplied with adjacency matrix $A_{T_k}$ is

$$\frac{|A_{T_k} \cdot v_k|}{|v_k|} = \frac{(n+1-k) \cdot |v_{k-1}|}{|v_k|} = (n+1-k) \cdot \sqrt{\frac{\binom{n}{k-1}}{\binom{n}{k}}} = \sqrt{k \cdot (n+1-k)}. \tag{13}$$

To prove the upper bound, we will use the result by Aaronson et al. [3],

$$\lambda(f) \leq \sqrt{s_0(f) \cdot s_1(f)}.$$

For $T_k$, the sensitivity of an input $x$ is $k$ if $|x| = k$ and $n + 1 - k$ if $|x| = k - 1$ (it is 0 everywhere else). We get the required upper bound by noticing that $s_0(T_k) = n + 1 - k$ and $s_1(T_k) = k$. Since the same lower bound has already been proved,

$$\lambda(T_k) = \sqrt{k \cdot (n + 1 - k)}.$$

On plotting the spectral sensitivity against the threshold $k$, we see that the spectral sensitivity is minimum when $k$ is 1(OR) or $n$(AND), and maximum when $k = \frac{n}{2}$ (MAJORITY).

**Total Symmetric Functions.** We observe that the sensitivity graph of any symmetric function $f$ can be written as sum of the sensitivity graphs of a subset of threshold functions. Define $S_f = \{1 \leq k \leq n : f(k) \neq f(k - 1)\}$.

**Lemma 6.** *For a symmetric function* $f : \{0,1\}^n \to \{0,1\}$, *the adjacency matrix of the sensitivity graph of* $f$ *can be written as* $A_f = \sum_{S_f} A_{T_k}$

*Proof.* Let $B = \sum_{S_f} A_{T_k}$. Since the support of $A_{T_k}$ where $k \in S_f$ is disjoint, $B$ is also a $\{0,1\}$ matrix. We need to prove that $B = A_f$.

From the definition of sensitivity graph, $A_f[x,y] = 1$ if and only if the Hamming distance between $x$ and $y$ is 1 and $f(x) \neq f(y)$. Without loss of generality ($A_f$ and $B$ are symmetric), assume $|x| > |y|$, then $|x| \in S_f$ implying $B[x,y] = 1$.

For the reverse direction, if $B[x,y] = 1$ and $|x| > |y|$, then $|x| \in S_f$. This means $f(x) \neq f(y)$, and the Hamming distance between $x$ and $y$ has to be 1 from the definition of $A_{T_k}$. Again, $A_f[x,y] = 1$.

Consider a symmetric function $f : \{0,1\}^n \to \{0,1\}$. As defined earlier, $t_f$ is the smallest value such that function value is a constant for the range of Hamming weights $\{t_f, .., n - t_f\}$. We capture the spectral sensitivity of $f$ using $t_f$.

*Proof (Proof of Theorem 2).*
From Lemma 6, the adjacency matrix of the sensitivity graph of $f$ can be written as $A_f = \sum_{S_f} A_{T_k}$. Since each $A_{T_k}$ has only non negative values, Lemma 1 gives us

$$\lambda(f) = \|A_f\| \geq \|A_{T_k}\| = \lambda(T_k) \quad \forall k \in S_f.$$

There is a change in function value of $f$ at Hamming weight $t_f$ or $(n + 1 - t_f)$ and from Theorem 6

$$\lambda(T_{t_f}) = \lambda(T_{n+1-t_f}) = \sqrt{t_f.(n + 1 - t_f)}.$$

We get $\lambda(f) \geq \sqrt{t_f.(n + 1 - t_f)}$. Since $(n + 1 - t_f) = \Theta(n)$,

$$\lambda(f) = \Omega(\sqrt{t_f \cdot n}).$$

The upper bound follows from $\text{Adv}^+(f)$, Theorem 5.

**Acknowledgements.** We would like to thank Sourav Chakraborty, Manaswi Paraashar and Swagato Sanyal for the discussions.

# References

1. Aaronson, S.: Quantum certificate complexity. In: 18th Annual IEEE Conference on Computational Complexity (Complexity 2003), Aarhus, Denmark, 7–10 July 2003, pp. 171–178 (2003). https://doi.org/10.1109/CCC.2003.1214418
2. Aaronson, S., Ambainis, A.: The need for structure in quantum speedups. Theory Comput. **10**, 133–166 (2014). https://doi.org/10.4086/toc.2014.v010a006
3. Aaronson, S., Ben-David, S., Kothari, R., Rao, S., Tal, A.: Degree vs. approximate degree and quantum implications of Huang's sensitivity theorem. In: STOC 2021: 53rd Symposium on Theory of Computing, Italy, 21–25 June 2021, pp. 1330–1342 (2021). https://doi.org/10.1145/3406325.3451047
4. Aaronson, S., Rall, P.: Quantum approximate counting, simplified. In: 3rd Symposium on Simplicity in Algorithms, SOSA 2020, Salt Lake City, UT, USA, 6–7 January 2020, pp. 24–32 (2020). https://doi.org/10.1137/1.9781611976014.5
5. Ambainis, A.: Quantum lower bounds by quantum arguments. In: Proceedings of the 32nd Symposium on Theory of Computing, Portland, OR, USA, 21–23 May 2000, pp. 636–643 (2000). https://doi.org/10.1145/335305.335394
6. Ambainis, A.: Polynomial degree vs. quantum query complexity. In: 44th Symposium on Foundations of Computer Science (FOCS 2003), pp. 230–239 (2003). https://doi.org/10.1109/SFCS.2003.1238197
7. Barnum, H., Saks, M.E., Szegedy, M.: Quantum query complexity and semi-definite programming. In: 18th Annual IEEE Conference on Computational Complexity (Complexity 2003), Aarhus, Denmark, 7–10 July 2003, pp. 179–193 (2003). https://doi.org/10.1109/CCC.2003.1214419
8. Beals, R., Buhrman, H., Cleve, R., Mosca, M., de Wolf, R.: Quantum lower bounds by polynomials. In: 39th Annual Symposium on Foundations of Computer Science, FOCS 1998, Palo Alto, California, USA, 8–11 November 1998, pp. 352–361 (1998). https://doi.org/10.1109/SFCS.1998.743485
9. Ben-David, S., Blais, E.: A tight composition theorem for the randomized query complexity of partial functions: extended abstract. In: 61st Symposium on Foundations of Computer Science, 2020, pp. 240–246 (2020). https://doi.org/10.1109/FOCS46700.2020.00031
10. Brassard, G., Høyer, P., Mosca, M., Tapp, A.: Quantum amplitude amplification and estimation. In: Quantum Computation and Information, pp. 53–74 (2002). https://doi.org/10.1090/conm/305/05215
11. Buhrman, H., de Wolf, R.: Complexity measures and decision tree complexity: a survey. Theor. Comput. Sci. **288**(1), 21–43 (2002). https://doi.org/10.1016/S0304-3975(01)00144-X
12. Chakraborty, S., Gál, A., Laplante, S., Mittal, R., Sunny, A.: Certificate games. In: Tauman Kalai, Y. (ed.) 14th Innovations in Theoretical Computer Science Conference (ITCS 2023), vol. 251, pp. 32:1–32:24. Schloss Dagstuhl - Leibniz-Zentrum für Informatik (2023). https://doi.org/10.4230/LIPIcs.ITCS.2023.32
13. Chakraborty, S., Kayal, C., Paraashar, M.: Separations between combinatorial measures for transitive functions. In: Bojańczyk, M., Merelli, E., Woodruff, D.P. (eds.) 49th International Colloquium on Automata, Languages, and Programming (ICALP 2022), vol. 229, pp. 36:1–36:20. Schloss Dagstuhl - Leibniz-Zentrum für Informatik (2022). https://doi.org/10.4230/LIPIcs.ICALP.2022.36

14. Gavinsky, D., et al.: Quadratically tight relations for randomized query complexity. CoRR abs/1708.00822 (2017). http://arxiv.org/abs/1708.00822
15. Grover, L.K.: A fast quantum mechanical algorithm for database search. In: Proceedings of the 28th Symposium on the Theory of Computing, 22–24 May 1996, pp. 212–219 (1996). https://doi.org/10.1145/237814.237866
16. Hoyer, P., Lee, T., Spalek, R.: Negative weights make adversaries stronger. In: Proceedings of the Thirty-Ninth Annual ACM Symposium on Theory of Computing - STOC 2007, p. 526 (2007). https://doi.org/10.1145/1250790.1250867
17. Høyer, P., Spalek, R.: Lower bounds on quantum query complexity. Bull. EATCS **87**, 78–103 (2005)
18. Huang, H.: Induced subgraphs of hypercubes and a proof of the sensitivity conjecture. CoRR abs/1907.00847 (2019). http://arxiv.org/abs/1907.00847
19. Kulkarni, R., Tal, A.: On fractional block sensitivity. Chic. J. Theor. Comput. Sci. **2016**, 1–16 (2016). http://cjtcs.cs.uchicago.edu/articles/2016/8/contents.html
20. Laplante, S., Magniez, F.: Lower bounds for randomized and quantum query complexity using Kolmogorov arguments. In: 19th Conference on Computational Complexity (CCC 2004), pp. 294–304 (2004). https://doi.org/10.1109/CCC.2004.1313852
21. Lee, T., Mittal, R., Reichardt, B.W., Spalek, R., Szegedy, M.: Quantum query complexity of state conversion. In: 2011 Symposium on Foundations of Computer Science, pp. 344–353 (2011). https://doi.org/10.1109/FOCS.2011.75
22. Mittal, R., Nair, S.S., Patro, S.: Lower bounds on quantum query complexity for symmetric functions (2021). https://arxiv.org/abs/2110.12616
23. Nayak, A., Wu, F.: The quantum query complexity of approximating the median and related statistics. In: Proceedings of the Thirty-First Annual ACM Symposium on Theory of Computing, STOC 1999, pp. 384–393. Association for Computing Machinery, New York (1999). https://doi.org/10.1145/301250.301349
24. Nisan, N., Szegedy, M.: On the degree of Boolean functions as real polynomials. Comput. Complex. **4**(4), 301–313 (1994). https://doi.org/10.1007/BF01263419
25. Paturi, R.: On the degree of polynomials that approximate symmetric Boolean functions (preliminary version). In: Proceedings of the Symposium on Theory of Computing, pp. 468–474 (1992). https://doi.org/10.1145/129712.129758
26. Simon, D.R.: On the power of quantum computation. In: 35th Annual Symposium on Foundations of Computer Science, Santa Fe, New Mexico, USA, 20–22 November 1994, pp. 116–123 (1994). https://doi.org/10.1109/SFCS.1994.365701
27. Spalek, R., Szegedy, M.: All quantum adversary methods are equivalent. Theory Comput. **2**(1), 1–18 (2006). https://doi.org/10.4086/toc.2006.v002a001
28. Tal, A.: Properties and applications of Boolean function composition. In: Proceedings of the 4th Conference on Innovations in Theoretical Computer Science - ITCS 2013, p. 441. ACM Press (2013). https://doi.org/10.1145/2422436.2422485
29. de Wolf, R.: A note on quantum algorithms and the minimal degree of $\epsilon$-error polynomials for symmetric functions. Quantum Inf. Comput. **8**(10), 943–950 (2008). https://doi.org/10.26421/QIC8.10-4
30. Zhang, S.: On the power of Ambainis lower bounds. Theor. Comput. Sci. **339**(2–3), 241–256 (2005). https://doi.org/10.1016/j.tcs.2005.01.019

# Computational Geometry

# Growth Rate of the Number of Empty Triangles in the Plane

Bhaswar B. Bhattacharya[1], Sandip Das[2], Sk. Samim Islam[2],
and Saumya Sen[2(✉)]

[1] University of Pennsylvania, Philadelphia, USA
bhaswar@wharton.upenn.edu
[2] Indian Statistical Institute, Kolkata, India
{sandipdas,sksamimislam_r,saumyasen_r}@isical.ac.in

**Abstract.** Given a set $P$ of $n$ points in the plane, in general position, denote by $N_\Delta(P)$ the number of empty triangles with vertices in $P$. In this paper we investigate by how much $N_\Delta(P)$ changes if a point $x$ is removed from $P$. By constructing a graph $G_P(x)$ based on the arrangement of the empty triangles incident on $x$, we transform this geometric problem to the problem of counting triangles in the graph $G_P(x)$. We study properties of the graph $G_P(x)$ and, in particular, show that it is kite-free. This relates the growth rate of the number of empty triangles to the famous Ruzsa-Szemerédi problem.

**Keywords:** Discrete geometry · Empty triangles · Kite-free graph

## 1   Introduction

Let $P$ be a set of $n$ points in the plane in general position, that is, no three are on a line. We define $N_\Delta(P)$ as the number of empty triangles in $P$, that is, the number of triangles with vertices in $P$ with no other point of $P$ in the interior. Counting the number of empty triangles in planar point sets is a classical problem in discrete geometry (see [1–3,5,6,8,10,11,14] and the references therein). Specifically, Bárány and Füredi [1] showed that $N_\Delta(P) \geq n^2 - O(\log n)$, for any set of points $P$, with $|P| = n$, in general position. On the other hand, a set of $n$ points chosen uniformly and independently at random from a convex set of area 1 contains $2n^2 + o(n^2)$ empty triangles on expectation [11,14].

In this paper we study the growth rate of $N_\Delta(P)$ when a point $x$ is removed from $P$. For this, let $N_\Delta(P\backslash\{x\})$ denote the number of empty triangles in the set $P\backslash\{x\}$ and consider the difference:

$$\Delta(x, P) = |N_\Delta(P) - N_\Delta(P\backslash\{x\})|.$$

In the following theorem we bound the above difference in terms of number of triangles in $P$ with $x$ as a vertex, which we denote by $V_P(x)$. To this end, denote by $K_4\backslash\{e\}$ the *kite graph*, that is, the complete graph $K_4$ with one of its diagonals removed (see Fig. 1).

S. Kalyanasundaram and A. Maheshwari (Eds.): CALDAM 2024, LNCS 14508, pp. 77–87, 2024.
https://doi.org/10.1007/978-3-031-52213-0_6

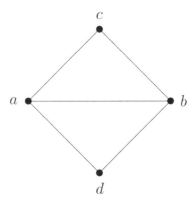

**Fig. 1.** The kite graph $K_4 \setminus \{e\}$.

**Theorem 1.** *For any set $P$, with $|P| = n$,*

$$\Delta(x, P) \leq V_P(x) + H(V_P(x), K_3, K_4 \setminus \{e\}), \tag{1}$$

*where $H(V_P(x), K_3, K_4 \setminus \{e\})$ is the maximum number of triangles in a $K_4 \setminus \{e\}$-free graph on $V_P(x)$ vertices. Moreover, there exists a set $P$, with $|P| = n$, and a point $x \in P$ such that $\Delta(x, P) \geq C V_P(x)^{\frac{3}{2}}$, for some constant $C > 0$.*

The proof of Theorem 1 is given in Sect. 2. To establish the upper bound in (1) we construct a graph $G_P(x)$ based on the arrangement of empty triangles in $P$ which have $x$ as a vertex and relate the problem of estimating $\Delta(x, P)$ to the problem of counting the number of triangles in the graph $G_P(x)$. The graph $G_P(x)$ has many interesting properties, specifically, we show that it is kite-free, which gives the upper bound in (1) (see Sect. 2.1). The lower bound construction is given in Sect. 2.2.

The extremal number $H(V_P(x), K_3, K_4 \setminus \{e\})$ that appear in the bound (1) is closely connected to the celebrated Ruzsa-Szemerédi problem, which asks for the maximum number of edges in a graph with $n$ vertices such that every edge belongs to a unique triangle [12]. Note that if a graph $G$ is kite-free, then every edge in $G$ has at most one triangle passing through it. Then removing the edges which have no triangles passing through them (which does not change the number of triangles), one can relate the problem of counting triangles in $G$ to the Ruzsa-Szemerédi problem. An application of the Szemerédi regularity lemma [13] shows that any solution to the Ruzsa-Szemerédi problem has at most $o(n^2)$ edges [12] (which is often also referred to as the diamond-free lemma). This can be improved to $n^2 / e^{\Omega(\log^* n)}$ by a stronger form of the graph removal lemma [7]. This, in particular, implies that $H(V_P(x), K_3, K_4 \setminus \{e\}) = o(V_P(x)^2)$. Hence, from Theorem 1 we get the upper bound: $\Delta(x, P) = o(V_P(x)^2)$. On the other hand, the lower bound of $\Delta(x, P)$ from Theorem 1 is $\Omega(V_P(x)^{\frac{3}{2}})$. We believe the correct order of magnitude of $\Delta(x, P)$ is closer to the lower bound, because the graph $G_P(x)$ has additional geometric structure. We collect some geometric properties of the graph $G_P(x)$ in Sect. 3, which we believe can be of independent interest.

## 2    Proof of Theorem 1

We prove the upper bound in (1) in Sect. 2.1. The lower bound construction is given in Sect. 2.2.

### 2.1    Proof of the Upper Bound

We begin with the following simple observation:

**Observation 1.** $\Delta(x, P) \leq V_P(x) + I_P(x)$, where $I_P(x)$ is the number of triangles in $P$ that contain only the point $x$ in the interior.

*Proof.* Let $U_P(x)$ denote the number of empty triangles in $P$ such that $x$ is not a vertex of the empty triangles. Note that $N_\triangle(P) = V_P(x) + U_P(x)$ and $N_\triangle(P\backslash\{x\}) = U_P(x) + I_P(x)$. This implies, $|N_\triangle(P) - N_\triangle(P\backslash\{x\})| = |V_P(x) - I_P(x)| \leq V_P(x) + I_P(x)$. $\qquad\square$

Given a set $P$, with $|P| = n$, and a point $x \in P$, define the graph $G_P(x)$ as follows: The vertex set of $G_P(x)$ is $V(G_P(x))$, the set of triangles in $P$ with $x$ as one of their vertices, and there should be edge between 2 vertices in $G_P(x)$ if the corresponding triangles, say $T_1$ and $T_2$, satisfy the following conditions:

- $T_1$ and $T_2$ share an edge,
- $T_1$ and $T_2$ are area disjoint,
- the sum of angles of $T_1$ and $T_2$ incident at $x$ is greater than $180°$.

We call the graph $G_P(x)$ the *empty triangle graph incident at $x$*. Figure 2 shows the graph $G_P(x)$ for a set of 4 points $P = \{x, a, b, c\}$. (It is worth noting that we use $\Delta$ in notations that count empty triangles in the point set $P$ and $K_3$ to denote a triangle in the graph $G_P(x)$.)

The following lemma shows that $I_P(x)$, as defined in Observation 1, is equal to the number of triangles in $G_P(x)$.

**Lemma 1.** *Suppose $P$ be a set of points in the plane, with $|P| = n$, in general position and $x \in P$. Then*

$$I_P(x) = N_{K_3}(G_P(x)),$$

*where $N_{K_3}(G_P(x))$ is the number of triangles in the graph $G_P(x)$.*

*Proof.* Let $(a, b, c)$ be a triangle in $I_P(x)$, that is, $(a, b, c)$ only has the point $x$ in the interior. This corresponds to a triangle in $G_P(x)$ as shown in Fig. 2.

Now, consider a triangle with vertices labeled $(1, 2, 3)$ in $G_P(x)$. The vertices 1, 2, and 3 in $G_P(x)$ correspond to 3 empty triangles in $P$ that have the point $x$ as one of their vertices, which we denote by $T_1$, $T_2$, and $T_3$, respectively. Since there is an edge between 1 and 2 in $G_P(x)$, the triangles $T_1$ and $T_2$ in $P$ share a common edge, are area disjoint, and the sum of angles of $T_1$ and $T_2$ incident at $x$ is greater than $180°$. Hence, we can assume, without loss of generality, $T_1$ and $T_2$

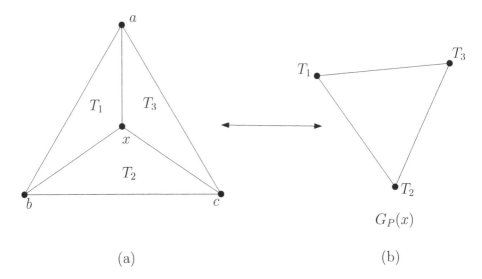

(a)                                   (b)

**Fig. 2.** A set of points $P = \{x, a, b, c\}$ and the empty triangle graph incident at $x$.

are arranged as in Fig. 3(a). Similarly, the pairs of triangles $(T_2, T_3)$ and $(T_1, T_3)$ share a common edge, are area disjoint, and the sum of their angles incident at $x$ is greater than $180°$. This implies, $T_1$, $T_2$, and $T_3$ are mutually area disjoint, and they cannot all share the same edge. Therefore, the triangles $T_1$, $T_2$, and $T_3$ have to be arranged as in Fig. 3(b). Note that the triangle formed by the union of the 3 triangles $T_1, T_2$ and $T_3$ only contains the point $x$ in $P$. Hence, for every triangle $(1, 2, 3)$ in $G_P(x)$ one gets a triangle in $I_P(x)$, formed by the union of the 3 triangles $T_1, T_2$ and $T_3$.                                                       □

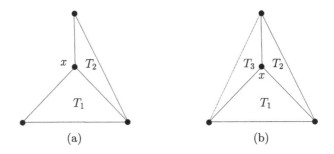

**Fig. 3.** Illustration for the proof of Lemma 1.

Applying Observation 1 and Lemma 1 it follows that

$$\Delta(x, P) \leq V_P(x) + I_P(x) = V_P(x) + N_{K_3}(G_P(x)).\tag{2}$$

Therefore, to prove the upper bound in (1), it remains to show that

$$N_{K_3}(G_P(x)) \leq H(V_P(x), K_3, K_4\backslash\{e\}) \tag{3}$$

This follows from the lemma below which shows that the graph $G_P(x)$ is kite-free.

**Lemma 2.** *The graph $G_P(x)$ does not contain $K_4\backslash\{e\}$ as a subgraph, that is, $G_P(x)$ is kite-free.*

*Proof.* Suppose $G_P(x)$ contains a kite $K_4\backslash\{e\}$, with vertices labeled $a, b, c, d$ as in Fig. 1. This corresponds to empty triangles $T_a, T_b, T_c, T_d$ with $x$ as a vertex such that $(T_a, T_b, T_c)$ and $(T_a, T_b, T_d)$ are mutually interior disjoint, and the pairs of triangles $(T_a, T_b)$, $(T_b, T_c)$, $(T_a, T_c)$, $(T_a, T_d)$, and $(T_b, T_d)$ share a common edge. This means the triangles $T_a, T_b, T_c$ must be arranged as in Fig. 2(a). Hence, it is impossible to place $T_d$ which share an edge with $T_a$ and $T_b$ and is interior disjoint from $T_a, T_b$, unless $T_d$ coincides with $T_c$. This shows $G_P(x)$ cannot contain a $K_4\backslash\{e\}$ as a subgraph. $\qquad\square$

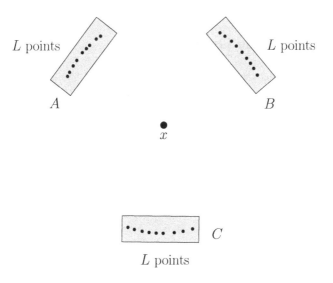

**Fig. 4.** Example showing the lower bound in Theorem 1.

Since $G_P(x)$ is kite-free by Lemma 2, the bound in (3) follows. This together with (2) gives the upper bound in (1).

## 2.2  Lower Bound Construction

To prove the lower bound in Theorem 1 consider the set of points $P$, with $|P| = n = 3L + 1$, as shown in Fig. 4. Specifically, $P$ consists of three sets points $A$, $B$, and $C$, with $|A| = |B| = |C| = L$, arranged along 3 disjoint convex chains and a point $x$ at the middle. Note that $N_\Delta(P\backslash\{x\}) = \binom{3L}{3} \sim \frac{9}{2}L^3$. Also,

$$V_P(x) = 3\binom{L}{2} + 3L^2 = \Theta(L^2). \tag{4}$$

To compute $N_\Delta(P)$ recall that $N_\Delta(P) = V_P(x) + U_P(x)$, where $U_P(x)$ is as defined in Observation 1. Now, note that

$$U_P(x) = 3\binom{L}{3} + 6L\binom{L}{2} \sim 3.5L^3.$$

Hence, $N_\Delta(P) \sim 3.5L^3 + \Theta(L^2)$ and

$$\Delta(x, P) = |N_\Delta(P) - N_\Delta(P\backslash\{x\})| = \Theta(L^3) = \Theta(V_P(x)^{\frac{3}{2}}),$$

from (4). This completes the proof of the lower bound in Theorem 1.

## 3  Properties of the Graph $G_P(x)$

In this section we collect some geometric properties of the graph $G_P(x)$. First, we show that $G_P(x)$ can contain arbitrarily large bipartite graphs.

**Lemma 3.** *Fix $r, s \geq 1$. Then there exists a set of points $P$ and $x \in P$ such that the graph $G_P(x)$ contains the complete bipartite graph $K_{r,s}$.*

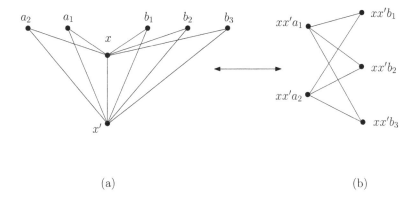

(a)                                      (b)

**Fig. 5.** Illustration for the proof of Lemma 3.

*Proof.* Consider the set of $r + s + 2$ points

$$P = \{x, x', a_1, a_2, \ldots, a_r, b_1, b_2, \ldots, b_s\},$$

as shown in Fig. 5(a) (with $r = 2$ and $s = 3$). Specifically, the points $\{a_1, a_2, \ldots, a_r\}$ and $\{b_1, b_2, \ldots, b_s\}$ lie on 2 disjoint convex chains and the points $x$ and $x'$ are in the middle. Note that, for all $1 \le i \le r$ and $1 \le j \le s$, the triangles $xx'a_i$ and $xx'b_j$ share a common edge, are interior disjoint, and the sum of the angles incident at $x$ is greater than $180°$. This implies that the graph $G_P(x)$ contains the complete bipartite graph $K_{r,s}$. $\qquad \square$

The reason it is worthwhile to know whether or not $\Delta(x, P)$ contains complete bipartite graphs as subgraphs, is because of a possible approach to improve the $o(n^2)$ upper bound on $\Delta(x, P)$ through the Kővári-Sós-Turán theorem [9]. Recall that the Kővári-Sós-Turán theorem states that any graph which is $K_{r,s}$-free, where $r \le s$, has at most $O(n^{2-\frac{1}{r}})$ edges. Therefore, if $\Delta(x, P)$ did not contain some complete bipartite graph as a subgraph, then it would have led to a polynomial improvement over the $o(n^2)$ upper bound on $\Delta(x, P)$. Lemma 3 shows that this is not the case, hence, one cannot directly apply the Kővári-Sós-Turán theorem to improve the upper bound on $\Delta(x, P)$.

Although the Kővári-Sós-Turán result cannot be directly applied to the graph $G_P(x)$, we believe a polynomial improvement over the Ruzsa-Szemerédi upper bound on $\Delta_P(x)$ is possible, because the graph $G_P(x)$ has additional geometric structure. To illustrate this we show that the well-known Behrend's construction [4], which gives a nearly quadratic lower bound on Ruzsa-Szemerédi problem, is not geometric realizable. We begin recalling Behrend's construction.

**Definition 1 (Behrend's graph).** *Suppose $p$ is an odd prime and $A \subseteq \mathbb{Z}/p\mathbb{Z}$ is a set with no 3-term arithmetic progression. The Behrend's graph $G(p, A)$ is a tripartite graph with vertices on each side of the tripartition numbered $\{0, 1, \ldots, p - 1\}$ and triangles of the form $(z, z + a, z + 2a)$ modulo $p$, for $z \in \{0, 1, \ldots, p - 1\}$ and $a \in A$.*

It is easy to check that the graph $G(p, A)$ has $3p$ vertices $3|A|p$ edges and each edge belongs to a unique triangle. For example, when $p = 3$ and $A = \{1, 2\}$ one gets the 9 vertex Paley graph shown in Fig. 6. Behrend [4] constructed a set $A$ of size $p/e^{O(\sqrt{\log p})}$ with no 3-term arithmetic progression. Using this set in the Behrend's graph in Definition 1 one gets a lower bound of $\Omega(p^2/e^{O(\sqrt{\log p})})$ for the Ruzsa-Szemerédi problem and, hence, for $H(n, K_3, K_4 \backslash \{e\})$.

The following result shows that the graph in Fig. 6 cannot be geometrically realized, that is, it is not possible to find a set of points $P$ and $x \in P$ such that $G_P(x)$ is isomorphic to the graph in Fig. 6.

**Proposition 1.** *The graph in Fig. 6 is not geometrically realizable.*

Proposition 1 shows that Behrend's graphs are not geometrically realizable. This, in particular, illustrates that the graph $G_P(x)$ has a richer geometric structure than the collection of kite-free graphs.

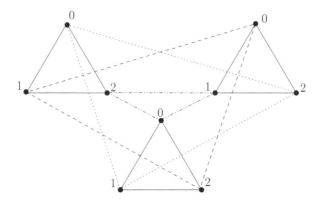

**Fig. 6.** The Paley graph with 9 vertices, 18 edges, and 6 triangles.

## 3.1    Proof of Proposition 1

We proceed by contradiction. Suppose there exists a point set $P$ and $x \in P$ such that $G_P(x)$ is isomorphic to the graph in Fig. 6. This implies $V_P(x) = 9$ and $I_P(x) = N_{K_3}(G_P(x)) = 6$ (by Lemma 1). This, in particular, means that there are 6 triangles in $P$ which only contains the point $x$ in the interior. Denote these triangles by $\mathcal{T} = \{T_1', T_2', \ldots, T_6'\}$.

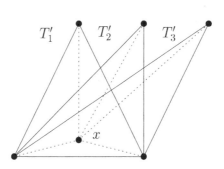

**Fig. 7.** Illustration for the proofs of Lemma 4.

**Lemma 4.** *There cannot be 3 triangles in the set $\mathcal{T}$ which share a common edge.*

*Proof.* Let, if possible, there exists 3 triangles $T_1', T_2', T_3'$ in $\mathcal{T}$ which share one edge (see Fig. 7(a)). As all the triangles contain $x$ as the only interior point they must have a common shareable area that contains $x$. Therefore, it is not possible to create a new triangle here by joining vertices of existing triangles which will contain $x$. Therefore, to create a new triangle which will contain $x$ as the only interior point, the following two cases may occur.

- *Case* 1: There is a triangle $T_4' \in \mathcal{T}$ that does not share any edge and vertex with $T_1', T_2', T_3'$ (see Fig. 8(a)) or there is a triangle $T_4' \in \mathcal{T}$ that does not share any edge with $T_1', T_2', T_3'$ but shares one vertex with $T_1', T_2', T_3'$. In both case $V_P(x) \geq 10$, which is a contradiction.
- *Case* 2: There is a triangle $T_4' \in \mathcal{T}$ which shares an edge with the triangles $T_1', T_2', T_3'$. This means $T_4'$ has a vertex that is not common with the vertices of $T_1', T_2', T_3'$ (see Fig. 8(b)) In this case $V_P(x) = 9$ and $I_P(x) = N_{K_3}(G_P(x)) = 4$. Consider the triangle $T_5'$ in $\mathcal{T} \setminus \{T_1', T_2', T_3', T_4'\}$. Note that, there is at least one edge in $T_5'$ which does not belong to the edge set of the triangles $T_1', T_2', T_3', T_4'$. This edge forms an empty triangle whose one vertex is $x$. This implies $V_P(x) \geq 10$, which is a contradiction.

Thus, if there are 3 triangles in the set $\mathcal{T}$ which share an edge, then it is not possible to place another 3 triangles satisfying the required geometric constraints for the graph in Fig. 6. Hence, there cannot be 3 triangle in $\mathcal{T}$ which share an edge. □

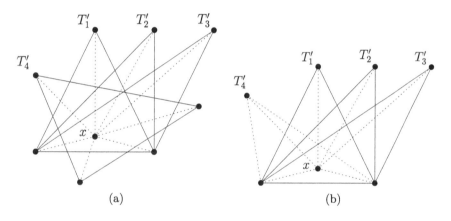

**Fig. 8.** Illustrations for the proofs of Lemma 4: (a) Case 1 and (b) Case 2.

Observe that each edge of a triangle in $\mathcal{T}$ generates an empty triangle with vertex $x$. Since there are 9 empty triangles with vertex $x$ in $G_P(x)$, the number of distinct edges generated by the triangles in $\mathcal{T}$ will be 9. Moreover, by Lemma 4 no edge of a triangle in $\mathcal{T}$ can be shared by three triangles of $\mathcal{T}$. Hence, each of the 9 edges generated by the triangles in $\mathcal{T}$ must belong to exactly 2 triangles in $\mathcal{T}$ (since, counting with repetitions, there are a total of $6 \times 3 = 18$ edges in the triangles in $\mathcal{T}$). The following lemma shows that this is not geometrically realizable.

**Lemma 5.** *All 3 edges of any triangle in $\mathcal{T}$ cannot be shared by other triangles in $\mathcal{T}$.*

*Proof.* Let, if possible, there exists a triangle $T'_1 \in \mathcal{T}$ whose all 3 edges are shared by the triangles $T'_2, T'_3, T'_4$ and $T'_2, T'_3, T'_4$ do not share edges in between. (see Fig. 9(a)). In this case $V_P(x) = 9$ and $I_P(x) = N_{K_3}(G_P(x)) = 4$, which is impossible by arguments similar to Case 2 of Lemma 4.

Alternatively, suppose there exists a triangle $T'_1 = (a, b, c) \in \mathcal{T}$, such that the edge $(a, b)$ is shared by the triangle $T'_2$, the edge $(b, c)$ is shared by the triangle $T'_3$, the edge $(c, a)$ is shared by the edge $T'_4$, and $T'_2$ and $T'_3$ share a common edge. This implies, there is a vertex $v$ such that $T'_2 = (a, b, v)$ and $T'_3 = (b, c, v)$ (see Fig. 9(b)). Note that $v$ cannot be inside the triangle $(a, b, c)$, since $(a, b, c)$ has only $x$ as the interior point. Also, $v$ cannot be in region $A$, region $C$, and region $E$, because then either the triangle $T'_2$ or the triangle $T'_3$ will have more than one point in the interior. Hence, $v$ has to be in region $B$, region $D$, or region $F$. In this case, the triangle $T'_2$ and the triangle $T'_3$ are area disjoint, hence only one of them can contain $x$ in the interior (see Fig. 9(b)). This gives a contradiction and completes the proof of the lemma.  □

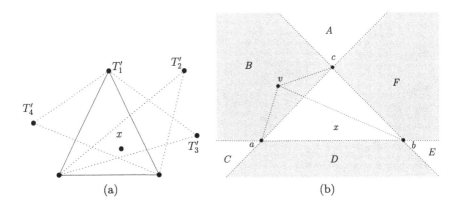

(a)                                    (b)

**Fig. 9.** Illustrations for the proofs of Lemma 5.

Combining Lemma 4 and Lemma 5, the result in Proposition 1 follows.

## 4    Conclusions

In this paper, we initiate the study of the growth rate of the number of empty triangles in the plane, by proving upper and lower bounds on the difference $\Delta(x, P)$. We relate the upper bound to the well-known Ruzsa-Szemerédi problem and study geometric properties of the triangle incidence graph $G_P(x)$. Our results show that $\Delta(x, P)$ can range from $O(V_P(x)^{\frac{3}{2}})$ and $o(V_P(x)^2)$. Understanding additional properties of the graph $G_P(x)$ is an interesting future direction, which can be useful in improving the bounds on $\Delta(x, P)$.

# References

1. Bárány, I., Füredi, Z.: Empty simplices in Euclidean space. Can. Math. Bull. **30**(4), 436–445 (1987)
2. Bárány, I., Marckert, J.-F., Reitzner, M.: Many empty triangles have a common edge. Discrete Comput. Geom. **50**, 244–252 (2013). https://doi.org/10.1007/s00454-013-9506-0
3. Bárány, I., Valtr, P.: Planar point sets with a small number of empty convex polygons. Stud. Sci. Math. Hung. **41**(2), 243–269 (2004)
4. Behrend, F.A.: On sets of integers which contain no three terms in arithmetical progression. Proc. Natl. Acad. Sci. **32**(12), 331–332 (1946)
5. Brass, P., Moser, W.O.J., Pach, J.: Research Problems in Discrete Geometry, vol. 18. Springer, New York (2005). https://doi.org/10.1007/0-387-29929-7
6. Erdos, P.: On some unsolved problems in elementary geometry. Mat. Lapok **2**(2), 1–10 (1992)
7. Fox, J.: A new proof of the graph removal lemma. Ann. Math. **174**(1), 561–579 (2011)
8. García, A.: A note on the number of empty triangles. In: Márquez, A., Ramos, P., Urrutia, J. (eds.) EGC 2011. LNCS, vol. 7579, pp. 249–257. Springer, Heidelberg (2012). https://doi.org/10.1007/978-3-642-34191-5_24
9. Kővári, T., Sós, V.T., Turán, P.: On a problem of Zarankiewicz. In: Colloquium Mathematicum, vol. 3, pp. 50–57. Polska Akademia Nauk (1954)
10. Pinchasi, R., Radoičić, R., Sharir, M.: On empty convex polygons in a planar point set. J. Comb. Theory Ser. A **3**(113), 385–419 (2006)
11. Reitzner, M., Temesvari, D.: Stars of empty simplices. arXiv preprint arXiv:1808.08734 (2018)
12. Ruzsa, I.Z., Szemerédi, E.: Triple systems with no six points carrying three triangles. Combinatorics (Keszthely, 1976) Coll. Math. Soc. J. Bolyai **18**(939–945), 2 (1978)
13. Szemerédi, E.: Regular partitions of graphs. Computer Science Department, Stanford University (1975)
14. Valtr, P.: On the minimum number of empty polygons in planar point sets. Stud. Sci. Math. Hung. **30**, 155–163 (1995)

# Geometric Covering Number: Covering Points with Curves

Arijit Bishnu$^{(\boxtimes)}$, Mathew Francis, and Pritam Majumder

Indian Statistical Institute, Kolkata, India
arijit@isical.ac.in, mathew@isichennai.res.in

**Abstract.** Given a point set, mostly a grid in our case, we seek upper and lower bounds on the number of curves that are needed to *cover* the point set. We say a curve *covers* a point if the curve passes through the point. We consider such coverings by monotonic curves, lines, orthoconvex curves, circles, etc. We also study a problem that is converse of the covering problem – if a set of $n^2$ points in the plane is covered by $n$ lines then can we say something about the configuration of the points?

**Keywords:** Discrete geometry · Incidence · Covering

## 1 Introduction

Let $S$ be a set of curves satisfying some fixed property (e.g., circle, convex curves, etc.) and $P$ be a set of points in $\mathbb{R}^d$. A curve $c \in S$ *covers* a point $p \in P$ if $p$ lies on the curve $c$. We say that $S$ covers $P$ if all points in $P$ are covered by the union of all members of $S$. We will be interested in the minimum cardinality of $S$, satisfying the given property, that covers $P$ (where the point set $P$ is fixed).

To start with, let $P$ be a set of points in $\mathbb{R}^2$ in general position and the goal is to figure out the number of simple curves needed to cover $P$. The solution is trivial – sort the points based on their $x$-coordinates and join them from left to right; i.e., we need just one simple curve to cover $P$. As we move from a simple curve with no restrictions whatsoever, to a straight line, the problem becomes hard and deserves non-trivial solutions [1,5,17,19]. This obviously gives rise to a natural question about what happens to this problem if we consider point sets with some special configuration, like grids vis-a-vis different kinds of simple curves like circles, convex curves, orthoconvex curves, etc. To bring the variety of different point sets and curves under a unifying framework, we propose the following definition of *geometric covering number*.

**Definition 1** *(Geometric covering number). The geometric covering number of a point set $P$ in $\mathbb{R}^d$ with respect to a curve type $C$ (like circle, convex curve, orthoconvex curve, etc.), denoted as $\mathcal{G}_C(P, d)$, is the minimum number of curves of type $C$ needed to* cover *all points in $P$. A curve* covers *a point if the point lies on the curve. If the dimension is understood, we just write $\mathcal{G}_C(P)$ instead of $\mathcal{G}_C(P, d)$.*

S. Kalyanasundaram and A. Maheshwari (Eds.): CALDAM 2024, LNCS 14508, pp. 88–102, 2024.
https://doi.org/10.1007/978-3-031-52213-0_7

The notion of covering a point set with different geometric structures have been studied in the literature [7,9,11,13,16]. The common theme running through all such problems is about figuring out the minimum number of structures, e.g., trees, paths, line segments, etc., needed to form a cover of the point set. Given a set of points, a *covering path* is a polygonal path that visits all the points and similarly a *covering tree* is a tree whose edges are line segments that jointly cover all the points. Covering paths and trees for planar grids have been studied in [16], where bounds on the minimum number of line segments of such paths and trees are given. Analogous questions on covering paths and trees for higher dimensional grids have been studied in [11]. Given a set $S$ of $n$ points in the plane, the problem of finding the smallest number $l$ of straight lines needed to cover all $n$ points in $S$ have been studied in [13], where bounds on the time complexity of this problem in terms of $n$ and $l$ (assuming $l$ to be small) is given.

On the other hand, incidence problems in geometry [20,21] studies questions about finding the maximum possible number of pairs $(p, \ell)$ such that $p$ is a point belonging to a set of points and $\ell$ is a line belonging to a set of lines and $p$ lies on $\ell$. Incidence between points and other geometric structures like circles, planes, algebraic curves, etc. have also been studied. We do not intend to go into all of them as an interested reader can find them in [20,21]. On the other hand, researchers have studied the problems of *point line cover*, or its more general form of *point curve cover* [1,5,17,19]. These problems consist of a set $P$ of $n$ points on the plane and a positive integer $k$, and the question is whether there exists a set of at most $k$ lines/hyperplanes/curves which cover all points in $P$. They are computationally hard problems, motivated from SET COVER, and the effort has been mostly in parametrized complexity where researchers focussed on finding tight kernels [8] for the problems [1,5,17,19].

*Notations:* We will use $[x]$ to denote the set of natural numbers $\{1, 2, \ldots, x\}$. $P$ will denote a set of $n$ points in dimension $d$. Unless otherwise stated, $P$ will be finite.

*Organization of the Paper:* In this paper, we study the notion of geometric covering number for a few types of curves. For most of the cases, our point set is a grid that we want to cover with a particular kind of curve. For completeness sake, we start with lines, the simplest curve, covering a finite grid in Sect. 2. We also investigate a converse question of covering in Sect. 2.2. Very simply put, the converse question deals with the following notion – if there is a guarantee that some lines cover an "unknown" point set, then can we say something about the configuration of the point set? From lines, we move onto monotone curves in Sect. 3. Section 4 considers three types of closed curves – circles, convex curves, and orthoconvex curves. Finally, Sect. 5 sums up the findings in this work. The Appendix is in Sect. A where we have put all the missing proofs and remarks. We feel our work will motivate studying the *geometric covering number* for more point set and curve pairs.

*Our Contributions:* Two of our major contributions in this paper are the following. As a converse to the covering by lines problem, we show in Theorem 4 that for a set $P$ of $n^2$ points covered by $n$ lines, it's not true that there always exists a subset of $P$ of size $\Theta(n^2)$ that can be put inside a grid of size $\Theta(n^2)$, possibly after a projective transformation. Regarding covering by orthoconvex curves, we proved in Theorem 15 that at least $2n/5$ (which is achieved for $n = 5$) orthoconvex curves with at most one inner corner and $2n/7$ curves with at most two inner corners are required to cover an $n \times n$ grid (Theorem 18). We also make the following observations regarding covering by other types of curves that are not very difficult to obtain. We noted in Proposition 10 that the answer to question of covering a grid by minimum number of monotonic curves can be obtained by applying Dilworth's Theorem on posets. For algebraic curves, the answer (Theorem 7) came as a consequence of the Combinatorial Nullstellensatz. For circles, the existing results in the literature imply very close upper and lower bounds (as noted in Proposition 11) and the case of convex curves is settled by an easy argument in Theorem 13.

## 2   Covering by Lines and Its Converse Problem

In the first part of this section, we consider covering grids by lines (the bounds are easy to obtain; we include it for the sake of completeness). In the next part, we consider a "converse" question – if a set of $n^2$ points in $\mathbb{R}^2$ is covered by $n$ lines, then can we say something about the configuration of the points?

### 2.1   Covering by Lines

Note that for any two points there exists a line covering them. Therefore, $\mathcal{G}_C(P) \leq \frac{|P|}{2}$ (the equality is achieved for any set of points in general position). Now let $\ell(P)$ denote the maximum number of points in $P$ any line can cover. Then we have $\mathcal{G}_C(P) \geq |P|/\ell(P)$. Therefore, we get $\frac{|P|}{\ell(P)} \leq \mathcal{G}_C(P) \leq \frac{|P|}{2}$. Now we consider the case when $P = [k_1] \times \cdots \times [k_d]$. We state the following whose proof is in Appendix A.1:

**Proposition 2.** $\ell(P) = \max\{k_1, \ldots, k_d\}$.

Proposition 2 implies that $\mathcal{G}_C(P) \geq \frac{\prod_{i=1}^{d} k_i}{\ell(P)} \geq \min\left\{\prod_{i \neq 1} k_i, \ldots, \prod_{i \neq d} k_i\right\} := N$. On the other hand, $\mathcal{G}_C(P) \leq N$ since there clearly exists an explicit covering of $P$ by $N$ lines (namely, by the lines parallel to the coordinate axis $i_0$, where $\ell(P) = k_{i_0}$). Therefore, we get that $\mathcal{G}_C(P) = \min\left\{\prod_{i \neq 1} k_i, \ldots, \prod_{i \neq d} k_i\right\}$.

**Remark 3 (Skew lines).** *We say that a line is skew if it is not parallel to $x$ or $y$-axis. We look at the question of covering an $n \times n$ grid by the minimum number of skew lines.*

*Note that the boundary of the $n \times n$ grid contains $4n - 4$ points. Now any skew line can contain at most 2 points from the boundary. So we need at least*

$2n - 2$ *skew lines to cover the grid. Also note that the $n \times n$ grid can be covered by $2n - 2$ skew lines (consider the $2n - 3$ lines parallel to the off-diagonal except the ones which pass through the bottom-left and top-right corners and these two corners are covered by the main diagonal).*

It is an open problem to find the minimum number of skew hyperplanes required to cover the $d$-dimensional hypercube. Current (2023) best known lower bound for the above problem is $d/2$, as observed in [22] (see Proposition 1.3).

## 2.2  On the Converse of the Covering Problem

Since $d = 2$ for an $n \times n$ grid, from the above discussion, it can be covered using $n$ lines. Here we look at the converse question, namely, if a set of $n^2$ points in $\mathbb{R}^2$ is covered by $n$ lines then can we say something about the configuration of the points?

Suppose a set of $n^2$ points is covered by $n$ lines. Then there exists a line containing $\Omega(n)$ points, since otherwise the total number of points is less than $n^2$. Now if this line contains $o(n^2)$ points, then there exists another line containing $\Omega(n)$ points. By continuing this, we can say that there exists a set of lines each containing $\Omega(n)$ points such that the total number of points in the union of these lines is $\Theta(n^2)$.

Now the following question seems natural. If a set $P$ of $n^2$ points is covered by $n$ lines, then does there always exist a subset of $P$ of size $\Theta(n^2)$ which can be put inside a grid of size $\Theta(n^2)$, possibly after applying a projective transformation? We show that the answer is no.

**Theorem 4.** *There exists a finite set $P$ of $n^2$ points in $\mathbb{R}^2$ which can be covered with $n$ lines but no subset of $P$ of size $\Omega(n^2)$ can be contained in a projective transformation of a rectangular grid of size $o(n^3)$.*

*Proof.* Given any two distinct points $p, p' \in \mathbb{R}^2$, we denote by $\ell(p, p')$ the unique line in $\mathbb{R}^2$ that contains both $p$ and $p'$. By an $s \times t$ grid, we mean a point set that can be obtained by a projective transform $f$ of the set $[t] \times [s]$. By a "horizontal line" of the grid, we mean a line $\ell(f(1, j), f(t, j))$ for some $j \in [s]$, and by a "vertical line" of the grid, we mean a line $\ell(f(i, 1), f(i, s))$, for some $i \in [t]$. The "size" of an $s \times t$ grid is $st$, i.e., the number of points in it. Note that every horizontal line of a grid intersects every vertical line of the grid (since there is a point of the grid that is contained in both of them).

For each $i \in [n]$, let $L_i$ denote the line with equation $y = i$ and let $\mathcal{L} = \{L_i\}_{1 \leq i \leq n}$. Let $\mathcal{P}$ be the set of points defined as follows. Define $P_1$ to be some set of $n$ distinct points from the line $L_1$. For each $1 < i \leq n$, we define $P_i$ to be a set of $n$ distinct points from $L_i$ that do not lie on any of the lines formed by points on other lines, i.e. in $\{\ell(p, p') : p \neq p' \text{ and } p, p' \in \bigcup_{1 \leq j \leq i-1} P_j\}$. Let $\mathcal{P} = \bigcup_{1 \leq i \leq n} P_i$. Let $m = |\mathcal{P}|$. Note that we have $|\mathcal{L}| = n$. We claim that for any $\mathcal{P}' \subseteq \mathcal{P}$ such that $|\mathcal{P}'| = \Omega(m) = \Omega(n^2)$, any grid that contains all the points of $\mathcal{P}'$ has size $\Omega(n^3)$.

Note that by our construction, if any line contains two points $p, p' \in \mathcal{P}$ such that $p \in P_i$ and $p' \in P_j$, where $i \neq j$, then $p$ and $p'$ are the only points in $\mathcal{P}$ that are contained in that line. This implies that the following property is satisfied by $\mathcal{P}$ and $\mathcal{L}$.

(*) Any line in $\mathbb{R}^2$ that contains more than two points in $\mathcal{P}$ belongs to $\mathcal{L}$.

Since every line in $\mathcal{L}$ contains exactly $n$ points of $\mathcal{P}$, we then have another property.

(+) Any line in $\mathbb{R}^2$ contains at most $n$ points in $\mathcal{P}$.

Let $\mathcal{P}' \subseteq \mathcal{P}$ be such that $|\mathcal{P}'| = \Omega(n^2)$. Consider any grid $\mathbb{G}$ that contains all the points of $\mathcal{P}'$. Let $\mathbb{G}$ be an $s \times t$ grid. Let $h_1, h_2, \ldots, h_s$ denote the horizontal lines of $\mathbb{G}$ and let $v_1, v_2, \ldots, v_t$ denote the vertical lines of $\mathbb{G}$. Suppose for the sake of contradiction that there exist $i \in [s]$ and $j \in [t]$ such that both the lines $h_i$ and $v_j$ contain at least 3 points of $\mathcal{P}$ each. Then by property (*), $h_i$ and $v_j$ are both lines in $\mathcal{L}$. But as $h_i$ and $v_j$ intersect, they are two lines in $\mathcal{L}$ that intersect, which is a contradiction, since the lines in $\mathcal{L}$ are all parallel to each other (note that parallel lines under a projective transformation may not be parallel but they do not intersect at any of the $s \times t$ grid points defined). Thus, we can conclude without loss of generality that for each $i \in [s]$, the horizontal line $h_i$ of $\mathbb{G}$ contains at most two points from $\mathcal{P}$, and hence at most two points from $\mathcal{P}'$. Since every point in $\mathcal{P}'$ is contained in at least one horizontal line of $\mathbb{G}$, we have that $s \geq |\mathcal{P}'|/2$ and therefore $s = \Omega(n^2)$. By property (+), each vertical line of $\mathbb{G}$ can contain at most $n$ points of $\mathcal{P}'$, and therefore, $t \geq |\mathcal{P}'|/n$, which implies that $t = \Omega(n)$. Thus the size of the grid $\mathbb{G}$ is $st = \Omega(n^3)$.    ◄

**Remark 5.** *The above construction also provides a counter-example[1] to the Conjecture 1.17 as stated in [23]. The formal statement of the conjecture is: Consider sufficiently large positive integers $m$ and $n$ that satisfy $m = O(n^2)$ and $m = \Omega(\sqrt{n})$. Let $P$ be a set of $m$ points and $L$ be a set of $n$ lines, both in $\mathbb{R}^2$, such that $I(P, L) = \Theta(m^{2/3}n^{2/3})$ (the number of incidences). Then there exists a subset $P' \subset P$ such that $|P'| = \Theta(m)$ and $P'$ is contained in a section of the integer lattice of size $\Theta(m)$, possibly after applying a projective transformation to it.*

## 2.3   Covering by Algebraic Curves

In this subsection, we address the question of covering a grid by algebraic curves. The answer comes as a direct application of the famous Combinatorial Nullstellensatz Theorem due to Noga Alon.

**Lemma 6 (Combinatorial Nullstellensatz [2]).** *Let $f = f(x_1, \ldots, x_d)$ be a polynomial in $\mathbb{R}[x_1, \ldots, x_d]$. Suppose the degree $\deg(f)$ of $f$ is $\sum_{i=1}^{d} t_i$ where each $t_i$ is a non-negative integer, and suppose the coefficient of $\prod_{i=1}^{d} x_i^{t_i}$ in $f$ is non-zero. Then, if $S_1, \ldots, S_n$ are subsets of $\mathbb{R}$ with $|S_i| > t_i$, there are $s_1 \in S_1, s_2 \in S_2, \ldots, s_d \in S_d$ so that $f(s_1, \ldots, s_d) \neq 0$.*

---

[1] This was communicated to Prof. Adam Sheffer who told us that this only exposes a typo in the statement of the conjecture which is more interesting and challenging when $m = o(n^2)$. Note that in our construction we have $m = n^2$.

**Theorem 7.** *Suppose the $n \times n$ grid is covered by $m$ algebraic curves of degree at most $k$. Then $m \geq n/k$.*

*Proof.* Suppose $m < n/k$. Let the algebraic curves defined by $f_1(x, y) = 0$, ..., $f_m(x, y) = 0$ cover the $n \times n$ grid, where $\deg(f_i) \leq k$. Then the polynomial $f(x, y) := \prod_{i=1}^{m} f_i(x, y)$ vanishes at each grid point. Suppose $\deg(f) = t_1 + t_2$ with the coefficient of $x^{t_1} y^{t_2}$ in $f$ being non-zero. Now note that $t_i \leq t_1 + t_2 = \deg(f) \leq mk < n$, for each $i = 1, 2$. So by Lemma 6, there exists a grid point $(s_1, s_2)$ so that $f(s_1, s_2) \neq 0$ and we arrive at a contradiction. Therefore, we conclude that $m \geq n/k$. ◄

**Corollary 8.** $\mathcal{G}_C(P) = \lceil n/k \rceil$, *where $P$ is an $n \times n$ grid and $C$ denotes algebraic curves of degree at most $k$.*

*Proof.* The lower bound follows from the previous theorem and the upper bound follows from covering by lines and then considering a set of $k$ lines as one curve of degree $k$. ◄

See Appendix A.2 for a discussion on irreducible algebraic curves.

## 3 Covering by Monotonic Curves

In this section, we consider the case when the the curve is *monotonic*.

**Definition 9.** *Let $f : [0, 1] \to \mathbb{R}^d$ be a curve and suppose $f(t) = (f_1(t), \ldots, f_d(t))$ for $t \in [0, 1]$. Then $f$ is called **monotonic** if it satisfies the following property: $t_1 \leq t_2 \Rightarrow f_i(t_1) \leq f_i(t_2)$ for each $i = 1, \ldots, d$.*

Given a finite subset $P$ of $\mathbb{R}^d$, we define the poset $\mathcal{P} := (P, \leq)$ as follows. For $x := (x_1, \ldots, x_d) \in \mathbb{R}^d$ and $y := (y_1, \ldots, y_d) \in \mathbb{R}^d$, we define $x \leq y$ if $x_i \leq y_i$ for $i = 1, \ldots, d$.

We say that two elements $a$ and $b$ of a poset $P$ are *comparable* if either $a \geq b$ or $b \leq a$. An *antichain* in a poset is a set of elements no two of which are comparable to each other, and a *chain* is a set of elements every two of which are comparable. A chain decomposition is a partition of the elements of the poset into disjoint chains. Size of an antichain is its number of elements, and the size of a chain decomposition is its number of chains.

**Proposition 10.** *Let $w(\mathcal{P})$ denote the size of the largest antichain, called the width, of $\mathcal{P}$. Then $\mathcal{G}_C(P) = w(\mathcal{P})$, where $P$ is any point set and $C$ denotes monotonic curves.*

*Proof.* Let $x_i \in \mathcal{P}$ for $i = 1, \ldots, r$. Then note that $x_1 \leq \cdots \leq x_r$ is a chain if and only if $x_1, \ldots, x_r$ lie on the same curve (which is monotonic). Therefore, $\mathcal{G}_C(P)$ equals the number of chains in a chain decomposition of smallest size of $\mathcal{P}$, which by Dilworth's theorem [10] equals the size of the largest antichain of $\mathcal{P}$. Hence $\mathcal{G}_C(P) = w(\mathcal{P})$. ◄

Note that the poset $\mathcal{P}$ can be decomposed into $w(\mathcal{P})$ many disjoint chains. Therefore, the points in $P$ can be covered by $\mathcal{G}_C(P)$ many monotonic curves such that *no two curves intersect at a point of $P$*.

## 4   Covering by Closed Curves

In this section, we consider covering grids by circles, convex curves and ortho-convex curves. Notice that the curves need not be of the same size, e.g., when we are considering covering by circles, all the circles need not be of the same size.

### 4.1   Covering by Circles and Convex Curves

*Covering by Circles.* A circle contains at most $O(n^\epsilon)$ points from an $n \times n$ grid for every $\epsilon > 0$ (see e.g. [14]). Therefore, the minimum number of circles required to cover an $n \times n$ grid is $\Omega(n^{2-\epsilon})$, for every $\epsilon > 0$. Regarding upper bound, note that there is a covering of the $n \times n$ grid by $O(n^2/\sqrt{\log n})$ circles. This is obtained by choosing a corner of the grid and drawing all concentric circles such that each of them is incident to at least one grid point (see Fig. 1). The number of such circles is $O(n^2/\sqrt{\log n})$ by a well known theorem of Ramanujan and Landau [4,18]. The theorem says that the number of positive integers that are less than $n$ that are the sum of two squares is $\Theta(n/\sqrt{\log n})$. We sum it up as the following.

**Proposition 11.** $\Omega(n^{2-\epsilon}) \leq \mathcal{G}_C(P) \leq O(n^2/\sqrt{\log n})$, *where $P$ is an $n \times n$ grid and $C$ denotes circles.*

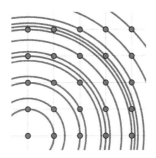

**Fig. 1.** Covering of $5 \times 5$ grid by circles

*Covering by Convex Curves.* A closed convex curve intersects non-trivially with a horizontal grid line if it contains more than two points from the line. Note that, any closed convex curve can intersect at most two horizontal grid lines non-trivially. This follows from the following lemma whose proof is in Appendix A.3.

**Lemma 12.** *If a closed convex curve intersects a horizontal grid line non-trivially, then it must lie entirely on one side of that line.*

**Theorem 13.** *The points of the $n \times n$ grid cannot be covered with less than $n/2$ closed convex curves, i.e. $\mathcal{G}_C(P) \geq n/2$ where $P$ is an $n \times n$ grid and $C$ denotes closed convex curves.*

*Proof.* Suppose, for the sake of contradiction, that $C_1, C_2, \ldots, C_k$ are $k$ closed convex curves such that they together cover every point of the $n \times n$ grid and that $k < n/2$. Then, since there are $n$ horizontal grid lines, and by Lemma 12 above, each $C_i$ can have a non-trivial intersection with at most 2 horizontal grid lines, we can conclude that there is some horizontal grid line such that no curve in $C_1, C_2, \ldots, C_k$ has a non-trivial intersection with that line. Now consider the points on that horizontal line. There are $n$ points on this line. Each curve in $C_1, C_2, \ldots, C_k$ can cover at most two points from that line and none of them intersects non-trivially with this horizontal line. But then, since $k < n/2$, there must be some point on this horizontal line that is not covered by any curve in $C_1, C_2, \ldots, C_k$, which is a contradiction.    ◄

Almost same argument can be used to get an answer for an $m \times n$ grid and this will be $\min\{\lceil m/2 \rceil, \lceil n/2 \rceil\}$.

## 4.2  Covering by Orthoconvex Curves

A set $K \subseteq \mathbb{R}^2$ is defined to be *orthogonally convex* if, for every line $\ell$ that is parallel to one of standard basis vectors $(1, 0)$ or $(0, 1)$, the intersection of $K$ with $\ell$ is empty, a point, or a single segment. The *orthogonal convex hull* of a point set $P \subseteq \mathbb{R}^2$ is the intersection of all connected orthogonally convex supersets of $P$. If the boundary of orthogonal convex hull (of a set of points) is a simple closed curve then we call it an *orthoconvex* curve. An orthoconvex curve has only two types of angles, namely 90° and 270°. By *inner corner* of an orthoconvex curve, we mean a point where the curve turns by 270°. See Fig. 2 for an example of an orthoconvex where the red points are its inner corners.

If an orthoconvex curve (with $k$ inner corners) covers a set of points, then there is also an orthoconvex curve (with $k$ inner corners) covering the same points which is not self-intersecting and all the corners are grid points. This can be done by pushing the sides/edges of the curve "outwards" (instead of inwards which corresponds to taking orthoconvex hull) until we hit a grid line. So w.l.o.g., we may impose the following assumptions of 'non-self-intersecting' and 'corners are grid points'.

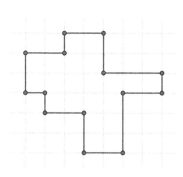

**Fig. 2.** An othoconvex curve and its inner corners (in red) (Color figure online)

In the following, by *curve*, we mean an orthoconvex curve having at most one inner corner (Fig. 3 shows examples of such curves). We say that a curve *hits* a (horizontal or vertical) grid line if the curve has a non-trivial intersection with that grid line (i.e., the curve follows that grid line for some distance, rather than just crossing it). We say that a collection of curves $C$ *hits* a (horizontal or vertical) grid line if there is some curve in $C$ that hits that grid line. Given a collection of curves $C$, we say that a grid point is *exposed* (by $C$) if the grid point is not covered by any curve in $C$, but it lies on a horizontal grid line and a vertical grid line both of which are hit by $C$. Given a collection of curves $C$, a *corner* of $C$ is a corner of the (minimum size) bounding box of $C$. So every collection $C$ of curves has exactly 4 corners. If a corner of $C$ is an exposed grid point, then we call it an *exposed corner*. We say that a sequence of curves $c_1, c_2, \ldots, c_t$ is *good* if for every $i \in \{2, 3, \ldots, t\}$, $c_i$ hits a grid line that is hit by $\{c_1, c_2, \ldots, c_{i-1}\}$. Clearly, every prefix of a good sequence is also a good sequence.

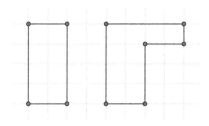

**Fig. 3.** Orthoconvex curves with at most one inner corner

**Lemma 14.** *Let $c_1, c_2, \ldots, c_t$ be a good sequence of curves. Then $\{c_1, c_2, \ldots, c_t\}$ either: (a) hits at most $5t$ grid lines, or (b) hits $5t + 1$ grid lines and has an exposed corner.*

(See Fig. 4 for an illustration of case (b), whereas Fig. 5 shows an example of cae (a))

*Proof.* We prove this by induction on $t$. It is not difficult to see that the lemma is true when $t = 1$. Let $i > 1$ and suppose that the lemma is true for the good sequence $c_1, c_2, \ldots, c_{i-1}$. Let $C = \{c_1, c_2, \ldots, c_{i-1}\}$. Then either $C$ hits (a) at most $5i - 5$ grid lines, or (b) hits $5i - 4$ grid lines and has an exposed corner.

In case (a), since the curve $c_i$ can hit at most 5 grid lines that are not hit by $C$ (recall that $c_i$ hits at least one grid line that is also hit by $C$), we have that $C \cup \{c_i\}$ can hit at most $5i$ grid lines, and we are done. Next, let us consider case (b). Note that if $c_i$ is a rectangle, then it can hit at most 3 grid lines that are not hit by $C$ (note that, a rectangle has four sides and $c_i$ hits at least one grid line that is also hit by $C$), and therefore, $C \cup \{c_i\}$ hits at most $5i - 4 + 3 = 5i - 1$ grid lines, and we are done. So we can assume that $c_i$ is not a rectangle. Also, if there are two grid lines that are hit by both $C$ and $c_i$, then $C \cup \{c_i\}$ hits at most $5i$ grid lines, and we are done. So we can assume that $c_i$ hits exactly one grid line that is hit by $C$, and therefore, $C \cup \{c_i\}$ hits exactly $5i+1$ grid lines. In this case, we have to show that one of the corners of $C \cup \{c_i\}$ is exposed. Let $B$ be the bounding box of $C \cup \{c_i\}$. Let $g_0, g_1, g_2, g_3$ be the grid lines on which the top, right, bottom, and left borders of $B$ lie. Clearly, each of $g_0, g_1, g_2, g_3$ is hit by either $C$ or $c_i$ or both. Since $c_i$ hits exactly one grid line that is hit by $C$, we have that at most one of $g_0, g_1, g_2, g_3$ is hit by both $C$ and $c_i$. This implies that $C$ and $\{c_i\}$ do not have shared corners. Note that a corner $v$ of $C$ is exposed, and a corner $v'$ of $\{c_i\}$ is exposed. If each of $g_0, g_1, g_2, g_3$ is hit by $C$, then $v$ is an exposed corner of $C \cup \{c_i\}$ (observe that $v$ cannot be covered by $c_i$, because if it is, it has to be a corner of $\{c_i\}$, which would mean that $C$ and $\{c_i\}$ have a shared corner) and we are done. Similarly, if each of $g_0, g_1, g_2, g_3$ is hit by $c_i$, then $v'$ is an exposed corner of $C \cup \{c_i\}$ and we are again done. Thus we can assume that neither $C$ nor $c_i$ hits all the grid lines $g_0, g_1, g_2, g_3$. Recall that all grid lines except at most one in $g_0, g_1, g_2, g_3$ are hit by exactly one of $C$ or $c_i$. Then there exists some $j \in \{0, 1, 2, 3\}$ such that one of $g_j, g_{j+1 \mod 4}$ is hit by $C$ and not by $c_i$, and the other is hit by $c_i$ and not by $C$. Then the grid point that is contained in both the grid lines $g_j$ and $g_{j+1 \mod 4}$ is an exposed corner of $C \cup \{c_i\}$. This completes the proof.     ◀

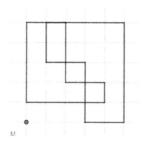

**Fig. 4.** Two curves that hit 11 grid lines and has an exposed corner (M)

**Theorem 15.** *If $m$ orthoconvex curves with at most one inner corner cover the $n \times n$ grid, then $m \geq 2n/5$.*

*Proof.* Let $C$ be a collection of $m$ curves that cover the $n \times n$ grid. For two curves $c$ and $d \in C$, we say that $cRd$ if there is a grid line that is hit by both $c$ and $d$. Let $R^*$ be the transitive closure of $R$. Clearly, $R^*$ is an equivalence relation. Let $S_1, S_2, \ldots, S_p$ be the equivalence classes of $R^*$. We need the following claims for the proof.

**Claim 16.** *For each $i \in [p]$, $S_i$ does not expose any grid point.*

*Proof.* Suppose for some $i \in [p]$, $S_i$ exposes a grid point $v$. That is, $v$ is not covered by $S_i$, but both the horizontal grid line as well as the vertical grid line that contains $v$ are hit by $S_i$. Since $C$ covers the whole grid, there is a curve $c \in C$ that covers $v$. As $S_i$ does not cover $v$, we have that $c \in C - S_i$. As $c$ covers $v$, $c$ hits either the horizontal grid line containing $v$ or the vertical grid line containing $v$. Since both these grid lines are hit by $S_i$, it follows that there exists some $d \in S_i$ such that $c$ and $d$ hit a common grid line. Then $d\mathcal{R}c$, which implies that $c \in S_i$, which is a contradiction. This proves the claim. ◄

**Claim 17.** *The curves of each equivalence class $S_i$ can be arranged in a good sequence.*

*Proof.* Let $G$ be the graph with vertex set $S_i$ and edge set $R$ restricted to $S_i$. By enumerating the curves of $S_i$ in the order in which they are visited by a graph traversal algorithm starting from an arbitrary vertex, we get a sequence of the curves in $S_i$ such that before a curve $c$ is encountered in the sequence, we encounter some curve $d$ such that $d\mathcal{R}c$ (except for the first curve in the sequence). This sequence is clearly a good sequence of the curves in $S_i$. This proves the claim. ◄

By Lemma 14 and Claims 16 and 17, we know that for each $i \in [p]$, $S_i$ hits at most $5|S_i|$ grid lines. Thus the total number of grid lines that are hit by $C$ is at most $5(|S_1| + |S_2| + \cdots + |S_p|) = 5|C| = 5m$. If the the curves in $C$ hit $2n$ grid lines, we then have $5m \geq 2n$, which gives $m \geq 2n/5$. Otherwise, suppose that the collection $C$ of $m$ curves, where $m \leq 2n/5$, hits less than $2n$ grid lines. That is, there is some (horizontal or vertical) grid line that is not hit by any curve in $C$. Then every curve in C can cover at most two points on this grid line (if it covers more than two, then the curve hits this grid line). So at most $2m \leq 4n/5$ points on this grid line can be covered by the collection of curves $C$, which means that some points on this grid line are not covered by any curve in $C$, which is a contradiction. So we conclude that $m \geq 2n/5$ and this proves the theorem. ◄

Note that, the inequality of the above theorem is tight for $n = 5$ since the $5 \times 5$ grid can be covered by 2 curves (shown in Fig. 5). As a consequence of the above theorem, we also get the following theorem on orthoconvex curves with *at most 2 inner corners*.

**Fig. 5.** Covering of $5 \times 5$ grid by two orthoconvex curves (with at most one inner corner)

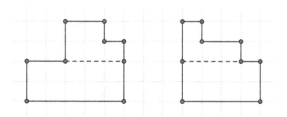

**Fig. 6.** Decomposition of orthoconvex curves with 2 inner corners

**Theorem 18.** *We need at least $2n/7$ orthoconvex curves with at most two inner corners to cover an $n \times n$ grid.*

*Proof.* Suppose we have a covering by $m$ such curves. Note that we can decompose each orthoconvex curve with two inner corners into an orthoconvex curve with at most one inner corner and a rectangle (see Fig. 6). Hence we obtain a covering by $m$ orthoconvex curves with at most one inner corner and $m$ rectangles. These $m$ orthoconvex curves with at most one inner corner can together hit at most $5m$ grid lines (see proof of Theorem 15) and the rectangles together hit at most $2m$ extra grid lines (since each rectangle hit at most two extra grid lines). So the total number of grid lines hit by our original curves is at most $7m$. Since the curves have to hit $2n$ grid lines (by the same reasoning as in proof of Theorem 15), we then have $7m \geq 2n$. Hence, we conclude that $m \geq 2n/7$. ◀

See Appendix A.4 for a remark on covering by orthoconvex curves.

## 5 Conclusion and Discussion

In this paper, we mainly discussed the problem of covering a grid (mostly planar) by minimum number of curves of various types. An interesting open problem in this direction is to cover the hypercube by minimum number of skew hyperplanes. We leave it as an open problem to figure out what happens when there are more inner corners for covering by an orthoconvex curve. Lastly, we mention that in this article we only considered 1-fold covering where every grid point was covered at least once. But, in general, we could ask analogous questions for $r$-fold covering (i.e., every point is covered at least $r$ times) for $r \geq 2$.

## A Appendix

### A.1 Proof of Proposition 2

*Proof.* Let $M := \max\{k_1, \ldots, k_d\}$. First we show that $\ell(P) \leq M$ by induction on $d$. The base case $d = 1$ is obvious. Now we proceed to the induction step. Let $L$ be a line segment that lies inside the rectangular parallelepiped $[1, k_1] \times \cdots \times [1, k_d]$. Then $L$ has length at most $\sqrt{\sum_{i=1}^{d}(k_i - 1)^2}$. Now let $x := (x_1, \ldots, x_d)$ and

$y := (y_1, \ldots, y_d)$ be two distinct points of $P$ lying on $L$. If $x_i = y_i$ for some $i$, then $L$ lies inside a lower dimensional rectangular parallelepiped and therefore, by induction hypothesis, $L$ covers at most $\max\{k_j \mid j \neq i\} \leq M$ many points. So let us assume $x_i \neq y_i$ for all $i = 1, \ldots, d$. Then the distance between $x$ and $y$ is at least $\sqrt{d}$. Suppose $L$ covers a total of $t$ points of $P$. Then we have

$$(t-1)\sqrt{d} \leq \sqrt{\sum_{i=1}^{d}(k_i - 1)^2} \leq \sqrt{d} \cdot \max\{k_1 - 1, \ldots, k_d - 1\}$$

and this implies $t \leq \max\{k_1, \ldots, k_d\}$. Therefore, we conclude that $\ell(P) \leq M$. On the other hand, there clearly exist lines covering $M$ points, namely the lines parallel to the coordinate axis $i_0$, where $M = k_{i_0}$. Hence, we have shown that $\ell(P) = M$.                                                                ◀

### A.2   A remark on Irreducible Algebraic Curves

**Remark 19 (Irreducible algebraic curves).** *By a result of Bombieri and Pila [6], an irreducible algebraic curve of degree $k$ can contain at most $O(n^{1/k})$ points from an $n \times n$ grid and hence, the minimum number of irreducible algebraic curves of degree $k$ to cover the $n \times n$ grid is at least $\Omega(n^{2-1/k})$.*

Using the same reasoning as in the previous theorem and corollary, one also has the following result on covering the $n_1 \times \cdots \times n_d$ grid by algebraic hypersurfaces.

**Theorem 20.** *The minimum number of algebraic hypersurfaces of degree at most $k$ needed to cover the $n_1 \times \cdots \times n_d$ grid is equal to $\lceil n/k \rceil$, i.e., $\mathcal{G}_C(P) = \lceil n/k \rceil$, where $P$ is an $n_1 \times \cdots \times n_d$ grid and $C$ denotes algebraic hypersurfaces of degree at most $k$.*

### A.3   Proof of Lemma 12

*Proof.* Suppose the curve intersects a horizontal line at three points $p, q, r$, where $q$ lies in the interior of line segment $[p, r]$. Since the curve is convex, there exists a line $L$ through $q$ such that the curve lies entirely on one side of $L$ (hyperplane separation theorem). Now if $L$ is different from the horizontal line, then $p$ and $r$ lie on different sides of $L$. But since the curve lies on one side of $L$, it can not pass through both $p$ and $r$, a contradiction. Therefore, $L$ is same as the horizontal line and the curve lies entirely on one side of this line.                    ◀

### A.4   Remark on Covering by Orthoconvex Curves

**Remark 21.** *We think that the bound $2n/7$ of Theorem 18 is probably not tight. So a natural problem is to obtain a tight bound for covering by orthoconvex curves with at most 2 inner corners. The next natural follow up question would be:*

*what happens for orthoconvex curves with at most $k$ inner corners for $k = 3, 4$ etc. It seems our arguments for $k = 1, 2$ can not be extended to these cases to obtain non-trivial bounds and hence require new ideas. Another question of interest is to find the minimum number of general orthoconvex curves (with no restrictions on the number of inner corners) required to cover an $n \times n$ grid. One can check that for $n = 4, 5, 6, 7, 8, 9$ and $10$, the $n \times n$ grid can be covered by $2, 2, 2, 3, 3, 3$ and $4$ orthoconvex curves, respectively. To us, the general problem of orthoconvex curves seems difficult. Note that we have obvious lower and upper bounds of $\lceil (n + 1)/4 \rceil$ and $\lfloor n/2 \rfloor$ respectively, since, any orthoconvex curve can contain at most $4n - 4$ grid points (the number of grid points on the boundary of an $n \times n$ grid) and on the other hand, an $n \times n$ grid can be covered by $\lfloor n/2 \rfloor$ orthoconvex curves. Any improvement over these bounds would be interesting.*

# References

1. Afshani, P., Berglin, E., van Duijn, I., Nielsen, J.S.: Applications of incidence bounds in point covering problems. In: 32nd SoCG, 14–18 June 2016. LIPIcs, vol. 51, pp. 60:1–60:15 (2016)
2. Alon, N.: Combinatorial Nullstellensatz. Comb. Probab. Comput. **8**(1–2), 7–29 (1999)
3. Andrews, G.E.: An asymptotic expression for the number of solutions of a general class of Diophantine equations. Trans. Am. Math. Soc. **99**, 272–277 (1961)
4. Berndt, B.C., Rankin, R.A.: Ramanujan: Letters and Commentary. American Mathematical Society, Providence (1995)
5. Boissonnat, J.-D., Dutta, K., Ghosh, A., Kolay, S.: Tight kernels for covering and hitting: POINT HYPERPLANE COVER and POLYNOMIAL POINT HITTING SET. In: Bender, M.A., Farach-Colton, M., Mosteiro, M.A. (eds.) LATIN 2018. LNCS, vol. 10807, pp. 187–200. Springer, Cham (2018). https://doi.org/10.1007/978-3-319-77404-6_15
6. Bombieri, E., Pila, J.: The number of integral points on arcs and ovals. Duke Math. J. **59**(2), 337–357 (1989)
7. Brass, P., Moser, W., Pach, J.: Research Problems in Discrete Geometry. Springer, Heidelberg (2005)
8. Cygan, M., et al.: Parameterized Algorithms (2015)
9. Dillencourt, M.B., Eppstein, D., Hirschberg, D.S.: Geometric thickness of complete graphs. In: Graph Algorithms and Applications, vol. 2, pp. 39–51. World Scientific (2004)
10. Dilworth, R.P.: A decomposition theorem for partially ordered sets. Ann. Math. **51**(1), 161–166 (1950)
11. Dumitrescu, A., Tóth, C.D.: Covering grids by trees. In: CCCG (2014)
12. Engel, K.: Sperner Theory. Encyclopedia of Mathematics and its Applications, Cambridge University Press, Cambridge (1997)
13. Grantson, M., Levcopoulos, C.: Covering a set of points with a minimum number of lines. In: Calamoneri, T., Finocchi, I., Italiano, G.F. (eds.) CIAC 2006. LNCS, vol. 3998, pp. 6–17. Springer, Heidelberg (2006). https://doi.org/10.1007/11758471_4
14. Guth, L.: Polynomial Methods in Combinatorics. University Lecture Series, vol. 64. American Mathematical Society (2016)
15. Har-Peled, S., Lidický, B.: Peeling the grid. SIAM J. Discret. Math. **27**, 650–655 (2013)

16. Keszegh, B.: Covering paths and trees for planar grids. arXiv preprint arXiv:1311.0452 (2013)
17. Kratsch, S., Philip, G., Ray, S.: Point line cover: the easy kernel is essentially tight. ACM Trans. Algorithms **12**(3):40:1–40:16 (2016)
18. Landau, E.: Uber die einteilung der positiven ganzen zahlen in vier klassen l nach der mindeszahl der zu ihrer additiven zusammensetzung erforderlichen quadrate. Arch. Math. Phys. **13**, 305–312 (1908)
19. Langerman, S., Morin, P.: Covering things with things. Discret. Comput. Geom. **33**(4), 717–729 (2005)
20. Matousek, J.: Lectures on Discrete Geometry. Graduate Texts in Mathematics, vol. 212. Springer, Cham (2002)
21. Pach, J., Agarwal, P.K.: Combinatorial Geometry. Wiley-Interscience Series in Discrete Mathematics and Optimization, Wiley, Hoboken (1995)
22. Sauermann, L., Xu, Z.: Essential covers of the hypercube require many hyperplanes (2023)
23. Sheffer, A.: Polynomial Methods and Incidence Theory. Cambridge Studies in Advanced Mathematics, Cambridge University Press, Cambridge (2022)

# Improved Algorithms
# for Minimum-Membership Geometric Set
# Cover

Sathish Govindarajan[(✉)] and Siddhartha Sarkar

Indian Institute of Science, Bengaluru, India
{gsat,siddharthas1}@iisc.ac.in

**Abstract.** Bandyapadhyay et al. introduced the generalized minimum-membership geometric set cover (GMMGSC) problem [SoCG, 2023], which is defined as follows: We are given two sets $P$ and $P'$ of points in $\mathbb{R}^2$, $n = \max(|P|, |P'|)$, and a set $S$ of $m$ axis-parallel unit squares. The goal is to find a subset $S^* \subseteq S$ that covers all the points in $P$ while minimizing $\mathsf{memb}(P', S^*)$, where $\mathsf{memb}(P', S^*) = \max_{p \in P'} |\{s \in S^* : p \in s\}|$.

We study GMMGSC problem and give a 16-approximation algorithm that runs in $O(m^2 \log m + m^2 n)$ time. Our result is a significant improvement to the 144-approximation given by Bandyapadhyay et al. that runs in $\tilde{O}(nm)$ time.

GMMGSC problem is a generalization of another well-studied problem called Minimum Ply Geometric Set Cover (MPGSC), in which the goal is to minimize the ply of $S^*$, where the ply is the maximum cardinality of a subset of the unit squares that have a non-empty intersection. The best-known result for the MPGSC problem is an 8-approximation algorithm by Durocher et al. that runs in $O(n + m^8 k^4 \log k + m^8 \log m \log k)$ time, where $k$ is the optimal ply value [WALCOM, 2023].

**Keywords:** Computational Geometry · Minimum-Membership Geometric Set Cover · Minimum Ply Covering · Approximation Algorithms

## 1 Introduction

Set Cover is a fundamental and well-studied problem in combinatorial optimization. Given a range space $(X, \mathcal{R})$ consisting of a set $X$ and a family $\mathcal{R}$ of subsets of $X$ called the ranges, the goal is to compute a minimum cardinality subset of $\mathcal{R}$ that covers all the points of $X$. It is NP-hard to approximate the minimum set cover below a logarithmic factor [10,17]. When the ranges are derived from geometric objects, it is called the geometric set cover problem. Computing the minimum cardinality set cover remains NP-hard even for simple 2D objects, such as unit squares on the plane [11]. There is a rich literature on designing approximation algorithms for various geometric set cover problems (see [1,5,7,12,15]). Several variants of the geometric set cover problem such as unique cover, red-blue cover, etc. are well-studied [6,13].

S. Kalyanasundaram and A. Maheshwari (Eds.): CALDAM 2024, LNCS 14508, pp. 103–116, 2024.
https://doi.org/10.1007/978-3-031-52213-0_8

In this paper, we study a natural variant of the geometric set cover called the Generalized Minimum-Membership Geometric Set Cover (GMMGSC). This is a generalization of two well-studied problems: minimum ply geometric set cover and minimum-membership geometric set cover, which were motivated by real-world applications in interference minimization in wireless networks and have received the attention of researchers [3,4,8,9]. We define the problem below.

**Definition 1 (Membership).** *Given a set $P$ of points and a set $\mathcal{S}$ of geometric objects, the membership of $P$ with respect to $\mathcal{S}$, denoted by* memb$(P, \mathcal{S})$, *is* $\max_{p \in P} |\{s \in \mathcal{S} : p \in s\}|$.

**Definition 2 (GMMGSC problem).** *Given two sets $P$ and $P'$ of points in $\mathbb{R}^2$, $n = \max(|P|, |P'|)$, and a set $\mathcal{S}$ of $m$ axis-parallel unit squares, the goal is to find a subset $\mathcal{S}^* \subseteq \mathcal{S}$ that covers all the points in $P$ while minimizing* memb$(P', \mathcal{S}^*)$.

## 1.1 Related Work

Bandyapadhyay et al. introduced the generalized minimum-membership geometric set cover (GMMGSC) problem and gave a polynomial-time constant-approximation algorithm for unit squares [3]. Specifically, they consider the special case when all the points lie within a unit grid cell and all the input unit squares intersect the grid cell. They use linear programming techniques to obtain a 16-approximation in $\tilde{O}(nm)$ time for GMMGSC problem for this special case. Here, $\tilde{O}(\cdot)$ hides some polylogarithmic factors. This implies a 144-approximation for GMMGSC problem for unit squares.

We note that GMMGSC problem is a generalization of two well-studied problems: (1) Minimum-Membership Geometric Set Cover problem where $P' = P$, and (2) Minimum Ply Geometric Set Cover problem where $P'$ is obtained by picking a point from each distinct *region* in the *arrangement* $\mathcal{A}(\mathcal{S})$ of $\mathcal{S}$.

Minimum-Membership Set Cover (MMSC) problem is well-studied in both abstract [14] and geometric settings [9]. Kuhn et al. showed that the abstract MMSC problem admits an $O(\log m)$-approximation algorithm, where $m$ is the number of ranges. They also showed that, unless P = NP, this is the best possible approximation ratio. Erlebach and van Leeuwen introduced the geometric version of the MMSC problem [9]. They showed NP-hardness for approximating the problem with ratio less than 2 on unit disks (i.e., disks with diameter 1) and unit squares. They gave a 5-approximation algorithm for unit squares that runs in $n^{O(k)}$ time, where $k$ is the minimum membership.

Biedl et al. introduced the Minimum Ply Geometric Set Cover (MPGSC) problem [4]. They gave 2-approximation algorithms for unit squares and unit disks that run in $(nm)^{O(k)}$ time, where $k$ is the optimal ply of the input instance. Durocher et al. presented the first constant approximation algorithm for the MPGSC problem with unit squares [8]. They divide the problem into subproblems by using a standard grid decomposition technique. They solve almost optimally the subproblem within a square grid cell using a dynamic programming scheme. Specifically, they give an algorithm that runs in $O(n + m^8 k^4 \log k + m^8 \log m \log k)$ time and outputs a solution with ply $\leq 8k + 32$, where $k$ is the optimal ply. Bandyapadhyay et al. also gave a $(36 + \epsilon)$-approximation algorithm for the MPGSC problem for unit squares that runs in $n^{O(1/\epsilon^2)}$ time [3,16].

## 1.2 Our Contribution

We first consider a special case of the GMMGSC problem called the line instance of GMMGSC, where the input squares are intersected by a horizontal line and the input points lie on only one side of the line. Refer to Definition 4. We design a polynomial-time algorithm (i.e., Algorithm 1) for this problem where the solution has some desirable properties.

Next, we consider the slab instance of GMMGSC, where the input squares are intersected by a unit-height horizontal slab and the points lie within the slab. Refer to Definition (3). As far as we know, there are no known approximation results for this problem. We adapt the linear programming techniques in [3] to decompose a slab instance of GMMGSC into two line instances of GMMGSC. We use Algorithm 1 to solve them. Then we merge the two solutions to obtain the final solution. A major challenge was finding a solution for the line GMMGSC which respects a key lemma (i.e., Lemma 1). This key lemma enables us to obtain a solution with membership at most $(8 \cdot OPT + 18)$ for the slab instance.

Finally, we give an algorithm for GMMGSC problem for unit squares that runs in $O(m^2 \log m + m^2 n)$ time and outputs a solution whose membership is at most $16 \cdot OPT + 36$. We divide GMMGSC instance into multiple line instances. Then we use Algorithm 1 on the line instances. Finally, we merge the solutions of the line instance to obtain the final solution.

For GMMGSC problem, we note that our result is a significant improvement in the approximation ratio as compared to the best-known result of Bandyapadhyay et al. [3]. For MPGSC problem, our result is a significant improvement in the running time as compared to the best-known result of Durocher et al. while achieving a slightly worse approximation ratio [8].

## 2 Generalized Minimum-Membership Set Cover for Unit Squares

Let $P$ and $P'$ be two sets of points in $\mathbb{R}^2$ and $S$ be a set of axis-parallel unit squares. We want to approximate the minimum-membership set cover (abbr. MMSC) of $P$ using $S$ where membership is defined with respect to $P'$. First, we divide the plane into horizontal slabs of unit height. Each slab is defined by two horizontal lines $L_1$ and $L_2$, unit distance apart, where $L_2$ is above $L_1$. We define an instance for the slab subproblem below. For an illustration, refer to Fig. 1.

**Definition 3 (Slab instance).** *Consider a set $S$ of unit squares where each square intersects one of the boundaries of a unit-height horizontal slab $\alpha$. The points of the set $P$ to be covered are located within $\alpha$, each point lying inside at* *least one of the squares in $\mathcal{S}$. Let $P'$ be a set of points with respect to which the membership is to be computed. The instance $(P, P', \mathcal{S})$ is called a slab instance.*

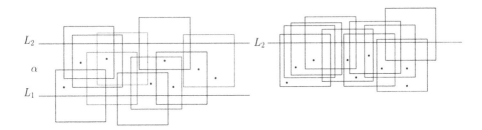

**Fig. 1.** A slab instance and a line instance of GMMGSC.

In Sect. 2.2, we solve the slab instance by decomposing it into two *line instances*. In the following section, we define and discuss the line instance.

### 2.1    GMMGSC for the Line Instance

**Definition 4 (Line instance).** *Consider a set $\mathcal{S}$ of unit squares where each square intersects a horizontal line $\ell$. The points of the set $P$ to be covered are located only on one side (above or below) of $\ell$, each point lying inside at least one of the squares in $\mathcal{S}$. Let $P'$ be a set of points with respect to which the membership is to be computed. The instance $(P, P', \mathcal{S})$ is called a line instance.*

In the rest of this section, we design an algorithm for the line instance where the input points lie below the defining horizontal line. For the slab, it would be the instance corresponding to the top boundary $L_2$ of the slab. Refer to Fig. 1 for an example. The algorithm for the line instance corresponding to the line $L_1$ is symmetric.

Let us introduce some notation first. For a unit square $s \in \mathcal{S}$, denote by $x(s)$ and $y(s)$ the $x$-coordinate and $y$-coordinate of the bottom-left corner of $s$, respectively. For a horizontal line $\ell$, denote by $y(\ell)$ the $y$-coordinate of any point on $\ell$.

We make the following non-degeneracy assumptions. First, no input square has its top boundary coinciding with the slab boundary lines. Second, $x$- and $y$-coordinates of the input squares are distinct. Note that a set $Q$ of intersecting unit squares also forms a clique in the intersection graph of $Q$.

In a set of unit squares $\mathcal{S}$, two squares $s_1, s_2 \in \mathcal{S}$ are **consecutive** (from left to right) if there exists no square $t \in \mathcal{S}$ such that $x(s_1) < x(t) < x(s_2)$. If a point $p$ is contained in exactly one square $s$ in a set cover $\mathcal{S}^*$ of a set of points $P$, then $p$ is said to be an **exclusive point** of the square $s$ with respect to $\mathcal{S}^*$. For $s \in \mathcal{S}^*$, the region in the plane, denoted by $\mathsf{Excl}(s)$, which is covered exclusively by $s$ is called the **exclusive region** of $s$ with respect to a set cover $\mathcal{S}^*$. For

$s_i, s_j \in \mathcal{S}^*$, the region in the plane, denoted by $\mathsf{Excl}(s_i, s_j)$, which is contained exclusively in $s_i \cap s_j$, is called the **pairwise exclusive region** of $s_i$ and $s_j$ with respect to a set cover $\mathcal{S}^*$. A square $s$ in a set cover $\mathcal{S}^*$ of a set of points $P$ is called **redundant** if it covers no point of $P$ exclusively. Refer to Fig. 2 for an illustration of these terms.

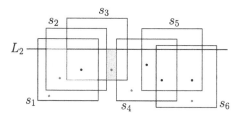

**Fig. 2.** A set cover for a line instance. For $i \in [5]$, $s_i, s_{i+1}$ are consecutive squares. The red points are exclusive points. The purple region is $\mathsf{Excl}(s_3)$. The yellow region is $\mathsf{Excl}(s_1, s_2)$. The squares $s_2, s_5$ are redundant. (Color figure online)

A set of unit squares having a common intersection is said to form a **geometric clique**. A set of unit squares containing a point of $P'$ in their common intersection region is said to form a **discrete clique**. The common intersection region of a set of unit squares forming a clique $Q$ is called the **ply region** of $Q$. The ply region of a clique $Q$ is always rectangular. For a clique $Q$, denote by $x_l(Q)$ (resp. $x_r(Q)$) the $x$-coordinate of the left (resp. right) boundary of the ply region of $Q$. Unless specified otherwise, a clique refers to a discrete clique.

**Types of Legal Cliques.** First, we classify a clique with respect to a line instance. A set of intersecting squares in a line instance is called a **top-anchored clique** when the points to be covered lie below the line with respect to which the line instance is defined. A set of intersecting squares in a line instance is called a **bottom-anchored clique** when the points to be covered lie above the line with respect to which the line instance is defined.

Let a line instance be defined with respect to a horizontal line $\ell$. Let $s_1, \ldots, s_k$ be a sequence of squares from left to right having a common intersection. This set of squares is called a **monotonic ascending clique** if $i < j$ implies $y(s_i) < y(s_j)$. We use the abbreviation $ASC$ to denote such a clique. Let a line instance be defined with respect to a horizontal line $\ell$. On the other hand, if $i < j$ implies $y(s_i) > y(s_j)$, then this set of squares is called a **monotonic descending clique**. We use the abbreviation $DESC$ to denote such a clique.

Let a line instance be defined with respect to a horizontal line $l$. Let $s_1, \ldots, s_k$, $s_{k+1}, \ldots s_{k+r}$ be a sequence of squares from left to right having a common intersection. This set of squares is called a **composite clique** if the following holds.

– The sequence of squares $s_1, \ldots, s_k$ forms a monotonic clique.
– Either $y(s_{k+1}) > y(s_k) < y(s_{k-1})$, or $y(s_{k-1}) < y(s_k) > y(s_{k+1})$, and

– The sequence of squares $s_{k+1}, \ldots, s_{k+r}$ forms a monotonic clique.

The square $s_k$ is called the *transition square*. We use the abbreviation $DESC|ASC$ to denote a composite clique where the sequence $s_1, \ldots, s_k$ is descending but the sequence $s_{k+1}, \ldots, s_{k+r}$ is ascending. For other types of composite cliques, the abbreviation would be self-explanatory (Fig. 3).

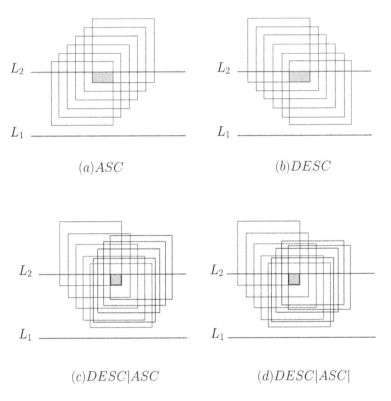

$(a)ASC$                                    $(b)DESC$

$(c)DESC|ASC$                         $(d)DESC|ASC|$

**Fig. 3.** (a), (b), and (c) show three different types of legal top-anchored cliques. (d) Shows an invalid clique where a top-anchored composite $DESC|ASC$ clique is followed by a *transition square*.

*Claim.* In a set cover for the line instance, where none of the constituent squares are redundant, a top-anchored composite clique must be of type $DESC|ASC$.

*Proof.* Suppose not. Let $Q$ be a top-anchored composite clique of type $ASC|ASC$, where the left ascending sequence be the squares $s_1, \ldots, s_k$ from left to right. Then the square $s_k$ would become redundant since the two squares $s_{k-1}, s_{k+1}$ would cover all the points covered by $s_k$ lying below the defining horizontal line. This contradicts the non-redundancy condition. For an example, refer to Fig. 4. Similar arguments apply to rule out the existence of top-anchored composite cliques of type $ASC|DESC$ and $DESC|DESC$.

In the rest of the paper, whenever we consider a top-anchored clique in the solution (containing no redundant squares) of a line instance of the GMMGSC problem, we assume that the clique is of one of the following legal types: (i) monotonic $ASC$, (ii) monotonic $DESC$, or (iii) composite $DESC|ASC$.

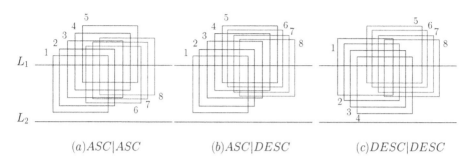

(a)$ASC|ASC$          (b)$ASC|DESC$          (c)$DESC|DESC$

**Fig. 4.** Different types of a forbidden top-anchored clique. In all the figures, the square 5 is redundant since the squares 4 and 6 fully cover the area of square 5 below the line $L_1$.

**The Algorithm.** In this subsection, we describe an algorithm for the line instance $(P, P', S)$ that produces a feasible set cover $S^*$ with some desirable structural properties. Multiple maximum cliques may exist in the intersection graph of $S^*$. We order the maximum cliques of $S^*$ in the increasing order of the $x_r(\cdot)$ values of their ply regions.

**Definition 5 (Leftmost maximum clique).** *The leftmost maximum clique $Q$ of a set of unit squares $S^*$ refers to that maximum clique in the intersection graph of $S^*$ for which $x_r(\cdot)$ value of the corresponding ply region is the minimum among all the maximum cliques in $S^*$.*

The procedure $\mathsf{Swap}(\mathcal{A}, \{s\}, S^*)$ consists of the deletion of a set of squares $\mathcal{A}$ from a set cover $S^*$ and the addition of a square $s \in S \setminus S^*$ into $S^*$.

**Definition 6 (Profitable swap).** *The operation $\mathsf{Swap}(\mathcal{A}, \{s\}, S^*)$ is a profitable swap if $\mathcal{A}$ is a set of two or more consecutive squares in the leftmost maximum clique $Q \subseteq S^*$, and $s \in S \setminus S^*$ such that $S^* \cup \{s\} \setminus \mathcal{A}$ is a feasible set cover for $P$.*

The procedure $\mathsf{RemoveRedundancy}(S)$ ensures that each square $s \in S^*$ contains at least one point $p \in P$ exclusively. We proved that in a set cover for the line instance, where none of the constituent squares are redundant, any maximum clique can be of only 3 types, as shown in Fig. 5.

The algorithm for the line instance is given in Algorithm 1 and has two steps. In the first step, $\mathsf{RemoveRedundancy}(\cdot)$ is applied on the set $S$ of input squares

---

**Algorithm 1:** Algorithm for line instance of GMMGSC

---

    **Input**  : A horizontal line $\ell$, a set $\mathcal{S}$ of $m$ unit squares intersecting $\ell$, a set $P$
                of $n$ points below $\ell$ such that each of them lies in at least one square
                in $\mathcal{S}$, and a set of points $P'$.
    **Output:** Returns a set of squares $\mathcal{S}^* \subseteq \mathcal{S}$ covering $P$.

1  $\mathcal{S}^* \leftarrow$ RemoveRedundancy$(\mathcal{S})$
2  Let $Q$ be the leftmost maximum clique in $\mathcal{S}^*$.
3  **while** *there is a* profitable swap *in* $Q$ **do**
4      |  $\mathcal{S}^* \leftarrow$ Swap$(\mathcal{A}, \{s\}, \mathcal{S}^*)$
5      |  $\mathcal{S}^* \leftarrow$ RemoveRedundancy$(\mathcal{S}^*)$
6      |  Update $Q$ to be the leftmost maximum clique in $\mathcal{S}^*$.
7  **end**
8  **return** $\mathcal{S}^*$

---

to obtain a feasible set cover $\mathcal{S}^*$ containing no redundant squares. The second step performs a set of *profitable swaps* on $\mathcal{S}^*$. This step aims to obtain a feasible solution with a maximum clique $Q$ with some desirable properties. A profitable swap is defined on the leftmost maximum clique of a feasible solution. Refer to Definition 6. We have the following observation about the solution.

**Observation 1.** *There exist no profitable swaps on the leftmost maximum clique $Q$ of $\mathcal{S}^*$ returned by Algorithm 1.*

Our algorithm implicitly implies the following lemma.

**Lemma 1 (Key Lemma).** *Let $Q$ be the leftmost maximum clique in the solution $\mathcal{S}^*$ returned by Algorithm 1. Let $|Q| = k$ and $s_1, \ldots, s_k$ be the squares of $Q$ from left to right. No input square contains* Excl$(s_i) \cup$ Excl$(s_{i+1})$, *where $1 \leq i \leq k - 1$.*

*Proof.* Fix an index $i$ with $1 \leq i \leq k - 1$. Suppose for the sake of contradiction that there exists an input square $t$ that covers Excl$(s_i) \cup$ Excl$(s_{i+1})$. Observe that Swap$(\{s_i, s_{i+1}\}, \{t\}, \mathcal{S}^*)$ is a profitable swap in $\mathcal{S}^*$, which is a contradiction to Observation 1.

First, we will show how to implement the algorithm and analyze the running time.

**Observation 2.** *The procedure* RemoveRedundancy$(\mathcal{S})$ *can be implemented in $O(nm)$ time.*

*Proof.* Consider an arbitrary ordering of the $n$ points in $P$ and an arbitrary ordering of the $m$ squares in $\mathcal{S}$. In $O(nm)$ time, we construct an $n \times m$ matrix $T$ where the $(i, j)$-th entry is 1 if the $i$-th point is contained in the $j$-th square. Also, construct an array count of length $n$. For every point $p_i$, count$[i]$ stores the number of squares in $\mathcal{S}$ that contain $p_i$. Initialize each entry of an $m$-length array removed to 0. Run a loop that iterates over each square $s_j \in \mathcal{S}$. If $s_j$ does not contain any point $p_i$ with count$[i] = 1$, remove $s_j$, i.e., set removed$[j] = 1$. Set each entry in the $j$-th column of $T$ to zero. For each point $p_{i'} \in s_j$, decrease count$[i']$ by one. At the end of the loop, output the squares with removed$[\cdot] = 0$. Naively, the time required to perform all the operations is $O(nm)$.

**Theorem 1.** *The running time of Algorithm 1 is $O(m^2 \log m + m^2 n)$.*

*Proof.* To compute the leftmost maximum clique $Q$ in $\mathcal{S}^*$, we first obtain the set of squares containing each point of $P'$. This takes $O(nm)$ time.

We need to check if there exists a profitable swap in $Q$. There are at most $m$ choices for the swapped-in square $s \in \mathcal{S} \setminus \mathcal{S}^*$. While computing $Q$, one could also obtain the left-to-right ordering of the squares in $Q$. For each candidate swapped-in square $s$, we find a set of consecutive squares $\mathcal{A} \subseteq Q$ that can lead to a profitable swap of $\mathcal{A}$ by $s$. We can use binary search on the squares of $Q$ to do this. This requires $O(\log |Q|) = O(\log m)$ time. Additionally, we may need to check if the extreme two squares of $\mathcal{A}$, say $s_i$ and $s_j$, can be swapped out safely. This is equivalent to checking if all points in Excl$(s_i, s_j)$ are covered by $s$. Naively, the time required to check this is $O(n)$. Thus, if one exists, we can execute a profitable swap in $Q$ in $O(m \log m + mn)$ time.

We need to remove from $\mathcal{S}^*$ those squares in $\mathcal{S}^* \setminus Q$ that may have become redundant because of swapping in the square $s$. This can be done by re-invoking RemoveRedundancy$(\cdot)$ on $\mathcal{S}^*$ in $O(nm)$ time (due to Observation 2). The leftmost maximum clique in $\mathcal{S}^*$ can be determined in $O(nm)$ time.

Every profitable swap decreases the size of the set cover by at least one. Therefore, at most $m$ profitable swaps are performed. Thus, there are at most $m$ iterations of the *while* loop. Hence, the total running time of the *while* loop is $O(m \cdot (m \log m + nm)) = O(m^2 \log m + m^2 n)$.

**Structural Properties of the Solution.** We state two properties about the structure of the solution returned by Algorithm 1. Let $Q$ be the leftmost maximum clique in the solution $\mathcal{S}^*$ returned by Algorithm 1 for the line instance $(P, P', \mathcal{S})$. Let $s_1, \ldots, s_k$ be the squares of $Q$ from left to right. For $1 \leq j \leq k$, let $p_j$ be the bottom-most exclusive point in $s_j$.

**Lemma 2.** *Let $p \in P'$ be an arbitrary point contained in the common intersection region of $Q$. For $k > 13$, there exists a set $J \subset [k]$ with $|J| \geq k - 9$ such that every input square $t \in \mathcal{S}$ containing $p_j$ also contains $p$, for $j \in J$.*

*Proof.* There are two cases to consider.

**Case 1:** $Q$ is a monotonic clique. There are two subcases.

- $Q$ is a monotonic descending clique. Define the set $J = \{3, \ldots, k - 3\}$. By definition $y(s_{j-1}) > y(s_j) > y(s_{j+1})$, for each $j \in J$. Let $t$ be an input square that contains $p_j$ but not $p$. Again, there are two cases.

- $p$ lies to the right of $t$, i.e. $x(p) > x(t) + 1$: Observe that $x(p_j) < x(s_k)$. Since $p \in s_1$ and $t$ does not contain $p$, therefore $t$ starts before $s_1$ starts, i.e., $x(t) < x(s_1)$. Since $t$ contains $p_j \in s_j$, hence $t$ must end below the bottom boundary of $s_{j-1}$ and should end to the right of where $s_j$ starts, i.e., $y(t) < y(s_{j-1})$ and $x(t) + 1 > x(s_j)$. Combining these with the fact that $t$ is a unit square, $t$ must cover $\mathsf{Excl}(s_{j-2}) \cup \mathsf{Excl}(s_{j-1})$ as shown in Fig. 5(b). This is a violation of Lemma 1.
- $p$ does not lie to the right of $t$ but is above $t$, i.e., $x(t) + 1 > x(p)$ but $y(p) > y(t)+1$: The square $t$ must end after $s_k$ starts, i.e., $x(t)+1 > x(s_k)$. Since $p \in s_k$, the square $t$ must end below $s_k$, i.e., $y(t) < y(s_k)$. Since $p_j \in t$, therefore $x(t) < x(p_j)$. So, $t$ covers $\mathsf{Excl}(s_{j+1}) \cup \mathsf{Excl}(s_{j+2})$. This is a violation of Lemma 1.

- $Q$ is a monotonic ascending clique. Define the set $J = \{4, \ldots, k-2\}$. By definition $y(s_{j-1}) < y(s_j) < y(s_{j+1})$, for each $j \in J$. Let $t$ be an input square that contains $p_j$ but not $p$. By an analogous argument, $t$ would be forced to cover either $\mathsf{Excl}(s_{j+1}) \cup \mathsf{Excl}(s_{j+2})$ (when $p$ lies to the left of $t$ as shown in Fig. 5(a)) or $\mathsf{Excl}(s_{j-1}) \cup \mathsf{Excl}(s_{j-2})$ (when $p$ does not lie to the left of $t$ but lies above $t$). This is again a violation of Lemma 1.

**Case 2:** $Q$ is a composite clique (of type $DESC|ASC$). Let $b$ be the index (in the left-to-right ordering) of the bottom-most square in $Q$. Define the set $J = \{3, \ldots, b-3, b+3, \ldots, k-2\}$. Using the two subcases in the previous case, we can arrive at a violation of Lemma 1 for $j \in J$. Refer to Fig. 5(c) for an illustration.

**Lemma 3.** *For $k > 13$, there exists a set $J \subset [k]$ with $|J| \geq k-5$ such that no input square $t \in S$ can contain $p_j, p_{j+1}, p_{j+2}$ for $j \in J$.*

*Proof.* There are two cases to consider.

**Case 1:** $Q$ is a monotonic clique. There are two subcases.

- $Q$ is a monotonic descending clique. Define the set $J = \{1, \ldots, k-3\}$.
- $Q$ is a monotonic ascending clique. Define the set $J = \{2, \ldots, k-2\}$. For $j \in J$, assume that $t \in S$ contains the bottom-most exclusive points $p_j, p_{j+1}, p_{j+2}$ of three consecutive squares of $Q$, namely $s_j, s_{j+1}, s_{j+2}$ respectively. Then $t$ would contain either $\mathsf{Excl}(s_j) \cup \mathsf{Excl}(s_{j+1})$ or $\mathsf{Excl}(s_{j+1}) \cup \mathsf{Excl}(s_{j+2})$. This implies a violation of Lemma 1, as can be seen in Fig. 5. Thus, we have a contradiction.

**Case 2:** $Q$ is a composite clique (of type $DESC|ASC$). Let $b$ be the index (in the left-to-right ordering) of the bottom-most square in $Q$. We define the set $J = \{1, \ldots, b-3, b+1, \ldots, k-2\}$. If $1 \leq j \leq b-3$, then $s_j, s_{j+1}, s_{j+2}$ are in a monotonic descending sequence and the corresponding subcase from Case 1 applies. If $b+1 \leq j \leq k-2$, then $s_j, s_{j+1}, s_{j+2}$ are in a monotonic ascending sequence and the corresponding subcase from Case 1 applies.

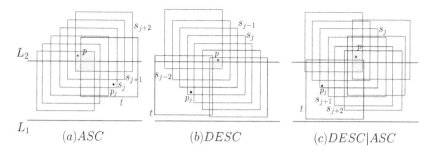

**Fig. 5.** $p \in P'$ is contained in the ply region of the cliques. The red square $t$ does not contain $p$. (a) The green squares constitute a monotonic $ASC$ clique. (b) The green squares constitute a monotonic $DESC$ clique. In (c), the green squares form a composite clique of type $DESC|ASC$. In (a), the red square $t$ can swap out $s_{j+1}, s_{j+2}$. In (b), the red square $t$ can swap out $s_{j-1}, s_{j-2}$. (Color figure online)

## 2.2   GMMGSC for the Slab Instance

In this section, we present a constant approximation algorithm for the slab instance. We will use an LP relaxation (adapted from Bandyapadhyay et al. [3]) to partition the slab instance into two line instances.

For each unit square $s \in \mathcal{S}$, we create a variable $x_s$, indicating whether $s$ is included in our solution. In addition, create another variable $y$, which indicates the maximum number of times a point in $P'$ is covered by our solution. Then, we formulate the following linear programming relaxation.

$$\min \; y$$

$$\text{s.t.} \sum_{s \in \mathcal{S}, p \in s} x_s \geq 1 \quad \text{for all } p \in P$$

$$\sum_{s \in \mathcal{S}, p' \in s} x_s \leq y \quad \text{for all } p' \in P'$$

$$0 \leq x_s \leq 1 \quad \text{for all } s \in \mathcal{S}$$

The input set $\mathcal{S}$ of squares is partitioned naturally into two parts $\mathcal{S}_1, \mathcal{S}_2$ where $\mathcal{S}_1$ (resp. $\mathcal{S}_2$) consists of the input squares intersecting the bottom (resp. top) boundary line $L_1$ (resp. $L_2$) of the horizontal slab, say $\alpha$.

We partition the set $P$ of points within $\alpha$ into $P_1$ and $P_2$ using the LP. Let $(\{x_s^*\}_{s \in \mathcal{S}}, y^*)$ be an optimal solution of the above linear program computed using a polynomial-time LP solver. For a point $p \in \mathbb{R}^2$ and $i \in \{1, 2\}$, define $\delta_{p,i}$ as the sum of $x_s^*$ for all $s \in \mathcal{S}_i$ satisfying $p \in s$. Then we assign each point $p \in P$ to $P_i$, where $i \in \{1, 2\}$ is the index that maximizes $\delta_{p,i}$.

Now we solve the two line instances $(P_i, P', \mathcal{S}_i)$, for $i \in \{1, 2\}$ using Algorithm 1. Finally, we output the union of the solutions of these two line instances. We discard redundant squares from the solution, if any. For $i \in \{1, 2\}$, denote by $\mathcal{S}_i^*$ the solution returned by Algorithm 1 for the line instance $(P_i, P', \mathcal{S}_i)$. We state and prove the following useful lemma.

**Lemma 4.** *For every $i \in \{1, 2\}$, $\mathsf{memb}(P', \mathcal{S}_i^*) \leq 4 \cdot OPT + 9$, where $\mathcal{S}_i^*$ is the solution returned by Algorithm 1 for the instance $(P_i, P', \mathcal{S}_i)$ and $OPT$ is the minimum membership for the slab instance.*

*Proof.* We consider the case when $i = 2$, i.e., all the squares intersect the top boundary line $L_2$ of the slab $\alpha$. The argument for the case of $i = 1$ is identical and is not duplicated. Consider the leftmost maximum clique $Q$ in the intersection graph of the squares in $\mathcal{S}_2^*$. Suppose, $\mathsf{memb}(P', \mathcal{S}_2^*) = k$ and the squares in $Q$ from left to right are $s_1, \ldots, s_k$.

If $k \leq 13$, then $\mathsf{memb}(P', \mathcal{S}_i^*) \leq 4 \cdot OPT + 9$ is satisfied trivially since $OPT \geq 1$.

Assume that $k > 13$. Denote by $p_j$ the bottom-most exclusive point of $s_j$. Let $p$ be any point in $P'$ contained in the ply region of $Q$. By Lemma 2, for each $j \in J$, every input square containing $p_j$ also contains $p$. By Lemma 3, no input square $s$ contains $p_j, p_{j+1}, p_{j+2}$ for $j \in J$. Thus we can write

$$\sum_{s \in \mathcal{S}_2, p \in s} x_s \geq \frac{1}{2} \sum_{\forall j \in J} \sum_{s \in \mathcal{S}_2, p_j \in s} x_s$$

Since a variable $x_s$ may appear at most twice in the double-sum on the right-hand side, we have multiplied by the factor $1/2$. The left-hand side of the above inequality is bounded above by the LP optimal $y^*$. Since every $p_j$ belongs to $P_2$, we have $\sum_{s \in \mathcal{S}_2, p_j \in s} x_s \geq 1/2$ from the partitioning criteria of $P$ into $P_1$ and $P_2$. From the proofs of Lemma 2 and Lemma 3, we observe that $|J| \geq k - 9$, irrespective of the type of clique $Q$. So we can write

$$y^* \geq \frac{1}{2} \cdot (k - 9) \cdot \frac{1}{2}$$
$$\implies k \leq 4y^* + 9$$

By definition, the optimal ply value, $OPT$, is at least $y^*$. Therefore, $k \leq 4 \cdot OPT + 9$.

**Lemma 5.** *For $i \in \{1, 2\}$, let $\mathcal{S}_i^*$ be the solution for the line instance $(P_i, P', \mathcal{S}_i)$ obtained via Algorithm 1. Then $\mathcal{S}^* = \mathcal{S}_1^* \cup \mathcal{S}_2^*$ is a feasible solution to the slab instance $(P, P', \mathcal{S})$ with $\mathsf{memb}(P', \mathcal{S}^*) \leq 8 \cdot OPT + 18$.*

*Proof.* Since $\mathcal{S} = \mathcal{S}_1 \cup \mathcal{S}_2$, so $\mathcal{S}^*$ is a feasible set cover for the slab instance $(P, P', \mathcal{S})$. Consider an arbitrary point $p \in P'$. Using Lemma 4 for $i \in \{1, 2\}$, we get that the number of unit squares in $\mathcal{S}_i^*$ containing $p$ is at most $4 \cdot OPT + 9$. Thus, $\mathsf{memb}(P', \mathcal{S}^*) \leq 8 \cdot OPT + 18$.

### 2.3    Putting Everything Together

**Theorem 2.** *GMMGSC problem admits an algorithm that runs in $O(m^2 \log m + m^2 n)$ time, and computes a set cover whose membership is at most $16 \cdot OPT + 36$, where $OPT$ denotes the minimum membership.*

*Proof.* We divide the plane into unit-height horizontal slabs. For each non-empty slab $\alpha$, we partition the slab instance into two subproblems, namely the line instances corresponding to the boundary lines of $\alpha$ using the LP-relaxation technique described in Sect. 2.2. We solve each line instance using Algorithm 1. Then we output the union of the solutions thus obtained while discarding redundant squares, if any. Consider any point $p \in P'$. Suppose $p$ lies within a slab $\alpha$ whose boundary lines are $L_1$ and $L_2$. Since the squares containing $p$ must intersect either $L_1$ or $L_2$, $p$ can be contained in squares from at most 4 subproblems. One is a line instance corresponding to $L_1$ where the points lie above $L_1$. The other is a line instance corresponding to $L_1$ where the points lie below $L_1$. The other two subproblems correspond to $L_2$. Thus, due to Lemma 4, the number of squares of our solution containing $p$ is at most $16 \cdot OPT + 36$, where $OPT$ is the optimal membership value for the instance $(P, P', \mathcal{S})$.

Formulating and solving the LP takes $\tilde{O}(nm)$ time [2]. The overall running time of the algorithm is dominated by the running time of Algorithm 1. There are at most $O(\min(n, m))$ line instances to solve. By a standard trick, the total running time remains $O(m^2 \log m + m^2 n)$ using Theorem 1.

# References

1. Agarwal, P.K., Pan, J.: Near-linear algorithms for geometric hitting sets and set covers. In: Proceedings of the Thirtieth Annual Symposium on Computational Geometry, SOCG 2014, pp. 271–279. Association for Computing Machinery, New York (2014). https://doi.org/10.1145/2582112.2582152
2. Allen-Zhu, Z., Orecchia, L.: Nearly linear-time packing and covering LP solvers. Math. Program. **175**(1), 307–353 (2019). https://doi.org/10.1007/s10107-018-1244-x
3. Bandyapadhyay, S., Lochet, W., Saurabh, S., Xue, J.: Minimum-membership geometric set cover, revisited. In: 39th International Symposium on Computational Geometry (SoCG 2023). Leibniz International Proceedings in Informatics (LIPIcs), vol. 258, pp. 11:1–11:14 (2023). https://doi.org/10.4230/LIPIcs.SoCG.2023.11
4. Biedl, T., Biniaz, A., Lubiw, A.: Minimum ply covering of points with disks and squares. Comput. Geom. **94**, 101712 (2021). https://doi.org/10.1016/j.comgeo.2020.101712
5. Chan, T.M., Grant, E.: Exact algorithms and APX-hardness results for geometric packing and covering problems. Comput. Geom. **47**(2, Part A), 112–124 (2014). https://doi.org/10.1016/j.comgeo.2012.04.001. Special Issue: 23rd Canadian Conference on Computational Geometry (CCCG11)
6. Chan, T.M., Hu, N.: Geometric red-blue set cover for unit squares and related problems. Comput. Geom. **48**(5), 380–385 (2015). https://doi.org/10.1016/j.comgeo.2014.12.005. Special Issue on the 25th Canadian Conference on Computational Geometry (CCCG)
7. Clarkson, K.L., Varadarajan, K.: Improved approximation algorithms for geometric set cover. Discret. Comput. Geom. **37**(1), 43–58 (2007). https://doi.org/10.1007/s00454-006-1273-8
8. Durocher, S., Keil, J.M., Mondal, D.: Minimum ply covering of points with unit squares. In: Lin, C.C., Lin, B.M.T., Liotta, G. (eds.) WALCOM 2023. LNCS,

vol. 13973, pp. 23–35. Springer, Cham (2023). https://doi.org/10.1007/978-3-031-27051-2_3

9. Erlebach, T., van Leeuwen, E.J.: Approximating geometric coverage problems. In: Proceedings of the Nineteenth Annual ACM-SIAM Symposium on Discrete Algorithms, SODA 2008, pp. 1267–1276. Society for Industrial and Applied Mathematics (2008)

10. Feige, U.: A threshold of ln n for approximating set cover. J. ACM **45**(4), 634–652 (1998). https://doi.org/10.1145/285055.285059

11. Fowler, R.J., Paterson, M.S., Tanimoto, S.L.: Optimal packing and covering in the plane are np-complete. Inf. Process. Lett. **12**(3), 133–137 (1981). https://doi.org/10.1016/0020-0190(81)90111-3

12. Hochbaum, D.S., Maass, W.: Approximation schemes for covering and packing problems in image processing and VLSI. J. ACM **32**(1), 130–136 (1985). https://doi.org/10.1145/2455.214106

13. Ito, T., et al.: A polynomial-time approximation scheme for the geometric unique coverage problem on unit squares. Comput. Geom. **51**, 25–39 (2016)

14. Kuhn, F., von Rickenbach, P., Wattenhofer, R., Welzl, E., Zollinger, A.: Interference in cellular networks: the minimum membership set cover problem. In: Wang, L. (ed.) COCOON 2005. LNCS, vol. 3595, pp. 188–198. Springer, Heidelberg (2005). https://doi.org/10.1007/11533719_21

15. Mustafa, N.H., Raman, R., Ray, S.: Settling the APX-hardness status for geometric set cover. In: 55th IEEE Annual Symposium on Foundations of Computer Science, FOCS 2014, Philadelphia, PA, USA, 18–21 October 2014, pp. 541–550. IEEE Computer Society (2014). https://doi.org/10.1109/FOCS.2014.64

16. Mustafa, N.H., Ray, S.: Improved results on geometric hitting set problems. Discret. Comput. Geom. **44**(4), 883–895 (2010). https://doi.org/10.1007/s00454-010-9285-9

17. Raz, R., Safra, S.: A sub-constant error-probability low-degree test, and a sub-constant error-probability PCP characterization of NP. In: Proceedings of the Twenty-Ninth Annual ACM Symposium on Theory of Computing, STOC 1997, pp. 475–484. Association for Computing Machinery, New York (1997). https://doi.org/10.1145/258533.258641

# Semi-total Domination in Unit Disk Graphs

Sasmita Rout and Gautam Kumar Das[(⊠)]

Indian Institute of Technology Guwahati, Guwahati, Assam, India
{sasmita18,gkd}@iitg.ac.in

**Abstract.** Let $G = (V, E)$ be a simple undirected graph with no isolated vertex. A set $D \subseteq V$ is a dominating set if each vertex $u \in V$ is either in $D$ or is adjacent to a vertex $v \in D$. A set $D_{t2} \subseteq V$ is said to be a semi-total dominating set if $(i)$ $D_{t2}$ is a dominating set, and $(ii)$ for every vertex $u \in D_{t2}$, there exists a vertex $v \in D_{t2}$ such that the distance between $u$ and $v$ in $G$ is within 2. Given a graph $G$, the semi-total domination problem is to find a semi-total dominating set of minimum cardinality. The semi-total domination problem is NP-complete for general graphs. It is also NP-complete on some special graph classes, such as planar, split, and chordal bipartite graphs. In this paper, we have shown that it is NP-complete for unit disk graphs. We propose a 6-factor approximation algorithm for the semi-total dominating set problem in unit disk graphs. The algorithm's running time is $O(nk)$, where $n$ and $k$ are the number of vertices and the size of the maximal independent set of the given UDG, respectively. In addition, we show that the minimum semi-total domination problem in a graph with maximum degree $\mathbb{D}$ admits a $2 + \ln(\mathbb{D} + 1)$-factor approximation algorithm which is an improvement over the best-known result $2 + 3\ln(\mathbb{D} + 1)$.

**Keywords:** Semi-total dominating set · NP-complete · Approximation algorithm

## 1 Introduction

Let $G = (V, E)$ be a simple undirected graph that may contain multiple components. However, no component in the graph is an isolated vertex. A set $N_G(v)$ denotes the open neighborhood of $v$ in $G$, and it is defined as $N_G(v) = \{u \in V : uv \in E\}$. On the other hand, the closed neighborhood $N_G[v]$ of $v$ is defined as $N_G[v] = N_G(v) \cup \{v\}$. For any subset $S \subseteq V$, $G[S]$ represents the subgraph induced by the vertex set $S$ in $G$ (i.e., for each $u, v \in S$, $uv \in E(G[S])$ if and only if $uv \in E$).[1] Given two vertices $u$ and $v$, the distance $d(u, v)$ between $u$ and $v$ is the minimum number of edges that connect $u$ with $v$ in $G$. A subset $D \subseteq V$ is said to be a dominating set (DS) of G if for each vertex $v \in V$, $|N_G[v] \cap D| \geq 1$. The dominating set with minimum cardinality

---

[1] $E(G)$ refers to the edge set of the graph $G$.

© The Author(s), under exclusive license to Springer Nature Switzerland AG 2024
S. Kalyanasundaram and A. Maheshwari (Eds.): CALDAM 2024, LNCS 14508, pp. 117–129, 2024.
https://doi.org/10.1007/978-3-031-52213-0_9

is called the minimum dominating set, and the size of the minimum dominating set is called the domination number, $\gamma(G)$. A vertex $v \in V$ dominates $N_G[v]$, and a subset $S \subseteq V$ dominates $\bigcup_{v \in S} N_G[v]$. A subset $D_t \subseteq V(G)$ is said to be a total dominating set (TDS) of $G$ if $D_t$ is a dominating set of $G$, and the vertices in $D_t$ induce a subgraph with no isolated vertex. The total dominating set with minimum cardinality is called the minimum total dominating set, and the size of the minimum total dominating set is called the total domination number, $\gamma_t(G)$. A subset $D_{t2} \subseteq V$ is said to be a semi-total dominating set of $G$ if $(i)$ $D_{t2}$ is a dominating set (domination property), and $(ii)$ for each $u \in D_{t2}$, there exists a vertex $v \in D_{t2}$ such that $d(u, v) \leq 2$ (semi-total property). The semi-total dominating set with minimum cardinality is called a minimum semi-total dominating set, and the corresponding cardinality is the semi-total domination number. We denote the semi-total domination number as $\gamma_{t2}(G)$. Given a graph $G$, the objective of the semi-total dominating set problem is to find a minimum semi-total dominating set.

## 1.1   Related Work

In computational complexity theory, the dominating set problem is a classical NP-complete problem [6]. Along with the domination problem, its variants are also generally hard in general graphs. So, researchers started exploring the behavior of the domination problem and its variants in different sub-classes of general graphs. The literature on domination and its variants can be found in [7–10,13,18]. In this paper, we focus on semi-total dominating set. The Semi-total dominating set was introduced by W. Goddard et al. [16] in 2015. Since every semi-total dominating set is a dominating set and every total dominating set is a semi-total dominating set, the semi-total domination number is squeezed between the domination number and the total domination number, i.e., for a given graph $G$, $\gamma(G) \leq \gamma_{t2}(G) \leq \gamma_t(G)$. In [16], authors showed that if $G$ is a connected graph with $n$ ($\geq 4$) vertices, then $\gamma_{t2} \leq \frac{n}{2}$. In the same paper, authors also showed that if $G$ is a graph with $n$ vertices and maximum degree $\mathbb{D}$, then $\gamma_{t2} \geq \frac{2n}{2\mathbb{D}+1}$. Subsequently, Henning and Marcon [13] showed that for a connected graph with at least 2 vertices, $\gamma_{t2}(G) \leq \alpha'(G) + 1$, where $\alpha'(G)$ is the matching number.[2] In [1], Asplund et al. studied the semi-total domination in cartesian product graphs and established that for any two graphs $G$ and $H$, $\gamma_{t2}(G \square H) \geq \frac{1}{3}\gamma_{t2}(G)\gamma_{t2}(H)$. In [4], authors showed that it is NP-complete to recognize the graphs that satisfy $\gamma_{t2}(G) = \gamma_t(G)$ and $\gamma(G) = \gamma_{t2}(G)$. In [11], authors showed that for every connected claw-free cubic graph $G$ of order $n$, $\gamma_{t2} \leq \frac{n}{3}$. In [12], authors showed that the semi-total domination problem remains NP-complete in planar graphs, chordal bipartite graphs and split graphs. They also gave a $2 + 3\ln(\mathbb{D} + 1)$-factor approximation algorithm for the minimum semi-total domination problem, where $\mathbb{D}$ is the maximum degree of $G$.

---

[2] Given a graph $G = (V, E)$, a set of edges $E' \subseteq E$ is said to be a matching of $G$ if no two elements of $E'$ are adjacent and the matching number is the size of the largest matching.

## 1.2   Our Contribution

The remaining part of this paper is organized as follows. In Sect. 2, we introduce the required preliminaries and notations. In Sect. 3, we prove that the semi-total dominating set problem is NP-complete in unit disk graphs. Next, in Sect. 4, we propose an $O(nk)$ time 6-factor approximation algorithm for the semi-total domination problem for unit disk graphs. In this section, we also propose a $2 + \ln(\mathbb{D} + 1)$-factor approximation algorithm for the semi-total domination problem for general graphs. Finally, we conclude the paper in Sect. 5.

## 2   Preliminaries

In this section, we will introduce the required notations and definitions. Let $P = \{p_1, p_2, \ldots, p_n\}$ be a set of $n$ points in $\mathbb{R}^2$. A graph $G = (V, E)$ is said to be a geometric UDG corresponding to the point set $P$ if there exists a one-to-one correspondence between each $v_i \in V$ with $p_i \in P$ and $v_i v_j \in E$ if and only if $\delta(p_i, p_j) \leq 1$, where $\delta(.,.)$ is the Euclidean distance between two points in $\mathbb{R}^2$. Let $\Delta(p)$ denote the unit disk centered at the point $p \in P$ and $\Delta(P) = \{\Delta(p) : p \in P\}$. The set of disks $\Delta(P)$ is considered independent if for every pair $p, q \in P$, $p \notin \Delta(q)$, i.e., $\delta(p, q) > 1$. This article often refers to a point as a vertex or node. Given a positive integer $i$ and a vertex $u$, $N_G^i[u]$ represents the set of all the vertices within distance $i$ from $u$ in $G$. We often refer to $N_G^1[.]$ as $N_G[.]$.

Next, we list some of the already proven lemmas, theorems, and observations useful in Sect. 3, and Sect. 4.

**Lemma 1** [19]. *Let $G = (V, E)$ be a planar graph of degree at most 3. The graph $G$ can be embedded in a grid of area $O(|V|^2)$ such that each $v \in V$ lies in a grid point with co-ordinate $(5i, 5j)$, where $i$ and $j$ are integers and each edge $e \in E$ is a finite sequence of consecutive segments of length 5 units along the grid lines.*

**Lemma 2** [15]. *Let $\mathcal{P}$ be a unit disk centered at point $p$ and let $S$ be a set of independent unit disks such that each disk in $S$ contains the point $p$, then $|S| \leq 5$.*

**Lemma 3** [3]. *Given a UDG $G$, there exists a $\frac{44}{9}$-approximation algorithm for the minimum dominating set problem with running time $O(n^2)$.*

**Observation 1** [5]. *For a given graph $G$, $\gamma(G) \leq \gamma_{t2}(G)$.*

## 3   NP-Completeness

In this section, we focus on the hardness result of the semi-total domination problem and prove that the decision version of the problem is NP-complete in unit disk graphs. We use a reduction from the **decision version of the vertex cover (VC) problem in planer graphs of degree at most** 3 to the **decision version of the semi-total dominating set problem in UDGs**. The corresponding decision problems are formally defined as follows:

The decision version of the VC problem in planar graphs of degree at most 3 (**D-VC-PGD3**): Given a positive integer $k$ and a planar graph $G$ of degree at most 3, does $G$ has a VC of size at most $k$?

The decision version of semi-total dominating set problem in UDGs (**D-T2DS-UDGs**): Given a positive integer $k$ and a UDG $G$, does $G$ has a semi-total dominating set of size at most $k$?

Lichtenstein and David [14] reduced the planar 3SAT problem to the planar vertex cover problem and proved that **D-VC-PGD3** is NP-complete. We prove the hardness result of the semi-total dominating set problem in UDGs by making a polynomial time reduction from an arbitrary instance of **D-VC-PGD3** to an instance of **D-T2DS-UDGs**. To prove this, we embed a planar graph $G = (V, E)$ of degree at most 3 in a grid of cell size $5 \times 5$ using Lemma 1.

**Lemma 4.** *If $G = (V, E)$ is an instance of* **D-VC-PGD3** *without any isolated vertex, then an instance $G' = (V', E')$ of* **D-T2DS-UDG** *can be constructed from $G$ in polynomial time.*

*Proof.* Let $V = \{v_1, v_2, \ldots, v_n\}$ and $E = \{e_1, e_2, \ldots, e_m\}$ be the vertex set and edge set of the given instance $G$. We construct a graph $G' = (V', E')$ from $G$ by the four steps as given below:

**Step 1 (Embedding):** We first embed the graph $G$ on a grid of cell size $5 \times 5$ using the algorithm proposed by Biedl et al. in [2]. In this embedding, each edge $e \in E$ is a consecutive sequence of line segment(s) in the grid, where the length of each segment is 5 unit. Let $\ell$ be the total number of line segments used in the embedding. For each vertex $v \in V$, a **node point** located at a grid having coordinate $(5i, 5j)$ for some integers $i$ and $j$. Let $p_i$ be the node point at the grid corresponding to vertex $v_i \in V$ for $1 \leq i \leq n$. Let the set of node points be $N$, i.e., $|N| = |V| = n$. Refer to Fig. 1(a) and (b) for an illustration of the embedding step.

**Step 2 (Inclusion of auxiliary points):** In this step, we add some auxiliary points on each segment of the graph after the embedding step as mentioned below: (*i*) for each $p_i p_j$ corresponding to the edge $v_i v_j \in E$, if the number of segments in $p_i p_j$ is *one* (length of $p_i p_j$ is exactly 5 unit), then add *six* points at distances 1, 1.3, 2.1, 2.6, 3.2 and 4 either from $p_i$ or $p_j$ as depicted in Fig. 2(a), (*ii*) if the number of segments is more than *one* (length of $p_i p_j$ is greater than 5 unit), then we add a point on each grid point along the line except the node points. We refer to those points as **grid points** (see the filled square points in Fig. 2). If both the endpoints of any segment are grid points, then add *four* points on the segment at distances 1, 2, 3, 4 from any of the endpoints of the segment (see Fig. 2(b)); otherwise, add *five* points at distances 1, 1.9, 2.5, 3 and 4 from node point $p_i$ (Fig. 2(c)). Let $A$ be the set representing the auxiliary points added in this step.

**Step 3 (Inclusion of gadgets):** Since each node in the planar graph has degree at most 3 and is embedded within a grid, at least one position at each node point exists to accommodate an extra edge. In this step, we add a gadget as shown in Fig. 2(d) at each node point $p_i$. The gadget at $p_i$ contains 4 points, namely $x_i$, $x_i'$, $y_i$ and $y_i'$. Here, the distances between $p_i$ and $x_i$, $x_i$ and $x_i'$, $x_i$ and $y_i$, $y_i$ and $y_i'$ are 0.9, 0.5, 0.9 and 0.5, respectively. Let $S$ be the number of points added in this step. Since each gadget contains 4 points, therefore, $|S| = 4|N| = 4n$.

**Step** 4 **(Construction of UDG):** Let $G' = (V', E')$ be the UDG constructed after applying the above 3 steps on graph $G$, where $V' = N \cup A \cup S$ and $E' = \{uv : u, v \in V' \text{ and } \delta(u, v) \le 1\}$.

From Lemma 1, we conclude that the number of segments $\ell = O(n^2)$. Therefore, the upper bound on the number of vertices and the number of edges in $G'$ is $O(n^2)$. Hence, $G'$ can be constructed from $G$ in polynomial time. For the complete illustration of the construction phases, refer to Fig. 1 and Fig. 2.    □

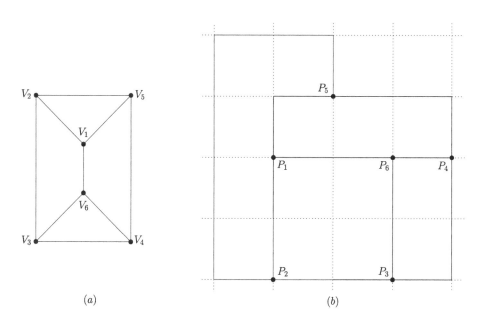

(a)                                                      (b)

**Fig. 1.** (a) A planar graph $G$, and (b) Embedding of $G$ in a grid

**Theorem 1.** *$D\text{-}T2DS\text{-}UDGs$ belongs to the class NP-complete.*

*Proof.* Let $G = (V, E)$ be a unit disk graph. Given a subset $S \subseteq V$ and a positive integer $k$, we can verify whether $S$ is a semi-total dominating set of $G$ of size at most $k$ or not in polynomial time. Therefore, $D\text{-}T2DS\text{-}UDGs \in NP$.

To prove $D\text{-}T2DS\text{-}UDGs$ is NP-hard, we will do a polynomial time reduction from **$D\text{-}VC\text{-}PGD3$** to **$D\text{-}T2DS\text{-}UDGs$**. Here, we use Lemma 4 to construct an instance $G' = (V', E')$ of $D\text{-}T2DS\text{-}UDGs$ from an arbitrary instance

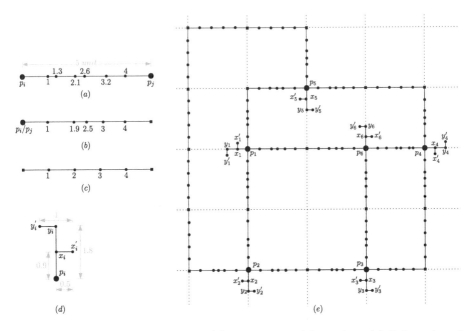

**Fig. 2.** (a) Orientation of six points, (b) Orientation of five points, (c) Orientation of four points, (d) Gadget, and (e) Graph $G'$

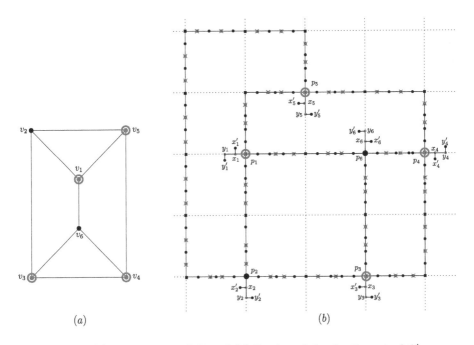

**Fig. 3.** (a) Vertex cover of $G$, and (b) Semi-total dominating set of $G'$

$G = (V, E)$ of $D$-$VC$-$PGD3$ in polynomial time. Next, we prove the following claim to complete the hardness result of $D$-$T2DS$-$UDGs$.

**Claim:** $G$ has a vertex cover $S_{vc}$ such that $|S_{vc}| \leq k$ if and only if $G'$ has a semi-total dominating set $D_{t2}$ such that $|D_{t2}| \leq k + 2\ell + 2n$.

($\Longrightarrow$) Let $S_{vc}$ be a vertex cover of $G$ such that $|S_{vc}| \leq k$ and $T_{vc}$ be the set of vertices in $G'$ corresponding to the vertices in $S_{vc}$, i.e., $T_{vc} = \{p_i \in V' : v_i \in S_{vc}\}$. Now, we construct two sets $T_a \subseteq A$ and $T_g \subseteq S$ such that $D_{t2} = T_{vc} \cup T_a \cup T_g$ is a semi-total dominating set with cardinality less than or equal to $k + 2\ell + 2n$. The construction of $T_a$ and $T_g$ are as follows. Since $S_{vc}$ is a vertex cover of $G$, at least one endpoint of every edge in $G$ is inside $S_{vc}$. Since every edge in $G$ corresponds to a sequence of segments in $G'$, we start from the endpoint, which is inside $T_{vc}$. For each $p_i p_j$ in $G'$ corresponding to each $v_i v_j \in E$, at least one out of $p_i$ and $p_j$ is in $T_{vc}$. Without loss of generality, let $p_i \in T_{vc}$. We traverse from $p_i$ towards $p_j$. While traversing, we leave two vertices next to $p_i$, add a single vertex to $T_a$, and then leave the next vertex and select the next one for $T_a$; again, we leave two vertices. This way, we repeat the process till we reach $p_j$. For each $v_i v_j \in E$, we apply this process to the corresponding $p_i p_j$ in $G'$ and observe that exactly two points from each segment are in $T_a$. So $|T_a| = 2\ell$, where $\ell$ is the number of segments in $G'$. In Fig. 3(b), the red cross points are in $T_a$. From each $p_i \in V'$, we choose $x_i$ and $y_i$ for $T_g$. So $|T_g| = 2n$. In Fig. 3(b), the red disks are in $T_g$.

Now, we are left to show that the set $D_{t2}$ is a semi-total dominating set of $G'$. Since $x_i, y_i \in T_g \subseteq D_{t2}$ and dominate $p_i$, $x_i'$ and $y_i'$, and $d(x_i, y_i) = 1$ for all $i = \{1, 2, \ldots, n\}$, and for each $p_i \in T_{vc}$, there exists a vertex $x_i \in T_g$ such that $d(p_i, x_i) = 1$. Hence, the sets $T_{vc}$ and $T_g$ satisfy the semi-total domination. There are three types of segments in $G'$. The types of segments are as follows: $(i)$ segments with two endpoints as node points as depicted in Fig. 2(a), $(ii)$ segments with one endpoint as a node point and another as a grid point as depicted in Fig. 2(b), and $(iii)$ segments with two endpoints as grid points as depicted in Fig. 2(c). We have to show that each segment from each segment type requires at least two points in $D_{t2}$ for semi-total domination.

$(i)$ Segments with two endpoints as node points: Since $S_{vc}$ is a vertex cover of $G$, at least one out of $v_i$ and $v_j$ is in $S_{vc}$, where $v_i v_j \in E$. Therefore, at least one out of $p_i$ and $p_j$ is in $T_{vc}$. Without loss of generality, let $p_i \in T_{vc}$ (pick any one if both $p_i, p_j \in T_{vc}$). The way graph $G'$ is constructed, there are 6 points on this type of segment excluding the two node points (i.e., $p_i$ and $p_j$). Let us refer to each point on the segment as $z_t$, where $1 \leq t \leq 6$ and $t$ is the position of the point from $p_i$. $p_i$ dominates $z_1$. The selection of $z_3$ and $z_5$ in $T_a$ ensure the domination of $z_2$, $z_4$ and $p_j$ and $d(z_3, z_5) \leq 2$ ensures the semi-total property. For a complete illustration, refer to $p_4 p_6$ in Fig. 3(b).

$(ii)$ Segments with one endpoint as a node point and another as a grid point: Let $p_i$ and $g_{ijt}$ be the node point and grid point, respectively, where the segment $p_i g_{ijt}$ is in $p_i p_j$, and $t$ represents the position of the grid point from $p_i$. This type of segment contains 5 points. Let the points are $z_1$, $z_2$, $\ldots$, $z_5$. $(a)$ If $p_i \in T_{vc}$, then the selection of $z_3$ and $z_5$ in $T_a$ ensures the domination of the segment. $(b)$ If $p_i \notin T_{vc}$, then a point in the other segment connected to $g_{ijt}$ dominates $g_{ijt}$

(since $p_j \in T_{vc}$). Hence, the selection of $z_2$ and $z_4$ ensures the domination of the remaining points in the segment $p_i g_{ijt}$. In either case, the selected two points are within a distance of 2. Hence, it satisfies the semi-total property.

($iii$) Segments with two endpoints as grid points: Let $g_{ijt}$ and $g_{ij(t+1)}$ be two grid points and $z_1$, $z_2$, $z_3$ and $z_4$ be the four points on the segment. If so, then at least one grid point is dominated by a point in the other segment connected either to $g_{ijt}$ or $g_{ij(t+1)}$ (since at least one out of $p_i$ or $p_j$ is in $T_{vc}$). Without loss of generality, if $p_i \in T_{vc}$, then the selection of $z_2$ and $z_4$ ensures domination of $z_1$, $z_2$, $z_3$, $z_4$ and $g_{ij(t+1)}$. Since the distance between $z_2$ and $z_4$ is 2, the selection satisfies the semi-total property for the segment.

From the above arguments, we conclude that $D_{t2} = T_{vc} \cup T_a \cup T_g$ is a semi-total dominating set of $G'$. Since $|T_{vc}| \leq k$, $|T_a| = 2l$ and $|T_g| = 2n$, $|D_{t2}| \leq k + 2\ell + 2n$.

($\Longleftarrow$) Let $D_{t2}$ be a semi-total dominating set of $G'$ such that $|D_{t2}| \leq k + 2\ell + 2n$. Then, we will show that $G$ has a vertex cover of size at most $k$. To prove this, we prove the following observations.

(i) Out of four points in the gadget associated with each $p_i$ in $G'$, at least two points belong to $D_{t2}$.
(ii) Each segment contributes at least two points to $D_{t2}$.
(iii) If $p_i$ and $p_j$ in $G'$ corresponds to the end vertices of an edge $v_i v_j \in E$ and if none of them ($p_i$ and $p_j$) is in $D_{t2}$, then there exists one segment whose 3 vertices are in $D_{t2}$ from the segment(s) representing the edge $v_i v_j$, i.e., if there are $\ell'$ number of segments in $G'$ corresponding to the edge $v_i v_j \in E$ such that $p_i, p_j \notin D_{t2}$, then out of $5\ell' + 1$ points (excluding $p_i$ and $p_j$) $2\ell' + 1$ points are in $D_{t2}$.

**Observation (i):** Correspond to each $p_i \in E'$, there are 4 points ($x_i$, $x_i'$, $y_i$ and $y_i'$) in the corresponding gadget. Since $x_i'$ and $y_i'$ are pendant vertices, the selection of any two vertices is sufficient for domination. However, for semi-total domination, it requires either the selection of ($x_i$, $y_i$), ($x_i$, $y_i'$), or ($x_i'$, $y_i$). Hence, in $G'$, $|S \cap D_{t2}| \geq 2n$.

**Observation (ii):** Since there are four types of segments. Let us consider a segment, say $s_1$ with 4 points (excluding the endpoints), say $q_1$, $q_2$, $q_3$, and $q_4$. On the contrary, suppose only one vertex is in $D_{t2}$. If so, then just for domination, it requires at least two vertices (since only consecutive points are adjacent). This implies that the segments with more than four vertices on the segment require at least two points for domination. Hence, $|D_{t2} \cap A| \geq 2\ell$, where $A$ and $\ell$ are the set of auxiliary points and the number of segments in $G'$, respectively.

**Observation (iii):** Let $\ell'$ be the number of segments in $G'$ correspond to an edge $v_i v_j \in E$. Since only consecutive points are adjacent, two points can dominate at most 5 points in semi-total domination. So the minimum number of points required to dominate (semi-total) $5\ell' + 1$ number of points on $\ell'$ segments is $\lceil \frac{5\ell'+1}{5} \rceil \times 2 = 2\ell' + 1$.

Given a semi-total dominating set $D_{t2}$ of $G'$ of size at most $k+2\ell+2n$, we are left to show that by deleting and/or replacing some of the vertices from $D_{t2}$, we can obtain a vertex cover $S_{vc}$ of $G$ of size at most $k$. Let define a set $S'_{vc} = D_{t2} \backslash S$, then $|S'_{vc}| \leq k + 2\ell$ (due to claim $(i)$). Then $S_{vc} = \{v_i \in V | p_i \in S'_{vc}\}$. For each edge $v_i v_j \in E$, if $v_i, v_j \notin S_{vc}$, then there exists a segment on $p_i p_j$ in $G'$, which has 3 points in $D_{t2}$ instead of 2 (refer to claim (ii) and $(iii)$). For each such edge, add $v_i$ (or $v_j$) to $S_{vc}$. Since every segment contributes at least 2 to $D_{t2}$ and there is $2\ell$ number of such points (refer to claim (ii)), $|S_{vc}| \leq k$. Since every edge in $G$ has at least one vertex in $S_{vc}$, $S_{vc}$ is a vertex cover of size at most $k$. This proves that $D\text{-}T2DS\text{-}UDGs \in NP\text{-}hard$.

Therefore, $D\text{-}T2DS\text{-}UDGs \in NP\text{-}complete$. □

# 4 Approximation Algorithms

In this section, we propose a 6-factor approximation algorithm for the semi-total domination problem in UDGs. We also propose a $2 + \ln(\mathbb{D} + 1)$-factor approximation algorithm for the semi-total domination problem for general graphs, where $\mathbb{D}$ is the maximum degree of the graph, which is an improvement over the approximation factor $2 + 3\ln(\mathbb{D} + 1)$ given in [12].

## 4.1 Algorithm for Semi-total Domination in UDGs

Given a geometric unit disk graph $G = (V, E)$ with $V = \{p_1, p_2, \ldots, p_n\} \subseteq \mathbb{R}^2$ as the set of disk centers, Algorithm 1 finds a semi-total dominating set $D_{t2}$ of $G$. Now, we describe the procedure for finding the set $D_{t2}$. First, we find a maximal independent set $D \subseteq V$ of $G$ to satisfy the domination property (see Lines 2–6 of Algorithm 1). Next, to satisfy the semi-total property, we choose a set of vertices $T \subseteq V$ such that for each $v \in D$, there exists a vertex $u \in D \cup T$ such that $d(u, v) \leq 2$. To find such a set $T$, first, we find each point $u \in D$, which satisfies the semi-total property (see Lines 8–13 of Algorithm 1) and next, we segregate the points (set $\mathcal{U}$) that do not satisfy the semi-total property in $D$ (see Line 14 of Algorithm 1) and then for each point $u \in \mathcal{U}$, we add a point $v \in N_G(u)$ into $T$ (see Lines 15–18 of Algorithm 1). Finally, we report $D_{t2} = D \cup T$ as a semi-total dominating set of $G$. Lemma 5 and Lemma 6 represent the algorithm's correctness and time complexity, respectively.

**Lemma 5.** *The set $D_{t2}$ in Algorithm 1 is a semi-total dominating set of $G$.*

*Proof.* In the first phase, we find a maximal independent set $D$ of $G$ to satisfy the domination property (see Lines 2–6 in Algorithm 1). Next, we segregate the points that do not satisfy the semi-total property in $D$. Note that, for each vertex $u \in V$, the algorithm finds $S_u = N_G(u) \cap D$. If $|S_u| > 1$, then the vertices in $S_u$ satisfy the semi-total property, and hence the vertices in the set $X \subseteq D$ also satisfy the semi-total property (see Lines 8–13). Since $\mathcal{U} = D \setminus X$, the vertices in the set $\mathcal{U}$ do not satisfy the semi-total property, so we choose a one-distance neighbor $v \in V \setminus D$ for each such $u \in \mathcal{U}$ and $T$ is the corresponding set (see

---

**Algorithm 1.** *T2DS-UDG(G)*

---

**Input:** A unit disk graph, $G = (V, E)$, with known disk centers
**Output:** A semi-total dominating set $D_{t2}$ for $G$

  1: $V' = V, D = \emptyset$
  2: **while** $V' \neq \emptyset$ **do**
  3:    choose a vertex $v \in V'$
  4:    $D = D \cup \{v\}$
  5:    $V' = V' \setminus N_G[v]$
  6: **end while**
  7: $T = \emptyset$, $X = \emptyset$
  8: **for** each $u \in V$ **do**
  9:    $S_u = N_G(u) \cap D$
10:    **if** $|S_u| > 1$ **then**         $\triangleright$ each vertex in $S_u$ satisfies the *semi-total* property
11:        $X = X \cup S_u$
12:    **end if**
13: **end for**
14: $\mathcal{U} = D \setminus X$         $\triangleright$ vertices in $\mathcal{U}$ do not satisfy the *semi-total* property
15: **for** each $u \in \mathcal{U}$ **do**
16:    choose a vertex $v \in N_G(u)$
17:    $T = T \cup \{v\}$
18: **end for**
19: $D_{t2} = D \cup T$
20: **return** $D_{t2}$

---

Lines 15–18). Since the set $T$ is the set of such one-distance neighbor of each point violating semi-total property in $D$, the inclusion of $T$ in $D_{t2}$ along with $D$ ensures that for each vertex $u \in D$, there exists another vertex $v \in D \cup T$ such that $d(u, v) \leq 2$. Therefore, combinedly, the nominated points in $D$ and $T$ satisfy the *domination* and *semi-total* properties. Hence, the set $D_{t2}$ is a semi-total dominating set of $G$.

**Lemma 6.** *Algorithm 1 runs in $O(nk)$ time.*

*Proof.* The complexity of Algorithm 1 is primarily dominated by the *three for* loops (see Lines 2–6, 8–13 and 15–18 of Algorithm 1). Let $V = \{p_1, p_2, \ldots, p_n\}$ be the set of disks' centers corresponding to graph $G = (V, E)$. Let all the disks lie on a plane's rectangular region $\mathbb{R}$. Let the rectangle's extreme left and bottom arms represent the $x$- and $y$-axis, respectively. Then, we split the plane $\mathbb{R}$ so that the region $\mathbb{R}$ becomes a grid with cell size $1 \times 1$. Let $[x, y]$ be the index associated with each cell, where $x, y \in \mathbb{N} \cup \{0\}$. If a point $p \in V$ is located at co-ordinate $(p_x, p_y)$ on $\mathbb{R}$, then the point belongs to a cell with index $[\lfloor p_x \rfloor, \lfloor p_y \rfloor]$.

In the first *for* loop (see Lines 2–6), Algorithm 1 constructs a maximal independent dominating set $D$ of the input graph $G$. To do so efficiently, each non-empty cell maintains a list that keeps the points of $V$ chosen for inclusion in $D$ located within that cell. While considering a point $p \in V$ as a candidate for the set $D$, it only probes into 9 cells surrounding the cell where $p$ lies. That means if $p$ is located at co-ordinate $(p_x, p_y)$, then it searches in each $[i, j]$ cell, where

$\lfloor p_x \rfloor - 1 \leqslant i \leqslant \lfloor p_x \rfloor + 1$ and $\lfloor p_y \rfloor - 1 \leqslant j \leqslant \lfloor p_y \rfloor + 1$.[3] If there does not exist any point $q \in D$ in those 9 cells such that $p \in \Delta(q)$, then $p$ is included in $D$. A height balance binary tree containing non-empty cells is used to store the points that are in $D$. Since each cell of size $1 \times 1$ can contain the centers of at most 3 independent unit disks (since placing the centers on the boundary maximizes the number of independent unit disks in a cell and one disk covers more than one edge in a cell), the processing time to decide whether a point is in $D$ or not requires $O(\log k)$ time, where $k = |D|$. Thus the time taken to process $|V| = n$ points is $O(n \log k)$.

In the second *for* loop (Lines 8–13), Algorithm 1 finds a set $X \subseteq D$ in which each vertex satisfy the semi-total property. To find the set $X$, it finds $S_u = N_G(u) \cap D$ for each vertex $u \in V$. Now, if $|S_u| > 1$, the vertices in $S_u$ satisfy the semi-total property; hence, these vertices collectively represent the set $X$. Thus, finding a set $X$ requires $O(nk)$ time.

Since each vertex $u$ in $\mathcal{U}$ does not satisfy the semi-total property (i.e., $|S_u| \leq 1$), we add a vertex $v \in N_G(u)$ to $T$ (see Lines 15–18). Thus, in the worst case, the time to construct the set $T$ is $O(k)$.

Therefore, in worst case, Algorithm 1 executes in $O(nk)$ time.    □

**Lemma 7.** *In Algorithm 1, $|T| \leq |D^*|$, where $D^*$ is an optimal DS of $G$.*

*Proof.* On contrary assume that $|T| > |D^*|$, i.e., $|\mathcal{U}| > |D^*|$. So, at least *one* vertex in $D^*$ dominates two or more vertices in $\mathcal{U}$, which leads to a contradiction that there is no vertex $v \in V$ that has more than one neighbor in $\mathcal{U}$.    □

**Analysis:** The set $D_{t2}$ in Algorithm 1 is a semi-total dominating set of $G$, where $D_{t2} = D \cup T$ (see Lemma 5). Let $D^*$ and $D_{t2}^*$ be the optimal dominating set and optimal semi-total dominating set of $G$, respectively. Since $D$ is a maximal independent set of $G$, from Lemma 2, we have $|D| \leq 5|D^*|$. The set $T$ in Algorithm 1 satisfies the semi-total property when added to the independent set $D$. Note that from Lemma 7, we have $|T| \leq |D^*|$. Therefore, using Lemma 2, Lemma 5, Lemma 7 and Observation 1, we conclude the approximation factor of Algorithm 1 as follows:

$$|D_{t2}| = |D \cup T| \leq |D| + |T| \leq 5|D^*| + |D^*| \leq 6 \times |D^*| \leq 6 \times |D_{t2}^*| \quad (1)$$

**Theorem 2.** *The proposed algorithm (T2DS-UDG) gives a 6-factor approximation result for the semi-total domination problem in UDGs. The algorithm runs in $O(nk)$ time, where $n$ is the number of vertices in the given UDG and $k$ is the size of the maximal independent set.*

*Proof.* The approximation factor and the time complexity result follow from Eq. 1 and Lemma 6, respectively.    □

---

[3] Any point outside these 9 cells is independent from $p$.

**Corollary 1.** *The semi-total domination problem achieves a $\frac{53}{9}$-factor approximation result in UDGs with running time $O(n^2)$, where $n$ is the number of vertices in the given UDG.*

*Proof.* From Lemma 3 and Lemma 7, we have $|D| \leq \frac{44}{9}|D^*|$ and $|T| \leq |D^*|$, respectively. Therefore,

$$
\begin{aligned}
|D_{t2}| = |D \cup T| &\leq |D| + |T| \\
&\leq \frac{44}{9}|D^*| + |D^*| = \frac{53}{9}|D^*| \\
&\leq \frac{53}{9}|D_{t2}^*|
\end{aligned}
\tag{2}
$$

$\square$

**Note:** In [12], the authors proposed a $2 + 3\ln(\mathbb{D}+1)$-factor approximation algorithm for the semi-total dominating set problem in general graphs. Here, the authors used two sets, namely $D$ and $T$, to find the semi-total dominating set of the given graph $G$, where $D$ is a DS and $T$ is a set of vertices such that $D \cup T$ is a semi-total dominating set. The authors used the approximation algorithm for the minimum DS problem to find the set $D$ and the approximation algorithm for the minimum set cover problem to find the set $T$. Since the approximation factors of the minimum DS and the minimum set cover problems are $1 + \ln(\mathbb{D}+1)$ and $1 + 2\ln\mathbb{D}$, respectively, the approximation factor of the algorithm in [12] is $2 + 3\ln(\mathbb{D}+1)$. However, to have an improvement over the approximation factor, we can modify the algorithm in [12] by selecting the set $T$ as in Algorithm 1 (the selection of the set $T$ needs the set $D$ to be a dominating set, not necessarily a maximal independent set). Then, by Lemma 7 and Observation 1, the approximation factor of the semi-total domination problem in general graphs is as follows:

$$
\begin{aligned}
|D_{t2}| = |D \cup T| &\leq |D| + |T| \leq (1 + \ln(\mathbb{D}+1))|D^*| + |D^*| \\
&\leq (2 + \ln(\mathbb{D}+1))|D^*| \leq (2 + \ln(\mathbb{D}+1))|D_{t2}^*|
\end{aligned}
\tag{3}
$$

**Note:** Since there exists a polynomial-time approximation scheme (PTAS) for the domination problem in UDGs with approximation factor $(1 + \epsilon)$ and time $n^{O(\frac{1}{\epsilon}\frac{1}{\log \epsilon})}$, for any $\epsilon > 0$ [17]. We have the following corollary.

**Corollary 2.** *The semi-total dominating set problem in UDGs admits a PTAS with approximation factor $(2 + \epsilon)$ in time $n^{O(\frac{1}{\epsilon}\frac{1}{\log \epsilon})}$.*

## 5   Conclusion

In this paper, we have introduced the concept of semi-total domination to UDGs and shown that the semi-total domination problem in UDGs is NP-complete.

Then, we proposed a 6-factor approximation algorithm for the same, with time complexity $O(nk)$, where $k$ is the size of the maximal independent set of the given UDG. In addition, we also proposed a $2 + \ln(\mathbb{D} + 1)$-factor approximation algorithm for general graphs, where $\mathbb{D}$ is the maximum degree of the given graph.

# References

1. Asplund, J., Davila, R., Krop, E.: A Vizing-type result for semi-total domination. Discret. Appl. Math. **258**, 8–12 (2019)
2. Biedl, T., Kant, G.: A better heuristic for orthogonal graph drawings. Comput. Geom. **9**(3), 159–180 (1998)
3. da Fonseca, G.D., de Figueiredo, C.M., de Sá, V.G.P., Machado, R.C.: Efficient sub-5 approximations for minimum dominating sets in unit disk graphs. Theoret. Comput. Sci. **540**, 70–81 (2014)
4. Galby, E., Munaro, A., Ries, B.: Semitotal domination: new hardness results and a polynomial-time algorithm for graphs of bounded mim-width. Theoret. Comput. Sci. **814**, 28–48 (2020)
5. Goddard, W., Henning, M.A., McPillan, C.A.: Semitotal domination in graphs. Utilitas Math. **94** (2014)
6. Hartmanis, J.: Computers and intractability: a guide to the theory of NP-completeness (M. R. Garey and D. S. Johnson). SIAM Rev. **24**(1), 90 (1982)
7. Haynes, T.W., Hedetniemi, S., Slater, P.: Fundamentals of Domination in Graphs. CRC Press, Boca Raton (1998)
8. Haynes, T.W., Hedetniemi, S.T., Henning, M.A.: Topics in Domination in Graphs. Springer, Cham (2020)
9. Hedetniemi, S.T., Laskar, R.C.: Bibliography on domination in graphs and some basic definitions of domination parameters. Ann. Discret. Math. **48**, 257–277 (1991)
10. Henning, M.A.: A survey of selected recent results on total domination in graphs. Discret. Math. **309**(1), 32–63 (2009)
11. Henning, M.A., Marcon, A.J.: Semitotal domination in claw-free cubic graphs. Ann. Comb. **20**, 799–813 (2016)
12. Henning, M.A., Pandey, A.: Algorithmic aspects of semitotal domination in graphs. Theoret. Comput. Sci. **766**, 46–57 (2019)
13. Henning, M.A., Yeo, A.: Total Domination in Graphs. Springer, New York (2013)
14. Lichtenstein, D.: Planar formulae and their uses. SIAM J. Comput. **11**(2), 329–343 (1982)
15. Marathe, M.V., Breu, H., Hunt, H.B., III., Ravi, S.S., Rosenkrantz, D.J.: Simple heuristics for unit disk graphs. Networks **25**(2), 59–68 (1995)
16. Marcon, A.J.: Semitotal domination in graphs. University of Johannesburg (2015)
17. Nieberg, T., Hurink, J.: A PTAS for the minimum dominating set problem in unit disk graphs. In: Erlebach, T., Persinao, G. (eds.) WAOA 2005. LNCS, vol. 3879, pp. 296–306. Springer, Heidelberg (2006). https://doi.org/10.1007/11671411_23
18. Poureidi, A.: Total roman domination for proper interval graphs. Electron. J. Graph Theory Appl. (EJGTA) **8**(2), 401–413 (2020)
19. Valiant, L.G.: Universality considerations in VLSI circuits. IEEE Trans. Comput. **100**(2), 135–140 (1981)

# Discrete Applied Mathematics

# An Efficient Interior Point Method for Linear Optimization Using Modified Newton Method

Sajad Fathi Hafshejani[✉], Daya Gaur, and Robert Benkoczi

Department of Math and Computer Science, University of Lethbridge, Lethbridge, Canada
{sajad.fathihafshejan,daya.gaur,robert.benkoczi}@uleth.ca

**Abstract.** Interior point methods are one of the popular iterative approaches for solving optimization problems. Search direction plays a vital role in the performance of the interior point methods. This paper uses a modification to the Newton method and proposes a new way to find the search direction. We introduce a two-step interior point algorithm for solving linear optimization problems based on the new search direction. We present theoretical results for the convergence of the algorithm. Finally, we evaluate the algorithm on some test problems from the Netlib collection and show that the proposed algorithm reduces the number of iterations and CPU time by 30.97% and 20.46%, respectively.

**Keywords:** Interior point method · Linear optimization · Experimental evaluation

## 1 Introduction

In this paper, we consider Linear Optimization (LO) problems in standard form as in [12].

$$(P) \qquad \min\{c^T x : Ax = b, x \geq 0\},$$

and its dual is given by:

$$(D) \qquad \max\{b^T y : A^T y + s = c, s \geq 0\},$$

where $x, c, s \in \mathbb{R}^n$, $b \in \mathbb{R}^m$, $y \in \mathbb{R}^m$ and $A \in \mathbb{R}^{m \times n}$ with $m \leq n$. Without loss of generality, we make the following assumptions:

- The matrix $A$ is full row rank, i.e., rank $(A) = m \leq n$.
- Both problems (P) and (D) satisfy the Interior Point Condition (IPC), i.e., there exists $x^0 > 0$ and $(y^0, s^0)$ with $s^0 > 0$ such that:

$$Ax^0 = b, \qquad A^T y^0 + s^0 = c.$$

© The Author(s), under exclusive license to Springer Nature Switzerland AG 2024
S. Kalyanasundaram and A. Maheshwari (Eds.): CALDAM 2024, LNCS 14508, pp. 133–147, 2024.
https://doi.org/10.1007/978-3-031-52213-0_10

Optimization problems can be solved efficiently with Interior Point Methods (IPMs). The first practical method was introduced by Karmarkar in 1984 for LO problems, but since then, several approaches have been introduced to improve complexity bounds and the implementation [2,4,5,9].

The Newton method is commonly used for computing the search direction of the IPMs. This method is used to move from the current iteration to the next point. Recently, McDougall and Wotherspoon [6] proposed a new two-step Newton method. Their algorithm computes the Newton direction by using an auxiliary point, and they demonstrated that it can significantly reduce the number of iterations required to reach an optimal solution.

CONTRIBUTIONS: Motivated by the works mentioned above, our paper contributes in several ways. First, we propose a new two-step interior point algorithm for LO problems based on the extension of the method suggested in [6]. Our new algorithm calculates the matrix inverse only once in each iteration and uses information from the previous iteration to determine the search direction. Second, we extend the theoretical results presented in [1] and prove the convergence of our proposed algorithm. Third, we evaluate the efficiency of our algorithm by comparing it with the classical search direction on test problems from the Netlib collection. Our results demonstrate that our new algorithm can reduce the number of iterations and CPU times by 30.97% and 20.46%, respectively. Finally, the theoretical and practical contributions of our work can be applied to other IPMs, potentially improving their efficiency as well.

The paper is structured in the following manner: Sect. 2 provides a brief overview of the interior point methods and goes on to describe the new search direction in detail. Theoretical results for the convergence of the algorithm are presented in Sect. 3, while Sect. 4 showcases some numerical results. Finally, concluding remarks are provided in Sect. 5.

In this paper, we will use the following notations. First, the Euclidean norm of a vector is denoted by $\|.\|$. We also use $\mathbb{R}^n_+$ and $\mathbb{R}^n_{++}$ to represent the nonnegative and positive orthants, respectively. When we refer to vectors $x$ and $s$, we use $xs$ and $\frac{x}{s}$ to indicate coordinate-wise operations on the vectors, meaning the components are $x_i s_i$ and $\frac{x_i}{s_i}$, respectively.

We denote the auxiliary point by $(\tilde{x}, \tilde{y}, \tilde{s})$. The average of the auxiliary and current points in each iteration of the algorithm is denoted by $(\hat{x}, \hat{y}, \hat{s})$. To refer to a specific iteration, we use subscripts, such as $x_n$, for the value of $x$ after $n$ updates. For simplicity, we use $v$ instead of $v_n$ and $v_+$ for $v_{n+1}$ for $v \in \{x, y, s, \tilde{x}, \tilde{y}, \tilde{s}, \hat{x}, \hat{y}, \hat{s}\}$. So, $x$ refers to the current point, and $x_+$ refers to the updated value of $x$.

Finally, we use the notation $f(t) = \Theta(g(t))$ to indicate that there exist positive constants $\omega_1$ and $\omega_2$ such that $\omega_1 g(t) \leq f(t) \leq \omega_2 g(t)$ for all $t > 0$. We also use the notation $f(t) = O(g(t))$ to indicate that there exists a positive constant $\omega$ such that $f(t) \leq \omega g(t)$ for all $t > 0$.

## 2   Mid-Point Algorithm

In this section, we will be discussing the new interior point algorithm and its fundamental idea. To start, we will take a look at the Karush-Kuhn-Tucker (KKT) conditions for problems (P) and (D). These conditions play a crucial role in the IPMs and will serve as the foundation for our discussions. KKT conditions for (P) and (D) are:

$$
\begin{aligned}
Ax &= b, & x &\geq 0, \\
A^T y + s &= c, & s &\geq 0, \\
xs &= 0,
\end{aligned}
\tag{1}
$$

where the coordinate-wise product of vectors $x$ and $s$ is denoted as $xs$. The first and second equalities in Eq. (1) represent the primal and dual feasibility constraints, respectively. The complementarity condition is given by the third equality in the same system of equations.

To obtain the optimal solution, the IPM solves the system by replacing the complementarity condition with the parameterized nonlinear equation $xs = \mu e$, where $\mu$ is a positive real parameter and $e$ is the all-one vector of length $n$. The resulting system is shown below.

$$
\begin{aligned}
Ax &= b, & x &\geq 0, \\
A^T y + s &= c, & s &\geq 0, \\
xs &= \mu e,
\end{aligned}
\tag{2}
$$

Based on the IPC condition and the full row rank property of matrix $A$, it can be concluded that the system (2) has a unique solution $(x(\mu), y(\mu), s(\mu))$. The terms $x(\mu)$ and $(y(\mu), s(\mu))$ are called the $\mu$-centers of (P) and (D), respectively, as stated in [2]. Additionally, [8,11] define the *central path* as the collection of all $\mu$-centers, where $\mu$ covers all positive real numbers. They have also demonstrated that the central path has a limit that exists and converges to an analytic center of the optimal solutions set of (P) and (D) as $\mu$ goes to zero.

Now, we are in a position to extend the method proposed by [6] to the interior point method. To this end, note that the optimal solution of (2) is equivalent to finding the roots of the following function:

$$
\xi = \begin{bmatrix} x \\ y \\ s \end{bmatrix} \text{ and } F(\xi) = \begin{bmatrix} Ax - b \\ A^T y + s - c \\ \mu e - xs \end{bmatrix}
\tag{3}
$$

where the operator $F$ is defined on the Banach space $B_1$ with values in a Banach space $B_2$. We can find the root of Eq. (3) denoted by $\xi^*$ where $F(\xi^*) = 0$. One way to calculate this root is to use iterative methods like Newton's method, which can find the difference $\Delta\xi$ given an initial value $\xi$ in the vicinity of the root. To do this, we can use a linear approximation using the Taylor series expansion around $\xi$.

$$F(\xi) + F'(\xi)\Delta\xi \simeq 0$$

where Jacobian $F' = \begin{bmatrix} \frac{\partial F_1}{\partial \xi_1} & \frac{\partial F_1}{\partial \xi_2} & \frac{\partial F_1}{\partial \xi_3} \\ \frac{\partial F_2}{\partial \xi_1} & \frac{\partial F_2}{\partial \xi_2} & \frac{\partial F_2}{\partial \xi_3} \\ \frac{\partial F_3}{\partial \xi_1} & \frac{\partial F_3}{\partial \xi_2} & \frac{\partial F_3}{\partial \xi_3} \end{bmatrix} = \begin{bmatrix} A & 0 & 0 \\ 0 & A^T & I \\ S & 0 & X \end{bmatrix}$ and $\Delta\xi = \begin{bmatrix} \Delta x \\ \Delta y \\ \Delta s \end{bmatrix}$,

where $X, S$ are diagonal matrices constructed from $x$ and $s$. It yields an iterative algorithm, the Newton-Raphson method, for finding roots. To start, we need an initial estimate $\xi_0$. Then, using a linear approximation with the Taylor series expansion around $\xi$, we can update the current estimate $\xi_n$ of the root using the following rule for some appropriate step size $\alpha$:

$$\xi_{n+1} = \xi_n - \alpha[F'(\xi_n)]^{-1}F(\xi_n)$$

The Newton-Raphson method's efficiency can be improved through a two-step approach suggested in [6]. Let $F$ be any function. The first step involves updating an auxiliary point $\tilde{\xi}_0 = \xi_0$ using a linear approximation around the current root estimate $\xi_n$. In the second step, the average $\hat{\xi}_n$ of the auxiliary point and the current estimate is calculated. Finally, in the third step, the current estimate is updated by employing the linear approximation to $F$ around $\xi_n$, but instead of using the Jacobian at point $\xi_n$, the Jacobian is evaluated at the average $\hat{\xi}_n$. The update rules used in the $n^{th}$ iteration can be concisely summarized as:

$$\tilde{\xi}_{n+1} = \xi_n - \alpha[F'(\hat{\xi}_n)]^{-1}F(\xi_n)$$
$$\hat{\xi}_{n+1} = \frac{1}{2}(\tilde{\xi}_{n+1} + \xi_n)$$
$$\xi_{n+1} = \xi_n - \alpha[F'(\hat{\xi}_{n+1})]^{-1}F(\xi_n)$$

We extend the approach of [6] to linear optimization problems. The computations proceed as follows:

**Initialization:**

1. Start with a feasible point $(x, y, s)$.
2. Set $\mu \leftarrow \mu(1 - \theta)$.
3. Compute the proximity function by:[1]

$$\delta(x, s, \mu) = \|\mu\mathbf{e} - xs\|_1.$$

Check the condition $\delta \geq \tau$. If $\delta < \tau$ go back to step 2.
4. Set $(\tilde{x}, \tilde{y}, \tilde{s}) = (x, y, s)$
5. Compute the average point $(\hat{x}_0, \hat{y}_0, \hat{s}_0) = (\frac{x+\tilde{x}}{2}, \frac{y+\tilde{y}}{2}, \frac{s+\tilde{s}}{2})$

---

[1] $\|x\|_1 = \sum_{i=1}^{n} |x_i|.$

6. Find the search direction using the following system:

$$A\Delta x = 0$$
$$A^T \Delta y + \Delta s = 0$$
$$\hat{S}\Delta x + \hat{X}\Delta s = \mu\mathbf{e} - XS\mathbf{e} \tag{4}$$

Note the $\hat{X}, \hat{S}$ are diagonal matrices constructed from $\hat{x}, \hat{s}$.

7. Find the maximum value for step size $\beta$ such that the new point is feasible.
8. Update the current point by using the following role:

$$(x_+, y_+, s_+) \leftarrow (x + \beta\Delta x, y + \beta\Delta y, s + \beta\Delta s) \tag{5}$$

To start the iterative process, we will use the same method as in the initialization phase for all values of $k$ greater than or equal to 1. However, we will incorporate a new step to update the auxiliary point in a specific manner.

**Iterative Update:**

1. Compute $\delta(x, s, \mu)$ and check Step 3 in the Initialization phase and compute $\delta(x, s, \mu)$.
2. Compute the search direction by solving the following system:

$$A\Delta\tilde{x} = 0$$
$$A^T \Delta\tilde{y} + \Delta\tilde{s} = 0$$
$$\hat{S}\Delta\tilde{x} + \hat{X}\Delta\tilde{s} = \mu\mathbf{e} - XS\mathbf{e} \tag{6}$$

Note that, in practice, we use the information of the previous iteration to calculate the search direction $(\Delta\tilde{x}, \Delta\tilde{y}, \Delta\tilde{s})$.

3. Find the maximum value for $\alpha$ such that the new auxiliary point will remain feasible.
4. Update the auxiliary point by:

$$(\tilde{x}_+, \tilde{y}_+, \tilde{s}_+) \leftarrow (x + \alpha\Delta\tilde{x}, y + \alpha\Delta\tilde{y}, s + \alpha\Delta\tilde{s}) \tag{7}$$

5. Compute the average.

$$(\hat{x}_+, \hat{y}_+, \hat{s}_+) \leftarrow \frac{1}{2}((\tilde{x}_+, \tilde{y}_+, \tilde{s}_+) + (x, y, s)) \tag{8}$$

6. Solve the following system of equations to obtain the search direction.

$$A\Delta x = 0$$
$$A^T \Delta y + \Delta s = 0$$
$$\hat{S}_+\Delta x + \hat{X}_+\Delta s = \mu\mathbf{e} - XS\mathbf{e} \tag{9}$$

Note the $\hat{X}_+, \hat{S}_+$ are diagonal matrices constructed from $\hat{x}_+, \hat{s}_+$.

7. Find the maximum value for step size $\beta$ such that the new point is feasible.

---

**Algorithm 1: A two-step feasible IPM algorithm for LPs.**

**Input:** $x_0$, $y_0$, $s_0$, $\mu > 0$, $\tau > 0$, $\varepsilon > 0$, $\theta \in (0,1)$, and $\delta(x,s,\mu)$

1  $(x,y,s) \leftarrow (x_0, y_0, s_0)$
2  $k, m \leftarrow 0$
3  **while** *stopping criteria is not met* **do**
4      $\mu \leftarrow \mu(1 - \theta)$
5      **while** $\delta(x,y,s,\mu) \geq \tau$ **do**
6         **if** $m==0$ **then**
7            $(\tilde{x}, \tilde{y}, \tilde{s}) \leftarrow (x,y,s)$
8            Update $(\hat{x}_+, \hat{y}_+, \hat{s}_+) \leftarrow \frac{1}{2}((\tilde{x}, \tilde{y}, \tilde{s}) + (x,y,s))$
9            Find search direction $(\Delta\tilde{x}, \Delta\tilde{y}, \Delta\tilde{s})$ using (4)
10          Find the value for $\beta$ and update
            $(x,y,s) \leftarrow (x + \beta\Delta x, y + \beta\Delta y, s + \beta\Delta s)$
11        **if** $m \geq 1$ **then**
12          Find the search direction using (6)
13          Find the maximum step size $\alpha$ and update the auxiliary point using (7)
14          Update the average point by using (8)
15          Find search direction $(\Delta x, \Delta y, \Delta s)$ by solving (9)
16          Find the maximum value for step size $\beta$ and update the current point by using (10)
17        $m \leftarrow m + 1$
18      $k \leftarrow k + 1$

---

8. Update the current point by using the following role:

$$(x_+, y_+, s_+) \leftarrow (x + \beta\Delta x, y + \beta\Delta y, s + \beta\Delta s) \qquad (10)$$

9. Check Step 1. If $\delta(x,s,\mu) < \tau$ stop the inner loop. If the terminating criteria are not met update $\mu$ and go to Step 1.

The pseudo-code is listed in Algorithm 1. Algorithm 1, begins with an initial point $(x_0, y, s_0)$ and some given parameters such as $\mu$, $\epsilon$, $\theta$, and $\tau$. The algorithm consists of two loops, the inner and outer loops. In the outer loop, the value of the barrier parameter is updated by factor $(1 - \theta)\mu$. Next, the value of the proximity function is computed. If $\delta < \tau$, then we again update the parameter $\mu$; otherwise, the algorithm begins the inner loop. In each inner loop iteration, the search direction is obtained by information from the previous iteration, and then the best value for step size is computed. Using this information, the auxiliary point is updated. Then the average of the principle point and the auxiliary point is computed. Next, the search direction for the principle point is obtained by solving (9). After that, the algorithm finds the maximum value for step size to make the new point feasible. Then the principle point is updated using (10). This procedure is stopped when we find an iterate with $\delta < \tau$. This means that the iterates are in a small enough neighbourhood of the $\mu$-center $(x(\mu), y(\mu), s(\mu))$. Again, the parameter $\mu$ is reduced by the factor $1 - \theta$, and we apply the inner loop targeting the new $mu$ center. This procedure is repeated until we obtain an iterate $(x, y, s)$ such that $x^T s < \epsilon$. In this case, an $\epsilon$-approximate solution of problems (P) and (D) is found.

When it comes to IPMs, the choice of the barrier update parameter $\theta$ is crucial both in theory and practice. In general, if $\theta$ remains constant regardless of the problem's dimension $n$, for instance, when it is set to $\frac{1}{2}$ or even $\theta = 0.99$, it is referred to as a "large-update" method. On the other hand, the "small-update" methods utilize much smaller values of $\theta$, such as $\theta = \frac{1}{\sqrt{n}}$, which depends on the problem's size.

## 3    Convergence

To prove the convergence of Algorithm 1, we first recall some Lemmas without proof from [1].

**Lemma 1 (Lemma 3 in [1]).** *Suppose that $h_0 : [0, \infty) \to \mathbb{R}$ is a continuous and non-decreasing function. Moreover, we assume that the equation*

$$h_0(t) - 1 = 0$$

*has the smallest positive solution $\kappa$. Moreover, suppose that there exists a function $h : [0, \kappa) \to R$, which is continuous and nondecreasing. Let also parameters $t_0$, $\kappa_0$, $t_1$ be such that $t_0 = 0$, $\kappa_0 > 0$ and $\kappa_0 < t_1$. Define the sequences $\{t_m\}$, $\{\kappa_m\}$ by*

$$p_m = h_0 \left( \frac{t_m + \kappa_m}{2} \right) \tag{11}$$

$$t_{m+1} = \kappa_m + \frac{h \left( \frac{\kappa_m - \kappa_{m-1} + t_m - t_{m-1}}{2} \right)}{1 - p_m} (\kappa_m - t_m) \tag{12}$$

$$\omega_{m+1} = \int_0^1 h \left( \frac{\kappa_m - t_m}{2} + \theta(t_{m+1} - t_m) \right) d\theta(t_{m+1} - t_m) \tag{13}$$

$$\kappa_{m+1} = t_{m+1} + \frac{\omega_{m+1}}{1 - p_m} \tag{14}$$

*If for each $m = 0, 1, 2, \dots$ and some $\beta > 0$*

$$p_m < 1 \quad and \quad t_m \leq \beta. \tag{15}$$

*Then, the following holds:*

$$0 \leq t_m \leq \kappa_m \leq t_{m+1} \leq \beta \tag{16}$$

*and*

$$\lim_{m \to \infty} t_m = \omega \leq \beta \tag{17}$$

The conditions connecting the $h$ functions to the operators $F$ and $F'$ are:

A1 There exists an initial guess $\xi_0 \in D$, $\kappa_0 \geq 0$, $t_1 \geq \kappa_0$ such that $F'(\xi_0) \in \mathcal{L}(B_2, B_1)$, $\|F'(\xi_0)^{-1}F(\xi_0)\| \leq \kappa_0$, and for $\hat{\xi}_0 = \xi_0 - F'(\xi_0)^{-1}F(\xi_0)$, $F'(\frac{\xi_0 + \hat{\xi}_0}{2})^{-1} \in \mathcal{L}(B_2, B_1)$ with

$$\left\| F'(\frac{x_0 + y_0}{2})^{-1}F(x_0) - F'(x_0)^{-1}F(x_0) \right\| \leq t_1 - \kappa_0$$

A2 $\|F'(\xi_0)^{-1}(F'(\xi) - F'(\xi_0))\| \leq h_0(\|\xi - \xi_0\|)$ for all $\xi \in D$. Set

$$D_3 = \bigcup(\xi_0, \kappa) \cap D.$$

A3 $\|F'(\xi_0)^{-1}(F'(\hat{\xi}) - F'(\xi))\| \leq h(\|\hat{\xi} - \xi\|)$ for all $\xi, \hat{\xi} \in D_3$
A4 $p_m < 1$ and $t_m \leq \beta$
A5 $\bigcup[\xi, \omega] \subseteq D$

As we can see from the next lemma, the inner loop of Algorithm 1 is able to converge to the optimal solution when a fixed value of $\mu$ is used.

**Lemma 2 (Theorem 3 in [1]).** *Suppose that the conditions (A1)–(A5) hold. Then, for a fixed $\mu$ the sequence $\{\xi_m\}$ converges to a solution $\xi_\mu^* \in \bigcup[\xi_0, \alpha]$ such that*

$$\|\xi_m - \xi_\mu^*\| \leq \omega - t_m.$$

The following lemma shows that the outer loop of the algorithm converges at most $O\left(\frac{1}{\theta} \log \frac{n}{\epsilon}\right)$ iteration, i.e., after performing $O\left(\frac{1}{\theta} \log \frac{n}{\epsilon}\right)$ iterations, we have $\mu \leq \epsilon$.

**Lemma 3 (Lemma I.36 in [10]).** *The total number of outer iterations to obtain $n\mu \leq \epsilon$ are $O\left(\frac{1}{\theta} \log \frac{n}{\epsilon}\right)$.*

Based on Lemma 2, we can see that the inner iteration for a fixed $\mu$ can converge to the optimal solution. Meanwhile, based on Lemma 3, we can conclude that the outer loop is also able to converge to the optimal solution, meaning that $\mu$ will be less than or equal to $\epsilon$ after a finite number of iterations. Combining these two conclusions, we can say that Algorithm 1 converges to the optimal solution.

**Theorem 1.** *Suppose that the outer loop updates the barrier parameter by factor $\theta \in (0, 1)$ and $k \to \infty$. Then one has:*

$$\|\xi_k - \xi^*\| \leq \epsilon.$$

**Remark.** While practical evidence supports the idea that the two-step strategy significantly speeds up Newton's method, its theoretical complexity hasn't been proven [6]. Moreover, the convergence rate has only been demonstrated through multiple examples. By considering the extension of this method to the interior point algorithm, the structure of the system (9) prevents us from calculating the complexity bound of the algorithm. Specifically, the complexity of IPMs involves multiplying the iterations within the inner and outer loops of Algorithm 1. Deriving the number of outer iterations is straightforward based on the value of the parameter $\theta$. However, calculating the number of inner iterations necessitates consideration of Newton's system. To compute these number of inner iterations, introducing an auxiliary variable, such as $v := \sqrt{\frac{xs}{\mu}}$, is imperative to reformulate the right side of the system (9). Unfortunately, due to the concurrent utilization of $x, s$ and $\hat{x}, \hat{s}$ defining such a variable is precarious, rendering the computation of the interior point method's complexity unattainable using the frameworks outlined in previous literature.

# 4    Numerical Results

## 4.1    Details About the Experimentation

- **Algorithms:**
  In order to demonstrate the effectiveness of our proposed method, we conducted a comparison with the algorithm from [12]. For this comparison, we analyzed the number of iterations and CPU times on various Netlib instances taken by the two algorithms.
- **Device's detail:**
  We programmed the two algorithms, Algorithm 1 and the Classical algorithm [12], in Python 3.10. For the equation-solving step in both algorithms, we used BLAS routines. Both algorithms successfully solved 46 instances within a total time limit of 24 h. We conducted the tests on the Alliance Canada cluster CEDAR (https://alliancecanada.ca).
- **Stopping condition:**
  For both algorithms, we stopped if the number of iterations exceeded 700 or if the relative gap was less than $10^{-6}$. The relative gap is the absolute difference between $c^T x$ and $b^T y$ divided by $1 + |c^T x| + |b^T y|$.
- **Test problems:**
  We have selected 46 test problems of varying sizes from the Netlib collection. (For more information on each test problem, please see Appendix, Table 2). Note that $nnz$ denotes the number of non-zeros elements in matrix $A$.
- **Barrier parameter**
  As for the barrier parameter, we have set the initial value to $\mu_0 = 1$ for both algorithms. In each iteration of the outer loop of the algorithm, we reduce the value of $\mu$ by $\mu = (1 - \theta)\mu$.

– **Proximity function:**
  For our algorithms, we rely on the proximity function specified as follows:

$$\delta(x, s, \mu) = \|\mu e - xs\|_1,$$

where $\|x\|_1 = \sum_{i=1}^n |x_i|$.

– **Threshold parameter:**
  We are currently working on a large-update method, and as referenced in a previous study [12], $\tau$ value should be set to $O(n)$. As a result, we have decided to choose $\tau$ to be equal to $n$ for both the algorithms.

– **Barrier update parameter:**
  In all experiments, we use $\theta = 0.6$ to update $\mu$.

– **Step size:**
  For all of our experiments, we follow a specific strategy for computing the step size in an inner iteration, as described in [3]. To calculate the step sizes $\alpha$ during the algorithm, we use the following equations:

$$\alpha_x^{\max} = \frac{1}{\max_{i=1,2,\cdots,n}\left\{1, -\frac{x_i}{\Delta x_i}\right\}}, \qquad \alpha_s^{\max} = \frac{1}{\max_{i=1,2,\cdots,n}\left\{1, -\frac{s_i}{\Delta s_i}\right\}}.$$

To ensure we don't hit the boundary, we reduce the maximum allowable step sizes by a fixed factor of $0 < \alpha_0 < 1$. Therefore, our final step sizes are given by $\alpha_x = \alpha_0.\alpha_x^{\max}$ and $\alpha_s = \alpha_0.\alpha_s^{\max}$.

### 4.2  Initialization

To find an initial feasible solution, we use the following strategy suggested in [7]. We first compute:

$$\tilde{y} = (AA^T)^{-1}Ac, \quad \tilde{s} = c - A^T y, \qquad \tilde{x} = A^T(AA^T)^{-1}b$$

and set

$$\delta_x = \max\{-1.5 * \min(\tilde{x}_i), 0\}, \quad \delta_s = \max\{-1.5\min *(\tilde{s}_i), 0\} \tag{18}$$

Then, we compute:

$$\tilde{\delta}_x = \delta_x + 0.5 * \frac{(\tilde{x} + \delta_x e)^T(\tilde{s} + \delta_s e)}{\sum_{i=1}^n (\tilde{s}_i + \delta_s)} \tag{19}$$

$$\tilde{\delta}_s = \delta_s + 0.5 * \frac{(\tilde{x} + \delta_x e)^T(\tilde{s} + \delta_s e)}{\sum_{i=1}^n (\tilde{x}_i + \delta_x)} \tag{20}$$

and generate $y^0 = \tilde{y}$ and $s_i^0 = \tilde{s}_i + \tilde{\delta}_s$ for $i = 1, 2, ..., n$ and $x_i^0 = \tilde{x}_i + \tilde{\delta}_x$ for $i = 1, 2, ..., n$.

**Remark.** Let's recall the concerns highlighted by Argyros et al. [1] regarding the use of the two-step Newton method, particularly focusing on the crucial task of choosing the initial point for their algorithm. They provided conditions for selecting this initial point, considering scenarios where the function $F$ behaves as a non-linear function with unbounded first or second derivatives within specific intervals. However, it's important to emphasize that all the problems we're dealing with here are linear in nature. In simpler terms, any feasible point we start with works perfectly well for Algorithm 1. Therefore, we adopt the structure previously described in [7] to decide on the starting point for our algorithm.

## 4.3    Results

**Number of Iterations and CPU Time(s).** Please refer to Table 3 for a detailed summary of the results on select Netlib instances for the two algorithms. The second column lists the names of the test problems, while the third and fourth columns show the number of iterations needed for the classical algorithm and Algorithm 1 respectively. The fifth and sixth columns display the CPU times for the classical algorithm and Algorithm 1.

**Speed-Up and Relative Reduction.** In order to determine the effectiveness of the new proposed method, we analyzed two criteria - the CPU times and the number of iterations achieved by Algorithms 1 and the classical algorithm. We calculated the speed-up and relative reduction for each algorithm on a given instance. The speed-up of Algorithm 1 is defined as the time it takes for the classical algorithm to complete divided by the time it takes for Algorithm 1 to complete. Similarly, the relative reduction in the number of iterations is defined as the number of iterations required by the classical algorithm divided by the number of iterations required by Algorithm 1. These measurements allowed us to accurately compare the performance of the two algorithms and determine the most effective approach.

Figure 1 displays the reduction and speed-up of CPU times and the number of iterations. The instances are ordered according to their size in an array, and the average value of speedup and relative reduction is calculated for each suffix of the array. The x-coordinate is equal to the number of instances, and the y-coordinate shows the average speedup and relative reduction for the 46-$n$ largest instances. Algorithm 1 has an average speedup between 1.25 and 1.17, and an average relative reduction between 1.44 and 1.18 over all 46 instances.

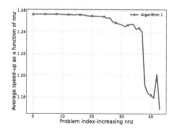

**Fig. 1.** Left: Relative Reduction. Right, Speedup vs. the number of instances.

**Average Number of Iterations and CPU Times.** Based on the obtained results, (see Appendix, Table 3), we have computed the average number of iterations and CPU time for both algorithms. As can be seen in Table 1, the results show that the new proposed approach can significantly reduce the number of iterations and CPU times by %30.97 and %20.46, respectively. This is a promising development that could have implications for future research in this field.

**Table 1.** The average number of iterations and CPU time

| Methods | Aver. Iter. | Aver. CPU |
|---|---|---|
| Classical Algorithm | 95.29 | 132.46 |
| Algorithm 1 | **65.77** | **105.43** |

## 5   Conclusions

We introduce a midpoint approach to tackle linear optimization problems. The approach consists of two steps and employs an auxiliary point to modify the Newton method. In the first step, the auxiliary point is updated, and in the second step, the algorithm utilizes this point to calculate the search direction. We establish the convergence of this novel algorithm. The algorithm is also tested on select LO instances from the Netlib collection. A comparison with a classical method for computing the search direction demonstrated a significant reduction in the number of iterations and CPU time by 30.97% and 20.46% respectively.

## Appendix

The Appendix section contains two tables. Table 2 specifies information related to each test problem, including the name, the number of non-zero elements, and the number of rows and columns of matrix A. Table 3 provides information on the number of iterations and the CPU time for performing Algorithm 1 and the classical algorithm proposed in [12]. Additionally, the averages of CPU time and the number of iterations from Table 3 are presented in Table 1.

**Table 2.** LP instances from NETLIB

| i | Name | m | n | nnz |
|---|------|---|---|-----|
| 1 | adlittle | 56 | 138 | 424 |
| 2 | afiro | 27 | 51 | 102 |
| 3 | agg | 488 | 615 | 2,862 |
| 4 | agg2 | 516 | 758 | 4,740 |
| 5 | agg3 | 516 | 758 | 4,756 |
| 6 | bandm | 305 | 472 | 2,494 |
| 7 | beaconfd | 173 | 295 | 3,408 |
| 8 | blend | 74 | 114 | 522 |
| 9 | bnl2 | 2,324 | 4,486 | 14,996 |
| 10 | capri | 271 | 482 | 1,896 |
| 11 | chemcom | 288 | 744 | 1,590 |
| 12 | czprob | 929 | 3,562 | 10,708 |
| 13 | e226 | 223 | 472 | 2,768 |
| 14 | fffff8000 | 524 | 1028 | 6401 |
| 15 | fit1p | 627 | 1,677 | 9,868 |
| 16 | forest6 | 66 | 131 | 246 |
| 17 | ganges | 1,309 | 1,706 | 6,937 |
| 18 | gfrd_pnc | 616 | 1,160 | 2,445 |
| 19 | israel | 174 | 316 | 2443 |
| 20 | lotfi | 153 | 366 | 1136 |
| 21 | pilot_we | 722 | 2,928 | 9,265 |
| 22 | sc50a | 50 | 78 | 160 |
| 23 | sc50b | 50 | 78 | 148 |
| 24 | sc105 | 105 | 163 | 340 |
| 25 | sc205 | 205 | 317 | 665 |
| 26 | scagr25 | 471 | 671 | 1,725 |
| 27 | scfxm1 | 330 | 600 | 2,732 |
| 28 | scfxm2 | 660 | 1,200 | 5,469 |
| 29 | scfxm3 | 990 | 1,800 | 8,206 |
| 30 | scrs8 | 490 | 1,275 | 3,288 |
| 31 | scagr7 | 129 | 185 | 465 |
| 32 | scsd1 | 77 | 760 | 2388 |
| 33 | scsd6 | 147 | 1,350 | 4,316 |
| 34 | scsd8 | 397 | 2,750 | 8,584 |
| 35 | sctap1 | 300 | 660 | 1,872 |
| 36 | sctap2 | 1,090 | 2,500 | 7,334 |
| 37 | sctap3 | 1,480 | 3,340 | 9,734 |
| 38 | share1b | 117 | 253 | 1179 |
| 39 | share2b | 96 | 162 | 777 |
| 40 | stair | 356 | 614 | 4,003 |
| 41 | standata | 359 | 1274 | 3230 |
| 42 | standmps | 467 | 1,274 | 3,878 |
| 43 | stocfor1 | 117 | 165 | 501 |
| 44 | stocfor2 | 2,157 | 3,045 | 9,357 |
| 45 | truss | 1,000 | 8,806 | 27,836 |
| 46 | lp-vtp-base | 198 | 346 | 1051 |

**Table 3.** Number of iterations and CPU times for classical algorithm and Algorithm 1.

| $i$ | Name | Number of iterations | | Time(s) | |
|---|---|---|---|---|---|
| | | Classical Algorithm | Algorithm 1 | Classical Algorithm | Algorithm 1 |
| 1 | adlittle | 53 | 41 | 0.251 | 0.263 |
| 2 | afiro | 10 | 12 | 0.020 | 0.118 |
| 3 | agg | 91 | 64 | 7.540 | 5.727 |
| 4 | agg2 | 95 | 52 | 12.595 | 7.521 |
| 5 | agg3 | 124 | 52 | 16.229 | 6.985 |
| 6 | bandm | 106 | 73 | 4.497 | 3.305 |
| 7 | beaconfd | 91 | 50 | 1.290 | 1.972 |
| 8 | blend | 61 | 24 | 0.340 | 0.184 |
| 9 | bnl2 | 205 | 166 | 1346.310 | 1086.060 |
| 10 | capri | 88 | 56 | 3.642 | 2.518 |
| 11 | chemcom | 65 | 41 | 6.288 | 3.715 |
| 12 | czprob | 174 | 161 | 8531.635 | 495.363 |
| 13 | e226 | 122 | 77 | 4.557 | 3.178 |
| 14 | ffff8000 | 130 | 109 | 28.523 | 25.609 |
| 15 | fit1p | 53 | 37 | 38.135 | 25.724 |
| 16 | forest6 | 43 | 24 | 0.221 | 0.150 |
| 17 | ganges | 51 | 41 | 41.833 | 33.589 |
| 18 | gfrd_pnc | 700 | 43 | 38.531 | 12.079 |
| 19 | israel | 170 | 95 | 5.896 | 3.000 |
| 20 | lotfi | 700 | 54 | 18.409 | 1.340 |
| 21 | pilot_we | 393 | 238 | 742.771 | 448.402 |
| 22 | sc50a | 22 | 20 | 0.073 | 0.111 |
| 23 | sc50b | 28 | 17 | 0.150 | 0.125 |
| 24 | sc105 | 33 | 24 | 0.346 | 0.510 |
| 25 | sc205 | 43 | 25 | 2.225 | 0.543 |
| 26 | scagr7 | 35 | 21 | 0.418 | 0.438 |
| 27 | scagr25 | 43 | 50 | 5.720 | 6.048 |
| 28 | scfxm1 | 84 | 53 | 5.426 | 3.908 |
| 29 | scfxm2 | 78 | 68 | 24.357 | 24.037 |
| 30 | scfxm3 | 86 | 72 | 63.983 | 54.576 |
| 31 | scrs8 | 155 | 108 | 51.426 | 35.941 |
| 32 | scsd1 | 34 | 19 | 5.217 | 2.000 |
| 33 | scsd6 | 47 | 29 | 16.232 | 9.223 |
| 34 | scsd8 | 39 | 24 | 53.510 | 32.258 |
| 35 | sctap1 | 77 | 49 | 5.619 | 3.823 |
| 36 | sctap2 | 85 | 58 | 127.099 | 86.693 |
| 37 | sctap3 | 104 | 86 | 329.885 | 274.156 |
| 38 | share1b | 127 | 95 | 2.703 | 2.200 |
| 39 | share2b | 34 | 30 | 0.405 | 0.467 |
| 40 | stair | 144 | 102 | 11.808 | 7.268 |
| 41 | standata | 118 | 85 | 37.289 | 28.812 |
| 42 | standmps | 214 | 127 | 72.569 | 43.784 |
| 43 | stocfor1 | 91 | 45 | 1.146 | 1.037 |
| 44 | stocfor2 | 217 | 177 | 684.794 | 556.632 |
| 45 | truss | 64 | 55 | 1531.321 | 1309.819 |
| 46 | lp-vtp-base | 66 | 42 | 2.1741 | 1.408 |

# References

1. Argyros, I.K., Deep, G., Regmi, S.: Extended Newton-like midpoint method for solving equations in Banach space. Foundations **3**(1), 82–98 (2023)
2. Bai, Y.Q., El Ghami, M., Roos, C.: A comparative study of kernel functions for primal-dual interior-point algorithms in linear optimization. SIAM J. Optim. **15**(1), 101–128 (2004)
3. Cai, X., Wang, G., Zhang, Z.: Complexity analysis and numerical implementation of primal-dual interior-point methods for convex quadratic optimization based on a finite barrier. Numer. Algorithms **62**(2), 289–306 (2013)
4. Fathi-Hafshejani, S., Fakharzadeh Jahromi, A., Peyghami, M.R.: A unified complexity analysis of interior point methods for semidefinite problems based on trigonometric kernel functions. Optimization **67**(1), 113–137 (2018)
5. Fathi-Hafshejani, S., Mansouri, H., Reza Peyghami, M., Chen, S.: Primal-dual interior-point method for linear optimization based on a kernel function with trigonometric growth term. Optimization **67**(10), 1605–1630 (2018)
6. McDougall, T.J., Wotherspoon, S.J.: A simple modification of Newton's method to achieve convergence of order $1 + \sqrt{2}$. Appl. Math. Lett. **29**, 20–25 (2014)
7. Mehrotra, S.: On the implementation of a primal-dual interior point method. SIAM J. Optim. **2**(4), 575–601 (1992)
8. Monteiro, R.D., Adler, I.: Interior path following primal-dual algorithms. Part I: Linear programming. Math. Program. **44**(1–3), 27–41 (1989)
9. Roos, C.: A full-newton step $O(n)$ infeasible interior-point algorithm for linear optimization. SIAM J. Optim. **16**(4), 1110–1136 (2006)
10. Roos, C., Terlaky, T., Vial, J.P.: Theory and Algorithms for Linear Optimization: An Interior Point Approach. Wiley, Chichester (1997)
11. Sonnevend, G.: An "analytical centre" for polyhedrons and new classes of global algorithms for linear (smooth, convex) programming. In: Prékopa, A., Szelezsáan, J., Strazicky, B. (eds.) System Modelling and Optimization. LNCIS, vol. 84, pp. 866–875. Springer, Heidelberg (2006). https://doi.org/10.1007/BFb0043914
12. Terlaky, T., Roos, C., Peng, J.: Self-regularity: A New Paradigm for Primal-Dual Interior-Point Algorithms (Princeton Series in Applied Mathematics). Princeton University Press, Princeton (2002)

# Unique Least Common Ancestors
# and Clusters in Directed Acyclic Graphs

Ameera Vaheeda Shanavas[1] , Manoj Changat[1]([⊠]) , Marc Hellmuth[2] ,
and Peter F. Stadler[3,4,5,6,7]

[1] Department of Futures Studies, University of Kerala, Trivandrum 695 581, India
ameerasv@gmail.com, mchangat@keralauniversity.ac.in
[2] Department of Mathematics, Faculty of Science, Stockholm University,
10691 Stockholm, Sweden
marc.hellmuth@math.su.se
[3] Bioinformatics Group, Department of Computer Science and Interdisciplinary
Center for Bioinformatics, Universität Leipzig, Härtelstrasse 16-18, 04107 Leipzig,
Germany
studla@bioinf.uni-leipzig.de
[4] Max Planck Institute for Mathematics in the Sciences, Leipzig, Germany
[5] Institute for Theoretical Chemistry, University of Vienna, Vienna, Austria
[6] Faculty of Ciencias, Universidad Nacional de Colombia, Bogotá, Colombia
[7] Santa Fe Institute, Santa Fe, NM, USA

**Abstract.** We investigate the connections between clusters and least
common ancestors (LCAs) in directed acyclic graphs (DAGs). We focus
on the class of DAGs having unique least common ancestors for certain
subsets of their minimal elements since these are of interest, particularly
as models of phylogenetic networks. Here, we use the close connection
between the canonical $k$-ary transit function and the closure function on
a set system to show that pre-$k$-ary clustering systems are exactly those
that derive from a class of DAGs with unique LCAs. Moreover, we show
that $k$-ary $\mathscr{T}$-systems and $k$-weak hierarchies are associated with DAGs
that satisfy stronger conditions on the existence of unique LCAs for sets
of size at most $k$.

**Keywords:** Monotone transit function · closure function · clustering
system · $k$-weak hierarchy

## 1  Introduction

Directed acyclic graphs (DAGs) play an increasing role in mathematical phylogenetics as models of more complex evolutionary relationships that are not
adequately represented by rooted trees. The set $X$ of minimal vertices of a DAG
$G = (V, E)$ corresponds to the extant taxa and thus generalizes the leaf set of
a phylogenetic tree. Inner vertices $u \in V$ are interpreted as ancestral states and
are naturally associated with the sets $C(u)$ of the descendant genes. These sets

S. Kalyanasundaram and A. Maheshwari (Eds.): CALDAM 2024, LNCS 14508, pp. 148–161, 2024.
https://doi.org/10.1007/978-3-031-52213-0_11

are often called the "hardwired clusters" [11,12]. A least common ancestor of a set $A$ of taxa is a minimal vertex $v$ in $G$ such that $A \subseteq \mathsf{C}(v)$, i.e., all taxa in $A$ are descendants of $v$. In phylogenetics, least common ancestors play a key role in understanding evolutionary relationships and processes. The clusters of $G$, on the other hand, are often accessible from data. Basic relations between clustering systems of (rooted) DAGs and the uniqueness of least common ancestors were explored recently in [10]. Here, we elaborate further on this theme, making use in particular of the fact that the canonical transit function of set systems is a restriction of the closure function to small spanning sets. In particular, we are interested here in characterizing DAGs in which least common ancestors of certain sets of "leaves" are uniquely defined.

Section 2 contains basic definitions, some useful properties of DAGs, and a characterization of $k$-weak hierarchies. Section 3 is about lca and $k$-lca-property, and the connection of the latter with the pre-$k$-ary clustering systems. Section 4 discusses the correspondence of the strict and the strong $k$-lca properties to the $k$-ary $\mathscr{T}$-systems and the $k$-weak hierarchies, respectively.

## 2    Background and Preliminaries

***Transit Functions and k-ary Transit Functions.*** Let $X$ be a non-empty, finite set. We write $X^k$ for the $k$-fold Cartesian set product of $X$ and $X^{(k)}$ for the set of all non-empty subsets of $X$ with cardinality at most $k$.

Following [6], a *$k$-ary transit function* on $X$ is a function $R : X^k \mapsto 2^X$ satisfying the axioms

(**t1**)  $u_1 \in R(u_1, u_2, \ldots, u_k)$;
(**t2**)  $R(u_1, u_2, \ldots, u_k) = R(\pi(u_1, u_2, \ldots, u_k))$ for all $u_i \in X$ and all permutations $\pi$ of $(u_1, u_2, \ldots, u_k)$;
(**t3**)  $R(u, u, \ldots, u) = \{u\}$ for all $u \in X$.

The "symmetry" axiom (**t2**) allows us to interpret a $k$-ary transit function also as a function over subsets $U \in X^{(k)}$. Then axiom (**t2**) becomes void, (**t3**) becomes $R(\{x\}) = \{x\}$ for all $x \in X$, and condition (**t1**) reads "$u \in U$ implies $u \in R(U)$ for all $U \in X^{(k)}$".

Given a $k$-ary transit function $R$ on $X$, we denote its system of *transit sets* by $\mathscr{C}_R := \{R(U) \mid U \in X^{(k)}\}$. A set system $\mathscr{C} \subset 2^X$ *is identified* by a $k$-ary transit function if $\mathscr{C} = \mathscr{C}_{R_\mathscr{C}}$ where $R_\mathscr{C} : X^{(k)} \to 2^X$ defined by $R_\mathscr{C}(U) := \bigcap\{C \in \mathscr{C} \mid U \subseteq C\}$ for all $U \in X^{(k)}$. As shown in [3,7], a system of non-empty sets $\mathscr{C} \subset 2^X$ is identified by a $k$-ary transit function, $k \geq 2$, if and only if $\mathscr{C}$ is a *($k$-ary) $\mathscr{T}$-system*, satisfying the following three axioms

(**KS**)  $\{x\} \in \mathscr{C}$ for all $x \in X$
(**KR**)  For all $C \in \mathscr{C}$ there is a set $T \subseteq C$ with $|T| \leq k$ such that $T \subseteq C'$ implies $C \subseteq C'$ for all $C' \in \mathscr{C}$.
(**KC**)  For every $U \subseteq X$ with $|U| \leq k$ holds $\bigcap\{C \in \mathscr{C} \mid U \subseteq C\} \in \mathscr{C}$.

Conversely, a $k$-ary transit function $R$ identifies a set system if and only if it satisfies the *monotone* axiom

**(m)** For every $w_1, \ldots, w_k \in R(u_1, \ldots, u_k)$ holds $R(w_1, \ldots, w_k) \subseteq R(u_1, \ldots, u_k)$.

That is, for all $U, W \in X^{(k)}$ holds: $W \subseteq R(U)$ implies $R(W) \subseteq R(U)$. The correspondence of monotone $k$-ary transit functions and $k$-ary $\mathscr{T}$-systems is then mediated by the *canonical transit function* $R_{\mathscr{C}}$ and $\mathscr{C}_R$, respectively.

For a general set system $\mathscr{C}$ on $X$, the *closure* function cl $: 2^X \rightarrow 2^X$, sometimes also called the "convex hull", is defined as $\mathrm{cl}(A) := \bigcap \{ C \in \mathscr{C} \mid A \subseteq C \}$ for all $A \in 2^X$. The canonical $k$-ary transit function $R_{\mathscr{C}}$ of a set system is the restriction of its closure function to small sets as arguments: $R_{\mathscr{C}}(U) = \mathrm{cl}(U)$ for all non-empty sets $U$ with $|U| \leq k$.

A set system is *closed* if for all non-empty set $A \in 2^X$ holds $A \in \mathscr{C} \iff \mathrm{cl}(A) = A$. By [10, L. 16], this is equivalent to the condition that for all $A, B \in \mathscr{C}$ with $A \cap B \neq \emptyset$ we have $A \cap B \in \mathscr{C}$, i.e., $\mathscr{C}$ is closed under pairwise intersection. A set system $\mathscr{C}$ consisting of *non-empty* subsets of $X$ is called a *clustering system* if it satisfies **(KS)** and **(K1)**: $X \in \mathscr{C}$. Note that axiom **(KS)** translates to $R_{\mathscr{C}}$ satisfying **(t3)**. A $k$-ary $\mathscr{T}$-system is thus a clustering system if and only it satisfies **(K1)** or, equivalently [7], if its canonical transit function $R = R_{\mathscr{C}}$ satisfies

**(a')** there is $U \in X^{(k)}$ such that $R(U) = X$.

A set system is called *pre-$k$-ary* if it satisfies **(KC)** for a given parameter $k$. The 2-ary case has received considerable attention in the literature for the special case of clustering systems, see [3]. A 2-ary transit function is called a *transit function*. A clustering system $\mathscr{C}$ is called pre-binary in [3] if **(KC)** with $k = 2$ is satisfied, i.e., if $R_{\mathscr{C}}(x, y) \in \mathscr{C}$ for all $x, y \in X$, and *binary* if in addition **(KR)** holds with $k = 2$. Binary clustering systems are therefore identified by monotone (2-ary) transit function satisfying **(a')** with $k = 2$; that is, there is $p, q \in X$ such that $R(p, q) = X$.

***Weak and k-Weak Hierarchies.*** Generalizations of hierarchies are important in the clustering literature. Recall that a clustering system $\mathscr{C}$ is a

**weak hierarchy** if for any three sets $A, B, C \in \mathscr{C}$ holds $A \cap B \cap C \in \{ A \cap B, A \cap C, B \cap C \}$ [1];

**k-weak hierarchy** if for any $k + 1$ sets $A_1, A_2, \ldots, A_{k+1} \in \mathscr{C}$ there is $1 \leq j \leq k + 1$ such that $\bigcap\limits_{i=1}^{k+1} A_i = \bigcap\limits_{i=1, i \neq j}^{k+1} A_i$ [2].

We write $A \between B$ if $A \cap B \notin \{A, B, \emptyset\}$ and say that $A$ and $B$ overlap. It is well known that weak hierarchy = 2-weak hierarchy $\implies$ $k$-weak hierarchy $\implies$ $(k+1)$-weak hierarchy for all $k \geq 3$. As outlined in [9], weak hierarchies always satisfy **(KR)** for $k = 2$. More generally, Lemma 6.3 of [7] ensures that $k$-weak hierarchies satisfy **(KR)** for the parameter $k$. For weak hierarchies, furthermore, axiom **(KC)** with $k = 2$ is equivalent to $\mathscr{C}$ being closed under pairwise intersection.

The characterization of $k$-weak hierarchies by condition (kW') in [5], together with the fact that every $k$-weak hierarchy is also a $k'$-weak hierarchy for all $k' \geq k$, can be rephrased as follows:

**Observation 1.** *A set system $\mathscr{C}$ is a $k$-weak hierarchy if and only if for every $A \in 2^X$ with $|A| > k$ there is $z \in A$ such that $z \in \mathrm{cl}(A \setminus \{z\})$.*

For our purposes, the following characterization of $k$-weak hierarchies in terms of their closure functions will be particularly useful:

**Proposition 1.** *A set system $\mathscr{C}$ on $X$ is a $k$-weak hierarchy if and only if for every $\emptyset \neq A \subseteq X$ there exists $U \subseteq A$ with $|U| \leq k$ such that $\mathrm{cl}(A) = \mathrm{cl}(U)$.*

*Proof.* First, assume that $\mathscr{C}$ is a $k$-weak hierarchy. If $|A| \leq k$, then $A = U$ trivially satisfies $\mathrm{cl}(A) = \mathrm{cl}(U)$. Hence, assume $|A| > k$. By Observation 1, there is $z \in A$ such that $z \in \mathrm{cl}(A \setminus \{z\})$, which implies $A \subseteq \mathrm{cl}(A \setminus \{z\})$. Together with isotony and idempotency of the closure function, we obtain

$$\mathrm{cl}(A \setminus \{z\}) \subseteq \mathrm{cl}(A) \subseteq \mathrm{cl}(\mathrm{cl}(A \setminus \{z\})) = \mathrm{cl}(A \setminus \{z\}).$$

Thus, there is $z \in A$ such that $\mathrm{cl}(A) = \mathrm{cl}(A \setminus \{z\})$. Repeating this argument for $A' := A \setminus \{z\}$, we observe that we can stepwise remove elements of $A$ while preserving $\mathrm{cl}(A)$ until we arrive at a residual set $U \subset A$ with $|U| = k$ that still satisfies $\mathrm{cl}(U) = \mathrm{cl}(A)$.

Now assume that $\mathscr{C}$ is *not* a $k$-weak hierarchy. Hence, there are $k + 1$ sets $A_1, A_2, \ldots, A_{k+1} \in \mathscr{C}$ such that for all $1 \leq j \leq k + 1$ it holds that $\cap_{i=1}^{k+1} A_i \subsetneq \cap_{i=1, i \neq j}^{k+1} A_i$. Thus, there are $k + 1$ (distinct) elements $x_1, \ldots, x_{k+1} \in X$ such that $x_i \in A_j$ if and only $i \neq j$. Set $A = \{x_1, x_2, \ldots, x_{k+1}\}$ and consider any subset $U \subset A$ with $|U| \leq k$. Then there is at least one set $A_h$, $1 \leq h \leq k + 1$, such that $U \subseteq A_h$. By the previous arguments, $x_h \notin A_h$. Since $A_h \in \mathscr{C}$, we have $\mathrm{cl}(U) \subseteq A_h$, and thus $x_h \notin \mathrm{cl}(U)$. Since $x_h \in A$ and $A \subseteq \mathrm{cl}(A)$, we have $x_h \in \mathrm{cl}(A)$ and, thus, $\mathrm{cl}(U) \neq \mathrm{cl}(A)$. $\qquad \square$

*Clusters*, **LCA** *and* lca *in DAGs.* Let $G$ be a directed acyclic graph (DAG) with an associated partial order $\preceq$ on its vertex set $V(G)$ defined by $v \preceq_G w$ if and only if there is a directed path from $w$ to $v$. In this case, we say that $w$ is an ancestor of $v$ and $v$ is a descendant of $w$. If the context is clear, we may drop the subscript and write $\preceq$. Two vertices $u, v \in V(G)$ are *incomparable* if neither $u \preceq v$ nor $v \preceq u$ is true. We denote by $X = L(G) \subseteq V(G)$ the $\preceq$-minimal vertices of $G$, and we call $x \in X$ a leaf of $G$. For every $v \in V(G)$, the set of its descendant leaves

$$\mathsf{C}(v) := \{x \in X \mid x \preceq v\} \tag{1}$$

is a cluster of $G$. We write $\mathscr{C}_G := \{\mathsf{C}(v) \mid v \in V(G)\}$. By construction, $\mathsf{C}(x) = \{x\}$ for $x \in X$, hence $\mathscr{C}_G$ satisfies (KS). For $v \in V(G)$, we write $\mathrm{Anc}(v) = \{w \in V(G) \mid v \preceq w\}$ for the ancestors of $v$. For every leaf $x \in X$, we have $\mathrm{Anc}(x) = \{v \mid x \in \mathsf{C}(v)\}$. We write $\mathrm{Anc}(Y) := \bigcap_{w \in Y} \mathrm{Anc}(w)$ for the set of common ancestors of all $w \in Y$. In general, not every set $Y \subseteq V$ has a common ancestor in a DAG: Consider the DAG with three leaves $\{x, y, z\}$, two maximal vertices $\{p, q\}$, $C(p) = \{x, y\}$, and $C(q) = \{x, z\}$. Then $\text{Anc}(\{y, z\}) = \emptyset$. A *(rooted) network* $G$ is a DAG such that there is a unique vertex $\rho \in V(G)$, called the root, with indegree 0. In a network, we have $x \preceq \rho$ for all $x \in V(G)$ and, thus, in particular, $C(\rho) = X$, i.e., $X \in \mathscr{C}_G$, and thus $\mathscr{C}_G$, satisfying **(K1)**, is a clustering system.

**Definition 1.** *[4] Let $G$ be a DAG. A vertex $w \in V(G)$ is a* least common ancestor *(LCA) of $Y \subseteq V(G)$ if it is a $\preceq$-minimal element in $\text{Anc}(Y)$. The set $\text{LCA}(Y)$ comprises all LCAs of $Y$ in $G$.*

An LCA of $Y$ thus is an ancestor of all vertices in $Y$ that is $\preceq$-minimal w.r.t. this property. Clearly, $\text{LCA}(\{v\}) = \{v\}$ for all $v \in V(G)$ and $\text{LCA}(Y) = \emptyset$ if and only if $\text{Anc}(Y) = \emptyset$. In a network, the root vertex is a common ancestor for any set of vertices, and thus $\text{LCA}(Y) \neq \emptyset$.

We will, in particular, be interested in situations where the LCA of certain sets of leaves is uniquely defined. More precisely, we are interested in DAGs where $|\text{LCA}(Y)| = 1$ holds for certain subsets $Y \subseteq X$; the most obvious examples are DAGs that satisfy the *2-lca-property* (also known as the *pairwise lca-property* [10]), i.e., for every pair of leaves $x, y \in L(G)$ there is a unique least common ancestor $\text{lca}(x, y)$. For simplicity, we will write $\text{lca}(Y) = q$ instead of $\text{LCA}(Y) = \{q\}$ whenever $|\text{LCA}(Y)| = 1$ and say that $\text{lca}(Y)$ *is defined*; otherwise, we leave $\text{lca}(Y)$ *undefined*.

The following result for networks [10, L. 17] remains valid for all DAGs.

**Lemma 1.** *Let $G$ be a DAG. Then $v \preceq_G w$ implies $C(v) \subseteq C(w)$ for all $v, w \in V(G)$.*

Consequently, [10, Obs. 12 & 13] also hold for DAGs in general:

**Observation 2.** *Let $G$ be a DAG with leaf set $X$, $\emptyset \neq A \subseteq X$, and suppose $\text{lca}(A)$ is defined. Then the following is satisfied:*

*(i) $\text{lca}(A) \preceq_G v$ for all $v$ with $A \subseteq C(v)$.*
*(ii) $C(\text{lca}(A))$ is the unique inclusion-minimal cluster in $\mathscr{C}_G$ containing $A$.*
*(iii) $\text{lca}(C(\text{lca}(A))) = \text{lca}(A)$.*

Note that the existence of $\text{lca}(A)$ for all $A \subseteq X$ does not imply that $G$ is a network since we could expand any network "upward" for $\rho$ by attaching an arbitrary DAG that has $\rho$ as its unique leaf. Clearly, the vertices "above" $\rho$ cannot be least common ancestors of any leaves.

Consider a set system $\mathscr{Q} \subseteq 2^X$. Then the Hasse diagram $\mathfrak{H}(\mathscr{Q})$ is the DAG with vertex set $\mathscr{Q}$ and directed edges from $A \in \mathscr{Q}$ to $B \in \mathscr{Q}$ if (i) $B \subsetneq A$ and (ii) there is no $C \in \mathscr{Q}$ with $B \subsetneq C \subsetneq A$. As we shall see later, Hasse diagrams are of interest here because they guarantee "well-behaved" least common ancestors.

The correspondence between Hasse diagrams that are networks and $k$-ary transit functions is summarized in the following:

**Lemma 2.** *Let $R$ be a $k$-ary transit function. Then $\mathfrak{H}(\mathscr{C}_R)$ is a network if and only if $R$ satisfies **(a')** for $k$.*

*Proof.* If $N := \mathfrak{H}(\mathscr{C}_R)$ is a network, it contains a unique vertex $\rho$ with indegree 0; the root of $N$. Since $R$ satisfies **(t3)**, all singletons $\{x\}$ with $x \in X$ are contained as vertices of $N$. Since $N$ has a unique root, it follows that $X$ is a vertex of $N$ and, in particular, $C(\rho) = X \in \mathscr{C}_R$. This implies that there must be a subset $U \in X^{(k)}$ such that $R(U) = X$. Hence, $R$ satisfies **(a')**. Conversely, if **(a')** with parameter $k$ holds, there is a subset $U \in X^{(k)}$ with $R(U) = X$ and thus $X \in \mathscr{C}_R$. Let $v_X$ be the vertex in $\mathfrak{H}(\mathscr{C}_R)$ for which $C(v_X) = X$ holds. Since $v \preceq v_X$ for every vertex $v$ in $\mathfrak{H}(\mathscr{C}_R)$, this is, in particular, true for the singletons, and thus $v_X$ serves as the unique root of $\mathfrak{H}(\mathscr{C}_R)$. □

Following [10], we say that a DAG $G = (V, E)$ has the *path-cluster-comparability* **(PCC)** property if it satisfies, for all $u, v \in V$: $u$ and $v$ are $\preceq_G$-comparable if and only if $C(u) \subseteq C(v)$ or $C(v) \subseteq C(u)$. By [10, Cor. 11 & Prop. 3], the Hasse diagram $G$ of a clustering system $\mathscr{C}$ satisfies **(PCC)** and [10, Prop. 2] implies that $\mathscr{C}_G = \mathscr{C}$.

# 3  DAGs with lca- and $k$-lca-Property

In the following, we consider the generalization of lca-networks introduced in [10] for arbitrary (not necessarily rooted) DAGs.

**Definition 2.** *A DAG with leaf set $X$ has the* lca-*property if* $\mathrm{lca}(A)$ *is defined for all non-empty $A \subseteq X$.*

By definition, every DAG with the lca-property also has the pairwise lca-property. The converse is, in general, not satisfied. An example of a network (rooted DAG) that satisfies the pairwise lca-property but that is not an lca-network, can be found in [10, Fig. 13(A)].

**Lemma 3.** *If a DAG $G$ has the* lca-*property, then its clustering system $\mathscr{C}_G$ is closed.*

*Proof.* To show that $\mathscr{C}_G$ is closed, we use the equivalent condition that $\mathscr{C}_G$ is closed under pairwise intersection. Thus, let $C(u), C(v) \in \mathscr{C}_G$ for some $u, v \in V(G)$. If $C(u) \subseteq C(v)$, $C(v) \subseteq C(u)$ or $C(u) \cap C(v) = \emptyset$, there is nothing to show. Hence, assume that $C(u) \between C(v)$ and set $A := C(u) \cap C(v) \neq \emptyset$. Since $G$ has the lca-property, there is $w \in V(G)$ such that $w = \mathrm{lca}(A)$, and thus $A \subseteq C(w)$. The contraposition of Lemma 1 shows that $u$ and $v$ are two incomparable common ancestors of $A$. Since $w$ is the unique $\preceq$-minimal common ancestor of $A$, we have $w \preceq u$ and $w \preceq v$, which together with Lemma 1 implies $C(w) \subseteq C(u)$ and $C(w) \subseteq C(v)$. Therefore $C(w) \subseteq A$. Hence $A = C(w) \in \mathscr{C}_G$ and thus, $\mathscr{C}_G$ is closed. □

The converse of Lemma 3 is not true. A counter-example can be found in Fig. 1. The following connection between the clusters, the least common ancestors, and the closure function will be useful in the remainder of this contribution:

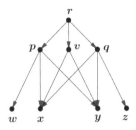

**Fig. 1.** The cluster system $\mathscr{C}_G = \big\{\{w\},\{x\},\{y\},\{x,y\},\{w,x,y\},\{x,y,z\},\{w,x,y,z\}\big\}$ of the network $G$ is closed and satisfies **(KC)** for every $k \in \{1,2,3,4\}$. However, we have $\mathrm{LCA}(\{x,y\}) = \{p,v,q\}$ and thus $G$ does not have the pairwise lca-property. By Proposition 2, if $G$ has the pairwise lca-property, then $\mathscr{C}_G$ is pre-binary. The example in this Figure shows that the converse is not true. In particular, the equivalence between pre-$k$-ary and the $k$-lca-property in Proposition 3 requires **(PCC)**, which is not satisfied by $G$.

**Observation 3.** *If $G$ is a DAG with leaf set $X$ and the* lca-*property, then* $\mathsf{C}(\mathrm{lca}(Y)) = \mathrm{cl}(Y)$ *for all $\emptyset \neq Y \subseteq X$.*

*Proof.* The argument follows the proof of [10, L. 41], observing that [10, L.17] remains true for arbitrary DAGs, and substituting Observation 2(i) and Lemma 3 for [10, Cor. 18] and [10, P. 11], respectively. □

The observations above can be extended to networks where more least common ancestors exist and are unique for all leaf sets of size at most $k$. Naturally, we start from the cluster system $\mathscr{C} := \mathscr{C}_G := \{\mathsf{C}(v); v \in V(G)\}$ and consider the map $R_{\mathscr{C}} : X^k \to 2^X$ defined by

$$R_{\mathscr{C}}(u_1, u_2, \ldots, u_k) := \bigcap\{\mathsf{C}(v) \mid v \in V(G), u_1, u_2, \ldots, u_k \in \mathsf{C}(v)\}$$

One easily verifies that $R_{\mathscr{C}}$ satisfies **(t2)**, and thus, we can again interpret $R_{\mathscr{C}}$ as a function over sets, in which case we have $R_{\mathscr{C}}(U) = \mathrm{cl}(U)$ for all $U \in X^{(k)}$. In this setting, we are interested in cases where $\mathrm{lca}(U)$ is defined at least for all sets of cardinality $|U| \leq k$. We formalize this idea in

**Definition 3.** *A DAG $G$ with leaf set $X$ has the $k$-lca-property if $\mathrm{lca}(A)$ is defined for all $A \in X^{(k)}$.*

Now, we define the $k$-ary map $R_G : X^k \to 2^X$ by $R_G(u_1, \ldots, u_k) := \mathsf{C}(\mathrm{lca}(u_1, \ldots, u_k))$; in set notation this reads $R_G(U) = \mathsf{C}(\mathrm{lca}(U))$ for all $U \in X^{(k)}$.

**Proposition 2.** *Let $G$ be a DAG with $k$-lca-property. Then $R_G$ is a monotone, $k$-ary transit function that satisfies $R_G = R_{\mathscr{C}_G}$. Moreover, $\mathscr{C}_G$ is pre-$k$-ary.*

*Proof.* Let $G$ be a DAG with $k$-lca-property and leaf set $X$. It follows directly from the definition and uniqueness of $\mathrm{lca}(U)$ for $U \in X^{(k)}$ that $R_G$ satisfies **(t1)**, **(t2)**, and **(t3)**, i.e., $R_G$ is a $k$-ary transit function. If $u_1, \ldots, u_k \in$

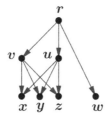

**Fig. 2.** Consider the DAG $G$ with leaf set $X = L(G)$ where $\mathscr{C}_G = \{\{x\}, \{y\}, \{z\}, \{w\}, \{x, y, z\}, X\}$. Here, $\mathscr{C}_G$ satisfies **(KS)** and **(KC)** for $k = 2$. By definition, $\mathscr{C}_G$ is thus pre-binary. However, $G$ is not a pairwise lca-network since $\mathrm{lca}(x, y)$, $\mathrm{lca}(x, z)$, and $\mathrm{lca}(y, z)$ are not defined. Moreover, $\mathscr{C}_G$ also satisfies **(KC)** for $k = 3$ but $G$ is not a $3$-lca-network since $\mathrm{lca}(x, y, z)$ is not defined.

$R_G(x_1, \ldots, x_k) = \mathtt{C}(\mathrm{lca}(x_1, \ldots, x_k))$, then $\{u_1, \ldots, u_k\} \subseteq \mathtt{C}(\mathrm{lca}(x_1, \ldots, x_k))$ and we can apply Observation 2(i) to conclude that $\mathrm{lca}(u_1, \ldots, u_k) \preceq \mathrm{lca}(x_1, \ldots, x_k)$. Applying Lemma 1 yields $\mathtt{C}(\mathrm{lca}(u_1, \ldots, u_k)) \subseteq \mathtt{C}(\mathrm{lca}(x_1, \ldots, x_k))$, and thus $R_G(u_1, \ldots, u_k) \subseteq R_G(x_1, \ldots, x_k)$, i.e., $R_G$ is monotone. It follows from Observation 2(ii) that $\mathtt{C}(\mathrm{lca}(U))$ is the unique inclusion minimal cluster in $\mathscr{C}_G$ containing $U$, i.e., $\mathrm{cl}(U) = \mathtt{C}(\mathrm{lca}(U))$ for all $U \in X^{(k)}$. Consequently, $R_G = R_{\mathscr{C}_G}$.

Since $G$ has the $k$-lca-property, $\mathrm{lca}(U)$ is defined for all $U \in X^{(k)}$. Thus $\mathtt{C}(\mathrm{lca}(U))$ is the unique inclusion minimal cluster in $\mathscr{C}_G$ containing $U$ for all $U \in X^{(k)}$ by Observation 2(ii); hence $\mathscr{C}_G$ satisfies **(KC)** for $k$, i.e., $\mathscr{C}_G$ is pre-$k$-ary. □

Note that a DAG $G$ for which $\mathscr{C}_G$ is pre-$k$-ary does not necessarily have the $k$-lca-property, see Fig. 2 and Fig. 1 for a counter-example. Moreover, **(KR)** with parameter $k$ is not necessarily satisfied since the $k$-lca-property does not claim the existence of clusters that are not associated with least common ancestors of a set $U \in X^{(k)}$. Therefore, $R_G$ need not identify $\mathscr{C}_G$. At least for an important subclass of DAGs there is a simple correspondence between the uniqueness of LCAs and a property of the $\mathscr{T}$-system.

**Proposition 3.** *Let $G$ be a DAG that satisfies* **(PCC)**. *Then, $G$ satisfies the $k$-lca-property if and only if $\mathscr{C}_G$ is pre-$k$-ary.*

*Proof.* Suppose that $G$ is a DAG with leaf set $X$ that satisfies **(PCC)**. If $G$ satisfies the $k$-lca-property, then Proposition 2 implies that $\mathscr{C}_G$ is pre-$k$-ary. Assume now that $\mathscr{C}_G$ is pre-$k$-ary. Hence, for all $U \in X^{(k)}$ we have $R_{\mathscr{C}_G}(U) = \bigcap \{C \in \mathscr{C}_G \mid U \subseteq C\} \in \mathscr{C}_G$. Therefore, $\mathrm{Anc}(\{U\}) \neq \emptyset$ and thus, $\mathrm{LCA}(\{U\}) \neq \emptyset$. Assume, for contradiction, that there are two distinct vertices $v, w \in \mathrm{LCA}(U)$. Note that $U \subseteq R_{\mathscr{C}_G}(U) \subseteq \mathtt{C}(v) \cap \mathtt{C}(w)$. By Definition 1, both $v$ and $w$ are $\preceq$-minimal ancestors of the vertices in $U$ and, therefore, $v$ and $w$ are incomparable in $G$. This, together with the fact that $G$ satisfies **(PCC)** implies that neither $\mathtt{C}(v) \subseteq \mathtt{C}(w)$ nor $\mathtt{C}(w) \subseteq \mathtt{C}(v)$ can hold. Then, $\mathtt{C}(v) \between \mathtt{C}(w)$ since $\mathtt{C}(v) \cap \mathtt{C}(w) \neq \emptyset$. Since $\mathscr{C}_G$ is pre-$k$-ary, $R_{\mathscr{C}_G}(U) \in \mathscr{C}_G$, i.e., there is a vertex $z \in V(G)$ such that $\mathsf{C}(z) = R_{\mathscr{C}_G}(U)$. Hence, $\mathsf{C}(z) \subseteq \mathsf{C}(w) \cap \mathsf{C}(v)$. Since $\mathsf{C}(v) \between \mathsf{C}(w)$ it must hold that $\mathsf{C}(z) \subsetneq \mathsf{C}(w)$ and $\mathsf{C}(z) \subsetneq \mathsf{C}(v)$. Since $G$ satisfies **(PCC)**, $z$ and $v$ must be $\preceq$-comparable. If, however, $v \preceq z$, then Lemma 1 implies that $\mathsf{C}(v) \subseteq \mathsf{C}(z)$; a contradiction. Hence, $z \prec v$ and, by similar arguments, $z \prec w$ must hold. This, however, contradicts the fact that $v$ and $w$ are $\prec$-minimal ancestors of all the vertices in $U$. Hence, $|\mathrm{LCA}(U)| = 1$ must hold for all $U \in X^{(k)}$. Consequently, $G$ satisfies the $k$-lca-property. $\qquad\square$

Consequently, we obtain a characterization of pre-$k$-ary clustering systems in terms of the DAGs from which they derive.

**Theorem 1.** *A clustering system $\mathscr{C}$ is pre-$k$-ary if and only if there is a DAG $G$ with $\mathscr{C} = \mathscr{C}_G$ and $k$-lca-property.*

*Proof.* Suppose that $\mathscr{C}$ is a pre-$k$-ary clustering system and consider the Hasse diagram $G := \mathfrak{H}(\mathscr{C})$. It satisfies **(PCC)** and $\mathscr{C}_G = \mathscr{C}$. Consequently, $\mathscr{C}_G$ is pre-$k$-ary. Thus, we can apply Proposition 3 to conclude that $G$ satisfies the $k$-lca-property. Conversely, suppose that $G$ is a DAG with the $k$-lca-property and $\mathscr{C} = \mathscr{C}_G$. By Proposition 2, $\mathscr{C}$ is pre-$k$-ary. $\qquad\square$

Next, we show that $k$-ary transit functions give rise to DAGs with the $k$-lca-property in a rather natural way:

**Lemma 4.** *Let $R$ be a monotone $k$-ary transit function. Then, the Hasse diagram of its transit sets $\mathfrak{H}(\mathscr{C}_R)$ satisfies the $k$-lca-property.*

*Proof.* Let $R$ be a monotone transit function on $X$ and $U \in X^{(k)}$. Considering $R$ as a function over subsets, conditions **(t1)** and **(t2)** imply that $U \subseteq R(U)$. In the following, let $v_C$ denote the unique vertex in $\mathfrak{H}(\mathscr{C}_R)$ that corresponds to the cluster $C \in \mathscr{C}_R$. For all $W \in X^{(k)}$ with $U \subseteq R(W)$ it holds, by condition **(m)**, that $R(U) \subseteq R(W)$. This, together with the definition of the Hasse diagram, implies that $v_{R(U)} \preceq v_{R(W)}$ for all $W \in X^{(k)}$ with $U \subseteq R(W)$. Thus, $v_{R(U)}$ is the unique $\preceq$-minimal vertex in $\mathfrak{H}(\mathscr{C}_R)$ satisfying $x \preceq v_{R(U)}$ for all $x \in U$, and thus $v_{R(U)} = \mathrm{lca}(U)$. $\qquad\square$

As an immediate consequence of the correspondence between monotone $k$-ary transit functions and $k$-ary $\mathscr{T}$-systems, we also conclude that the Hasse diagram of $k$-ary $\mathscr{T}$-systems has the $k$-lca-property.

The converse of Lemma 4, however, need not be true: A Hasse diagram $\mathfrak{H}(\mathscr{C}_R)$ with the $k$-lca-property for some $k$ is not sufficient to imply that $R$ is monotone:

*Example 1.* Let $R$ on $X = \{a, b, c, d\}$ be symmetric and defined by $R(a, b) = X$, $R(a, c) = \{a, b, c\}$, and all other sets are singletons or $X$ in such a way that **(t1)** and **(t3)** is satisfied. One easily verifies that $R$ is a transit function satisfying **(a')** and that $\mathfrak{H}(\mathscr{C}_R)$ is a network with root $X$. In fact, $\mathfrak{H}(\mathscr{C}_R)$ is a rooted tree having pairwise lca-property. However, $R$ is not monotone since $R(a, b) = X \nsubseteq R(a, c)$.

**Theorem 2.** *Let $R$ be a $k$-ary transit function. Then $R$ is monotone if and only if there is a DAG $G$ with $k$-lca-property, which satisfies $\mathscr{C}_G = \mathscr{C}_R$ and $R_{\mathscr{C}_G} = R$.*

*Proof.* If $R$ is a monotone $k$-ary transit function, then $G = \mathfrak{H}(\mathscr{C}_R)$ satisfies $\mathscr{C}_G = \mathscr{C}_R$. By Lemma 4, $G$ has the $k$-lca-property. Moreover, since $\mathscr{C}_G = \mathscr{C}_R$ it follows that $R_{\mathscr{C}_G} = R_{\mathscr{C}_R}$. Since $R$ is monotone, $R = R_{\mathscr{C}_R} = R_{\mathscr{C}_G}$.

Conversely, let $G$ be a DAG with $\mathscr{C}_G = \mathscr{C}_R$ and $k$-lca-property. By Proposition 2, $R_G = R_{\mathscr{C}_G}$ is monotone. Therefore, $R$ is a monotone $k$-ary transit function. □

## 4   DAGs with Strict and Strong $k$-lca-Property

In general, lca($\mathtt{C}(w)$) is not necessarily defined for all $w \in V(G)$, see e.g. the DAG in Fig. 2. As discussed in [10], it is, however, a desirable property:

**(CL)** For every $v \in V(G)$, lca($\mathtt{C}(v)$) is defined.

By definition, every DAG $G$ that has the lca-property satisfies **(CL)**.

Since lca($\mathtt{C}(v)$) is defined, Observation 2(i) implies lca($\mathtt{C}(v)$) $\preceq v$. Then by Lemma 1, $\mathtt{C}(\text{lca}(\mathtt{C}(v))) \subseteq \mathtt{C}(v)$. The reverse inclusion is trivial. Hence, we have the following Observation:

**Observation 4.** *Let $G$ be a DAG satisfying* **(CL)**. *Then* $\mathtt{C}(\text{lca}(\mathtt{C}(v))) = \mathtt{C}(v)$ *for all $v \in V(G)$.*

**Definition 4.** *Let $G$ be a DAG with leaf set $X$ and $k$-lca property. Then, $G$ has the* strict $k$-lca-property *if $G$ satisfies* **(CL)**, *and for every $w \in V(G)$, there is $U \in X^{(k)}$ such that* lca($\mathtt{C}(w)$) = lca($U$).

**Proposition 4.** *Let $G$ be a DAG with leaf set $X$ and $k$-lca-property. Then $G$ has the strict $k$-lca-property if and only if $\mathscr{C}_G$ is a $k$-ary $\mathscr{T}$-system. In this case, $\mathscr{C}_G$ is identified by $R_G$.*

*Proof.* The $k$-lca-property of $G$ implies that $\mathscr{C}_G$ is pre-$k$-ary, from Proposition 2. Assume that $G$ has the strict $k$-lca-property. Consider $\mathtt{C}(w) \in \mathscr{C}_G$. By definition, there exists $U \in X^{(k)}$ such that lca($\mathtt{C}(w)$) = lca($U$). Since $G$ satisfies **(CL)**, Observation 4 implies $\mathtt{C}(w) = \mathtt{C}(\text{lca}(\mathtt{C}(w))) = \mathtt{C}(\text{lca}(U))$. Moreover, by Observation 2(ii), $\mathtt{C}(\text{lca}(U)) = \mathtt{C}(w)$ is the unique inclusion minimal cluster in $\mathscr{C}_G$ containing $U$. This implies both $U \subseteq \mathtt{C}(w)$ and $C(w) \subseteq C(v)$ for every $v \in V(G)$ with $U \subseteq C(v)$. Hence, $\mathscr{C}_G$ satisfies **(KR)**. Hence, $\mathscr{C}_G$ is a $k$-ary $\mathscr{T}$-system.

Conversely, assume that $G$ holds $k$-lca-property and $\mathscr{C}_G$ satisfies **(KR)**. Thus, for every $w \in V(G)$, there is $U \in X^{(k)}$ such that $U \subseteq C(w)$ and $U \subseteq C(v)$ implies $C(w) \subseteq C(v)$ for all $v \in V(G)$. Hence, $C(w)$ is an inclusion minimal set in $\mathscr{C}_G$ containing $U$. Since $G$ has the $k$-lca-property, lca($U$) is defined and, by Observation 2(ii), $\mathtt{C}(\text{lca}(U))$ is the unique inclusion minimal set in $\mathscr{C}_G$ containing $U$, and thus $\mathtt{C}(w) = \mathtt{C}(\text{lca}(U))$ must hold. By Observation 2(iii) we have lca($\mathtt{C}(w)$) = lca($\mathtt{C}(\text{lca}(U))$) = lca($U$). Therefore, $G$ has the strict $k$-lca-property.

Since a set system is identified by a $k$-ary transit function if and only if it is a $k$-ary $\mathscr{T}$-system and its canonical transit function identifies it, we have $\mathscr{C}_G$ is identified by $R_{\mathscr{C}_G}$. Moreover, $R_{\mathscr{C}_G} = R_G$ from Proposition 2. Hence the result. □

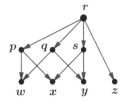

**Fig. 3.** The network $G$ with leaf set $X = \{w, x, y, z\}$ has the clustering systems $\mathscr{C}_G = \{\{x\}, \{y\}, \{z\}, \{w\}, \{w, x\}, \{w, y\}, \{x, y\}, X\}$ satisfying **(KS)**, **(KC)** and **(KR)** for $k = 2, 3, 4$. Moreover, $\mathrm{lca}(u, v)$ is defined for all $u, v \in X$. However, the sets $\{w, x\}, \{w, y\}, \{x, y\}$ violates the condition of weak hierarchy. $G$ is an lca-network but not a strong-2-lca-network since $\mathrm{lca}(\{w, x, y\}) = r \neq \mathrm{lca}(\{u, v\})$ for any $u, v \in \{w, x, y\}$.

In [10], networks with the *strong* lca-*property* were introduced. These satisfy (i) the lca-property, and (ii) for every non-empty subset $A \subseteq X$, there are $x, y \in A$ such that $\mathrm{lca}(x, y) = \mathrm{lca}(A)$. As it turns out, these networks are characterized by their clustering systems: $G$ is a strong lca-network if and only if $G$ has the lca-property and $\mathscr{C}_G$ is a weak hierarchy [10, Prop. 13]. In the following, we generalize these results to DAGs in general and spanning sets for $\mathrm{lca}(A)$ that are larger than a pair of points:

**Definition 5.** *Let $G$ be DAG with leaf set $X$ and* lca-*property. Then, $G$ has the strong $k$-lca-property if, for every non-empty subset $A \subseteq X$, there is $U \in X^{(k)}$ such that $\mathrm{lca}(U) = \mathrm{lca}(A)$.*

Figure 3 shows that the lca property does not imply the strong $k$-lca-property, i.e., the uniqueness of LCAs for all $A \subseteq X$ does not imply that these are spanned by small subsets of leaves.

**Lemma 5.** *If a DAG $G$ has the strong $k$-lca-property, then it has the strict $k$-lca-property.*

*Proof.* Suppose that $G$ is a DAG with leaf set $X$ and that has the strong $k$-lca-property. By definition, $G$ has the lca-property. Hence, for all non-empty $A \in 2^X$, $\mathrm{lca}(A)$ is defined. This implies that $\mathrm{lca}(A)$ is defined for all $A \in X^{(k)} \subseteq 2^X$, and thus, $G$ has the $k$-lca-property. Furthermore, since $\mathsf{C}(v) \in 2^X$ for all $v \in V(G)$, the DAG $G$ satisfies **(CL)**. Let $w \in V(G)$. Since $A := \mathsf{C}(w) \subseteq X$ and since $G$ has the strong $k$-lca-property, there exists $U \in X^{(k)}$ such that $\mathrm{lca}(U) = \mathrm{lca}(A) = \mathrm{lca}(\mathsf{C}(w))$. In summary, $G$ has the strict $k$-lca-property.    □

**Proposition 5.** *Let $G$ be a DAG with leaf set $X$ and the* lca-*property. Then $G$ has the strong $k$-lca-property if and only if for every non-empty subset $A \subseteq X$ there exists $U \subseteq A$ with $|U| \leq k$ such that $\mathrm{cl}(A) = \mathrm{cl}(U)$ in $\mathscr{C}_G$.*

*Proof.* First, assume that $G$ has the strong $k$-lca-property and let $\emptyset \neq A \subseteq X$. Then $\mathrm{lca}(A) = \mathrm{lca}(U)$ for some $U \subseteq X$ with $|U| \leq k$. Applying Observation 3, we obtain $\mathrm{cl}(A) = \mathtt{C}(\mathrm{lca}(A)) = \mathtt{C}(\mathrm{lca}(U)) = \mathrm{cl}(U)$. Conversely, assume that for every non-empty subset $A \subseteq X$, there exists $U \subseteq A$ with $|U| \leq k$ such that $\mathrm{cl}(A) = \mathrm{cl}(U)$ in $\mathscr{C}_G$. Let $A \subseteq X$ be non-empty. Applying Observation 2(iii) and Observation 3 yields $\mathrm{lca}(A) = \mathrm{lca}(\mathtt{C}(\mathrm{lca}(A))) = \mathrm{lca}(\mathrm{cl}(A)) = \mathrm{lca}(\mathrm{cl}(U)) = \mathrm{lca}(\mathtt{C}(\mathrm{lca}(U))) = \mathrm{lca}(U)$. $\qquad\square$

Using Proposition 1 and 5, we can establish the correspondence between strong $k$-lca DAGs and $k$-weak hierarchies.

**Theorem 3.** *$G$ is a DAG with the strong $k$-lca-property if and only if $G$ has the lca-property and $\mathscr{C}_G$ is a $k$-weak hierarchy.*

*Proof.* For the *only if*-direction, suppose that $G$ is a DAG with the strong $k$-lca-property. By Definition 5, $G$ has the lca-property. Moreover, Proposition 5 implies that the following *"Condition $(U, A, X)$"* is satisfied: for every non-empty subset $A \subseteq X$, there exists $U \subseteq A$ with $|U| \leq k$ such that $\mathrm{cl}(A) = \mathrm{cl}(U)$ in $\mathscr{C}_G$. By Proposition 1, $\mathscr{C}_G$ is a $k$-weak hierarchy. For the *if*-direction, assume that $\mathscr{C}_G$ is a $k$-weak hierarchy. By Proposition 1, Condition $(U, A, X)$ is satisfied. This, together with the assumption that $G$ has the lca-property and Proposition 5 implies that $G$ has the strong $k$-lca-property. $\qquad\square$

## 5   Concluding Remarks

The connection between clusters and LCAs in DAGs is not limited to the relationships discussed so far and summarized in the following diagram:

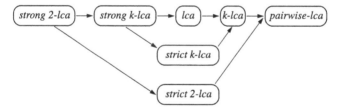

Note that there are no implications between the lca-, strict 2-lca- and strict $k$-lca-property.

*Weak pyramids* [8] are weak hierarchies that, in addition, satisfy a necessary (but not sufficient) condition for $\mathscr{C}$ to comprise intervals [13]:

**(WP)** If $A, B, C \in \mathscr{C}$ have pairwise non-empty intersections, then one set is contained in the union of the two others.

As an example for further close connections between least common ancestors and clustering properties, we mention the following result:

**Lemma 6.** *Let $G$ be a strong lca-network with leaf set $X$. Then, $\mathscr{C}_G$ is a weak pyramid if and only if there are no four distinct vertices $a, b, c, d \in X$ such that $b, c \not\preceq \mathrm{lca}(a, d)$, $a, c \not\preceq \mathrm{lca}(b, d)$, and $a, b \not\preceq \mathrm{lca}(c, d)$.*

*Proof.* We prove both directions through contraposition. First suppose that $\mathscr{C}_G$ is not weakly pyramidal. Then, there exists $A, B, C \in \mathscr{C}_G$ with pairwise non-empty intersection such that $A \nsubseteq B \cup C$, $B \nsubseteq A \cup C$, and $C \nsubseteq A \cup B$ and vertices $u', v', w' \in V(G)$ such that $A = \mathsf{C}(u')$, $B = \mathsf{C}(v')$ and $C = \mathsf{C}(w')$. Moreover, by Theorem 3, $\mathscr{C}_G$ is a weak hierarchy and hence $A \cap B \cap C \neq \emptyset$. By the latter arguments, there are four vertices $d \in A \cap B \cap C$, $a \in A \setminus (B \cup C)$, $b \in B \setminus (A \cup C)$ and $c \in C \setminus (A \cup B)$. For $u := \operatorname{lca}(a, d)$, we have, $u \preceq u'$ by Observation 2(i) since $a, d \in \mathsf{C}(u')$. This, and the fact that $b, c \notin A = \mathsf{C}(u')$ (and therefore, $b, c \nleq u'$) implies that $b, c \nleq u$. Analogous argumentation for $\operatorname{lca}(b, d)$ and $\operatorname{lca}(c, d)$ shows that $a, c \nleq \operatorname{lca}(b, d)$ and $a, b \nleq \operatorname{lca}(c, d)$.

Conversely, assume that there exist distinct $a, b, c, d \in L(G)$ such that $b, c \nleq u = \operatorname{lca}(a, d)$, $a, c \nleq v = \operatorname{lca}(b, d)$, and $a, b \nleq w = \operatorname{lca}(c, d)$. Consider $A := \mathsf{C}(u), B := \mathsf{C}(v)$ and $C := \mathsf{C}(w)$. Then $A, B, C \in \mathscr{C}_G$ have pairwise non-empty intersection since $d \in A \cap B \cap C$. However, since $a \in A \setminus (B \cup C)$, $b \in B \setminus (A \cup C)$ and $c \in C \setminus (A \cup B)$, none of the sets $A, B, C$ can be contained in the union of the two others, i.e., **(WP)** does not hold. □

Results like Lemma 6 suggest that the connection between LCAs and clusters in DAGs remains an interesting topic for future research.

In this contribution, we have restricted ourselves to least common ancestors of the minimal vertices, i.e., the "leaves" of a DAG $G$. In a more general setting, with applications beyond phylogenetics, one may ask analogous questions for arbitrary subsets of $V(G)$. At present, no characterization of the class of graphs with unique LCAs for all subsets of $V(G)$, or unique LCAs for all pairs of vertices in $V(G)$, appears to be known. These graphs could serve as interesting generalizations of rooted trees and may also be of interest from an algorithmic point of view as there are computational problems that are hard on DAGs in general, but become tractable if LCAs are unique, in particular also on trees.

**Acknowledgments.** AVS acknowledges the financial support from the CSIR-HRDG for the Senior Research Fellowship (09/0102(12336)/2021-EMR-I).

# References

1. Bandelt, H.J., Dress, A.W.M.: Weak hierarchies associated with similarity measures – an additive clustering technique. Bull. Math. Biol. **51**, 133–166 (1989). https://doi.org/10.1007/BF02458841
2. Bandelt, H.J., Dress, A.W.M.: An order theoretic framework for overlapping clustering. Discret. Math. **136**, 21–37 (1994). https://doi.org/10.1016/0012-365X(94)00105-R
3. Barthélemy, J.P., Brucker, F.: Binary clustering. Discret. Appl. Math. **156**(8), 1237–1250 (2008). https://doi.org/10.1016/j.dam.2007.05.024
4. Bender, M.A., Pemmasani, G., Skiena, S., Sumazin, P.: Finding least common ancestors in directed acyclic graphs. In: Proceedings of the 12th Annual ACM-SIAM Symposium on Discrete Algorithms, SODA 2001, pp. 845–853. Society for Industrial and Applied Mathematics, Washington, D.C., USA (2001). https://doi.org/10.5555/365411.365795

5. Bertrand, P., Diatta, J.: Multilevel clustering models and interval convexities. Discret. Appl. Math. **222**, 54–66 (2017). https://doi.org/10.1016/j.dam.2016.12.019
6. Changat, M., Mathews, J., Peterin, I., Narasimha-Shenoi, P.G.: $n$-ary transit functions in graphs. Discussiones Math. Graph Th. **30**(4), 671–685 (2010). https://eudml.org/doc/270794
7. Changat, M., Narasimha-Shenoi, P.G., Stadler, P.F.: Axiomatic characterization of transit functions of weak hierarchies. Art Discret. Appl. Math. **2**, P1.01 (2019). https://doi.org/10.26493/2590-9770.1260.989
8. Changat, M., Shanavas, A.V., Stadler, P.F.: Transit functions and pyramid-like binary clustering systems. Technical report. 2212.08721, arXiv (2023). https://doi.org/10.48550/arXiv.2212.08721
9. Dress, A.: Towards a theory of holistic clustering. In: Mirkin, B., McMorris, F.R., Roberts, F.S., Rzhetsky, A. (eds.) Mathematical Hierarchies and Biology. DIMACS Series in Discrete Mathematics and Theoretical Computer Science, vol. 37, pp. 271–290. American Mathematical Society (1996)
10. Hellmuth, M., Schaller, D., Stadler, P.F.: Clustering systems of phylogenetic networks. Theory Biosci. **142**(4), 301–358 (2023). https://doi.org/10.1007/s12064-023-00398-w
11. Huson, D.H., Scornavacca, C.: A survey of combinatorial methods for phylogenetic networks. Genome Biol. Evol. **3**, 23–35 (2011). https://doi.org/10.1093/gbe/evq077
12. Nakhleh, L., Wang, L.-S.: Phylogenetic networks: properties and relationship to trees and clusters. In: Priami, C., Zelikovsky, A. (eds.) Transactions on Computational Systems Biology II. LNCS, vol. 3680, pp. 82–99. Springer, Heidelberg (2005). https://doi.org/10.1007/11567752_6
13. Nebeský, L.: On a certain numbering of the vertices of a hypergraph. Czechoslovak Math. J. **33**, 1–6 (1983). https://doi.org/10.21136/CMJ.1983.101849

# The Frobenius Problem for the Proth Numbers

Pranjal Srivastava[1(✉)] and Dhara Thakkar[2]

[1] Indian Institute of Science Education and Research Bhopal, Bhopal, India
pranjal.srivastava194@gmail.com
[2] Indian Institute of Technology Gandhinagar, Gandhinagar, India
thakkar_dhara@iitgn.ac.in

**Abstract.** Let $n$ be a positive integer greater than 2. We define *the Proth numerical semigroup*, $P_k(n)$, generated by $\{k2^{n+i} + 1 \mid i \in \mathbb{N}\}$, where $k$ is an odd positive number and $k < 2^n$. In this paper, we introduce the Frobenius problem for the Proth numerical semigroup $P_k(n)$ and give formulas for the embedding dimension of $P_k(n)$. We solve the Frobenius problem for $P_k(n)$ by giving a closed formula for the Frobenius number. Moreover, we show that $P_k(n)$ has an interesting property such as being Wilf.

**Keywords:** Combinatorial techniques · Frobenius problem · Proth Number · Numerical semigroup · Apéry Set · pseudo-Frobenius number · type · Wilf's conjecture

## 1 Introduction

The mathematician Ferdinand Frobenius defines the problem that asks to find the largest integer that is not expressible as a non-negative integer linear combination of elements of $L$, where $L$ is a set of $m$ coprime positive integers.

The Frobenius problem is defined as follows: Given a set $L = \{l_1, l_2, ..., l_m\}$ of coprime positive integers and $l_i \geq 2$, find the largest natural number that is not expressible as a non-negative linear combination of $l_1, l_2, ..., l_m$. It is also known as the money exchange or coin exchange problem in number theory. In literature, the connection between graph theory, theory of computer science and Frobenius problem has been developed (see [10,11,14,15]). This is because the Frobenius problem has attracted mathematicians as well as computer scientists since the 19-th century (see [3], Chap. 1 in [6], Problem C7 in [9,28]).

For the special case e.g., $m = 2$, the explicit formula to find the Frobenius number is known, it is $l_1 l_2 - l_1 - l_2$ proved in [26]. In addition to that, for the case $m = 3$, semi-explicit formula is known to find the Frobenius number [17]. Moreover, Rödseth [24], Selmer [25] and Beyer [4] have developed algorithms to solve the Frobenius problem in the case $m = 3$. In 1996, Ramírez-Alfonsín showed that the Frobenius problem for variable $m$ is NP-hard [16].

The Frobenius problem has been studied for several special cases, e.g., numbers in a geometric sequence, arithmetic sequence, Pythagorean triples, three

S. Kalyanasundaram and A. Maheshwari (Eds.): CALDAM 2024, LNCS 14508, pp. 162–175, 2024.
https://doi.org/10.1007/978-3-031-52213-0_12

consecutive squares or cubes [7,13,29,30]. Moreover, the Frobenius problem is defined on some special structure like Numerical semigroup (see the definition below).

Let $\mathbb{N}$ and $\mathbb{Z}$ be the set of non-negative integers and set of integers, respectively. A subset $S$ of $\mathbb{N}$ containing 0 is a *numerical semigroup* if $S$ is closed under addition and has a finite complement in $\mathbb{N}$. If $S$ is a numerical semigroup and $S = \langle B \rangle$, then we call $B$, a system of generators of $S$. A system of generators $B$ of $S$ is minimal if no proper subset of $B$ generates $S$. In [18] Rosales et al. proved that every numerical semigroup admits a unique minimal system of generators and such a system is finite. The cardinality of a minimal system of generators of $S$ is called the *embedding dimension* of $S$ denoted by $e(S)$.

The Frobenius number of a numerical semigroup $S = \langle \{a_1, a_2, \ldots, a_n\} \rangle$ (denoted by $F(S)$) is the greatest integer that cannot be expressed as a sum $\sum_{i=1}^{n} t_i a_i$, where $t_1, \ldots, t_n \in \mathbb{N}$ [2,18].

To solve the Frobenius problem for numerical semigroups, several methods were introduced, e.g., see [5,18–20]. In particular, in recent articles, the method of computing the Apéry set (see Definition 1) and deduce the Frobenius number using the Apéry set has been presented. In literature, there exists a large list of publications devoted to solve the Frobenius problem for special classes of numerical semigroup, including the Frobenius problem for Fibonacci numerical semigroup [12], Mersenne numerical semigroup [21], Thabit numerical semigroup [22] and repunit numerical semigroup [23]. We note that the study of the Frobenius number for the mentioned numerical semigroups has been inspired by special primes such as Fibonacci, Mersenne, Thabit and repunit primes. In this paper, we introduce Proth numerical semigroup motivated by the Proth number. The main aim of this paper is to study the Proth numerical semigroup and its invariants like embedding dimension, Frobenius number, etc.

In number theory, the *Proth number* (named in honor of the mathematician François Proth) is a natural number of the form $k2^n + 1$, where $n$ and $k$ are positive numbers and $k < 2^n$ is an odd number. We say that a Proth number is a *Proth prime* if it is prime.

A numerical semigroup $S$ is the *Proth numerical semigroup* if $n \in \mathbb{N}$ such that $S = \langle \{k2^{n+i} + 1 \mid i \in \mathbb{N}\} \rangle$, where $n$ and $k$ are positive numbers and $k < 2^n$ is an odd number. We denote by $P_k(n)$ the numerical semigroup $\langle \{k2^{n+i} + 1 \mid i \in \mathbb{N}\} \rangle$. It is easy to see that when $k = 1$ the Proth numerical semigroup is the Cunningham numerical semigroup [27]. Hence, we can assume that $2^r < k < 2^{r+1}$ for some $r$.

In this paper, we first prove that $e(P_k(n))$ is $n + r + 1$ where $2^r < k < 2^{r+1}$. Later, we find the Frobenius number of the Proth numerical semigroup. More formally, we prove the following theorem.

**Theorem 1.** *Let $n > 2$ be a positive integer. Then $F(P_{2^r+1}(n)) = 2s_1 + s_n + s_{n+r} - s_0$, where $s_i = k2^{n+i} + 1$ for $i \in \mathbb{N}$.*

Let $S$ be a numerical semigroup. An integer $x$ is a *pseudo-Frobenius number* of $S$ if $x \in \mathbb{Z} \setminus S$ and $x + s \in S$ for all $s \in S \setminus \{0\}$. The set of pseudo-Frobenius numbers of $S$ is denoted by $\mathrm{PF}(S)$, and the cardinality of the set $\mathrm{PF}(S)$ is called the *type* of $S$ denoted by $\mathrm{t}(S)$ [2,18].

We find the set of pseudo-Frobenius numbers of the Proth numerical semigroup $P_{2^r+1}(n)$ and prove that its type is $n + r - 1$.

In the context of a numerical semigroup, it is reasonable to study the problems that connect the Frobenius number and other invariants of a numerical semigroup. One such problem posed by Wilf (known as Wilf's conjecture) in [31] is as follows: Let $S$ be a numerical semigroup and $\nu(S) = |\{s \in S \mid s \leq \mathrm{F}(S)\}|$, is it true that $\mathrm{F}(S) + 1 \leq e(S)\nu(S)$, where $e(S)$ is the embedding dimension and $\mathrm{F}(S)$ is the Frobenius number of $S$? Note that the numerical semigroups that satisfy Wilf's conjecture are called Wilf.

The conjecture is still open; in spite of it, an affirmative answer has been given for a few special classes of a numerical semigroup. In this paper, we prove that the Proth numerical semigroup $P_{2^r+1}(n)$ supports Wilf's conjecture.

This paper is an attempt to understand the Frobenius problem and Wilf conjecture for arbitrary embedding dimension through the Proth numerical semigroup. Our approach was inspired by the ideas discussed in [21,22]. However, it is worth noting that our techniques to find the Apéry set of the Proth numerical semigroups differ from the existing ones [21,22].

The reader not familiarized with the study of numerical semigroup and the terminologies like embedding dimension, pseudo-Frobenius numbers, type, etc., can refer to the literature [2,18].

## 2    The Embedding Dimension

We begin this section by proving that $P_k(n)$ is a numerical semigroup. Later, we prove that the embedding dimension of $P_k(n)$ is $n+r+1$. Some of the techniques used in this section are introduced earlier see, e.g., [21,22,27].

**Lemma 1.** *(Lemma 2.1 in [18]) Let $S$ be a nonempty subset of $\mathbb{N}$. Then $\langle S \rangle$ is a numerical semigroup if and only if $\gcd(S) = 1$.*

**Theorem 2.** *Let $n > 2$ be an integer, then $P_k(n)$ is a numerical semigroup.*

*Proof.* It is clear that $P_k(n) \subseteq \mathbb{N}$ is closed under addition and contains zero. Note that from Lemma 1 it is enough to show that $\gcd(P_k(n)) = 1$. Let $k2^n + 1$, $k2^{n+1} + 1 \in P_k(n)$. Then $\gcd(k2^n + 1, k2^{n+1} + 1) = \gcd(k2^n + 1, k2^{n+1} - k2^n) = \gcd(k2^n + 1, k2^n) = 1$. Therefore, $P_k(n)$ is a numerical semigroup. □

Next we give the minimal system of generators of the Proth numerical semigroup. To this purpose, we need some preliminary results.

**Lemma 2.** *(Lemma 2.1 in [27]) Let $S$ be a numerical semigroup generated by a non-empty set $M$ of positive integers. Then the following conditions are equivalent:*

*(i) $2m - 1 \in S$ for all $m \in M$;*

(ii) $2s - 1 \in S$ for all $s \in S \setminus \{0\}$.

**Theorem 3.** *Let $n > 2$ be an integer, then $P_k(n) = \langle\{k2^{n+i}+1 \mid i = 0,\ldots,n+r\}\rangle$.*

*Proof.* Let $P = \langle\{k2^{n+i}+1 \mid i \in \{0,1,\ldots,n+r\}\}\rangle$. It is clear that $P \subseteq P_k(n)$. To prove the other direction it is enough to prove that $k2^{n+i}+1 \in P$ for all $i \in \mathbb{N}$. Let $i \in \{0,1,\ldots,n+r-1\}$, then $2(k2^{n+i}+1)-1 = k2^{n+i+1}+1 \in P$. For $i = n+r$, $2(k2^{n+n+r}+1)-1 = ((k-2^r)2^{n+1}+3)(k2^n+1)+((2^{r+1}-k)2^n-2)(k2^{n+1}+1) \in P$. From Lemma 2, we get $2s - 1 \in P$ for all $s \in P \setminus \{0\}$. By induction, we can deduce that $k2^{n+i}+1 \in P$ for all $i \geq n+r+1$ and hence $P_k(n) = \langle\{k2^{n+i}+1 \mid i = 0,\ldots,n+r\}\rangle$. $\qquad\square$

Note that, Theorem 3 tells us that $\{k2^{n+i}+1 \mid i = 0,\ldots,n+r\}$ is a system of generators of $P_k(n)$.

**Lemma 3.** *Let $n > 2$ be an integer, then $k2^{n+n+r}+1 \notin \langle\{k2^{n+i}+1 \mid i \in \{0,1,\ldots,n+r-1\}\}\rangle$.*

*Proof.* Assume to the contrary that there exists $a_0, a_1, \ldots, a_{n+r-1} \in \mathbb{N}$ such that

$$k2^{n+n+r}+1 = \sum_{i=0}^{n+r-1} a_i(k2^{n+i}+1)$$

$$= k2^n \left(\sum_{i=0}^{n+r-1} 2^i a_i\right) + \sum_{i=0}^{n+r-1} a_i.$$

Hence, $\sum_{i=0}^{n+r-1} a_i = 1 \pmod{k2^n}$ and we get, $\sum_{i=0}^{n+r-1} a_i = tk2^n + 1$ for some $t \in \mathbb{N}$. Observe that $t \neq 0$. Thus, $\sum_{i=0}^{n+r-1} a_i \geq k2^n+1$. Therefore, $k2^{n+n+r}+1 = \sum_{i=0}^{n+r-1} a_i(k2^{n+i}+1) \geq (\sum_{i=0}^{n+r-1} a_i)(k2^n+1) \geq (k2^n+1)^2$. Since $2^r < k$ we get

$$2^{r+n} < 2^n k < 2^n k + 2 \Rightarrow k2^{r+n+n} < k^2 2^{2n} + 2k2^n$$
$$\Rightarrow k2^{r+n+n}+1 < k^2 2^{2n} + 2k2^n + 1$$
$$\Rightarrow k2^{n+n+r}+1 < (k2^n+1)^2.$$

Hence, $k2^{n+n+r}+1 \geq (k2^n+1)^2 > k2^{n+n+r}+1$, which is a contradiction. Therefore, $k2^{n+n+r}+1 \notin \langle\{k2^{n+i}+1 \mid i \in \{0,1,\ldots,n+r-1\}\}\rangle$. $\qquad\square$

**Theorem 4.** *Let $n > 2$ be an integer and let $P_k(n)$ be the Proth numerical semigroup associated to $n$, then $e(P_k(n)) = n+r+1$. Moreover, $\{k2^{n+i}+1 \mid i \in \{0,1,\ldots,n+r\}\}$ is the minimal system of generators of $P_k(n)$.*

*Proof.* By Theorem 3, we know that $\{k2^{n+i}+1 \mid i \in 0,1,\ldots,n+r\}$ is a system of generator for $P_k(n)$. Suppose that it is not minimal system of generators of $P_k(n)$. Then there exists $l \in \{1,2,\ldots,n+r-1\}$ such that $k2^{n+l}+1 \in \langle k2^{n+i}+1 \mid i \in \{0,1,\ldots,l-1\}\rangle$. Let $T = \langle k2^{n+i}+1 \mid i \in \{0,1,\ldots,l-1\}\rangle$. If $i \in \{0,1,\ldots,l-2\}$, then $2(k2^{n+i}+1)-1 = k2^{n+i+1}+1 \in T$ and $2(k2^{n+l-1}+1)-1 = k2^{n+l}+1 \in T$. From Lemma 2, we have $2t-1 \in T$ for all $t \in T \setminus \{0\}$. Hence, by induction we can obtain that $k2^{n+i}+1 \in T$ for all $i \geq l$, which is a contradiction as $k2^{n+n+r}+1 \notin T$ from Lemma 3. Therefore, $\{k2^{n+i}+1 \mid i \in \{0,1,\ldots,n+r\}\}$ is the minimal system of generators of $P_k(n)$ and $e(P_k(n)) = n+r+1$.  □

## 3   The Apéry Set

In this section, we study the notion of Apéry set and give the explicit description of the elements of the Apéry set of the Proth numerical semigroup $P_{2^r+1}(n)$ for all $r \geq 1$. We denote by $s_i$ the element $k2^{n+i}+1$ for all $i \in \mathbb{N}$. Thus, with this notation, $\{s_0, s_1, \ldots, s_{n+r}\}$ is the minimal system of generators of $P_k(n)$.

**Definition 1.** *[1, 18] Let S be a numerical semigroup and $n \in S \setminus \{0\}$. The Apéry set of S with respect to n is $\mathrm{Ap}(S,n) = \{s \in S \mid s-n \notin S\}$.*

It is clear from the following lemma that $|\mathrm{Ap}(S,n)| = n$.

**Lemma 4.** *(Lemma 2.4 in [18]) Let S be a numerical semigroup and let n be a nonzero element of S. Then $\mathrm{Ap}(S,n) = \{w(0), w(1), \ldots, w(n-1)\}$, where $w(i)$ is the least element of S congruent with i modulo n, for all $i \in \{0, \ldots, n-1\}$.*

Our next goal is to describe the elements of $\mathrm{Ap}(P_k(n), s_0)$.

**Lemma 5.** *Let $n > 2$ be an integer. Then:*

*(1) if $0 < i \leq j < n+r$ then $s_i + 2s_j = 2s_{i-1} + s_{j+1}$;*
*(2) if $0 < i \leq n+r$ then $s_i + 2s_{n+r} = 2s_{i-1} + \alpha s_0 + \beta s_1$, where $\alpha = (k-2^r)2^{n+1}+3$ and $\beta = (2^{r+1}-k)2^n - 2$.*

*Proof.* (1) If $0 < i \leq j < n+r$ then we have

$$\begin{aligned} s_i + 2s_j &= (k2^{n+i}+1) + 2(k2^{n+j}+1) \\ &= 2(k2^{n+i-1}+1) + (k2^{n+j+1}+1) = 2s_{i-1} + s_{j+1}. \end{aligned}$$

(2) If $0 < i \leq n+r$ then we get

$$\begin{aligned} s_i + 2s_{n+r} &= (k2^{n+i}+1) + 2(k2^{n+n+r}+1) \\ &= 2(k2^{n+i-1}+1) + k2^{2n+r+1}+1 \\ &= 2s_{i-1} + \alpha(k2^n+1) + \beta(k2^{n+1}+1) = 2s_{i-1} + \alpha s_0 + \beta s_1, \end{aligned}$$

where $\alpha = (k-2^r)2^{n+1}+3, \beta = (2^{r+1}-k)2^n - 2$.  □

Let $P(r,n)$ denotes the set of all $n+r$-tuple $(a_1, \ldots, a_{n+r})$ that satisfies the following conditions:

1. for every $i \in \{1, \ldots, n+r\}$, $a_i \in \{0,1,2\}$;
2. if $a_j = 2$ for some $j = 2, \ldots, n+r$ then $a_i = 0$ for $i < j$.

**Lemma 6.** *(Lemma 3.3 in [8]) The cardinality of $P(r,n)$ is equal to $2^{n+r+1}-1$.*

**Lemma 7.** *Let $n > 2$ be an integer and let $P_{2^r+1}(n)$ be the Proth numerical semigroup minimally generated by $\{s_0, s_1, \ldots, s_{n+r}\}$. If $s \in \mathrm{Ap}(P_{2^r+1}(n), s_0)$ then there exist $(a_1, \ldots, a_{n+r}) \in P(r,n)$ such that $s = a_1 s_1 + \cdots + a_{n+r} s_{n+r}$.*

*Proof.* Let $s \in \mathrm{Ap}(P_{2^r+1}(n), s_0)$. We prove the result of lemma using induction on $s$. When $s = 0$ then result follows trivially. Assume that $s > 0$ and $j$ be the smallest element from $\{0, 1, \ldots, n+r\}$ such that $s - s_j \in P_{2^r+1}(n)$. Since $s \in \mathrm{Ap}(P_{2^r+1}(n), s_0)$ we have $j \neq 0$ and $s - s_j \in \mathrm{Ap}(P_{2^r+1}(n), s_0)$. Now from induction hypothesis there exist $(a_1, \ldots, a_{n+r}) \in P(r,n)$ such that $s - s_j = a_1 s_1 + a_2 s_2 + \cdots + a_{n+r} s_{n+r}$, hence $s = a_1 s_1 + a_2 s_2 + \cdots + (a_j + 1)s_j + \cdots + a_{n+r} s_{n+r}$. Note that, to conclude the proof it suffices to prove that $(a_1, \ldots, a_j + 1, \ldots, a_{n+r}) \in P(r,n)$.

(1) To prove $(a_1, a_2, \ldots, a_j + 1, \ldots, a_{n+r}) \in \{0, 1, 2\}^{n+r}$, it is enough to show that $a_j + 1 \neq 3$. If $a_j + 1 = 3$ then from Lemma 5,

(i) for $j < n + r$, we have $s_j + 2s_j = 2s_{j-1} + s_{j+1}$. This implies that,
$$s - s_{j-1} = a_1 s_1 + \cdots + s_{j-1} + (a_{j+1} + 1)s_{j+1} + \cdots + a_{n+r} s_{n+r}.$$
(ii) for $j = n + r$, we have, $s_j + 2s_j = 2s_{j-1} + \alpha s_0 + \beta s_1$. This implies that,
$$s - s_{j-1} = \alpha s_0 + (a_1 + \beta)s_1 + a_2 s_2 + \cdots + (a_{n+r-1} + 1)s_{n+r-1}.$$

In both the cases, we get $s - s_{j-1} \in P_{2^r+1}$, which is a contradiction to the minimality of $j$. Hence, $a_j + 1 \neq 3$.

(2) From the minimality of $j$, we obtain that $a_i = 0$ for all $1 \leq i < j$. Now assume that there exist $l > j$ such that $a_l = 2$, then again from Lemma 5, we have

(i) for $l < n + r$, we have $s_j + 2s_l = 2s_{j-1} + s_{l+1}$;
(ii) for $l = n + r$, we have, $s_j + 2s_l = 2s_{j-1} + \alpha s_0 + \beta s_1$.

Again by the same argument as in (1), we have $s - s_{j-1} \in P_{2^r+1}$, which contradict the minimality of $j$.

Therefore, $(a_1, \ldots, a_j + 1, \ldots, a_{n+r}) \in P(r,n)$. $\qquad\square$

It follows from Lemma 7 that $\mathrm{Ap}(P_{2^r+1}(n), s_0) \subseteq \{a_1 s_1 + \cdots + a_{n+r} s_{n+r} \mid (a_1, \ldots, a_{n+r}) \in P(r,n)\}$.

The next remark tells that the equality in the above expression does not hold in general.

*Remark 1.* If possible suppose that, $\mathrm{Ap}(P_{2^r+1}(n), s_0) = \{a_1 s_1 + \cdots + a_{n+r} s_{n+r} \mid (a_1, \ldots, a_{n+r}) \in P(r,n)\}$. Then $|\mathrm{Ap}(P_{2^r+1}(n), s_0)| = |\{a_1 s_1 + \cdots + a_{n+r} s_{n+r} \mid (a_1, \ldots, a_{n+r}) \in P(r,n)\}| = 2^{n+r+1} - 1 \neq s_0$.

Thus, it remains to find the elements of the set $\{a_1 s_1 + \cdots + a_{n+r} s_{n+r} \mid (a_1, \ldots, a_{n+r}) \in P(r,n)\}$ which belongs to $\mathrm{Ap}(P_{2^r+1}(n), s_0)$. To do so, we first define the following sets:
$$F_1 = \big\{a_1 s_1 + \cdots + a_{n+r-1} s_{n+r-1} + s_{n+r} \mid a_i \in \{0, 1, 2\} \text{ for } 1 \leq i \leq n + r -$$
2, $a_{n+r-1} \in \{1, 2\}$ and if $a_j = 2$ for some $j$ then $a_i = 0$ for $i < j\big\}$; and

$$F_2 = \left(\bigcup_{l=0}^{r-2} E_l \cup \{2s_{n+r}\}\right) \setminus \{s_1 + s_n + s_{n+r}, \; 2s_1 + s_n + s_{n+r}, \; s_n + s_{n+r}\}, \text{ where}$$

$E_l = \{a_1 s_1 + \cdots + a_{n+l} s_{n+l} + s_{n+r} \mid a_i \in \{0,1,2\}$ for $1 \leq i \leq n+l-1, \; a_{n+l} \in \{1,2\}$ and if $a_j = 2$ then $a_i = 0$ for $i < j\}$. Take $F = F_1 \cup F_2$.

**Lemma 8.** *Under the standing hypothesis and notation, the following equalities hold.*

*(a)* $s_{n+l} + s_{n+r} - s_0 = ((2^{n+r} - 2^{n+l}) + 2^{n+1} + 4)s_0 + (2^{n+l} - 2^n - 3)s_1$, for $1 \leq l \leq r$;

*(b)* $s_i + s_n + s_{n+r} - s_0 = ((2^r + 1)2^n + 2 - (2^i - 4))s_0 + (2^i - 4)s_1$ for $2 \leq i \leq n$;

*(c)* $s_1 + s_i + s_n + s_{n+r} - s_0 = ((2^r + 1)2^n + 2 - (2^i - 4))s_0 + (2^i - 3)s_1$ for $2 \leq i \leq n$;

*Proof.* (a) Let $1 \leq l \leq r$. Consider

$$(2^{n+r} - 2^{n+l} + 2^{n+1} + 4)s_0 + (2^{n+l} - 2^n - 3)s_1$$
$$= (2^{n+r} - 2^{n+l} + 2^{n+1} + 4)((2^r + 1)2^n + 1) + (2^{n+l} - 2^n - 3)((2^r + 1)2^{n+1} + 1)$$
$$= (2^r + 1)2^n(2^{n+r} - 2^{n+l} + 2 \cdot 2^n + 4 + 2(2^{n+l} - 2^n - 3)) + 2^{n+r} + 2^n + 1$$
$$= (2^r + 1)2^n(2^{n+r} + 2^{n+l} - 2) + 2^{n+r} + 2^n + 1$$
$$= (2^r + 1)2^n(2^{n+r} + 2^{n+l} - 1) + 1$$
$$= (2^r + 1)(2^{n+n+r}) + 1 + (2^r + 1)2^{n+n+l} + 1 - (2^r + 1)2^n - 1$$
$$= s_{n+r} + s_{n+l} - s_0.$$

(b) Let $2 \leq i \leq n$. Consider

$$((2^r + 1)2^n + 2 - (2^i - 4))s_0 + (2^i - 4)s_1$$
$$= ((2^r + 1)2^n + 2 - (2^i - 4))((2^r + 1)2^n + 1) + (2^i - 4)((2^r + 1)2^{n+1} + 1)$$
$$= (2^r + 1)2^n((2^r + 1)2^n + 2 - 2^i + 4 + 2 \cdot 2^i - 8) + (2^r + 1)2^n + 2$$
$$= (2^r + 1)2^n((2^r + 1)2^n + 2^i - 2) + (2^r + 1)2^n + 2$$
$$= (2^r + 1)2^n((2^r + 1)2^n + 2^i - 1) + 2$$
$$= (2^r + 1)2^{n+n+r} + 1 + (2^r + 1)2^{n+n} + 1 + (2^r + 1)2^{n+i} + 1 - ((2^r + 1)2^n + 1)$$
$$= s_{n+r} + s_n + s_i - s_0.$$

(c) Follows from the proof of part (b).  □

The following lemmas give the explicit description of the elements in the Apéry set $\mathrm{Ap}(P_{2^r+1}(n), s_0)$.

**Lemma 9.** *Let $n > 2$ be an integer. Then $F \cap \mathrm{Ap}(P_{2^r+1}(n), s_0) = \phi$.*

*Proof.* Let $a_1 s_1 + \cdots + a_{n+r-1} s_{n+r-1} + s_{n+r} \in F_1$. From Lemma 8(a), we have $s_{n+r-1} + s_{n+r} - s_0 \in P_{2^r+1}(n)$. Since $a_{n+r-1} \in \{1, 2\}$, we have $a_1 s_1 + \cdots + a_{n+r-1} s_{n+r-1} + s_{n+r} - s_0 = a_1 s_1 + \cdots + (a_{n+r-1} - 1)s_{n+r-1} + s_{n+r-1} + s_{n+r} - s_0 \in P_{2^r+1}(n)$.

Let $a_1 s_1 + \cdots + a_{n+l} s_{n+l} + s_{n+r} \in F_2$ for $1 \leq l \leq r - 2$. From Lemma 8(a), we have $s_{n+l} + s_{n+r} - s_0 \in P_{2^r+1}(n)$. Similar argument as above implies that $a_1 s_1 + \cdots + a_{n+l} s_{n+l} + s_{n+r} - s_0 \in P_{2^r+1}(n)$.

Let $a_1 s_1 + \cdots + a_n s_n + s_{n+r} \in F_2$ (i.e. $l = 0$). Note that $a_i \neq 0$ for some $i \in \{2, \ldots, n-1\}$. From Lemma 8(b) and (c), we have $s_i + s_n + s_{n+r} - s_0 \in P_{2^r+1}(n)$ and $s_1 + s_i + s_n + s_{n+r} - s_0 \in P_{2^r+1}(n)$. Since $a_i \neq 0$ for $2 \leq i \leq n-1$, we have $a_1 s_1 + \cdots + a_n s_n + s_{n+r} - s_0 \in P_{2^r+1}(n)$.

Finally, consider $2 s_{n+r} \in F_2$. From Lemma 8(a), we have $2 s_{n+r} - s_0 \in P_{2^r+1}(n)$.

Thus, for any element of $F$ say $x$, we have $x - s_0 \in P_{2^r+1}(n)$ and hence $F \cap \mathrm{Ap}(P_{2^r+1}(n), s_0) = \phi$. □

**Lemma 10.** *Under the standing hypothesis and notation, we have* $|F| = 2^{n+r} - 2^n - 2$.

*Proof.* Consider the set $L_{11} = \{a_1 s_1 + \cdots + a_{n+r-1} s_{n+r-1} + s_{n+r} \mid a_i \in \{0,1\}$ for $1 \leq i \leq n + r - 2$ and $a_{n+r-1} = 1\}$. Clearly, $|L_{11}| = 2^{n+r-2}$. Now we construct a new set $L_{12}$ as follows: Let $a_1 s_1 + \cdots + a_{n+r-1} s_{n+r-1} + s_{n+r} \in L_{11}$. Take the least index $m \in \{1, 2, \ldots, n+r-1\}$ for which $a_m = 1$, add an element $b_1 s_1 + \cdots + b_{n+r-1} s_{n+r-1} + s_{n+r}$ in $L_{12}$ with $b_m = 2$ and $b_j = a_j$ for all $j \neq m$. Clearly, $|L_{12}| = 2^{n+r-2}$. Note that $F_1$ is the disjoint union of $L_{11}$ and $L_{12}$. Hence, $|F_1| = 2^{n+r-1}$.

Consider the set $L_{21} = \{a_1 s_1 + \cdots + a_{n+l} s_{n+l} + s_{n+r} \mid a_i \in \{0,1\}$ for $1 \leq i \leq n + l - 1$ and $a_{n+l} = 1\}$. Clearly, $|L_{21}| = 2^{n+l-1}$. Now we construct a new set $L_{22}$ as follows: Let $a_1 s_1 + \cdots + a_{n+l} s_{n+l} + s_{n+r} \in L_{21}$. Take the least index $m$ for which $a_m = 1$, add an element $b_1 s_1 + \cdots + b_{n+l} s_{n+l} + s_{n+r}$ in $L_{22}$ with $b_m = 2$ and $b_j = a_j$ for all $j \neq m$. Clearly, $|L_{22}| = 2^{n+l-1}$. Note that $E_l$ is the disjoint union of $L_{21}$ and $L_{22}$. Hence, $|E_l| = 2^{n+l}$. Thus we get, $|F_2| = \sum_{l=0}^{r-2} |E_l| + 1 - 3 = \sum_{l=0}^{r-2} 2^{n+l} - 2 = 2^{n+r-1} - 2^n - 2$. Therefore, $|F| = |F_1| + |F_2| = 2^{n+r-1} + 2^{n+r-1} - 2^n - 2 = 2^{n+r} - 2^n - 2$. □

**Theorem 5.** *Let $n > 2$ be an integer. Then*

$$\mathrm{Ap}(P_{2^r+1}(n), s_0) = \{a_1 s_1 + \cdots + a_{n+r} s_{n+r} \mid (a_1, \ldots, a_{n+r}) \in P(r, n)\} \setminus F.$$

*Proof.* Let $P'(r, n) = \{a_1 s_1 + \cdots + a_{n+r} s_{n+r} \mid (a_1, \ldots, a_{n+r}) \in P(r, n)\} \setminus F$. Now from Lemma 7 and Lemma 9, it is clear that $\mathrm{Ap}(P_{2^r+1}(n), s_0) \subseteq P'(r, n)$). Note that from Lemma 6 and Lemma 10, we have

$$|P'(r, n)| = 2^{n+r+1} - 1 - (2^{n+r} - 2^n - 2) = s_0 = |\mathrm{Ap}(P_{2^r+1}(n), s_0)|.$$

Thus, $\mathrm{Ap}(P_{2^r+1}(n), s_0) = \{a_1 s_1 + \cdots + a_{n+r} s_{n+r} \mid (a_1, \ldots, a_{n+r}) \in P(r, n)\} \setminus F$. □

# 4    The Frobenius Problem

In this section, we give the formula for the Frobenius number of the Proth numerical semigroup $P_{2^r+1}(n)$ for all $r \geq 1$. We recall Lemma 4 from Sect. 3.

Let us begin with some preliminary lemmas.

**Lemma 11.** *Let* $s \in P_{2^r+1}(n)$ *such that* $s \not\equiv 0 (\mathrm{mod}\, s_0)$, *then* $s + 1 \in P_{2^r+1}(n)$. *Moreover,* $w(i+1) \leq w(i) + 1$ *for* $1 \leq i \leq s_0 - 1$.

*Proof.* Since $s \in P_{2^r+1}(n)$, there exist $a_0, \ldots, a_{n+r} \in \mathbb{N}$ such that $s = a_0 s_o + \cdots + a_{n+r} s_{n+r}$. If $s \not\equiv 0 (\mathrm{mod}\, s_0)$ then there exist $i \in \{1, \ldots, n+r\}$ such that $a_i \neq 0$ and we get, $s + 1 = a_0 s_0 + \cdots + (a_i - 1)s_i + \cdots + a_{n+r}s_{n+r} + s_i + 1$.

Now, $s_i + 1 = k2^{n+i} + 1 + 1 = 2k^{n+i-1} + 2 = 2s_{i-1}$. Hence, $s + 1 = a_0 s_o + \cdots + (a_{i-1} + 2)s_{i-1} + (a_i - 1)s_i + \cdots + a_{n+r}s_{n+r} \in P_{2^r+1}(n)$.

Moreover, by definition, $w(i) \not\equiv 0 (\mathrm{mod}\, s_0)$ for $1 \leq i \leq s_0 - 1$. Thus, $w(i) + 1 \in P_{2^r+1}(n)$. Now, $w(i) + 1 \equiv i + 1 (\mathrm{mod}\, s_0)$. As $w(i + 1)$ is the least element of $P_{2^r+1}(n)$ which is congruent with $i + 1$ modulo $s_0$, we get $w(i + 1) \leq w(i) + 1$. □

**Lemma 12.** *Let* $n > 2$ *be an integer. Then*

1. $w(2) = s_1 + s_n + s_{n+r}$;
2. $w(1) = 2s_1 + s_n + s_{n+r}$. *Moreover,* $w(1) - w(2) = s_1$.

*Proof.* (1) Consider

$$s_1 + s_n + s_{n+r} - 2 = (2^r + 1)2^{n+1} + 1 + (2^r + 1)2^{n+n} + 1 + (2^r + 1)2^{n+n+r} - 1$$
$$= 2 \cdot (2^r + 1)2^n + (2^r + 1)2^{2n}(2^r + 1) + 1$$
$$= (2^r + 1) \cdot 2^n + 1)^2 = s_0^2.$$

Therefore, $s_1 + s_n + s_{n+r} \equiv 2 (\mathrm{mod}\, s_0)$. From Lemma 5 we have, $s_1 + s_n + s_{n+r} \in \mathrm{Ap}(P_{2^r+1}(n), s_0)$. Thus, $w(2) = s_1 + s_n + s_{n+r}$.

(2) Note that from (1) we have $s_1 + s_n + s_{n+r} - 2 = s_0^2$. Now

$$2s_1 + s_n + s_{n+r} - 1 = s_1 + s_n + s_{n+r} + 2(2^r + 1)2^n + 1 - 1$$
$$= s_1 + s_n + s_{n+r} - 2 + 2s_0 = s_0^2 + 2s_0.$$

Therefore, $2s_1 + s_n + s_{n+r} \equiv 1 (\mathrm{mod}\, s_0)$. Again From Lemma 5 we have, $2s_1 + s_n + s_{n+r} \in \mathrm{Ap}(P_{2^r+1}(n), s_0)$. Thus, $w(1) = 2s_1 + s_n + s_{n+r}$. Clearly, $w(1) - w(2) = s_1$. □

The next Lemma is due to Selmer [25] gives us the relation among the Frobenius number and Apéry Set.

**Lemma 13.** *([25], Proposition 5 in [2]) Let* $S$ *be a numerical semigroup and let* $n$ *be a non-zero element of* $S$. *Then* $\mathrm{F}(S) = \max(\mathrm{Ap}(S, n)) - n$.

**Lemma 14.** *Under the standing notation, we have*

$$w(1) = \max(\mathrm{Ap}(P_{2^r+1}(n), s_0)).$$

*Proof.* From Lemma 11, $w(i + 1) \leq w(i) + 1$, for $1 \leq i \leq s_0 - 1$. Thus, we get $w(3) \leq w(2) + 1, w(4) \leq w(3) + 1 \leq w(2) + 2$. In general, for $3 \leq j \leq s_0 - 1$, we have $w(j) \leq w(2) + (j - 2)$. Since $w(1) - w(2) = s_1$, we get $w(j) \leq w(1) - s_1 + (j - 2) = w(1) - (s_1 - (j - 2)) < w(1)$ as $s_1 - (j - 2) > 0$. Therefore, $w(1) \geq w(i)$ for $0 \leq i \leq s_0 - 1$ and $w(1) = \max(\mathrm{Ap}(P_{2^r+1}(n), s_0))$.    □

Thus, from Lemma 13 and 14 we obtain the following formula for the Frobenius number of $P_{2^r+1}(n)$.

**Theorem 6.** *Let $n > 2$ be a positive integer. Then* $\mathrm{F}(P_{2^r+1}(n)) = 2s_1 + s_n + s_{n+r} - s_0$.

Next we define the genus of a numerical semigroup.

**Definition 2.** *Let $S$ be a numerical semigroup then the set $\mathbb{N} \setminus S$ is called set of gaps of $S$ and its cardinality is said to be genus of $S$ denoted by $g(S)$.*

*Remark 2.* It is well known that (see Lemma 3 in [2]), $g(S) \geq \frac{\mathrm{F}(S)+1}{2}$.

**Corollary 1.** *Let $n > 2$ be a positive integer. Then, $g(P_{2^r+1}(n)) \geq k(2^{n+1} + 2^{2n-1} + 2^{2n+r-1} - 2^{n-1}) + 2$.*

# 5    Pseudo-Frobenius Numbers and Type

Our purpose in this section is to give the pseudo-Frobenius set and the formula for the type of the Proth numerical semigroup $P_{2^r+1}(n)$ for all $r \geq 1$. Let us recall the definition of pseudo-Frobenius numbers.

Let $S$ be a numerical semigroup. An integer $x$ is a *pseudo-Frobenius number* of $S$ if $x \in \mathbb{Z} \setminus S$ and $x + s \in S$ for all $s \in S \setminus \{0\}$.

Consider the following relation on the set of integers $\mathbb{Z}$: $a \leq_S b$ if $b - a \in S$. Note that this relation is an order relation i.e., it is reflexive, transitive and anti-symmetric (see [18]). The next lemma characterizes pseudo-Frobenius numbers in terms of the Apéry set using the relation defined above.

**Lemma 15.** *(Proposition 2.20 in [18]) Let $S$ be a numerical semigroup and let $n$ be a nonzero element of $S$. Then*

$$\mathrm{PF}(S) = \{w - n \mid w \in maximals_{\leq_S}(\mathrm{Ap}(S, n))\}.$$

*Remark 3.* [22] If $w, w' \in \mathrm{Ap}(S, x)$, then $w' - w \in S$ if and only if $w' - w \in \mathrm{Ap}(S, x)$. Hence $maximal_{\leq_S}(\mathrm{Ap}(S, x)) = \{w \in \mathrm{Ap}(S, x) \mid w' - w \notin \mathrm{Ap}(S, x) \setminus \{0\}$ for all $w' \in \mathrm{Ap}(S, x)\}$.

Let $n > 2$ be an integer. We define the set $X$ as follows: $X = \{(a_1, \ldots, a_{n+r}) \mid a_1 s_1 + \cdots + a_{n+r} s_{n+r} \in F\}$. Let us consider $M(n) = P(r, n) \setminus X$. It is clear that maximal elements in $M(n)$ (with respect to the product order) are

- $(2, 1, \ldots, 1, 1, 0), \ldots, (0, \ldots, 0, \overset{\overset{r}{\downarrow}}{2}, 1, \ldots, 1, 0), \ldots, (0, \ldots, 0, 2, 0)$;

$$\bullet (2, 1, \ldots, \overset{n-1}{\underset{\downarrow}{1}}, 0, \ldots, 0, 1), \ldots, (0, \ldots, 0, 2, \overset{n-1}{\underset{\downarrow}{1}}, 0, \ldots, 0, 1);$$

$$\bullet (0, \ldots, 0, \overset{n-1}{\underset{\downarrow}{2}}, 0, \ldots, 0, 1), (2, 0, \ldots, 0, \overset{n}{\underset{\downarrow}{1}}, 0, \ldots, 0, 1).$$

As a consequence of Theorem 5, we get the following lemma.

**Lemma 16.** *Under the standing notation, we have*

$$maximal_{\leq P_{2^r+1}(n)}(Ap(P_{2^r+1}(n), s_0)) = maximal_{\leq P_{2^r+1}(n)}\{\{2s_i + s_{i+1} + \cdots + s_{n+r-1} \mid 1 \leq i \leq n + r - 1\} \cup \{2s_j + s_{j+1} + \cdots + s_{n-1} + s_{n+r} \mid 1 \leq j \leq n - 2\} \cup \{2s_{n-1} + s_{n+r}, 2s_1 + s_n + s_{n+r}\}\}.$$

We are now already to give the main result of this section.

**Theorem 7.** *Let $n > 2$ be an integer and let $P_{2^r+1}(n)$ be the Proth numerical semigroup associated to $n$. Then $maximal_{\leq P_{2^r+1}(n)}(Ap(P_{2^r+1}(n), s_0)) = \{2s_i + s_{i+1} + \cdots + s_{n+r-1} \mid 1 \leq i \leq r\} \cup \{2s_j + s_{j+1} + \cdots + s_{n-1} + s_{n+r} \mid 1 \leq j \leq n - 2\} \cup \{2s_1 + s_n + s_{n+r}\}.$*

*Proof.* Let $i \in \{r + 1, \ldots, n + r - 1\}$, then

$$2s_i + s_{i+1} + \cdots + s_{n-1} + s_{n+r} - (2s_{r+i} + s_{r+i+1} + \cdots + s_n + s_{n+r-1})$$
$$= k2^{n+i} + k2^{n+i}(2^r - 1) + r + k2^{2n+r} + 2 - (k2^{2n}(2^r - 1) + r + k2^{n+r+i} + 1)$$
$$= (k2^{2n} + 1) = s_n.$$

Also, $2s_1 + s_n + s_{n+r} - (2s_{n-1} + s_{n+r}) = 2s_1 + k2^n + 1 - 2(k2^{n-1} + 1) = s_2$.

Hence, we get $2s_{r+i} + s_{r+i+1} + \cdots + s_n + s_{n+r-1} \leq_{P_{2^r+1}(n)} 2s_i + s_{i+1} + \cdots + s_{n-1} + s_{n+r}$ for $i \in \{r + 1, \ldots, n + r - 1\}$ and $2s_{n-1} + s_{n+r} \leq_{P_{2^r+1}(n)} 2s_1 + s_n + s_{n+r}$. From Lemma 16 we obtain that $maximal_{\leq P_{2^r+1}(n)}(Ap(P_{2^r+1}(n), s_0)) = maximal_{\leq P_{2^r+1}(n)}\{\{2s_i + s_{i+1} + \cdots + s_{n+r-1} \mid 1 \leq i \leq r\} \cup \{2s_j + s_{j+1} + \cdots + s_{n-1} + s_{n+r} \mid 1 \leq j \leq n - 2\} \cup \{2s_1 + s_n + s_{n+r}\}\}$.

Consider a set $L_1 = \{p_i = 2s_i + s_{i+1} + \cdots + s_{n+r-1} \mid 1 \leq i \leq r\}$ and $L_2 = \{q_j = 2s_j + s_{j+1} + \cdots + s_{n-1} + s_{n+r} \mid 1 \leq j \leq n - 2\}$. Take $L = L_1 \cup L_2 \cup \{2s_1 + s_n + s_{n+1}\}$. We show that $L = maximal_{\leq P_{2^r+1}(n)}(Ap(P_{2^r+1}(n), s_0))$.

Thus, to conclude the proof, it is enough to show that, for any $x, y \in L$, $x \not\leq_{P_{2^r+1}(n)} y$.

Let $p_i, p_{i+1} \in L_1$, then

$$p_{i+1} - p_i = 2s_{i+1} + s_{i+2} + \cdots + s_{n+r-1} - (2s_i + s_{i+1} + \cdots + s_{n+r-1})$$
$$= -2s_i + s_{i+1} = -1.$$

Thus, the difference between any two element of $L_1$ is smaller than $r < s_0$. Which implies that $p_i \not\leq_{P_{2^r+1}(n)} p_j$ for any $1 \leq i, j \leq r$ and $i \neq j$.

Similarly, one can check that for $q_i, q_{i+1} \in L_2$, $q_{i+1} - q_i = -1$ and $q_i \not\leq_{P_{2^r+1}(n)} q_j$ for any $1 \leq i, j \leq n - 2$ and $i \neq j$.

Let $p_i \in L_1$ and $q_j \in L_2$. Note that, $q_1 - p_1 = s_{n+r} - (s_n + \cdots + s_{n+r-1}) = k2^{2n} + 1 - r$. Now consider $q_j - p_i = q_1 - (j-1) - (p_1 - (i-1)) = q_1 - p_1 - (j-i) = k2^{2n} + 1 - r - j + i$.

Suppose that $k2^{2n}+1-r-j+i \in P_{2^r+1}(n)$, then there exists $\lambda_0, \lambda_1, ..., \lambda_{n+r} \in \mathbb{N}$ such that

$$k2^{2n} + 1 - r - j + i = \lambda_0 s_0 + \lambda_1 s_1 + \cdots + \lambda_{n+r} s_{n+r}$$
$$= (\lambda_0 + \cdots + \lambda_{n+r}) + k2^n(\lambda_0 + 2\lambda_1 + \cdots + 2^{n+r}\lambda_{n+r}).$$

We get, $(\lambda_0 + \cdots + \lambda_{n+r}) = 1 - r - j + i \leq 0$ which is a contradiction as $\lambda_i \in \mathbb{N}$. Thus, $q_j - p_i \notin P_{2^r+1}(n)$ and hence $p_i \nleq_{P_{2^r+1}(n)} q_j$ for $1 \leq i \leq r$, $1 \leq j \leq n-2$. Now consider,

$$2s_1 + s_n + s_{n+r} - p_i = 2s_1 + s_n + s_{n+r} - (p_1 - (i-1))$$
$$= -s_2 - \cdots - s_{n-1} - s_{n+1} - \cdots - s_{n+r-1} + s_{n+r} + i - 1$$
$$= (k2^{n+2} - (n-3)) + k2^{2n} - r + 1 + (i-1)$$
$$= k2^n(4 + 2^n) - n - r + 3 + i.$$

If possible suppose that $k2^{2n} + k2^{n+2} - n - r + 3 + i \in P_{2^r+1}(n)$, then there exists $\lambda_0, \lambda_1, ..., \lambda_{n+r} \in \mathbb{N}$ such that

$$k2^n(4 + 2^n) - n - r + 3 + i = \lambda_0 s_0 + \lambda_1 s_1 + \cdots + \lambda_{n+r} s_{n+r}$$
$$= (\lambda_0 + \cdots + \lambda_{n+r}) + k2^n(2^0\lambda_0 + \cdots + 2^{n+r}\lambda_{n+r}).$$

We get, $(\lambda_0 + \cdots + \lambda_{n+r}) = -(n + r - 3 - i) \leq 0$, which is a contradiction as $\lambda_i \in \mathbb{N}$. Therefore, $p_i \nleq_{P_{2^r+1}(n)} 2s_1 + s_n + s_{n+r}$ for $1 \leq i \leq r$.

Similarly, it is clear that $2s_1 + s_n + s_{n+r} - q_j = k2^{n+2} + (j - n + 2) \notin P_{2^r+1}(n)$. Therefore, $q_j \nleq_{P_{2^r+1}(n)} 2s_1 + s_n + s_{n+r}$ for $1 \leq j \leq n-2$.

Hence, difference between any two elements of $L$ do not belongs to $P_{2^r+1}(n)$. Thus, from Remark 3, we have $L = maximal_{\leq P_{2^r+1}(n)}(\text{Ap}(P_{2^r+1}(n), s_0))$.  □

By applying Lemma 15 and Theorem 7 we obtained the following theorem.

**Theorem 8.** *Let $n > 2$ be an integer and let $P_{2^r+1}(n)$ be the Proth numerical semigroup. Then*
$$\text{PF}(P_{2^r+1}(n)) = \{2s_i + s_{i+1} + \cdots + s_{n+r-1} - s_0 \mid 1 \leq i \leq r\} \cup \{2s_j + s_{j+1} + \cdots + s_{n-1} + s_{n+r} - s_0 \mid 1 \leq j \leq n-2\} \cup \{2s_1 + s_n + s_{n+r} - s_0\}$$
*and $t(P_{2^r+1}(n)) = |\text{PF}(P_{2^r+1}(n)| = r + n - 1$.*

## 6   Wilf's Conjecture

In this section, we prove that the Proth numerical semigroup $P_{2^r+1}(n)$ supports Wilf's conjecture. Let us begin with the statement of Wilf's conjecture.

*Conjecture 1.* [31] Let $S$ be a numerical semigroup, and $\nu(S) = |\{s \in S \mid s \leq F(S)\}|$, then
$$F(S) + 1 \leq e(S)\nu(S),$$

where $e(S)$ is the embedding dimension of $S$ and $F(S)$ is the Frobenius number of $S$.

**Lemma 17.** *(Corollary 5 in [2]) Let $S$ be a numerical semigroup. We have $F(S) + 1 \leq (t(S) + 1)\nu(S)$.*

From the previous lemma we obtain the following theorem.

**Theorem 9.** *The Proth numerical semigroup $P_{2^r+1}(n)$ satisfies Wilf's conjecture.*

*Proof.* Recall that $e(P_{2^r+1}(n)) = n + r + 1$ and from Lemma 17

$$
\begin{aligned}
F(P_{2^r+1}(n)) + 1 &\leq (t(P_{2^r+1}(n)) + 1)\,\nu(P_{2^r+1}(n)) \\
&= (n + r)\,\nu(P_{2^r+1}(n)) \\
&< (n + r + 1)\,\nu(P_{2^r+1}(n)) \\
&= e(P_{2^r+1}(n))\,\nu(P_{2^r+1}(n)).
\end{aligned}
$$

$\square$

## 7    Conclusion

In this work, we obtained the formula for the embedding dimension of the Proth numerical semigroup $P_k(n)$. As a main result, we solved the Frobenius problem for $P_{2^r+1}(n)$. Moreover, we also attained the pseudo-Frobenius set and the type of $P_{2^r+1}(n)$. We concluded the paper by examining that $P_{2^r+1}(n)$ supports Wilf's conjecture. The following is an immediate open question to investigate: Is there a formula to find the Frobenius number and other invariants of the Proth numerical semigroup $P_k(n)$ for arbitrary $k$?

## References

1. Apéry, R.: Sur les branches superlinéaires des courbes algébriques. C. R. Hebd. Seances Acad. Sci. **222**, 1198–1200 (1946)
2. Assi, A., D'Anna, M., García-Sánchez, P.A.: Numerical Semigroups and Applications. Springer, Cham (2020)
3. Alfonsín, J.L.: The Diophantine Frobenius Problem. OUP, Oxford (2005)
4. Beyer, Ö., Selmer, E.S.: On the linear Diophantine problem of Frobenius in three variables (1978)
5. Bras-Amorós, M.: Bounds on the number of numerical semigroups of a given genus. J. Pure Appl. Algebra **213**(6), 997–1001 (2009)
6. Beck, M., Robins, S.: Computing the Continuous Discretely: Integer-Point Enumeration in Polyhedra. Undergraduate Texts in Mathematics (2007)
7. Gil, B.K., et al.: Frobenius numbers of Pythagorean triples. Int. J. Number Theory **11**(02), 613–619 (2015)
8. Gu, Z., Xilin, T.: The Frobenius problem for a class of numerical semigroups. Int. J. Number Theory **13**(05), 1335–1347 (2017)
9. Guy, R.: Unsolved Problems in Number Theory. Springer, Heidelberg (2004)
10. Heap, B.R., Lynn, M.S.: On a linear Diophantine problem of Frobenius: an improved algorithm. Numer. Math. **7**(3), 226–231 (1965)

11. Hujter, M., Vizvári, B.: The exact solutions to the Frobenius problem with three variables. Ramanujan Math. Soc. **2**(2), 117–143 (1987)
12. Marín, J.M., Ramírez-Alfonsín, J.L., Revuelta, M.P.: On the Frobenius number of Fibonacci numerical semigroups. Integers. Electron. J. Comb. Number Theory (2007)
13. Lepilov, M., O'Rourke, J., Swanson, I.: Frobenius numbers of numerical semigroups generated by three consecutive squares or cubes. Semigroup Forum. **91**, 238–259 (2015)
14. Owens, R.W.: An algorithm to solve the Frobenius problem. Math. Mag. **76**(4), 264–275 (2003)
15. Raczunas, M., Chrzastowski-Wachtel, P.: A Diophantine problem of Frobenius in terms of the least common multiple. Discret. Math. (1996)
16. Ramirez-Alfonsin, J.L.: Complexity of the Frobenius problem. Combinatorica **16**(1), 143–147 (1996)
17. Rosales, J., Robles-Pérez, A.: The Frobenius problem for numerical semigroups with embedding dimension equal to three. Math. Comput. **81**(279), 1609–1617 (2012)
18. Rosales, J.C., García-Sánchez, P.A.: Numerical Semigroups. Springer, New York (2009)
19. Rosales, J.C.: Numerical semigroups with Apéry sets of unique expression. J. Algebra **226**, 479–487 (2000)
20. Rosales, J.C., García-Sánchez, P.A., García-García, J.I., Jiménez Madrid, J.A.: Fundamental gaps in numerical semigroups. J. Pure Appl. Algebra **189**, 301–313 (2004)
21. Rosales, J.C., Branco, M., Torrão, D.: The Frobenius problem for Mersenne numerical semigroups. Math. Z. **286**(1), 741–749 (2017)
22. Rosales, J.C., Branco, M., Torrão, D.: The Frobenius problem for Thabit numerical semigroups. J. Number Theory **155**, 85–99 (2015)
23. Rosales, J.C., Branco, M., Torrão, D.: The Frobenius problem for repunit numerical semigroups. Ramanujan J. **40**(2), 323–334 (2016)
24. Rödseth, Ö.J.: On a linear Diophantine problem of Frobenius (1978)
25. Selmer, E.S.: On the linear Diophantine problem of Frobenius (1977)
26. Sylvester, J.: Mathematical questions with their solutions. Educ. Times **41**(21), 171–178 (1884)
27. Song, K.: The Frobenius problem for numerical semigroups generated by the Thabit numbers of the first, second kind base b and the Cunningham numbers. Bull. Korean Math. Soc. **57**(3), 623–647 (2020)
28. Sylvester, J.: On subvariants, i.e. semi-invariants to binary quantics of an unlimited order. Am. J. Math. **5**(1), 79–136 (1882)
29. Tripathi, A.: On the Frobenius problem for geometric sequences. Integers **8**(1), i43 (2008)
30. Tripathi, A.: The Frobenius problem for modified arithmetic progressions. J. Integer Sequences **16**(2) (2013)
31. Wilf, H.: A circle-of-lights algorithm for the "money-changing problem". Am. Math. Mon. **85**, 562–565 (1978)

# Graph Algorithms

# Eternal Connected Vertex Cover Problem in Graphs: Complexity and Algorithms

Kaustav Paul[(✉)] and Arti Pandey[(✉)]

Department of Mathematics, Indian Institute of Technology Ropar, Nangal Road, Rupnagar 140001, Punjab, India
{kaustav.20maz0010,arti}@iitrpr.ac.in

**Abstract.** A variation of the vertex cover problem is the eternal vertex cover problem. This is a two-player (attacker and defender) game, where the defender must allocate guards at specific vertices in order for those vertices to form a vertex cover. The attacker can attack one edge at a time. The defender must move the guards along the edges so that at least one guard passes through the attacked edge (guard moves from one end point of the attacked edge to the another end point), and the new configuration still acts as a vertex cover. If the defender is unable to make such a maneuver, the attacker prevails. If a strategy for defending the graph against any infinite series of attacks emerges, the defender wins. The *eternal vertex cover problem* is to find the smallest number of guards with which the defender can develop a successful strategy. The same problem is referred as the *eternal connected vertex cover problem* if the following additional requirement is added: underlying vertices of each defensive configuration form a connected vertex cover. The smallest number of guards that can be used to create a successful defensive strategy, in this case, is known as the *eternal connected vertex cover number* and is denoted by the $ecvc(G)$. The decision version of the *eternal connected vertex cover problem* is NP-hard for general graphs and it also remains NP-hard for bipartite graphs. In this paper, we proved that the problem is polynomial-time solvable for chain graphs and cographs. In addition, we proved that the problem is NP-hard for Hamiltonian graphs, and proposed a polynomial-time algorithm to compute eternal connected vertex cover number for Mycielskian of a given Hamiltonian graph.

**Keywords:** Eternal Connected vertex cover · Chain graphs · Hamiltonian graphs · Mycielskian · Cographs · Distance-hereditary Graphs. Graph algorithms

## 1 Introduction

The dynamic variant of the *vertex cover problem*, referred to as the *eternal vertex cover problem*, was initially introduced by Klostermeyer et al. in 2009 [14]. Given

A. Pandey—Research supported by CRG project, Grant Number-CRG/2022/008333, Science and Engineering Research Board (SERB), India.

S. Kalyanasundaram and A. Maheshwari (Eds.): CALDAM 2024, LNCS 14508, pp. 179–193, 2024.
https://doi.org/10.1007/978-3-031-52213-0_13

a graph $G = (V, E)$, the problem is framed as a two-player game featuring an attacker and a defender. This game is played in rounds. Before starting the game, the defender is tasked with strategically assigning guards to specific vertices of $G$ so that a guard is assigned to at least one endpoint of every edge. Now in the first round, the attacker's job is to attack one edge.

After the attack, the defender faces a decision dilemma for every guard: either relocate it to an adjacent vertex by traversing through the edge between them or leave it undisturbed. This reallocation of guards must be executed in such a manner that at least one guard from either endpoint of the attacked edge can traverse the edge and take up residence at the other endpoint and the new set of vertices where guards are assigned, still form a vertex cover. If such a reallocation is possible, the attack is successfully defended; if not then the attacker wins. After the reallocation, the second round begins, and the attacker gets to attack an edge, and the defender gets to defend the attack, and the game goes on. If the defender fails to establish such an arrangement, the attacker prevails. But, if the defender's allocation of guards proves capable of defending any infinite sequence of attacks, victory is secured by the defender.

Note that if at any round of the game, after the reconfiguration of guards, the underlying vertices of the newly allocated guards do not form a vertex cover, then in the next round, the attacker can simply attack the edge that does not have a guard assigned in either of its endpoints and win.

In this context, the fundamental parameter of interest is the *eternal vertex cover number* of graph $G$, denoted as $evc(G)$. This parameter signifies the minimum number of guards required to formulate a winning strategy for the defender. In a winning strategy, every guard configuration at any round of the game is called an *eternal vertex cover*. The MINIMUM ETERNAL VERTEX COVER problem (MIN-EVC) is to find an eternal vertex cover of minimum cardinality. The decision version of the problem is denoted as DECIDE-EVC.

Another variation of the vertex cover problem is the *connected vertex cover problem* where the objective is to identify a minimum cardinality vertex cover $S$ of the given graph $G$, such that the subgraph induced on $S$, that is, $G[S]$ is connected. This concept was initially introduced by Garey et al. in their seminal work [12]. The *connected vertex cover number* denoted as $cvc(G)$, represents the minimum cardinality of a connected vertex cover of $G$, and such a vertex cover is termed a *minimum connected vertex cover*. The problem of finding a minimum connected vertex cover is denoted as MIN-CVC.

When the following additional constraint is imposed on the MINIMUM ETERNAL VERTEX COVER problem: each guard configuration constitutes a connected vertex cover rather than just a vertex cover, the problem is termed as the MINIMUM ETERNAL CONNECTED VERTEX COVER (MIN-ECVC) problem. In that context, the minimum number of guards needed to devise a successful defensive strategy for the defender is referred to as the *eternal connected vertex cover number*, and is denoted as $ecvc(G)$. The ETERNAL CONNECTED VERTEX COVER DECISION (DECIDE-ECVC) problem is the decision version of the MIN-ECVC problem.

The inception of the *eternal connected vertex cover problem* can be attributed to Fujito et al. [11]. In this work, a fundamental premise is that each vertex within a given guard allocation configuration can accommodate at most one guard.

## 1.1   Notations and Definitions

All the graphs used in this study are finite, undirected, and simple. Consider a graph $G = (V, E)$, where $V$ represents the set of vertices and $E$ represents the set of edges; $n$ denotes the cardinality of $V$ and $m$ denotes the cardinality of $E$. The notation $N(v)$ denotes the set of neighbouring vertices of a given vertex $v$ in $V(G)$. An independent set $I$ of a graph $G$ is defined as a subset of $V$ such that for any two vertices $u$ and $v$ in $I$, the edge $\{u, v\}$ is not present in the edge set $E$. The degree of a vertex $v \in V$ is determined by the number of its neighbours, denoted as $deg(v)$. Let $V'$ be a subset of $V$. We use the notation $deg_{V'}(v)$ to represent the number of neighbours that vertex $v$ has in $V'$. Similarly, we use the notation $N_{V'}(v)$ to represent the set of neighbours of $v$ in $V'$. A vertex $v$ is defined as a *cut vertex* if the number of components in the graph $G$ is strictly smaller than the number of components in the graph obtained by removing vertex $v$ from $G$. The set of all cut vertices in $G$ is represented as $Cut(G)$.

The *join* of two graphs $H_1$ and $H_2$ is a graph created by combining separate copies of $H_1$ and $H_2$ by adding edges between every vertex in $V(H_1)$ and every vertex in $V(H_2)$. The symbol $\oplus$ will be used to denote the join operation.

A *vertex cover* $S$ of a graph $G = (V, E)$ is defined as a subset of the vertex set $V$ such that for every edge in $E$, at least one of its endpoints is included in $S$. If $S$ is a vertex cover, then the complement of $S$ in $V$, denoted as $V \backslash S$, forms an independent set. A vertex cover with minimum cardinality is referred to as a *minimum vertex cover*. The minimum vertex cover number, denoted as $mvc(G)$, represents the cardinality of the minimum vertex cover in graph $G$. Let $B$ be a subset of vertices in a graph $G$. The minimum cardinality of a vertex cover that contains $B$ is denoted as $mvc_B(G)$. If the induced graph on set $S$ is connected, then set $S$ is defined as a connected vertex cover. A connected vertex cover with minimum cardinality is known as *minimum connected vertex cover* and the cardinality of a minimum connected vertex cover is denoted as $cvc(G)$. Let $B$ be a subset of the vertex set $V$ of a graph $G$. The quantity $cvc_B(G)$ is the cardinality of the smallest connected vertex cover that contains $B$.

An independent set of maximum cardinality is called *maximum independent set* of $G$ and its cardinality is denoted by $mis(G)$. A subset of vertices $X \subseteq V$ is defined as a *separating set* if the number of connected components in the graph $G$ after removing the vertices in $X$ is more than the number of connected components in the original graph $G$. The problem referred to as the *maximum non-separating independent set problem* involves identifying a maximum independent set that does not include any separating set. If $S$ is a connected vertex cover of a graph $G$, then the complement of $S$ in $V$, denoted as $V \backslash S$, forms a non-separating independent set in $G$.

For a given graph $G = (V, E)$, the *Mycielskian* of the graph is $\mu(G)$, where the vertex set of $\mu(G)$ is $\{p\} \cup N \cup V$, where $N = \{v_1, \ldots, v_n\}$ is a copy of $V = \{u_1, \ldots, u_n\}$. The edge set of $\mu(G)$ is $E \cup \{\{p, v\}|v \in N\} \cup \{\{u_i, v_j\}|\{u_i, u_j\} \in E\}$.

A *Hamiltonian cycle* in a graph $G = (V, E)$ is defined as a cycle in $G$ that traverses each vertex $v \in V$ exactly once. A graph containing a Hamiltonian cycle is called Hamiltonian graph. A graph $G = (V, E)$ is said to be $k$-regular if the degree of each vertex in $V$ is equal to $k$. *Distance-hereditary* graphs are graphs in which the distance between any two vertices in any connected induced subgraph is the same as in the original graph.

## 1.2   Related Works

The eternal vertex cover problem is polynomial-time solvable for chordal graphs [4], cactus graphs [5], generalized trees [1], co-bipartite graphs [3], cographs, split graphs, and chain graphs [16]. Still, whether the DECIDE-EVC problem is in NP is unknown. Fomin et al. [10] have shown that the problem belongs to the class PSPACE. Later Babu et al. [2] have shown that the problem is in NP for locally connected graphs, till now which is the largest graph class for which the problem is known to be in NP. The problem is known to be NP-hard for general graphs [10], bipartite graphs [3], and locally connected graphs [2]. Some combinatorial results on the problem can be found in [14].

The connected vertex cover problem remains NP-hard even when considering a variety of graph classes, such as planar bipartite graphs with maximum degree 4 [7], planar biconnected graphs with maximum degree 4 [17], 3-connected graphs [20], and $k$-regular graphs, $k \geq 4$ [15]. The only nontrivial cases in which the computation of $cvc(G)$ can be achieved in polynomial-time are restricted to chordal graphs [19] and graphs with maximum degree 3 [6].

Fujito et al. have established a bound for the eternal connected vertex cover number $(ecvc(G))$ as follows: $cvc(G) \leq ecvc(G) \leq cvc(G)+1$ [11]. Their contributions include the polynomial-time algorithm to solve the MIN-ECVC problem for chordal graphs and the proof of the NP-hardness of the DECIDE-ECVC problem when restricted to locally connected graphs [11]. Furthermore, Fujito et al. have presented a 2-approximation algorithm for the MIN-ECVC problem on general graphs. Additionally, they have provided a comprehensive characterization of $ecvc(G)$ for cactus graphs, block graphs, and other graphs where every block is either a cycle or a clique [11].

## 1.3   Our Results

The remaining sections of the paper are structured in the following manner: Sect. 2 of the paper presents a polynomial-time algorithm for calculating the connected vertex cover number and the eternal connected vertex cover number of chain graphs. Section 3 of the paper proves the NP-hardness of the DECIDE-ECVC problem for Hamiltonian graphs. This section also presents a polynomial-time algorithm for calculating the values of $cvc(\mu(G))$ and $ecvc(\mu(G))$, where

$\mu(G)$ denotes the Mycielskian of a Hamiltonian graph $G$, given that a Hamiltonian cycle is also provided. In Sect. 4, it is proved that there exists a linear-time algorithm for computing the connected vertex cover number of distance-hereditary graphs. In Sect. 5, a polynomial-time algorithm is proposed to compute $ecvc(G)$ for cographs. The work is concluded in Sect. 6.

### 1.4   Existing Theorems Used in the Paper

To facilitate the subsequent proofs presented in this paper, we will briefly enumerate some fundamental theorems.

**Theorem 1** *[11].* *For any connected vertex cover $C$ of a connected graph $G$, all the edge attacks of $G$ can be eternally defended by $(|C| + 1)$ guards in such a way that guards always occupy all the vertices in $C$. In particular, $ecvc(G) \leq cvc(G) + 1$.*

**Theorem 2** *[11].* *Let $G = (V, E)$ be a connected graph. If $ecvc(G) = cvc(G)$, then $cvc(G) = cvc_{\{v\}}(G)$ for each $v \in V$.*

**Theorem 3** *[11].* *Let $G = (V, E)$ be a connected graph with $|V| \geq 2$. Suppose every vertex cover of size $mvc_{Cut(G)}(G)$ containing $Cut(G)$ is connected. Then $ecvc(G) = cvc(G)$ if and only if $cvc_{\{v\}}(G) = cvc(G)$ for each $v \in V$.*

## 2   Computing $cvc(G)$ and $ecvc(G)$ for Chain Graphs

Let $G = (X \cup Y, E)$ be a bipartite graph. $G$ is said to be a *chain graph* if the vertices in $X$ can be ordered $\{x_1, x_2, \ldots, x_{|X|}\}$, such that $N(x_1) \subseteq N(x_2) \subseteq \ldots \subseteq N(x_{|X|})$. Similarly, the vertices of $Y$ can be ordered $\{y_1, y_2, \ldots, y_{|Y|}\}$, such that $N(y_1) \supseteq N(y_2) \supseteq \ldots \supseteq N(y_{|Y|})$. We propose polynomial-time algorithms for computing $cvc(G)$ and $ecvc(G)$ for a chain graph $G$.

For complete bipartite graph $K_{n_1,n_2}$ (where $n_1 \leq n_2$), it is easy to observe that $cvc(K_{n_1,n_2}) = ecvc(K_{n_1,n_2}) = n_1 + 1$. So, throughout this section, it will be assumed that $G = (X \cup Y, E)$ is a chain graph such that $|X| \leq |Y|$ and $G$ is not a complete bipartite graph. Also, $|X| = p$ and $|Y| = q$. Now we state some lemmas that will directly imply the correctness of Algorithm 1.

**Lemma 1.** *For any minimum vertex cover $S$ of $G$, if $x_i \in S$, then $x_{i+1} \in S$, for each $i \in [p-1]$. Similarly if $y_i \in S$, then $y_{i-1} \in S$, for each $i \in \{2, 3, \ldots, q\}$.*

*Proof.* For the sake of contradiction let, $S$ be a minimum vertex cover of $G$ and $x_i \in G$ but $x_j \notin S$ for $j > i$. This implies that $N(x_j) \subseteq S$. But since $j > i$, $N(x_i) \subseteq N(x_j) \subseteq S$. Hence, $S \backslash \{x_i\}$ forms a vertex cover, which contradicts the minimality of $S$. Hence the lemma follows.                            □

**Lemma 2.** *If $mvc(G) < |X|$, then every minimum vertex cover $S$ contains $x_p$ and $y_1$, hence $G[S]$ is connected.*

*Proof.* The proof is easy, and hence is omitted.    □

**Lemma 3.** *If* $mvc(G) = |X|$, *then the following statements are true:*

a. *if there exists an index* $i \in [p-1]$, *such that* $deg(x_i) = i$, *then* $N(x_i) \cup \{x_{i+1}, ..., x_p\}$ *forms a connected vertex cover and* $cvc(G) = mvc(G)$.
b. *if such an index* $i$ *does not exist, then* $cvc(G) = mvc(G) + 1$. *Also,* $X \cup \{y_1\}$ *forms a minimum connected vertex cover.*

*Proof.* If such index $i$ exists, then it is easy to observe that $S = N(x_i) \cup \{x_{i+1}, ..., x_p\}$ forms a connected vertex cover of cardinality $p = mvc(G)$. Hence $S$ is also a minimum connected vertex cover.

If such index $i$ does not exist, then let there exists a connected vertex cover $S'$ of size $< p + 1$, that is $|S'| = p = mvc(G)$. Hence, this implies $S'$ is also a minimum vertex cover. So, $S'$ is of the form $N(x_j) \cup \{x_{j+1}, ..., x_p\}$ for some $j \in [p-1]$ by Lemma 1 and 2. This implies $deg(x_j) = j$, which is a contradiction to the assumption that no such index $i$ exists. Hence $cvc(G) = mvc(G) + 1$ and $X \cup \{y_1\}$ forms a minimum connected vertex cover.    □

---

**Algorithm 1.** An algorithm to compute the $cvc(G)$ for connected chain graph $G$

---

**Input**: A connected chain graph $G = (X \cup Y, E)$, with $|X| \leq |Y|$.
**Output**: A minimum connected vertex cover of $G$.
Find a minimum vertex cover $S$ of $G$, $j \leftarrow 0$;
**if** $S \cap X \neq \phi$ & $S \cap Y \neq \phi$ **then**
    **return** $S$;
**else**
    Initialize $A[\ ]$, such that $A[i] = deg(x_i)$, for all $i \in [p]$;
    $j \leftarrow min\{i \in [p] : A[i] = i\}$;
    **if** $j \neq 0$ and $j \neq p$ **then**
        $S' \leftarrow N(x_j) \cup \{x_{j+1}, ..., x_p\}$;
    **else**
        $S' \leftarrow X \cup \{y_1\}$;
    **end if**
**end if**
**return** $S'$;

---

The proof of correctness for Algorithm 1 can be deduced easily from Lemma 1,2 and 3. Hence the following theorem can be concluded.

**Theorem 4.** *Given a connected chain graph* $G = (V, E)$, $cvc(G)$ *can be computed in* $O(n + m)$ *time.*

Now for eternal connected vertex cover, we state and prove the following lemmas.

**Lemma 4.** *If* $cvc(G) < |X|$, *then* $ecvc(G) = cvc(G) + 1$.

*Proof.* Since $cvc(G) < |X|$, then $mvc(G) < |X|$. So, each minimum vertex cover is connected by Lemma 2. So, there does not exist any minimum connected vertex cover $S$ which contains $x_1$, otherwise $X \subseteq S$ by Lemma 1. So, by Theorem 2, $ecvc(G) \neq cvc(G)$. Also by Theorem 1, we can conclude that $ecvc(G) = cvc(G) + 1$ and $S \cup \{u\}$ (where $u$ is any vertex form $V \backslash S$) forms a minimum eternal connected vertex cover. □

**Lemma 5.** *If $cvc(G) = |X|$, then $ecvc(G) = cvc(G) + 1$.*

*Proof.* Since $cvc(G) = |X|$, then $mvc(G) = |X|$. If there exists a minimum connected vertex cover $C$ that contains $x_1$, then $C = X$ and it is not connected. Hence, a contradiction arises. So, by Theorem 1 and 2, $ecvc(G) = cvc(G) + 1$. □

**Lemma 6.** *Given $|X| < |Y|$, if $cvc(G) > |X|$, then $ecvc(G) = cvc(G) + 1$.*

*Proof.* If $cvc(G) \geq |X| + 1$, then $cvc(G) = |X| + 1$, as $X \cup \{y_1\}$ forms a connected vertex cover. Let there exist a minimum connected vertex cover $S$ that contains $y_q$. Note that either $x_p$ is contained in $S$ or $Y \subseteq S$. If $Y \subseteq S$, then $S = Y$, as $|Y| \geq |X| + 1 = cvc(G)$. But this must not be the case since the graph induced on $Y$ is not connected, hence $x_p \in S$. Let $k = max\{i : N(y_i) = X\}$. If there exists a $y_i$, with $i \leq k$, such that $y_i \notin S$, then $X \subseteq S$ and $S = X \cup \{y_q\}$, which contradicts the fact that the graph induced on $S$ is connected as $N(y_q) \neq X$. So, $y_i \in S$ for each $i \in [k]$.

Now, if there exists $j > k$, such that $y_j \notin S$, then $N(y_j) \subseteq S$, implying that $N(y_q) \subseteq S$. This implies $S \backslash \{y_q\}$ is a connected vertex cover, as $y_1 \in S$. This contradicts the minimality of $S$. So, $y_j \in S$, for each $j > k$, and hence $Y \subseteq S$, leading to a contradiction. So, there does not exist any minimum connected vertex cover that contains $y_q$, which implies $ecvc(G) = cvc(G) + 1$ by Theorem 1 and 2. □

**Lemma 7.** *Given $|X| = |Y|$, if $cvc(G) > |X|$, then $ecvc(G) = cvc(G)$.*

*Proof.* If $cvc(G) \geq |X| + 1$, then $cvc(G) = |X| + 1$. In this case, we will show that any attack can be defended by moving the guards from configuration $X \cup \{y_1\}$ to $Y \cup \{x_p\}$ (or $Y \cup \{x_p\}$ to $X \cup \{y_1\}$).

Assume $X \cup \{y_1\}$ is the initial configuration of guards and edge $\{x_i, y_j\}(i \neq p, j \neq p)$ is attacked. Move the guard at $x_p$ to $y_p$, $y_1$ to $x_p$ and $x_i$ to $y_j$ and rest of the guards through the matching mentioned in following claim.

**Claim 1.** *There exists a perfect matching in $G[(X \backslash \{x_i, x_p\}) \cup (Y \backslash \{y_j, y_p\})]$.*

*Proof.* Let $U \subseteq X \backslash \{x_i, x_p\}$ and $k = max\{l : x_l \in U\}$. So, $|U| \leq k$.

First assume that $i < k$, which implies $|U| \leq k - 1$. Note that if $|N(x_k)| \leq k$, then $\{y_1, \ldots, y_k\} \cup \{x_{k+1}, \ldots, x_p\}$ forms a connected vertex cover of size $p$, which is a contradiction. Hence $|N(x_k)| \geq k + 1$, which implies that $|N(x_k) \cap (Y \backslash \{y_j, y_p\})| \geq k - 1$. So, $|N(U)| \geq k - 1 \geq |U|$. So, by Hall's theorem, there exists a perfect matching in $G[(X \backslash \{x_i, x_p\}) \cup (Y \backslash \{y_j, y_p\})]$.

Now, if $i > k$, then $x_k \neq x_{p-1}$. If $|N(x_k)| = k+1$, then $y_p \notin N(x_k)$, implying $|N(U)| \geq k = |U|$. If $|N(x_k)| > k + 1$, then $|N(U)| > k = |U|$. So by Hall's theorem, there exists a perfect matching in $G[(X \setminus \{x_i, x_p\}) \cup (Y \setminus \{y_j, y_p\})]$.  □

By the above claim, any attack on $\{x_i, y_j\}$ can be defended when neither of $i, j$ is $p$. If $y_j = y_p$ and $x_i \neq x_p$, then move the guard at $x_i$ to $y_j$, and all the guards at $X \setminus \{x_i, x_p\}$ to $Y \setminus \{y_1, y_p\}$, which is possible since there exists a perfect matching in $G[(X \setminus \{x_i, x_p\}) \cup (Y \setminus \{y_1, y_p\})]$, which can be proved by similar arguments used in the above claim. If $x_i = x_p$, then move the guard from $x_p$ to $y_j$, $y_1$ to $x_p$ and all the guards from $X \setminus \{x_p\}$ to $Y \setminus \{y_j\}$, which can be done because of the existence of a perfect matching in $G[(X \setminus \{x_p\}) \cup (Y \setminus \{y_j\})]$. So, if the initial configuration is $X \cup \{y_1\}$, any attack on an edge $e \in E$ can be defended by exchanging the guards at both endpoints of $e$ or by moving the guards to the configuration $Y \cup \{x_p\}$. If $|X| = |Y|$, when the configuration is $Y \cup \{x_p\}$ (or $X \cup \{y_1\}$), any attack on edge $e$ (where at least one of the endpoints of $e$ does not have guard assigned) can be defended by moving the guards to configuration $X \cup \{y_1\}$ (or $Y \cup \{x_p\}$). So, $ecvc(G) = cvc(G)$.  □

So, combining the above lemmas and observations the following theorem can be concluded.

**Theorem 5.** *The MIN-ECVC problem is solvable for chain graphs in $O(n+m)$ time.*

## 3    Eternal Connected Vertex Cover for Mycielskian of Hamiltonian Graphs

The Mycielskian graphs are well-studied graphs. Here at first the iterated Mycielskian graph class which is generated from $K_2$, will be discussed, which is $G_1 = K_2$, $G_2 = \mu(G_1), G_3 = \mu(G_2) \ldots$ and so on. The vertex set of Mycielskian of $G$, that is $V(\mu(G))$, is denoted as $V(\mu(G)) = \{p\} \cup O \cup N$. The set $O$ denotes the "old" vertices (that is, vertices of $G$), $N$ denotes the "new" vertices which form an independent set and is a copy of $O$; $p$ is another new vertex. The edge set is definded as $E(\mu(G)) = \{u_i u_j \mid u_i, u_j \in O, u_i u_j \in E(G)\} \cup \{u_i u'_j \mid u_i \in O, u'_j \in N, u_i u_j \in E(G)\} \cup \{p u'_i \mid u'_i \in N\}$. We define the vertex set of $G_i$ as $\{p_i\} \cup O_i \cup N_i$ and the number of vertices in $G_i$ are $2k_i + 1$, where $k_i$ denotes the number of vertices in $G_{i-1}$; for $i \geq 2$. In this section, at first we will show that the DECIDE-ECVC problem is NP-hard for Hamiltonian graphs.

**Theorem 6.** *The DECIDE-ECVC problem is NP-hard for Hamiltonian graphs.*

*Proof.* It is known that the vertex cover problem is NP-hard for Hamiltonian graphs [9]. A reduction will be shown from the decision version of the vertex cover problem for Hamiltonian graphs to the DECIDE-ECVC problem for Hamiltonian graphs. Let $G$ be a Hamiltonian graph (and not a complete graph), with $|V(G)| \geq 3$. Construct $H$ in such a way that, $H = (V(G) \cup \{u, v, w\}, E(G) \cup \{lk | l \in \{u, v\}, k \in V(G)\} \cup \{uw, vw\})$. It is obvious that $H$ is also a Hamiltonian graph.

Now we show that $G$ has a vertex cover of size at most $k(< n - 1)$ if and only if $H$ has an eternal connected vertex cover of size at most $k + 3$.

Let $G$ has a vertex cover (say $S$) of size at most $k$. Note that $S \cup \{u, v\}$ forms a connected vertex cover of $H$. Hence it is easy to observe by Theorem 1, that $S \cup \{u, v, w\}$ forms an eternal connected vertex cover of $H$. Conversely, let $H$ has an eternal connected vertex cover of size at most $k + 3$ and $D$ be the initial configuration. If $|D \cap V(G)| \leq k$, then $D \cap V(G)$ is a vertex cover of $G$ of size at most $k$ and we are done. If not then $|D \cap V(G)| = k + 1$, which means $D$ contains $u, v$ but not $w$. Now attack the edge $uw$ and after defending the attack the new configuration is $D'$ which contains $u, v$ and $w$. So, $|D' \cap V(G)| \leq k$ and $D' \cap V(G)$ is a vertex cover of $G$ of size at most $k$. Hence the DECIDE-ECVC problem is NP-hard for Hamiltonian graphs. □

The following theorem will be used in upcoming proofs.

**Theorem 7** *[8]. If $G$ is Hamiltonian, then $\mu(G)$ is also Hamiltonian.*

Since $\mu(K_2) = C_5$, and $C_5$ is Hamiltonian, $G_i$ is Hamiltonian for each $i \geq 2$.

**Lemma 8.** *If $|V(G_i)| = 2k_i + 1$, then $mvc(G_i) = k_i + 1$, for $i \geq 2$.*

*Proof.* By Theorem 7, each $G_i$ is Hamiltonian for $i \geq 2$. So, a cycle of order $2k_i + 1$ has vertex cover of size at least $k_i + 1$ and $O_i \cup \{p_i\}$ forms a vertex cover of size $k_i + 1$ of $G_i$, which implies $mvc(G_i) = k_i + 1$. □

Coming to the eternal connected vertex cover, in the following lemma an upper bound for $ecvc(\mu(G))$ is given.

**Lemma 9.** *Given the Mycielskian $\mu(G)(= \{p\} \cup O \cup N)$ of a connected graph $G$; $\{p, v_i\} \cup O$ (where $v_i$ is any vertex of $N$) forms an initial configuration of an eternal connected vertex cover of $\mu(G)$.*

*Proof.* The proof of Lemma 9 is omitted due to space constraints. □

Note that by Lemma 8 and Lemma 9 it follows that $cvc(G_i) \in \{k_i + 1, k_i + 2\}$. Further, to calculate the exact value of $cvc(G_i)$ for each $i \geq 3$, existence of a polynomial-time algorithm is shown in Theorem 8. Before proving Theorem 8, we prove the following lemma.

**Lemma 10.** *Given a Hamiltonian graph $G$ with odd number of vertices and its Mycielskian $\mu(G)$, (where $\mu(G) = \{p\} \cup O \cup N$) and $|\mu(G)| = 2k + 1$, if $S$ is a vertex cover of $\mu(G)$ then $|S \cap O| \geq \frac{k+1}{2}$.*

*Proof.* Note that if $S$ is a vertex cover of $\mu(G)$, then $S \cap O$ is a vertex cover of $G$; if not then there is some edge $e$ in $G$ that is not covered by $S \cap O$, which implies $e$ will not be covered in $\mu(G)$ also by $S$, contradicting the fact that $S$ is a vertex cover. So, $|S \cap O| \geq mvc(G) \geq \frac{k+1}{2}$. □

**Theorem 8.** *Given a $G_i = (V_i, E_i)$, there exists a polynomial-time algorithm to compute $cvc(G_i)$.*

*Proof.* Let $cvc(G_i) = mvc(G_i) = k_i + 1$ and $C$ be a minimum connected vertex cover of $G_i$. Note that $C \setminus \{p_i\}$ forms a minimum vertex cover of $G_i[V_i \setminus \{p_i\}]$. Now, $G_i$ is Hamiltonian, which implies $G_i[V_i \setminus \{p_i\}]$ has a Hamiltonian path. So, if $p_i v_1 v_2 \ldots v_{2k_i} p_i$ is the Hamiltonian cycle of $G_i$, then $v_1 v_2 \ldots v_{2k_i}$ is a Hamiltonian path of $G_i[V_i \setminus \{p_i\}]$. So, the possible minimum vertex cover candidates for $G_i[V_i \setminus \{p_i\}]$ are the following since they are the vertex cover choices for a path of length $2k_i$.

(1) $\lambda_1 = \{v_1 v_3 \ldots v_{2k_i-1}\}$
(2) $\lambda_2 = \{v_2 v_4 \ldots v_{2k_i-2} v_{2k_i}\}$
(3) $\lambda_3 = \{v_2 v_3 v_5 v_7 \ldots v_{2k_i-1}\}$
(4) $\lambda_4 = \{v_2 v_4 v_5 v_7 \ldots v_{2k_i-1}\}$
(5) $\lambda_5 = \{v_2 v_4 v_6 v_7 \ldots v_{2k_i-1}\}$
$\vdots$
$(k_i+1)$ $\lambda_{k_i+1} = \{v_2 v_4 v_6 v_8 \ldots v_{2k_i-2} v_{2k_i-1}\}$

So, there exist $k_i + 1$ choices (because of the Hamiltonian path) and the collection of choices be $\Lambda = \{\lambda_1, \lambda_2, \ldots \lambda_{k_i+1}\}$. For each $\lambda_j \in \Lambda$, check if $\{p_i\} \cup \lambda_j$ is a connected vertex cover. If no such $\lambda_j$ is found, then $cvc(G_i) = mvc(G_i) + 1 = k_i + 2$. Otherwise $cvc(G_i) = mvc(G_i) = k_i + 1$. It is easy to observe that the whole algorithm takes polynomial-time. $\square$

**Theorem 9.** $ecvc(G_i) = k_i + 2$, *for each* $i \geq 2$.

*Proof.* The proof is omitted due to space constraints. $\square$

Now if a Hamiltonian graph $G$ with an odd number of vertices along with its Hamiltonian cycle is given, then $mvc(\mu(G)), cvc(\mu(G))$ and $ecvc(\mu(G))$ can be computed similarly like the proofs of Lemma 8, Lemma 10 and Theorems 8,9.

**Lemma 11.** *Let* $G$ *be an even cycle of size* $2k$, *for* $k > 1$. *Then* $mvc(\mu(G)) = 2k + 1$ *and* $ecvc(\mu(G)) = 2k + 2$.

*Proof.* $G = C_{2k}$ and it is easy to observe that $mvc(\mu(G)) = 2k+1$ from previous proofs. For the sake of contradiction, let us assume that $ecvc(\mu(G)) = 2k + 1$, then for each $v \in V(G)$, there exists a minimum connected vertex cover that contains $v$. Let the cycle orientation of $G$ be $x_1 x_2 \ldots x_{2k}$. Let $G' = \mu(G) = \{p\} \cup O \cup N$; where $O = \{x_1, x_2, \ldots, x_{2k}\}$ and $N = \{y_1, y_2, \ldots, y_{2k}\}$. where $y_i$ is a copy of $x_i$ for each $i \in [2k]$. Then the Hamiltonian cycle in $G'$ be $y_1 x_2 y_3 \ldots x_{2k} x_1 y_2 x_3 \ldots x_{2k-1} y_{2k} p$ by [8]. Then from the proof of Theorem 8 there is only one choice for minimum connected vertex cover which contains $y_1$, that is $y_1 y_3 \ldots y_{2k-1} x_1 x_3 \ldots x_{2k-1} p$ but this is not a connected vertex cover as $\{x_1, x_3, \ldots x_{2k-1}\}$ is an independent set, so there does not exist any edge from $\{x_1, x_3, \ldots x_{2k-1}\}$ to $\{y_1, y_3, \ldots y_{2k-1}\}$. So, there does not exist any minimum connected vertex cover that contains $y_1$, hence contradiction arises. So, $ecvc(\mu(G)) = k + 2$. $\square$

**Lemma 12.** *For a Hamiltonian graph $G = (V, E)$ for which $|V|$ is even, that is, $|V| = 2k$ and the Hamiltonian cycle $x_1 x_2 \ldots x_{2k}$ is given; $mvc(\mu(G)) = 2k + 1$ and $ecvc(\mu(G)) = 2k + 2$.*

*Proof.* The proof is omitted due to space constraints.    □

Theorem 8 can be generalized for any Hamiltonian graph $G$, as the proof does not depend on whether $G$ is odd or even. As a result, the following theorem follows.

**Theorem 10.** *Given a Hamiltonian graph $G = (V, E)$, there exists a polynomial-time algorithm to compute $cvc(\mu(G))$ if Hamiltonian cycle representation of $G$ is given.*

So, from the above lemmas and theorems, Theorem 11 can be concluded.

**Theorem 11.** *Given a Hamiltonian graph $G = (V, E)$ with $|V(\mu(G))| = 2k+1$; $mvc(\mu(G)) = k+1$, $ecvc(\mu(G)) = k+2$ and given at least one of the Hamiltonian cycle representation of $G$, $cvc(\mu(G))$ can be computed in linear-time.*

# 4    Connected Vertex Cover for Graphs with Bounded Cliquewidth

In this section, we will show that the minimum connected vertex cover problem can be solved for distance-hereditary graphs using the fact that the distance-hereditary graphs are of bounded cliquewidth. With this, we settle an open problem mentioned in [6], to resolve the complexity status of the minimum connected vertex cover problem for distance-hereditary graphs. The details of the proof are omitted due to space constraints.

# 5    Eternal Connected Vertex Cover for Cographs

*Cographs* are exactly $P_4$-free graphs. A graph $G = (V, E)$ is called a *cograph* if it can be generated from $K_1$ by complementation and disjoint union. Recursively, the class of cographs can be defined as follows

1. $K_1$ is a cograph.
2. Complement of a cograph is a cograph.
3. $G_1$ and $G_2$ are cographs, then $G_1 \cup G_2$ is a cograph.

Cographs can be represented as join of $k$ graphs, $G_1, G_2, \ldots, G_k$; where each $G_i$ is either $K_1$ or a disconnected graph and this representation can be found in time $O(n^2)$ [18]. For $k \geq 3$, every minimum vertex cover of $G$ is connected. At first an algorithm is proposed which takes cograph $G$ for which every minimum vertex cover is connected, as input and outputs $ecvc(G)$. The Algorithm 2 (named as ECVC_CHECK) is deduced from Theorem 3, combined with the fact that cographs are also distance-hereditary graphs and the class of distance-hereditary graphs is closed under pendant vertex addition.

Hammer et al. [13] proved that the minimum vertex cover problem is solvable in linear-time for distance-hereditary graphs. So, in Algorithm 2, for each $v \in V$, computing $mvc(G_v)$ takes $O(n)$-time. So, the overall time complexity of Algorithm 2 is $O(n^2)$.

Now since for $k \geq 3$, every minimum vertex cover is connected, ECVC_CHECK can solve the problem in polynomial-time. Now for the case when $k = 2$, that is $G$ is join of $G_1$ and $G_2$, several cases may arise. In this section, without loss of generality, we assume that $|G_1| \leq |G_2|$.

**Case 1 :** $|G_1| = |G_2|$

If $mis(G) < |G_1|$, then every minimum vertex cover is connected, so the $ecvc(G)$ can be calculated by ECVC_CHECK($G$). If $mis(G) = |G_1|$ and $G_1$, $G_2$ are both independent, then $G$ is a complete bipartite graph and $ecvc(G) = |G_1| + 1$. But if $mis(G) = |G_1|$ and $G_2$ is not independent then only possible maximum independent set is $G_1$, so only possible minimum vertex cover is $G_2$; implying $ecvc(G) = |G_2| + 1$, where $V(G_2) \cup \{x\}$ ($x$ is any vertex from $V(G_1)$) forms an initial configuration of minimum eternal connected vertex cover.

---

**Algorithm 2.** ECVC_CHECK(G)

---

**Input** : A connected cograph $G = (V, E)$ for which every minimum vertex cover is connected.
**Output** : $ecvc(G)$.
Compute $mvc(G)$;
$count = 0$;
**for** each $u \in V$ **do**
    Add a pendant vertex $v$ to $u$ in $G$, the new graph is $G_u$;
    Compute $mvc(G_u)$;
    **if** $mvc(G) = mvc(G_u)$ **then**
        $count + +$;
    **end if**
**end for**
**if** $count = |V|$ **then**
    $ecvc(G) = mvc(G)$;
**else**
    $ecvc(G) = mvc(G) + 1$;
**end if**

---

**Case 2 :** $|G_1| < |G_2|$

If $mis(G) < |G_1| < |G_2|$, then any maximum independent set is a proper subset of $V(G_1)$ or $V(G_2)$, which implies every minimum vertex cover is connected. So, $ecvc(G)$ can be computed by ECVC_CHECK.

If $|G_1| < mis(G) < |G_2|$, then any maximum independent set is a proper subset of $V(G_2)$; so each minimum vertex cover is connected; so $ecvc(G)$ can be computed by ECVC_CHECK($G$).

If $|G_1| < mis(G) = |G_2|$, then the only possible minimum vertex cover of $G$ is $G_1$, as the only maximum independent set is $G_2$; implying $ecvc(G) = |G_1|+1$, as $V(G_1) \cup \{v\}$, (for any $v \in V(G_2)$) forms an initial configuration of minimum eternal connected vertex cover.

If $|G_1| = mis(G) < |G_2|$ and $mis(G_2) < mis(G)$, then $V(G_2)$ is the only minimum vertex cover of $G$, which implies $ecvc(G) = |G_2| + 1$.

Now consider the last case, if $|G_1| = mis(G) < |G_2|$ and $mis(G_2) = mis(G)$, then $cvc(G) = mvc(G)$; if $V(G_1)$ is not an independent set then every minimum vertex cover contains $V(G_1)$ and at least one vertex form $V(G_2)$, which implies every minimum vertex cover is connected, so ECVC_CHECK can compute $ecvc(G)$. If $V(G_1)$ is an independent set, then there exists minimum vertex cover which is not connected, for example, $V(G_2)$. Now construct a new graph $G'$ as follows: $G' = G_1 \oplus G_2'$; where $G_2' = (V(G_2) \cup \{u, u'\}, E(G_2) \cup \{uu'\})$. Now $mis(G') = |G_1| + 1 = mis(G_2) + 1$, which implies every minimum vertex cover of $G'$ is connected and $mvc(G') = cvc(G')$. Now, here comes a lemma.

**Lemma 13.** $ecvc(G') = cvc(G')$ *if and only if* $ecvc(G) = cvc(G)$.

*Proof.* Let $ecvc(G') = cvc(G')$ and let $C$ is an initial configuration for a minimum eternal connected vertex cover; implying $C$ is a minimum connected vertex cover of $G'$. Note that, $V(G_1) \subseteq V(C)$ and $|V(C) \cap \{u, u'\}| = 1$. So, $C \backslash \{u, u'\}$ is a minimum connected vertex cover of $G$. Now let the guards be assigned at the configuration $C \backslash \{u, u'\}$ in $G$ and some edge $e = xy$ is attacked. If $x, y$ both are guarded then we are done. If not, then the attack on $e$ in $G'$ is defended by moving the guards to some configuration $C'$. Similarly the guards at $C \backslash \{u, u'\}$ can be moved to $C' \backslash \{u, u'\}$ by using the same moving pattern like $C$ to $C'$. The attack is defended. So, $C \backslash \{u, u'\}$ forms an initial configuration of minimum eternal connected vertex cover of $G$. So, $ecvc(G) = cvc(G)$.

Conversely, let $ecvc(G) = cvc(G)$ and $S'$ is the initial configuration of minimum eternal connected vertex cover of $G$. Then it is easy to observe that $S'' = S' \cup \{u\}$ is a minimum connected vertex cover of $G'$. Let an edge $e = xy$ is attacked in $G'$. If $x$ and $y$ are both covered by guards then we are done. If not, two cases may arise.

If $x \in V(G_1)$ and $y \in V(G_2')\backslash S''$; then either $y$ is in $G_2$ or $y = u'$. If $y = u'$, then move the guard at $x$ to $u'$ and the guard at $u$ to $x$. The attack is defended successfully. If $y \in V(G_2)$, then move the guards in $G'$ according to how the guards are moved in $G$ when $e$ was attacked in $G$.

$x \in V(G_2') \cap C''$ and $y \in V(G_2') \backslash C''$. If $x = u$ and $y = u'$, then shift the guard at $u$ to $u'$ and the attack is defended successfully. If $x \in V(G_2') \cap (C''\backslash\{u, u'\})$ and $y \in V(G_2')\backslash(C'' \cup \{u, u'\})$, then defend the attack by moving the guards in the same pattern in $G'$ as the guards were moved (except the guard at $u$ or $u'$) when $e$ was attacked in $G$. So, the attack is defended which implies $ecvc(G') = cvc(G')$. Hence the lemma is proved. □

So, a polynomial-time algorithm can be made for a cograph $G = G_1 \oplus G_2$ where $|G_1| = mis(G) = mis(G_2) < |G_2|$ and $V(G_1)$ is an independent set in the following way: first $G'$ is constructed from $G$ using the above description. Note that every minimum vertex cover of $G'$ is connected, which implies that we can compute $ecvc(G')$ by ECVC_CHECK. If $ecvc(G') = cvc(G')$, then $ecvc(G) = cvc(G)$. Otherwise $ecvc(G) = cvc(G) + 1$. So from the above lemmas and algorithm, we may conclude the following theorem.

**Theorem 12.** *The MIN-ECVC problem is solvable for cographs in* $O(n^2)$ *time.*

## 6   Conclusion and Future Aspects

In this paper, we have solved the MIN-ECVC problem for the chain graphs and cographs. To the best of our knowledge, the class of chain graphs is the biggest subclass of bipartite graphs for which a linear-time algorithm has been proposed to solve the MIN-ECVC problem. It will be interesting to look for efficient algorithms to solve the MIN-ECVC problem for other important subclasses of bipartite graphs. For distance-hereditary graphs, we have shown that the connected vertex cover problem can be solved in linear-time. But, the complexity status of the MIN-ECVC problem is open for distance-hereditary graphs. One may try to propose an algorithm to solve the MIN-ECVC problem for distance-hereditary graphs.

**Acknowledgements.** The authors would like to thank Prof. Ton Kloks for their invaluable inputs and suggestions which helped in improving the paper.

## References

1. Araki, H., Fujito, T., Inoue, S.: On the eternal vertex cover numbers of generalized trees. IEICE Trans. Fundam. Electron. Commun. Comput. Sci. **98**, 1153–1160 (2015)
2. Babu, J., Chandran, L.S., Francis, M., Prabhakaran, V., Rajendraprasad, D., Warrier, N.J.: On graphs whose eternal vertex cover number and vertex cover number coincide. Discret. Appl. Math. **319**, 171–182 (2022)
3. Babu, J., Misra, N., Nanoti, S.G.: Eternal vertex cover on bipartite graphs. In: Kulikov, A.S., Raskhodnikova, S. (eds.) CSR 2022. LNCS, vol. 13296, pp. 64–76. Springer, Cham (2022). https://doi.org/10.1007/978-3-031-09574-0_5
4. Babu, J., Prabhakaran, V.: A new lower bound for the eternal vertex cover number of graphs. J. Comb. Optim. **44**, 2482–2498 (2022)
5. Babu, J., Prabhakaran, V., Sharma, A.: A substructure based lower bound for eternal vertex cover number. Theor. Comput. Sci. **890**, 87–104 (2021)
6. Escoffier, B., Gourvès, L., Monnot, J.: Complexity and approximation results for the connected vertex cover problem in graphs and hypergraphs. J. Discrete Algorithms **8**, 36–49 (2010)
7. Fernau, H., Manlove, D.F.: Vertex and edge covers with clustering properties: complexity and algorithms. J. Discrete Algorithms **7**, 149–167 (2009)
8. Fisher, D.C., McKenna, P.A., Boyer, E.D.: Hamiltonicity, diameter, domination, packing, and biclique partitions of Mycielski's graphs. Discret. Appl. Math. **84**, 93–105 (1998)

9. Fleischner, H., Sabidussi, G., Sarvanov, V.I.: Maximum independent sets in 3- and 4-regular Hamiltonian graphs. Discret. Math. **310**, 2742–2749 (2010)
10. Fomin, F.V., Gaspers, S., Golovach, P.A., Kratsch, D., Saurabh, S.: Parameterized algorithm for eternal vertex cover. Inf. Process. Lett. **110**, 702–706 (2010)
11. Fujito, T., Nakamura, T.: Eternal connected vertex cover problem. In: Chen, J., Feng, Q., Xu, J. (eds.) TAMC 2020. LNCS, vol. 12337, pp. 181–192. Springer, Cham (2020). https://doi.org/10.1007/978-3-030-59267-7_16
12. Garey, M.R., Johnson, D.S.: The rectilinear Steiner tree problem is $np$-complete. SIAM J. Appl. Math. **32**, 826–834 (1977)
13. Hammer, P.L., Maffray, F.: Completely separable graphs. Discret. Appl. Math. **27**, 85–99 (1990)
14. Klostermeyer, W.F., Mynhardt, C.M.: Edge protection in graphs. Australas. J Comb. **45**, 235–250 (2009)
15. Li, Y., Wang, W., Yang, Z.: The connected vertex cover problem in k-regular graphs. J. Comb. Optim. **38**, 635–645 (2019)
16. Paul, K., Pandey, A.: Some algorithmic results for eternal vertex cover problem in graphs. In: Lin, C.C., Lin, B.M.T., Liotta, G. (eds.) WALCOM 2023. LNCS, vol. 13973, pp. 242–253. Springer, Cham (2023). https://doi.org/10.1007/978-3-031-27051-2_21
17. Priyadarsini, P.K., Hemalatha, T.: Connected vertex cover in 2-connected planar graph with maximum degree 4 is NP-complete. Int. J. Math. Comput. Phys. Electr. Comput. Eng, **1**, 570–573 (2007)
18. Stewart, L.: Cographs: a Class of Tree Representable Graphs. University of Toronto, Department of Computer Science (1978)
19. Ueno, S., Kajitani, Y., Gotoh, S.: On the nonseparating independent set problem and feedback set problem for graphs with no vertex degree exceeding three. Discret. Math. **72**, 355–360 (1988)
20. Watanabe, T., Kajita, S., Onaga, K.: Vertex covers and connected vertex covers in 3-connected graphs. In: 1991 IEEE International Symposium on Circuits and Systems (ISCAS), vol. 2, pp. 1017–1020 (1991)

# Impact of Diameter and Convex Ordering for Hamiltonicity and Domination

R. Mahendra Kumar[(⊠)] and N. Sadagopan

Indian Institute of Information Technology, Design and
Manufacturing,Kancheepuram, Chennai, India
{coe18d004,sadagopan}@iiitdm.ac.in

**Abstract.** A bipartite graph $G(X, Y)$ is called a star convex bipartite graph with convexity on $X$ if there is an associated star $T$ on $X$ such that for each vertex in $Y$, its neighborhood induces a subtree in $T$. A split graph $G(K, I)$ is a graph that can be partitioned into a clique ($K$) and an independent set ($I$). The objective of this study is twofold: (i) to strengthen the results presented in [1] for the Hamiltonian cycle (HCYCLE), the Hamiltonian path (HPATH), and Domination (DS) problems on star convex bipartite graphs (ii) to reinforce the results of [2] for HCYCLE, and HPATH on split graphs by introducing convex ordering on one of the partitions (clique or independent set). We establish the following dichotomy results on star convex bipartite graphs: (i) HCYCLE is NP-complete for diameter 3, and polynomial-time solvable for diameter 2, 5, and 6 (ii) HPATH is polynomial-time solvable for diameter 2, and NP-Complete, otherwise. Note that HCYCLE and HPATH are NP-complete on star convex bipartite graphs with diameter 4 1. Similarly, we present the following results on split graphs by imposing convexity on $K$ ($I$); HCYCLE and HPATH are NP-complete on star (comb) convex split graphs with convexity on $K$ ($I$). On the positive side, we show that for $K_{1,5}$-free star convex split graphs with convexity on $I$, HCYCLE is polynomial-time solvable.

We further show that the domination problem and its variants (Connected, Total, Outer-Connected, and Dominating biclique) are NP-complete on star convex bipartite graphs with diameter 3 (diameter 5, diameter 6). On the parameterized complexity front, we prove that the parameterized version of the domination problem, with the parameter being the solution size, is not fixed-parameter tractable on star convex bipartite graphs with a diameter at most 4, whereas it is fixed-parameter tractable when the parameter is the number of leaves in the associated star.

**Keywords:** Star (Bi-star) convex bipartite graph · Star (Comb) convex split graphs · Hamiltonian cycle (path) · Domination · Diameter · Dichotomy results

---

This work is partially supported by NBHM project, NBHM/02011/24/2023/6051.

S. Kalyanasundaram and A. Maheshwari (Eds.): CALDAM 2024, LNCS 14508, pp. 194–208, 2024.
https://doi.org/10.1007/978-3-031-52213-0_14

# 1   Introduction

The Hamiltonian cycle (path) problem is one of the most significant problems in graph theory. The complexity of the Hamiltonian cycle (path) problem for various graph classes has been well-studied in the literature. The Hamiltonian cycle (path) problem asks for a cycle (path) that visits each node exactly once. The Hamiltonian cycle (path) problem is NP-complete for chordal bipartite graphs [2], and split graphs [2]. Interestingly, this problem is polynomial-time solvable for distance-hereditary graphs, and bipartite permutation graphs [3].

Bipartite graphs are one of the well-studied graph classes in the literature. An important restricted bipartite graphs studied in the literature is tree convex bipartite graphs [4]. The Hamiltonian cycle (path) problem is NP-complete on star convex bipartite graphs [1] and comb convex bipartite graphs [1]. In this study, we aim to investigate the computational complexity of HPATH on star convex bipartite graphs from a different perspective. One can study the structural characterization of a graph in terms of graph parameters. Among the graph parameters, diameter is one of the popular parameters. Interestingly, for star convex bipartite graphs, we show that the diameter is at most 6. Further we establish a dichotomy result for star convex bipartite graphs, HCYCLE is NP-complete for diameter 3, and polynomial-time solvable for diameter 2, 5 and 6. Similarly, we show that HPATH is Polynomial-time solvable for diameter 2, and NP-Complete, otherwise. Also, we show that for bi-star convex bipartite graphs with diameter 3, HCYCLE and HPATH are NP-complete. All our results on HCYCLE and HPATH are also true for star convex chordal bipartite graphs. Note that HCYCLE and HPATH are NP-complete on star convex bipartite graphs with diameter 4 [1].

Similar to HCYCLE and HPATH, domination (DS) is well-studied in the literature. For a graph $G$, a set $D \subseteq V(G)$ is a dominating set if each vertex $v \in V(G) \backslash D$ is adjacent to at least one vertex in $D$. It is known that DS is NP-complete for star convex bipartite graphs [1], and comb convex bipartite graphs [1], star-convex split graphs with convexity on $I$ and comb-convex split graphs with convexity on I [12]. In this paper, we show that for star convex bipartite graphs with diameter 3, 5, and 6, DS is NP-complete. We consider some of the variants of DS and present hardness results. A set $D \subseteq V(G)$ of a graph is called a Connected Dominating set (CDS), Total Dominating set (TDS), respectively, if $D$ is a dominating set and the induced subgraph of $D$ is connected, has no isolated vertices respectively. A set $D \subseteq V(G)$ of a graph is called an outer-connected dominating set (OCDS) of $G$ if $D$ is a dominating set and the induced subgraph of $G$ on $V(G) \backslash D$ is connected. A set $D \subseteq V(G)$ of a graph is called a dominating biclique (DB) of $G$ if $D$ is a dominating set and the induced subgraph of $D$ is a biclique. The outer-connected dominating set problem is NP-complete for bipartite graphs [5], perfect elimination bipartite graphs [6], and split graphs [7]. On the positive side, OCD is linear-time solvable on chain graphs and bounded tree-width graphs [6].

Having imposed convex ordering on bipartite graphs, it is natural to explore this line of study on graphs having two partitions. A natural choice after bipartite graphs is the class of split graphs. A split graph is a graph that can be partitioned into a clique and an independent set. We wish to extend this line of study to split graphs by considering convex ordering with respect to the clique set, an independent set.

Interestingly, HCYCLE on $K_{1,5}$-free star convex split graphs with convexity on $I$ is polynomial-time solvable. This brings an interesting dichotomy for HCY-CLE on star convex split graphs with convexity on $I$. Note that for $K_{1,5}$-free split graphs, HCYCLE is NP-complete [10]. Further, we show that HCYCLE and HPATH are NP-complete for comb convex split graphs with convexity on $K$ ($I$). Also, we show that HCYCLE on split graphs with $|K| > |I|$ is NP-complete. Similarly, we show that for split graphs with $|K| = |I|$, and $|K| > |I|$ HPATH is NP-complete.

In this paper, we work with simple, connected, undirected, and unweighted graphs. We follow the notation and definitions as defined in [13,14]. We shall now define tree convex bipartite, and split graphs.

**Definition 1.** *A bipartite graph $G = (X, Y, E)$ is called tree convex bipartite if there is an associated tree $T = (X, F)$, such that for each vertex $u$ in $Y$, its neighborhood $N_G(u)$ induces a subtree of $T$.*

**Definition 2.** *A split graph $G$ is called $\Pi$-convex with convexity on $K$ if there is an associated graph $\Pi$ on $K$ such that for each $v \in I$, $N_G(v)$ induces a connected subgraph in $\Pi$.*

**Definition 3.** *A split graph $G$ is called $\Pi$-convex with convexity on $I$ if there is an associated graph $\Pi$ on $I$ such that for each $v \in I$, $N_G^I(v)$ induces a connected subgraph in $\Pi$.*

**Definition 4.** *A bipartite graph $G = (X, Y, E)$ is called bi-star convex bipartite if there is an associated star $T_1 = (X, F)$, and $T_2 = (Y, F')$ such that for each vertex $u$ in $Y$, its neighborhood $N_G(u)$ induces a subtree of $T_1$ and for each vertex $v$ in $X$, its neighborhood $N_G(v)$ induces a subtree of $T_2$.*

Due to the page constraint, we ignore some of the proofs. For complete details, readers can refer to arXiv.

## 2   Structural Results

In this section, we shall present a structural characterization of star convex bipartite graphs with respect to their diameter. We shall fix the following notation to present our results. For a star convex bipartite graph $G$ with bi-partition $(X, Y)$, let $X = \{u_1, u_2, \ldots, u_m\}$ and $Y = \{v_1, v_2, \ldots, v_n\}$. By $T$, we denote an associated star in $X$.

**Lemma 1.** *Let $G(X, Y)$ be a star convex bipartite graph with convexity on $X$. If $\delta(G_Y) \geq 2$, then there exists a vertex $u \in X$ universal to $Y$.*

*Proof.* We know that $G$ is a star convex bipartite graph with convexity on $X$. Since $\delta(G_Y) \geq 2$, each vertex in $Y$ must be adjacent to the root of $T$, say $u \in X$, to satisfy the convexity property on $X$. This shows that $u$ is universal to $Y$.    □

We show that the diameter of $G$ is bounded, in particular $diam(G)$ is at most 6.

**Theorem 1.** *Let $G(X,Y)$ be a connected star convex bipartite graph with convexity on $X$ $(Y)$. Then, $diam(G)$ is at most 6.*

*Proof.* Let $G$ be a star convex bipartite graph with convexity on $X$ and $u_t$ be the root of an associated star $T$. Assume on the contrary, that $diam(G)$ is at least 7. Without loss of generality, we assume that $diam(G)$ is 7. Since $diam(G)$ is 7, there exists a pair of vertices $u, v \in V(G)$ whose distance is 7. This implies that there exists an induced $P_8$ in $G$. Let the $P_8$ be $(u = z_1, z_2, z_3, z_4, z_5, z_6, z_7, z_8 = v)$. Without loss of generality, we partition the vertices of $P_8$ as follows; $X' = \{z_1, z_3, z_5, z_7\}$ and $Y' = \{z_2, z_4, z_6, z_8\}$. Case 1: $u_t \in X'$. Let $H(X', Y')$ be the graph induced on $V(P_8)$. We observe that $\delta(H_{Y'} - z_8) \geq 2$. Note that there is no $u \in V(H_{X'})$ such that $u$ is universal to $V(H_{Y'})$, which is a contradiction to Lemma 1. Case 2: $u_t \notin X'$. Since $G$ is a star convex bipartite graph with convexity on $X$ and the vertices of $H_{Y'} - z_8$ have minimum degree two, they must be adjacent to the root vertex $u_t$. Clearly, the distance between $u, v$ is at most four, which is a contradiction. Both cases clearly show that the diameter of $G$ is at most six.    □

**Theorem 2.** *A graph $G(X,Y)$ is star convex bipartite with diameter at most 2 if and only if $G(X,Y)$ is complete bipartite.*

**Corollary 1.** *For star convex bipartite graphs with diameter 2, the Hamiltonian cycle (path) problem is polynomial-time solvable.*

**Lemma 2.** *Let $G(X,Y)$ be a star convex bipartite graph with diameter 5. Then, there exists at least one pendant vertex in $Y$.*

*Proof.* Assume, on the contrary, that there does not exist a pendent vertex in $Y$. This implies that $\delta(G_Y) \geq 2$. We know from Lemma 1 that there exists a vertex $x \in X$ universal to $Y$. We now argue that the distance between any arbitrary pair of vertices $u, v$ in $G$ is at most 4. We observe that $u, v$ can be from the same partition or from different partitions. Case 1: $u, v \in Y$. It is easy to see that the distance between $u$ and $v$ is 2. Case 2: $u, v \in X$. Since $G$ is connected, there exists $u' \in Y$ such that $u' \in N_G(u)$. Similarly, there exists $v' \in Y$ such that $v' \in N_G(v)$. From Case 1, we know that the distance between $u'$ and $v'$ is 2. This shows that $d(u, v) \leq 4$. Case 3: $u \in X$ and $v \in Y$. Since $G$ is connected, there exists $u' \in Y$ such that $u' \in N_G(u)$. From Case 1, we know that the distance between $u'$ and $v$ is 2. This shows that $d(u, v) \leq 3$. We know that $diam(G)$ is 5. Since $diam(G)$ is 5, there must exist a pair $u, v$ in $G$ whose distance is 5. Since we arrive at a contradiction in all cases, it follows that our assumption is wrong. Therefore, there exists at least one pendant vertex in $Y$.    □

**Lemma 3.** *Let $G(X,Y)$ be a star convex bipartite graph with convexity on $X$. If there exists a universal vertex in $X$, then $diam(G)$ is at most 4.*

**Lemma 4.** *Let $G(X,Y)$ be a star convex bipartite graph with diameter 6. Then, there exist at least two pendant vertices $y_1, y_2$ in $Y$.*

*Proof.* On the contrary, assume that at most one pendant vertex exists in $Y$. Case 1: There does not exist a pendant vertex in $Y$. Proof of this case is similar to the proof of Lemma 2. Case 2: There exists exactly one pendant vertex in $Y$. Let the pendant vertex be $u$. On removing $u$ from $G$, $\delta(G_Y)$ becomes at least 2. An argument similar to the proof of Lemma 2 shows that the diameter of $G - u$ is at most 4. Adding $u$ to $G$ can increase the diameter of $G$ by at most one. This shows that $diam(G)$ is at most 5, a contradiction. □

**Lemma 5.** *Let $G(X,Y)$ be a star convex bipartite graph with convexity on $X$. If there exists a universal vertex in both $X$ and $Y$, then $diam(G)$ is at most 3.*

### 2.1   Hamiltonian Cycle in Star Convex Bipartite Graphs

It is known from [2] that the Hamiltonian path problem in chordal bipartite graph $G(X,Y)$ with $|X| = |Y|$ is NP-complete. Using this reduction, we show that the Hamiltonian cycle problem is NP-complete on star convex bipartite graphs with diameter 3. Note that chordal bipartite graphs which are bipartite graphs that forbid induced cycles of length at least six.

**Theorem 3.** *For star convex bipartite graphs with diameter 3, the Hamiltonian cycle problem is NP-complete.*

*Proof.* We present a deterministic polynomial-time reduction that reduces an instance of chordal bipartite graph $G(X,Y)$ with $|X| = |Y|$ to a corresponding star convex bipartite graph with diameter 3 instance $G'(X',Y')$. The mapping of an instance of $G$ to the corresponding instance of $G'$ is as follows: $V(G') = X' \cup Y'$, $X' = \{v_i \mid v_i \in X, 1 \leq i \leq |X|\} \cup \{v_{|X|+1}, v_{|X|+2}, v_{|X|+3}\}$, $Y' = \{u_j \mid u_j \in Y, 1 \leq j \leq |Y|\} \cup \{u_{|Y|+1}, u_{|Y|+2}, u_{|Y|+3}\}$ and $E(G') = E(G) \cup E'$, $E' = \{\{v_{|X|+2}, u_{|Y|+1}\}, \{v_{|X|+2}, u_{|Y|+3}\}, \{v_{|X|+3}, u_{|Y|+3}\}\} \cup \{\{v_{|X|+1}, u_j\} \mid v_{|X|+1} \in X', u_j \in Y', 1 \leq j \leq |Y'|\} \cup \{\{u_{|Y|+2}, v_i\} \mid u_{|Y|+2} \in Y', v_i \in X', 1 \leq i \leq |X'|\}$.

We show that $G'(X',Y')$ is a star convex bipartite graph with diameter 3 and convexity on $X'$. We first show that $diam(G')$ is 3. Note that $v_{|X|+1}$ is universal to $Y'$ and $u_{|Y|+2}$ is universal to $X'$. From Lemma 5, it is clear that $diam(G')$ is at most 3. In particular, the distance between $v_{|X|+3}$ and $u_{|Y|+1}$ is 3. Therefore, $diam(G')$ is 3. We now argue that $G'$ is a star convex bipartite graph with convexity on $X'$. The star $T'$ on $X'$ has root vertex $v_{|X|+1}$ and leaves $v_1, \ldots, v_{|X|}, v_{|X|+2}, v_{|X|+3}$. For any vertex $u_j \in Y'$ in $G'$, its neighborhood $N_{G'}(u_j) = N_G(u_j) \cup \{v_{|X|+1}\}$ is a subtree of $T'$. This implies that $G'$ is a star convex bipartite graph with convexity on $X'$. Clearly, it is a polynomial-time reduction as we add only six vertices. We claim that $G$ is a yes-instance of the

Hamiltonian path problem if and only if $G'$ is a yes-instance of the Hamiltonian cycle problem.

($\Rightarrow$) $|X| = |Y|$, any Hamiltonian path $P$ in $G$ must have one endpoint in $X$ and another in $Y$. Suppose that there exists a Hamiltonian path $P$ in $G$. Let $P = (v_1, u_1, v_2, u_2, \ldots, v_{|X|-1}, u_{|Y|-1}, v_{|X|}, u_{|Y|})$. By the construction of $G'$, we know that exists a path $P' = (v_{|X|+1}, u_{|Y|+1}, v_{|X|+2}, u_{|Y|+3}, v_{|X|+3}, u_{|Y|+2})$ such that it visits all the six newly added vertices. Since $v_{|X|+1}$ and $u_{|Y|+2}$ are universal to $Y$ and $X$, respectively, we join the paths $P$ and $P'$ to get the Hamiltonian cycle in $G'$. $C = (v_1, u_1, v_2, u_2, \ldots, v_{|X|-1}, u_{|Y|-1}, v_{|X|}, u_{|Y|}, v_{|X|+1}, u_{|Y|+1}, v_{|X|+2}, u_{|Y|+3}, v_{|X|+3}, u_{|Y|+2}, v_1)$.

($\Leftarrow$) We know that $d_{G'}(v_{|X|+3}) = 2$ and $d_{G'}(u_{|Y|+1}) = 2$. Since $N_{G'}(v_{|X|+3}) = \{u_{|Y|+2}, u_{|Y|+3}\}$ and $N_{G'}(u_{|Y|+1}) = \{v_{|X|+1}, v_{|X|+2}\}$, any Hamiltonian cycle must contain $P' = (v_{|X|+1}, u_{|Y|+1}, v_{|X|+2}, u_{|Y|+3}, v_{|X|+3}, u_{|Y|+2})$ as a subpath. Note that $V(G) = V(G') \backslash V(P')$. We remove $P'$ from $C$ to get a Hamiltonian path in $G$. □

Note that the constructed graph $G'$ has two vertices, $v_{|X|+1}$ is universal to $Y'$ and $u_{|Y|+2}$ is universal to $X'$. This shows that $G'$ is a bi-star convex bipartite graphs. Therefore, for bi-star convex bipartite graphs with diameter 3, the Hamiltonian cycle problem is NP-complete.

**Theorem 4.** *If $G$ is a star convex bipartite graph with diameter 5 (diameter 6), then $G$ has no Hamiltonian cycle.*

*Proof.* Follows from Lemma 2 and Lemma 4. Further, this can be answered in polynomial time. □

## 2.2   Hamiltonian Path in Star Convex Bipartite Graphs

In this section, we show that the Hamiltonian path problem is NP-complete for star convex bipartite graphs with diameter 3, diameter 5, and diameter 6. Since the necessity part is straight forward, we present the construction and the sufficiency part of the proof of the NP-complete reductions.

**Theorem 5.** *For star convex bipartite graphs with diameter 3, the Hamiltonian path problem is NP-complete.*

*Proof.* We present a reduction that reduces an instance of chordal bipartite graph $G(X, Y)$ with $|X| = |Y|$ to a corresponding instance of star convex bipartite graph with diameter 3 $G'(X', Y')$. The mapping of an instance of $G$ to the corresponding instance of $G'$ is as follows: $V(G') = X' \cup Y'$, $X' = \{v_i \mid v_i \in X, 1 \leq i \leq |X|\} \cup \{v_{|X|+1}, v_{|X|+2}\}$, $Y' = \{u_j \mid u_j \in Y, 1 \leq j \leq |Y|\} \cup \{u_{|Y|+1}, u_{|Y|+2}\}$ and $E(G') = E(G) \cup E'$, $E' = \{\{v_{|X|+1}, u_{|Y|+1}\}, \{v_{|X|+1}, u_{|Y|+2}\}, \{v_{|X|+2}, u_{|Y|+2}\}\} \cup \{\{v_{|X|+1}, u_j\} \mid v_{|X|+1} \in X', u_j \in Y', 1 \leq j \leq |Y|\} \cup \{\{u_{|Y|+2}, v_i\} \mid u_{|Y|+2} \in Y', v_j \in X', 1 \leq i \leq |X|\}\}$.

It is easy to see that the generated instances are star convex bipartite graphs with convexity on $X'$ and $diam(G')$ is three, due to the presence of the universal in $X$. We skip the proof of the necessity part as it is easy.

($\Leftarrow$) Since $v_{|X|+2}$ and $u_{|Y|+1}$ are pendant vertices, any Hamiltonian path in $G'$ must start at $v_{|X|+2}$ and ends at $u_{|Y|+1}$ or vice-versa. Without loss of generality, assume that the Hamiltonian path $P'$ starts at $v_{|X|+2}$ and ends at $u_{|Y|+1}$. In particular, $P' = (v_{|X|+2}, u_{|Y|+2}, v \ldots, u, v_{|X|+1}, u_{|Y|+1})$. Clearly, $P_{vu}$ is a Hamiltonian path in $G$.    $\square$

**Corollary 2.** *For bi-star convex bipartite graphs with diameter 3, the Hamiltonian path problem is NP-complete.*

**Theorem 6.** *For star convex bipartite graphs with diameter 5, the Hamiltonian path problem is NP-complete.*

**Remarks:** We know that from Lemma 2 that star convex bipartite graphs with diameter 5 have at least one pendant vertex. Note that the reduction instances have exactly one pendant vertex. This shows that the presence of pendant vertices makes the problem NP-hard.

**Theorem 7.** *For star convex bipartite graphs with diameter 6, the Hamiltonian path problem is NP-complete.*

### 2.3    Domination and Its Variants on Star Convex Bipartite Graphs

**2.3.1    Classical Complexity** In this section, we prove that the complexity of the dominating set problems for star convex bipartite graphs with diameter 3, 5 and 6 are NP-complete. Also, we present hardness results for some of the variants of dominating set problems, such as connected dominating set and total dominating set for star convex bipartite graphs with diameter 3, 5 and 6.

**Theorem 8.** *For star convex bipartite graphs with diameter 3, the dominating set problem is NP-complete.*

*Proof.* It is known that the vertex cover problem (VC) on general graphs is NP-complete, and this can be reduced in polynomial time to dominating set problem (DS) on star convex bipartite graphs with diameter 3 using the following reduction. We construct a star convex bipartite graph $G'(X', Y')$ with diameter 3 and convexity on $X'$ from the given graph $G(V, E)$ in polynomial-time. Let $n = |V(G)|$, $m = |E(G)|$ and $E(G) = \{e_1, e_1, \ldots, e_m\}$ . $V(G') = X' \cup Y'$, $X' = X'_1 \cup X'_2 \cup \{z_{m+1}\}$, $Y' = Y'_1 \cup Y'_2 \cup \{u_{n+1}\}$. $X'_1 = \{v_i \mid e_i \in E(G), 1 \le i \le m\}$, $X'_2 = \{z_i \mid e_i \in E(G), 1 \le i \le m\}$, $Y'_1 = \{u_j \mid x_j \in V(G), 1 \le j \le n\}$, $Y'_2 = \{u_{n+r+1} | 1 \le r \le 2m^2 - m\}$, and associate each vertex $u_{n+1+r}$ with a distinct subset of two elements from $X'_1 \cup X'_2$ arbitrarily. We now define the edges of $G'$. For each edge $e_k = \{x_i, x_j\} \in E(G)$, we add the following edges $\{u_i, v_k\}, \{u_i, z_k\}, \{u_j, v_k\}, \{u_j, z_k\}\}$. Also, for each vertex $u_{n+r+1}, 1 \le r \le 2m^2 - m$ in $Y'_2$ which we created for every pair of vertices $\{w_i, w_j\}$ in $X'_1 \cup X'_2$, we add the following edges $\{w_i, u_{n+r+1}\}$ and $\{w_j, u_{n+r+1}\}$. We make $z_{m+1}$ universal to $Y'$.

We show that the constructed graph $G'(X', Y')$ is a star convex bipartite graph with diameter 3 and convexity on $X'$. Since $z_{m+1} \in X'$ is universal to $Y'$, the distance between any pair of vertices in $Y'$ is two. Similarly, $d(v_i, u_j), v_i \in X', u_j \in Y'$ is at most 3. By our construction, we know that each pair of vertices in $X'$ has a common neighbor. Clearly, the distance between any pair of vertices in $X'$ is two. We know that $G'$ is not a complete bipartite graph (Theorem 2); therefore, $diam(G')$ is 3. We now show that $G'$ is a star convex bipartite graph with convexity on $X'$. The star $T'$ on $X'$ has root vertex $z_{m+1}$ and leaves $v_1, \ldots, v_m, z_1, \ldots, z_m$. For any vertex $u_j \in Y'$ and $d_{G'}(u_j) \geq 2$, its neighborhood contains the vertex $z_{m+1}$ which forms a subtree of $T'$. This implies that $G'$ is a star convex bipartite graph with convexity on $X'$. Clearly, it is a polynomial-time reduction. We claim that $G$ has a VC of size at most $k$ if and only if $G'$ has a DS of size at most $k' = k + 1$.

($\Rightarrow$) Let $S = \{x_j | 1 \leq j \leq k\}$ is a vertex cover of size $k$ in $G$. Then we construct the dominating set $D$ of size $k' = k + 1$ in $G'$ as follows: $D = \{u_j | 1 \leq j \leq k\} \cup \{z_{m+1}\}$. Since $S$ is a vertex cover, for any edge $e_k = \{x_i, x_j\} \in E(G)$, $x_i$ or $x_j$ is in $S$. Assume that $x_i \in S$, we include $u_i$ to $D$ ($v_k$ and $z_k$ are adjacent to $u_i$ in $G'$). By our construction, every vertex $v_i \in X_1'$ and $z_i \in X_2'$ are adjacent to at least one vertex from $\{u_i \mid 1 \leq j \leq k\}$ and every vertex in $Y_1'$ and $Y_2'$ is adjacent to $z_{m+1}$. Since $u_{n+1}$ is a pendant vertex, we include its neighbor $z_{m+1}$ to $D$. This implies that $G'$ has a DS of size at most $k' = k + 1$.

($\Leftarrow$) We know from our construction that any dominating set $D$ must contain either $z_{m+1}$ or $u_{n+1}$. Suppose $D$ contains $u_{n+1}$, then we replace $u_{n+1}$ by $z_{m+1}$ to get a dominating set of same size, since $u_{n+1}$ is a pendant vertex. Suppose $D$ contains any vertex from $Y_2'$ then we replace it with $z_{m+1}$, since $z_{m+1}$ is universal to $Y_2'$, and the vertices of $Y_2'$ can dominate only two vertices of $X_1'$. We now argue that any dominating set of $G'$ does not contain vertices from $X_1'$ and $X_2'$. For an edge $e_k = \{x_i, x_j\} \in E(G)$, $D$ can contain Case 1: only one vertex from $\{v_k, z_k\}$, Case 2: both the vertex from $\{v_k, z_k\}$. Case 1: $D$ contains any one of the vertices from $\{v_k, z_k\}$, say $v_k$. This implies that $D$ contain at most one vertex from $\{u_i, u_j\}$. Suppose $D$ contains both $u_i$ and $u_j$, then $v_k$ is redundant in $D$, a contradiction to the minimality of $D$. Without loss of generality, assume that $D$ contains $u_i$ to dominate $z_k$. Since $v_k$ is only adjacent to $u_i$ and $u_j$, we can replace $v_k$ by $u_j$ in $D$ to get a dominating set of the same size. Case 2: $D$ contains both the vertices from $\{v_k, z_k\}$. Since $v_k$ and $z_k$ are only adjacent to $u_i$ and $u_j$, we can replace $v_k(z_k)$ by $u_i(u_j)$ in $D$ to get a dominating set of the same size. It is clear from both cases that $D$ does not contain vertices from $X_1'$ and $X_2'$. We use $Y_1'$ vertices to dominate $X_1'$ and $X_2'$. Therefore, without loss of generality we assume that $D = \{u_j | 1 \leq j \leq k\} \cup \{z_{m+1}\}$ is a DS of size $k' = k+1$ in $G'$. It is easy to see that $S = \{x_j | 1 \leq j \leq k\}$ is a vertex cover of size $k$ in $G$. □

**Corollary 3.** *For star convex bipartite graphs with diameter 3, CDS (TDS) is NP-complete.*

For the following results we present only the construction as the proofs are similar to Theorem 8.

**Theorem 9.** *For star convex bipartite graphs with diameter 5, the dominating set problem is NP-complete.*

*Proof.* We construct a star convex bipartite graph $G'(X', Y')$ with diameter 5 and convexity on $X'$ from the given graph $G(V, E)$ in polynomial-time. Let $n = |V(G)|$, $m = |E(G)|$ and $E(G) = \{e_1, e_1, \ldots, e_m\}$. $V(G') = X' \cup Y'$, $X' = X'_1 \cup X'_2 \cup X'_3$, $Y' = Y'_1 \cup Y'_2$. $X'_1 = \{v_i \mid e_i \in E(G), 1 \leq i \leq m\}$, $X'_2 = \{z_i \mid e_i \in E(G), 1 \leq i \leq m\}$, $X'_3 = \{z_{m+1}, z_{m+2}, z_{m+3}\}$, $Y'_1 = \{u_j \mid x_j \in V(G), 1 \leq j \leq n\}$, $Y'_2 = \{u_{n+1}, u_{n+2}, u_{n+3}, u_{n+4}\}$ and $E(G') = E_1 \cup E_2$, $E_1 = \{\{u_i, v_k\}, \{u_i, z_k\}, \{u_j, v_k\}, \{u_j, z_k\}\} | e_k = \{x_i, x_j\} \in E(G), 1 \leq k \leq m, 1 \leq i \leq n, 1 \leq j \leq n\}$, $E_2 = \{\{z_{m+1}, u_{n+1}\}, \{z_{m+1}, u_{n+2}\}, \{z_{m+1}, u_{n+3}\}, \{z_{m+2}, u_{n+1}\}, \{z_{m+3}, u_{n+2}\}, \{z_{m+3}, u_{n+3}\}, \{z_{m+3}, u_{n+4}\}\} \cup \{\{z_{m+3}, u_j\} \mid z_{m+3} \in X'_3, u_j \in Y'_1, 1 \leq j \leq n\}$. We show that the constructed graph $G'(X', Y')$ is a star convex bipartite graph with diameter 5 and convexity on $X'$. By our construction, we know that $d(v_i, z_{m+3})$, $1 \leq i \leq m$ is two. Similarly, $d(z_{m+3}, u_{n+3})$ is three. This shows that $d(v_i, u_{n+3})$ is five. Therefore $diam(G')$ is 5. We now show that $G'$ is a star convex bipartite graph with convexity on $X'$. The star $T'$ on $X'$ has root vertex $z_{m+3}$ and leaves $v_1, \ldots, v_m, z_1, \ldots, z_m, z_{m+1}, z_{m+2}$. For any vertex $u_j \in Y'$ and $d_{G'}(u_j) \geq 2$, its neighborhood contains the vertex $z_{m+3}$ which forms a subtree of $T'$. This implies that $G'$ is a star convex bipartite graph with convexity on $X'$. Clearly, it is a polynomial-time reduction. We claim that $G$ has a VC of size at most $k$ if and only if $G'$ has a DS of size at most $k' = k + 3$.     □

**Corollary 4.** *For star convex bipartite graphs with diameter 5, CDS (TDS) is NP-complete.*

Observe that the dominating set $D$ obtained from Theorem 9 is also a connected (total) dominating set.

**Theorem 10.** *For star convex bipartite graphs with diameter 6, the dominating set problem is NP-complete.*

**Corollary 5.** *For star convex bipartite graphs with diameter 6, CDS (TDS) is NP-complete.*

From Corollary 2.3.1, 4, and 5, it is clear that the connected dominating set and total dominating set for star convex bipartite graphs with diameter 3,5 and 6 are NP-complete. We now show that the complexity of the outer-connected dominating set and dominating biclique problems for star convex bipartite graphs are NP-complete. Further, we show that for bounded degree star-convex bipartite graphs, the outer-connected dominating set problem is linear-time solvable.

**Theorem 11.** *For star convex bipartite graphs, the outer-connected dominating set problem is NP-complete.*

**Theorem 12.** *For star convex bipartite graphs, the dominating biclique problem is NP-complete.*

**2.3.2    Parameterized Complexity** In this section, we show that the parameterized version of dominating set problem with solution size as the parameter for star convex bipartite graphs is W[2]-hard.

**Theorem 13.** *For star convex bipartite graphs, the parameterized dominating set problem is W[2]-hard when the parameter is the solution size.*

*Proof.* We give a polynomial-time reduction from the parameterized version of dominating set problem in general graphs. We map an instance $(G, k)$ of the parameterized version of dominating set problem on general graphs to the corresponding star convex bipartite instance $(G', k')$ as follows: $V(G') = X' \cup Y'$, $X' = X_1' \cup \{z_1\}$, $Y' = Y_1' \cup \{z_2\}$. $X_1' = \{v_i \mid x_i \in V(G), 1 \leq i \leq n\}$, $Y_1' = \{u_i \mid x_i \in V(G), 1 \leq i \leq n\}$, and $E(G') = E_1 \cup E_2 \cup E_3 \cup \{\{z_1, z_2\}\}$, $E_1 = \{\{v_i, u_j\}, \{u_i, v_j\}, |\{x_i, x_j\} \in E(G), i \neq j, 1 \leq i \leq |V(G)|, 1 \leq j \leq |V(G)|\}$, $E_2 = \{\{v_i, u_i\} | x_i \in V(G)\}$ and $E_3 = \{\{z_1, u_i\} | u_i \in Y'\}$. From the above construction, we know that $z_1$ is universal to $Y'$. The star $T'$ on $X'$ has root vertex $z_1$ and leaves $v_1, \ldots, v_n$. For any vertex $u_j \in Y'$ and the degree is at least two, then its neighborhood contains the vertex $z_1$, which forms a subtree of $T'$. This implies that $G'$ is a star convex bipartite graph with convexity on $X'$. Clearly, it is a polynomial-time reduction. We show that $G$ has a dominating set of size at most $k$ if and only if $G'$ has a dominating set of size at most $k' = k + 1$.

($\Rightarrow$) Let $D = \{x_i | 1 \leq i \leq k\}$ be a DS in $G$. If $x_i$ is in $D$, then we include $u_i \in Y'$ to DS of $G'$. Since $z_1$ is universal, it dominates $Y'$. Thus $D' = \{u_i | x_i \in D, 1 \leq i \leq k\} \cup \{z_1\}$ is a DS of size $K' = k + 1$.

($\Leftarrow$) Since $z_2$ is a pendant vertex and it is adjacent to $z_1$, any dominating set $D'$ in $G'$ must contain either $z_1$ or $z_2$. Suppose $D'$ contains $z_2$, then we replace $z_2$ by $z_1$ to get a DS of the same size $k'$. Since $z_1$ is universal, $Y'$ is dominated. Now $D'$ includes any vertices of $Y'$ so as to dominate $X'$. Suppose $D'$ includes $v_i$. By our construction there exists an edge $\{x_i, x_j\}$ in $G$ and $v_i$ is adjacent to $u_i$ and $u_j$. Since $Y'$ is already dominated, we replace $v_i$ by either $u_i$ or $u_j$ to get a DS of the same size $k'$. We construct $D$ as follows: Suppose $D'$ contains $u_i$, then we include $x_i$ to $D$. Clearly, the size of $D$ is at most $k$, and $D$ is a DS in $G$. Therefore, the parameterized version of dominating set problem on star convex bipartite graphs with solution size as a parameter is W[2]-hard. □

**Theorem 14.** *For star convex bipartite graphs, the parameterized outer-connected dominating set problem (dominating biclique problem) is W[2]-hard when the parameter is the solution size.*

*Proof.* Proof is similar to Theorem 11 (Theorem 12). □

Having shown that the parameterized version dominating set problem on star convex bipartite graphs is W[2]-hard. A natural direction is to study the star convex bipartite graphs with some restrictions. One possible direction is to bound the number of leaves in a star. This direction has been addressed by Arti Pandey et al. in [8] for the dominating set problem, where the degree of a universal vertex is bounded. We adopt the algorithm presented in [9] to show the parameterized dominating set problem is FPT. Since the number of leaves is bounded by $l$, where $l$ is a parameter, $G$ does not contain $K_{l,l}$ as a subgraph. Thus the dominating set problem is FPT when the parameter is the number leaves in an associated star.

## 2.4    Hamiltonicity in Split Graphs

In this section, we present some hardness results of HCYCLE and HPATH for split graphs, and we use these results to show that HCYCLE (HPATH) is NP-complete for star (comb) convex split graphs. We know that for strongly chordal split graphs, HCYCLE is NP-complete [2]. It is easy to see that the reduction instances are strongly chordal split graphs with $|K| = |I|$. Since strongly chordal split graphs are split graphs, HCYCLE for split graphs $G(K, I)$ with $|K| = |I|$ is NP-complete. Similarly, for split graphs $G(K, I)$ with $|K| = |I| - 1$, HPATH is NP-complete [11]. In this paper, we consider other cases of $|K|$ and $|I|$ to study the complexity of HCYCLE and HPATH in split graphs. For HCYCLE we consider the case $|K| > |I|$ and for HPATH we consider $|K| = |I|$ and $|K| > |I|$. We show that for all these cases, HCYCLE and HPATH are NP-complete, and the results of this section shall be used in subsequent sections.

1. HPATH is NP-complete for split graphs $G(K, I)$ with $|K| = |I|$, and $|K| > |I|$
   We observe that reduction instances are $K_{1,5}$-free split graphs
2. HCYCLE is NP-complete for split graphs $G(K, I)$ with $|K| > |I|$. The reduction instances are $K_{1,5}$-free split graphs

We omit the proof of above two results (1) and (2).

## 2.5    Hamiltonicity in Star Convex Split Graphs

In this section, we shall prove the HPATH and the HYCLE problems are NP-complete on star convex split graphs with convexity on $K$ and $I$.

### 2.5.1    Star Convex Split Graphs with Convexity on $I$

**Theorem 15.** *For star convex split graphs $G(K, I)$ with convexity on $I$, the Hamiltonian cycle problem is NP-complete.*

*Proof.* We proved the Hamiltonian cycle problem in split graph $G(K, I)$ such that $|K| > |I|$ is NP-complete (Sect. 2.4). We construct a star convex split graph $H(K', I')$ with convexity on $I'$ from the given split graph $G(K, I)$ such that $|K| > |I|$ in polynomial-time. $V(H) = K' \cup I'$ and $K' = K$, $I' = \{y_{|I|+1}\} \cup \{y_j \mid y_j \in I,\ 1 \leq j \leq |I|\}$, and $E(H) = E' \cup E(G)$, $E' = \{\{y_{|I|+1}, x_i\} \mid y_{|I|+1} \in I',\ x_i \in K',\ 1 \leq i \leq |K'|\}$.

The star $T$ on $I'$ has root vertex $y_{|I|+1}$ and leaves $y_1, \ldots, y_{|I|}$. It is easy to see that, for any vertex $x_i \in K'$ in $H$, its neighborhood induces a subtree in $T$. We claim that $G$ is a yes-instance of the Hamiltonian cycle problem if and only if $H$ is a yes-instance of the Hamiltonian cycle problem.

($\Rightarrow$) Forward direction of the proof of correctness is trivial.

($\Leftarrow$) By our construction, the vertex $y_{|I|+1}$ is adjacent to all the vertices of $K'$ in $H$. This implies that any Hamiltonian cycle must contain $P_3 = (x_i, y_{|I|+1}, x_j)$ for some $i, j \leq |K|$ as a subpath. Suppose that there exists a Hamiltonian cycle in $H$. Since $\{x_i, x_j\} \in E(G)$, we remove the vertex $y_{|I|+1}$ from the cycle and join the vertices $x_i, x_j$ to obtain the Hamiltonian cycle in $G$.     □

**Insights into reduction instances of Theorem** 15
Note that the graph considered for the reduction is $K_{1,5}$-free split graph. We have added the vertex $y_{|I|+1}$ as part of the construction. Since $y_{|I|+1}$ is the root of the associated star $T$ on $I'$, each vertex in $K'$ is adjacent to $y_{|I|+1}$. A closer look at the construction reveals that $K_{1,5}$-free split graph instance becomes an instance of $K_{1,6}$-free star convex split graph due to the addition of $y_{|I|+1}$. This implies that for $K_{1,6}$-free star convex split graphs with convexity on $I$, HCYCLE is NP-complete. It is natural to study the complexity of HCYCLE in $K_{1,5}$-free star convex split graphs. Interestingly, HCYCLE on $K_{1,5}$-free star convex split graphs with convexity on $I$ is polynomial-time solvable, which we prove next.

**Theorem 16.** *For $K_{1,5}$-free star convex split graphs with convexity on $I$, the Hamiltonian cycle problem is polynomial-time solvable.*

*Proof.* We know that for $K_{1,4}$-free split graphs with $|K| = |I|$, HCYCLE is polynomial-time solvable [10]. We modify $K_{1,5}$-free star convex split graph instances to $K_{1,4}$-free split graph instances by removing the root vertex from the associated star $T$ on $I$. Note that $K_{1,4}$-free split graph instances has the property $|K| > |I|$.

*Claim.* Let $G(K, I)$ be a $K_{1,5}$-free star convex split graph and $H(K', I')$ be the modified instance of $G$ ($K_{1,4}$-free split graph). $G$ has a Hamiltonian cycle if and only if $H$ has a Hamiltonian cycle.

*Proof.* ($\Rightarrow$) Assume that $G$ has a Hamiltonian cycle. Let $u$ be the root of the associated star. Observe that any Hamiltonian cycle must contain $P_3 = (x_i, u, x_j)$ as a subpath. Note that $x_i$ and $x_j$ are clique vertices. We remove the vertex $u$, and join $x_i$ and $x_j$ to get Hamiltonian cycle in $H$.

($\Leftarrow$) We know that $H$ has a Hamiltonian cycle. Since $|K'| > |I'|$, any Hamiltonian cycle $C$ must contain a clique edge. Let the clique edge be $\{x_i, x_j\}$. Since the root $u$ is adjacent to both $x_i$ and $x_j$, we add $u$ to $C$ ($x_i, u, x_j$) to obtain a Hamiltonian cycle $C'$ in $G$.     □

Theorem 15 and Theorem 16 establishes a dichotomy for HCYCLE in star convex split graphs with convexity on $I$, i.e. for $K_{1,5}$-free star convex split graphs with convexity on $I$, HCYCLE is polynomial-time solvable whereas for $K_{1,6}$-free star convex split graphs with convexity on $I$, HCYCLE is NP-complete.

It is natural to investigate the complexity of HCYCLE with convexity on $K$ and HPATH with convexity on $K(I)$. Further, one can explore the complexity of HCYCLE and HPATH for other convex properties such as comb, triad, and circular convex. In this paper, we shall show for comb convex split graphs with convexity on $K(I)$, HCYCLE and HPATH are NP-complete. We first show that the Hamiltonian path problem in split graphs with two pendant vertices is NP-complete. We use this result to show that the Hamiltonian path problem in star convex split graphs with convexity on $I$ is NP-complete.

We use a construction similar to the construction presented in Theorem 15 to prove the following results.

1. For split graphs $G(K, I)$ such that $|K| = |I|$ with two pendant vertices, the Hamiltonian path problem is NP-complete.
2. For star convex split graphs $G(K, I)$ with convexity on $I$, the Hamiltonian path problem is NP-complete.
3. For star convex split graphs $G(K, I)$ with convexity on $K$, the Hamiltonian cycle (path) problem is NP-complete.
4. For comb convex split graphs $G(K, I)$ with convexity on $I$, the Hamiltonian cycle (path) problem is NP-complete and with convexity on $K$, the Hamiltonian cycle (path) problem is NP-complete.

We shall present the reduction for the the Hamiltonian path problem on star (comb) convex split graphs.

**Theorem 17.** *For star convex split graphs $G(K, I)$ with convexity on $K$, the Hamiltonian path problem is NP-complete.*

*Proof.* The Hamiltonian path problem in split graph $G(K, I)$ is such that $|K| = |I|$ is NP-complete. We now show that the Hamiltonian path problem in the star convex split graph $H(K, I)$ is NP-complete. We present a deterministic polynomial-time reduction that reduces an instance of split graph $G(K, I)$ such that $|K| = |I|$ to a corresponding star convex split graph instance $H(K', I')$. We map an instance of $G$ to the corresponding instance of $H$ as follows: $V(H) = K' \cup I'$, $K' = \{x_{|K|+1}\} \cup \{x_i \mid x_i \in K, \ 1 \le i \le |K|\}$, $I' = \{y_{|I|+1}\} \cup \{y_j \mid y_j \in I, \ 1 \le j \le |I|\}$ and $E(H) = E' \cup E(G)$, $E' = \{\{x_{|K|+1}, y_j\} \mid x_{|K|+1} \in K, \ y_j \in I, \ 1 \le j \le |I|+1\}$. We now show that $H$ is a star convex split graph $H(K', I')$ convexity on $K'$. The star $T$ on $K'$ has root vertex $x_{|K|+1}$ and leaves $x_1, \ldots, x_{|K|}$. For any vertex $y_j \in I'$ in $H$, its neighborhood $N_H(y_j) = N_G(y_j) \cup \{x_{|K|+1}\}$ is a subtree of $T$. This implies that $H$ is a star convex split graph with convexity on $K'$. ($\Rightarrow$) We skip the proof of necessity as it is easy. ($\Leftarrow$) Since $y_{|I|+1}$ is a pendant vertex, any Hamiltonian path in $H$ must start (end) at $y_{|I|+1}$. Since $x_{|K|+1}$ is the only neighbor of $y_{|I|+1}$, any Hamiltonian path that starts at $y_{(|I|+1)}$ must contain $P_2 = (y_{|I|+1}, x_{|K|+1})$ as a subpath. Without loss of generality, assume that the Hamiltonian path starts at $y_{|I|+1}$. We remove $P_2 = (y_{|I|+1}, x_{|K|+1})$ to obtain the Hamiltonian path in $G$.    □

**Theorem 18.** *For comb convex split graphs $G(K, I)$ with convexity on $K$, the Hamiltonian path problem is NP-complete.*

*Proof.* We reduce the Hamiltonian path problem in split graphs with $|K| = |I|$ to Hamiltonian path problem in comb convex split graphs with convexity on $K$ as follows. For a given instance of split graph $G(K, I)$ such that $|K| = |I|$, we add $2|K|$ vertices such that each partition receives $|K|$ vertices. $V(H) = K' \cup I'$, $K' = K \cup \{x_{|K|+1}, x_{|K|+2}, \ldots, x_{|K|+|K|-1}, x_{|K|+|K|}\}$, $I' = I \cup \{y_{|K|+1}, y_{|K|+2}, \ldots, y_{|K|+|K|-1}, y_{|K|+|K|}\}$ and $E(H) = E^1 \cup E^2 \cup E^3 \cup E^4 \cup E(G)$, $E^1 = \{\{x_{|K|+i}, y_j\} \mid x_{|K|+i} \in K, \ y_j \in I, \ 1 \leq i \leq |K|, \ 1 \leq j \leq |I|\}$, $E^2 = \{\{x_{|K|+i}, y_{|K|+i}\} \mid x_{|K|+i} \in K, \ y_{|K|+i} \in I, \ 1 \leq i \leq |K|\}$, $E^3 = \{\{x_{|k|+i}, y_{|k|+i-1}\} \mid x_{|K|+i} \in K, \ y_{|K|+i-1} \in I, \ 2 \leq i \leq |K|\}$, $E^4 = \{\{x_i, x_j\} \mid x_i \in K, x_j \in K, i \neq j, |K| + 1 \leq i \leq 2|K|, 1 \leq j \leq 2|K|\}$. The comb $T$ on $K'$ has backbone $\{x_{|K|+1}, x_{|K|+2}, \ldots, x_{|K|+|K|}\}$ and teeth $x_1, x_2, \ldots, x_{|K|}$. For any vertex $y_j \in I, 1 \leq j \leq |I|$, its neighborhood $N_H(y_j) = N_G(y_j) \cup \{x_{|K|+i} \mid 1 \leq i \leq |K|\}$ is a subtree on $T$. For a comb, any subset of teeth and the backbone form a subtree. Similarly, for any vertex $y_j \in I, |I|+1 \leq j \leq 2|I|-1$, its neighborhood is $N_H(y_j) = \{x_j, x_{j+1}\}$. We observe that each $y_j \in I, |I|+1 \leq j \leq 2|I|-1$ is adjacent only to two adjacent vertices in the backbone. This shows that this forms a subtree. It is easy to see that the neighborhood of $y_{2|I|}$ induces a subtree in $T$.

($\Rightarrow$) Since $|K| = |I|$, at least one endpoint of any Hamiltonian path in $G$ must be in the partition $I$. Suppose that there exists a Hamiltonian path $P$ in $G$. Let the path be $P = (x_1, y_1, x_2, y_2, \ldots, x_i, x_j, \ldots, x_{|K|-1}, y_{|I|-1}, x_{|K|}, y_{|I|})$. From the construction of $H$, we observe that the graph induced on the set $\{x_{|K|+1}, x_{|K|+2}, \ldots, x_{|K|+|K|}, y_{|K|+1}, y_{|K|+2}, \ldots, y_{|K|+|K|}\}$ is a path $P'$. In particular, the path $P'$ starts at the vertex $x_{|K|+1}$. Since $x_{|K|+1}$ is adjacent to all the vertices of $I$, we join the path $P'$ to $P$ to obtain a Hamiltonian path in $H$. The Hamiltonian path $Q = (x_1 P y_{|I|}, x_{|K|+1} P')$.

($\Leftarrow$) By our construction, the vertex $y_{|K|+|K|}$ is a pendant vertex. Any Hamiltonian path $Q$ in $H$ must start at $y_{|K|+|K|}$. Since the degree of $y_{|I|+1}, \ldots, y_{|I|+|I|}$ is two, any Hamiltonian path must contain $Q^1 = (y_{|I|+|I|}, x_{|K|+|K|}, y_{|I|+|I|-1}, x_{|K|+|K|-1}, \ldots, y_{|I|+2}, x_{|K|+2}, y_{|I|+1}, x_{|K|+1})$ as a subpath. Suppose if $H$ has a Hamiltonian path, then there exists a simple path $Q^2$ that visits $V(H) \backslash V(Q')$ exactly once. We know that the graph induced on $V(H) \backslash V(Q^1)$ is the same as $G$. Hence, $Q^2$ is a Hamiltonian path in $G$. □

# 3    Conclusion

In this paper, we made an attempt to reduce the gap between P vs. NPC for the problems Hamiltonian cycle (path), dominating set problem (DS), and its variants (CDS, TDS, OCDS, DB) for star convex bipartite graphs with diameter as a parameter. Also, we have shown NP-completeness of the Hamiltonian cycle and Hamiltonian path problem for bi-star convex bipartite graphs. On the parameterized front, we have shown that the dominating set problem is W[2]-hard for star convex bipartite graphs. We believe that the diameter can be used as a parameter to explore other well-known combinatorial problems restricted to star convex bipartite graphs. Further, for split graphs, one can consider other structures of the trees, such as path and triad, to investigate the complexity of the HCYCLE and HPATH.

# References

1. Chen, H., Lei, Z., Liu, T., Tang, Z., Wang, C., Xu, K.: Complexity of domination, hamiltonicity and treewidth for tree convex bipartite graphs. J. Comb. Optim. **32**(1), 95–110 (2016)
2. Müller, H.: Hamiltonian circuits in chordal bipartite graphs. Discret. Math. **156**(1–3), 291–298 (1996)
3. Spinrad, J., Brandstädt, A., Stewart, L.: Bipartite permutation graphs. Discret. Appl. Math. **18**(3), 279–292 (1987)
4. Jiang, W., Liu, T., Wang, C., Xu, K.: Feedback vertex sets on restricted bipartite graphs. Theor. Comput. Sci. **507**, 41–51 (2013)
5. Cyman, J.: The outer-connected domination number of a graph. Australas. J. Comb. **38**, 35–46 (2007)
6. Panda, B.S., Pandey, A.: Algorithm and hardness results for outer-connected dominating set in graphs. In: Pal, S.P., Sadakane, K. (eds.) WALCOM 2014. LNCS, vol. 8344, pp. 151–162. Springer, Cham (2014). https://doi.org/10.1007/978-3-319-04657-0_16
7. Pradhan, D.: On the complexity of the minimum outer-connected dominating set problem in graphs. J. Comb. Optim. **31**(1), 1–12 (2016)
8. Pandey, A., Panda, B.S.: Domination in some subclasses of bipartite graphs. Discrete Appl. Math. **252**, 51–66 (2019)
9. Telle, J.A., Villanger, Y.: FPT algorithms for domination in biclique-free graphs. In: Epstein, L., Ferragina, P. (eds.) ESA 2012. LNCS, vol. 7501, pp. 802–812. Springer, Heidelberg (2012). https://doi.org/10.1007/978-3-642-33090-2_69
10. Renjith, P., Sadagopan, N.: Hamiltonian cycle in $K_{1,r}$-free split graphs-a dichotomy. Int. J. Found. Comput. Sci. **33**(01), 1–32 (2022)
11. Renjith, P., Sadagopan, N.: Hamiltonian path in $K_{1,t}$-free split graphs- a dichotomy. In: Panda, B.S., Goswami, P.P. (eds.) CALDAM 2018. LNCS, vol. 10743, pp. 30–44. Springer, Cham (2018). https://doi.org/10.1007/978-3-319-74180-2_3
12. Mohanapriya, A., Renjith, P., Sadagopan, N.: P versus NPC: minimum Steiner trees in convex split graphs. In: Balachandran, N., Inkulu, R. (eds.) CALDAM 2022. LNCS, vol. 13179, pp. 115–126. Springer, Cham (2022). https://doi.org/10.1007/978-3-030-95018-7_10
13. Golumbic, M.C.: Algorithmic Graph Theory and Perfect Graphs. Academic Press, New York (1980)
14. West, D.B.: Introduction to Graph Theory, 2nd edn. Prentice hall, Upper Saddle River (2003)

# On Star Partition of Split Graphs

D. Divya and S. Vijayakumar[⊠]

Indian Institute of Information Technology, Design and Manufacturing (IIITDM)
Kancheepuram, Chennai 600127, India
{mat19d001,vijay}@iiitdm.ac.in

**Abstract.** A graph that is isomorphic to $K_{1,r}$ for some $r \geq 0$ is called a *star*. A partition $\{V_1, \ldots, V_k\}$ of the vertex set of a graph $G$ into $k$ sets is called a *star partition* of $G$ of *size* $k$ if each set in the partition induces a star. The minimum $k$ for which a graph $G$ admits a star partition of size $k$ is called the *star partition number* of $G$ and is denoted by $sp(G)$. Given a graph $G$, the problem MIN STAR PARTITION asks for a star partition of $G$ of minimum size. Given a graph $G$ and a positive integer $k$, its decision version STAR PARTITION asks whether $sp(G) \leq k$. STAR PARTITION is NP-complete for many natural graph classes [25]. In particular, it is NP-complete for $K_{1,5}$-free split graphs. In this paper, we study the star partition problems on split graphs, with a special focus on the degrees of vertices in the independent part. We call a split graph (a) an $r$-split graph if each vertex in the independent part has degree $r$ and (b) an $(r_1, \ldots, r_k)$-split graph if each vertex in the independent part has degree equal to one of $r_1, \ldots, r_k$. We obtain the following NP-completeness results: (1) STAR PARTITION is NP-complete even for $K_{1,5}$-free 2-split graphs. (2) Deciding whether $sp(G) = \lceil \omega(G)/2 \rceil$ is NP-complete even for $K_{1,6}$-free 2-split graph $(sp(G) \geq \lceil \omega(G)/2 \rceil$ for any graph $G)$. (3) STAR PARTITION is NP-complete even for $(1, r)$-split graphs ($r \geq 2$ and is fixed). We obtain the following fixed parameter (in)tractability results (in each case, $k$ stands for the parameter) (1) Given any connected split graph $G$ and an integer $k \geq 1$, deciding whether $sp(G) \leq k$ is fixed parameter tractable and has an $O((2k)^{2k+1}n)$ time algorithm. (2) Given a graph $G$ and an integer $k \geq 0$, deciding whether $sp(G) \leq \lceil \omega(G)/2 \rceil + k$ is para-NP-hard even when restricted to either (a) $K_{1,6}$-free $(0,2)$-split graphs or (b) $K_{1,6}$-free $(0,1,3)$-split graphs. (3) Given a graph $G$ and an integer $k \geq 0$, the problem of deciding whether $sp(G) \leq \omega(G) - k$ is $W[1]$-hard even for $(1,2)$-split graphs and lies in $W[3]$ for connected split graphs $(sp(G) \leq \omega(G)$ for any connected split graph $G)$. We also obtain the following polynomial time algorithms: (1) 3/2-approximation algorithms for several subclasses of 2-split graphs. (2) A linear time algorithm for $(0,1)$-split graphs; in particular, for any 1-split graph $G$, we prove that $sp(G) = \max(\lceil \omega(G)/2 \rceil, \alpha(G^2))$. Most of these results are obtained by an elegant framework that we have developed for the study of star partition on split graphs. Using this, we also obtain a simple characterization for any connected split graph $G$ having $sp(G) = \omega(G)$.

The second author is supported by DST-SERB MATRICS: MTR/2022/000870.

S. Kalyanasundaram and A. Maheshwari (Eds.): CALDAM 2024, LNCS 14508, pp. 209–223, 2024.
https://doi.org/10.1007/978-3-031-52213-0_15

**Keywords:** Star Partition · Split Graphs · NP-completeness ·
Polynomial Time Algorithms · Fixed Parameter Tractability

## 1   Introduction

A graph is called a *star* if it is isomorphic to the complete bipartite graph $K_{1,r}$
for some $r \geq 0$; please see Fig. 1. When a graph models a network, like a road or
computer network, each *induced* subgraph that is a star corresponds to a sub-
network that is a star network. The *center* of an induced star in such a graph
potentially corresponds, in the underlying network, to either a bottleneck or a
point that is desirable for locating some facilities; this is especially the case when
the star has two or more *non-center* vertices. These practical considerations as
well as their combinatorial appeal motivate a study of two optimization problems, namely MIN STAR COVER and MIN STAR PARTITION, that will be defined
below.

**Fig. 1.** Some examples of stars.

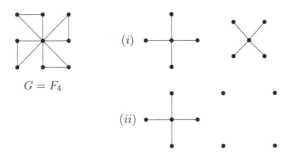

**Fig. 2.** The friendship graph $G = F_4$ along with (i) an optimal star cover and (ii) an
optimal star partition of it; thus $sc(G) = 2$ and $sp(G) = 5$.

By extension, any subset $S$ of the vertex set $V$ of a graph $G = (V, E)$ is called
a *star* of $G$ if the subgraph induced by $S$ is a star. A collection $\mathcal{C} = \{V_1, \ldots, V_k\}$
of stars in $G$ is called a *star cover* of $G$ if $V_1 \cup \ldots \cup V_k = V$. A star cover $\mathcal{C}$
of $G$ is called a *star partition* of $G$ if it is also a partition of $V$. A star cover $\mathcal{C}$
consisting of $k$ distinct stars of $G$ is called a *star cover* of $G$ of *size* $k$. The *size*

of a star partition is defined similarly. The minimum $k$ for which $G$ admits a star cover of size $k$ is called the *star cover number* of $G$ and is denoted by $sc(G)$. The minimum $k$ for which $G$ admits a star partition of size $k$ is called the *star partition number* of $G$ and is denoted by $sp(G)$. For instance, the graph $G$ in Fig. 2 has $sc(G) = 2$ and $sp(G) = 5$.

In this paper, we study the following optimization problems and their decision versions, especially on split graphs.

MIN STAR COVER
**Input:** A graph $G$.
**Goal:** A minimum star cover of $G$.

MIN STAR PARTITION
**Input:** A graph $G$.
**Goal:** A minimum star partition of $G$.

We study the above problems on split graphs, with a special focus on those split graphs for which each vertex in the independent part has degree at most $r$ for a small constant $r$, like $1 \le r \le 3$. An investigation in this direction is motivated by the NP-hardness of the problems for $K_{1,5}$-free split graphs [25] in which case the *independent* degrees of the vertices in the clique part are bounded by a small constant.

Incidentally, split graphs and stars are intimately related in that every split graph is the intersection graph of substars of some star [16].

Consider any graph $G$. Since a star partition of $G$ is also a star cover of $G$, we have $sc(G) \le sp(G)$. Also the centers of any star cover of $G$ form a dominating set of $G$. So, we further have $\gamma(G) \le sc(G) \le sp(G)$, where $\gamma(G)$ is the domination number of $G$. Thus, if $G$ is triangle-free, we also have $sp(G) = sc(G) = \gamma(G)$; this easily follows from Proposition 1. Consequently, MIN STAR COVER and MIN STAR PARTITION are both polynomially equivalent to the minimum dominating set problem on triangle-free graphs.

In general, $sp(G) \ne sc(G)$. For instance, for the friendship graph $F_n = K_1 \oplus (nK_2)$, where $n \ge 1$, $sc(F_n) = 2$ but $sp(G) = n + 1$; see Fig. 2. By Theorem 1 below, $sp(G) = sc(G)$ for any butterfly-free graph $G$. Interestingly, butterfly-free graphs include both bipartite graphs and split graphs.

**Theorem 1** ([17]). *If $G$ is a butterfly-free graph, then $sp(G) = sc(G)$. Moreover, given a star cover of a butterfly-free graph $G$ of size $k$, a star partition of $G$ of size at most $k$ can be computed in time $O(n^2 \log n)$.*

Thus MIN STAR COVER and MIN STAR PARTITION are polynomially equivalent for butterfly-free graphs. In particular, this holds for split graphs. Thus, in our study of split graphs, we will focus only on MIN STAR PARTITION. Nevertheless, we now provide a name to the decision versions of both the problems.

STAR COVER
**Input:** A graph $G$ and a positive integer $k$.
**Question:** Does $G$ have a star cover of size $k$?

STAR PARTITION
**Input:** A graph $G$ and a positive integer $k$.
**Question:** Does $G$ have a star partition of size $k$?

In [13], stars have been considered in the context of generalizing the well-known maximum matching problem. It considers the problem of covering a maximum number of vertices of the input graph by vertex-disjoint *induced* stars of the form $K_{1,i}, 1 \leq i \leq r$, where $r \geq 1$ is any fixed positive integer. This problem is shown to have a polynomial time algorithm in [13]. This algorithm leads to a simple $\frac{r}{2}$-approximation algorithm for MIN STAR PARTITION on $K_{1,r}$-free graphs, $r \geq 2$, that remains NP-hard even for line graphs, a subclass of claw-free graphs [25].

A summary of known results on STAR COVER and STAR PARTITION follows. Deciding whether an input graph can be covered by *or* partitioned into at most two stars has polynomial time algorithms but deciding whether an input graph can be covered by *or* partitioned into three stars is NP-complete [25]; these problems remain NP-complete even for $K_4$-free graphs. Both STAR COVER and STAR PARTITION are NP-complete for (a) chordal bipartite graphs [20], (b) $(C_4, C_6, \ldots, C_{2t})$-free bipartite graphs for every fixed $t \geq 2$ [9], (c) subcubic bipartite planar graphs [11,25], (d) $K_{1,5}$-free split graphs [25], (e) line graphs (and hence claw-free graphs and, more generally $K_{1,r}$-free graphs for any fixed $r \geq 3$) [7,25] and (f) co-tripartite graphs [15,25].

It is NP-hard to approximate MIN STAR PARTITION within $n^{1/2-\epsilon}$ for all $\epsilon > 0$ [25,29]. Also both MIN STAR COVER and MIN STAR PARTITION do not have any polynomial time $c \log n$-approximation algorithm for some constant $c > 0$ unless P = NP [27]. Both the problems have a polynomial time (a) 2-approximation algorithm for split graphs [25]; (b) $O(\log n)$-approximation algorithms for butterfly-free graphs [17]; (c) $(d + 1)$-approximation algorithm for triangle-free graphs of degree at most $d$ [27]. MIN STAR PARTITION has a polynomial time $r/2$-approximation algorithm for $K_{1,r}$-free graphs (which implies a $3/2$-approximation algorithm for, for instance, line graphs and cobipartite graphs as they are claw-free) [13,25]. MIN STAR COVER has a polynomial time $O(\log n)$-approximation algorithm for any hereditary graph class for which the maximum independent set problem has a polynomial time algorithm [17].

With the number of stars in a star cover/partition as the parameter, the problems are $W[2]$-complete for bipartite graphs and are fixed parameter tractable for graphs of girth at least five [24]. Recently, the star partition problem has been studied in the FPT framework with respect to structural parameters of graphs such as vertex cover, treewidth and cliquewidth in [21]. The following results are obtained: (1) with vertex cover number as the parameter, the star partition problem is fixed parameter tractable; (2) with treewidth as the parameter, the problem is fixed parameter tractable on bounded treewidth graphs. (For an introduction to parameterized complexity one may refer to [8].)

The problems have exact (a) $2^n n^{O(1)}$ time and exponential space algorithms and (b) $3^n n^{O(1)}$ time and polynomial space algorithms [3].

Both STAR COVER and STAR PARTITION have polynomial time algorithms for bipartite permutation graphs [4,10], convex bipartite graphs [2,6], doubly-convex bipartite graphs [2], trees [5], double-split graphs [18], trivially perfect graphs and complements of trivially perfect graphs [17].

*Related Literature:* The problem of covering a maximum number of vertices of an input graph by vertex-disjoint *induced* stars of the form $K_{1,i}, 1 \leq i \leq r$, where $r \geq 1$ is any fixed positive integer, has a polynomial time algorithm [13].

The problem of partitioning (the vertex set of) an input graph into induced paths of length $t$ ($t \geq 3$ is fixed) is NP-complete even for bipartite graphs with maximum degree three [19]. This result implies that the problem of partitioning a graph into induced $K_{1,2}$'s is NP-complete, even when restricted to bipartite graphs with maximum degree three.

The problem of partitioning an input graph into equal but fixed size stars, not necessarily induced, is investigated for many natural subclasses of perfect graphs in [26]. (This problem is also referred to as STAR PARTITION in that paper.) This problem had already been shown to be NP-complete even when the star size is fixed to be three in [14]. Given a graph $G$ and a positive integer $k$, the problem of deciding whether $G$ can be partitioned into exactly $k$ (not necessarily induced) stars, each of size at least two, is investigated in [1].

We may recall that a graph $G = (C \cup I, E)$ is called a *split graph* if its vertex set partitions into a clique $C$ and an independent set $I$.

**Definition 1.** *Let $G = (C \cup I, E)$ be a split graph and let $r$ and $r_1 \leq \ldots \leq r_k$ be non-negative integers. Then:*

1. *$G$ is called an $r$-split graph if $d(v) = r$ for each $v \in I$.*
2. *$G$ is called an $(r_1, \ldots, r_k)$-split graph if $d(v)$ equals one of $r_1, \ldots, r_k$ for each $v \in I$.*

*Our Results:* In this paper, we study the star partition problems on split graphs, with a special focus on degrees of vertices in the independent part; please see Definition 1 above.

We obtain the following NP-completeness results: (1) STAR PARTITION is NP-complete even for $K_{1,5}$-free 2-split graphs. (2) Deciding whether $sp(G) = \lceil \omega(G)/2 \rceil$ is NP-complete even for $K_{1,6}$-free 2-split graphs ($sp(G) \geq \lceil \omega(G)/2 \rceil$ for any graph $G$). (3) STAR PARTITION is NP-complete even for $(1, r)$-split graphs ($r \geq 2$ and is fixed).

We obtain the following fixed parameter (in)tractability results: (1) Given any connected split graph $G$ and an integer $k \geq 1$ as the parameter, deciding whether $sp(G) \leq k$ is fixed parameter tractable and has an $O((2k)^{2k+1}n)$ time algorithm. (2) Given a graph $G$ and an integer $k \geq 0$ as the parameter, deciding whether $sp(G) \leq \lceil \omega(G)/2 \rceil + k$ is para-NP-hard even when restricted to either (a) $K_{1,6}$-free $(0, 2)$-split graphs or (b) $K_{1,6}$-free $(0, 1, 3)$-split graphs. (3) Given a graph $G$ and an integer $k \geq 0$ as the parameter, the problem of deciding whether $sp(G) \leq \omega(G) - k$ is $W[1]$-hard even for $(1, 2)$-split graphs and lies in

$W[3]$ when restricted to connected split graphs ($sp(G) \leq \omega(G)$ for any connected split graph $G$).

We obtain the following polynomial time algorithms: (1) 3/2-approximation algorithms for several subclasses of 2-split graphs. (1) A linear time algorithm for $(0,1)$-split graphs; in particular, for any 1-split graph $G$, we prove that $sp(G) = \max(\lceil \omega(G)/2 \rceil, \alpha(G^2))$.

Most of these results are obtained by an elegant framework that we have developed for the study of star partition on split graphs. Using this, we also obtain a simple characterization for any connected split graph $G$ having $sp(G) = \omega(G)$.

## 2    Preliminaries

All graphs considered in this paper are finite. For basic graph theory terminology, we refer to [28]. Thus $K_n$, $P_n$ and $C_n$ denote the complete graph, the path and the cycle on $n$ (unlabeled) vertices. Also $K_{m,n}$ denotes the complete bipartite graph with independent bipartitions of sizes $m$ and $n$. In particular, the complete bipartite graph $K_{1,n}$ is the *star* on $n+1$ vertices, where $n \geq 0$.

A *clique (independent set)* of a graph $G$ is any set of pairwise adjacent (nonadjacent) vertices of $G$. The size of a maximum clique (maximum independent set) in $G$ is denoted by $\omega(G)$ ($\alpha(G)$).

For any set $X$ of vertices in a graph $G$, $G[X]$ denotes the graph induced by $X$ in $G$. Given a graph $H$, a graph $G$ is said to be $H$-free if it has no induced subgraphs isomorphic to $H$. More generally, given a family (finite or infinite) of graphs $\mathcal{F}$, a graph $G$ is said be $\mathcal{F}$-free if $G$ does not have any induced subgraph $H$ isomorphic to some graph $H \in \mathcal{F}$.

The complement of a graph $G$ is denoted by $\overline{G}$. Let $G_1 = (V_1, E_1)$ and $G_2 = (V_2, E_2)$ be two vertex-disjoint graphs. Then the graph $G = (V_1 \cup V_2, E_1 \cup E_2)$ is called the *union* of $G_1$ and $G_2$ and is denoted by $G_1 \cup G_2$. In particular, $pK_n$ denotes the union of $p$ vertex-disjoint complete graphs, each isomorphic to $K_n$. Also the graph $G$ obtained by joining each vertex of $G_1$ to each vertex of $G_2$ in the union $G_1 \cup G_2$ is called the *join* of $G_1$ and $G_2$ and is denoted by $G_1 \oplus G_2$.

If $G = (V, E)$ is any graph, then, as already defined in the introductory section, we also call a subset $S$ of $V$ a *star* of $G$ if the induced subgraph $G[S]$ is a star.

*Notation:* Let $G$ be any graph. Then any star $S$ in $G$ is *often* written as a disjoint union, namely $\{x\} \cup I$, where $x$ is a center vertex of the star $G[S]$ and $I = S\backslash\{x\}$ is the independent set consisting of the non-center vertices of $G[S]$ (where $|I| \geq 0$).

**Proposition 1** ([18]). *If a graph $G$ has a star cover of size $k$ such that the stars in it have their centers distinct, then it also has a star partition of size $k$.*

## 3   Structure of Star Partitions of Split Graphs

In this section, we prove a few basic facts about optimal star partitions of split graphs. We begin with the following lemma.

**Lemma 1.** *Let $G = (C \cup I, E)$ be a connected split graph. If $G$ has a star partition of size $k$, then it also has a star partition $S$ of size at most $k$ such that each star in $S$ has its center in $C$.*

*Proof.* Let $S$ be a star partition of $G$ of size $k$. If every star in $S$ has its center from $C$, then we are done. Otherwise consider any star $X = \{x\} \cup J$ of $G$ that has its center $x$ in $I$. Then, since $I$ is an independent set and $C$ is a clique, we have $J \subseteq C$ and $|J| \leq 1$. If $|J| = 1$, we can as well think that $X$ is a star with its center in $C$. So, suppose $|J| = 0$. Then $X = \{x\}$ and $x \in I$. Consider $N(x)$. Since $G$ is connected, $N(x) \neq \emptyset$. If one of the stars in $S$, say $Y$, has its center alone in $N(x)$, then $Z = Y \cup X = Y \cup \{x\}$ is also a star in $G$. In this case, $S' = \{Z\} \cup [S \setminus \{X, Y\}]$ is a star partition of $G$ of size at most $|S|$. Else $N(x)$ has a non-center vertex, say $z$, of some star $Y$ in $S$. In this case, $Y' = Y \setminus \{z\}$ and $X' = \{x, z\}$ are stars in $G$ with $X \cup Y = X' \cup Y'$. Thus, $S' = \{X', Y'\} \cup (S \setminus \{X, Y\})$ is a star partition of $G$ of size at most $|S|$.

Proceeding similarly with every star in $S$, if any, that has its center in $I$, we eventually obtain a star partition $S^*$ of $G$ of size at most $|S| = k$ with all its centers from $C$. □

*Important Note:* In this paper, in the light of Lemma 1, for any connected split graph $G = (C \cup I, E)$, without loss of generality, we shall assume that any star partition $S$ that we consider *always* has the centers of all its stars in $C$.

Since $C$ is a clique, any star of $G$ can have at most two vertices from $C$.

*Observation 1:* Consider a star partition $S$ of a connected split graph $G = (C \cup I, E)$ such that each star in $S$ has its center in $C$; such a star partition exists by Lemma 1. Suppose we partition $S$ into the collections $S_1$ and $S_2$ so that each star in $S_1$ has only its center from $C$ and each star in $S_2$ has its center as well as a non-center vertex from $C$. Then the following holds true:

(a) We have $|S_1| + 2|S_2| = |C|$.
(b) If $G$ has $|C| = q$, then $q/2 \leq sp(G) \leq q$. (We have either $\omega(G) = q$ or $\omega(G) = q + 1$.)
(c) If a star $X \in S_1$ has $u \in C$ as its center, then $X \subseteq \{u\} \cup N_I(u)$. Indeed $\{u\} \cup N_I(u)$ is in itself a star in $G$.
(d) If a star $X \in S_2$ has $v \in C$ as its center and $w \in C$ as a non-center, then $X \subseteq \{v, w\} \cup [N_I(v) \setminus N_I(w)]$. Indeed $\{v, w\} \cup [N_I(v) \setminus N_I(w)]$ is in itself a star in $G$.

In the definition below, we now encapsulate the natural transition between the partition $[S_1, S_2]$ of $S$ and an associated special partition of the clique part $C$ of $G$.

**Definition 2.** *Let $G = (C \cup I, E)$ be a split graph with $|C| = q$. Suppose $C$ partitions into three ordered sets $S = \{u_1, \ldots, u_s\}$, $T_1 = \{v_1, \ldots, v_t\}$ and $T_2 = \{w_1, \ldots, w_t\}$ such that the union of the $s$ stand-alone sets $N_I(u_1), \ldots, N_I(u_s)$ and $t$ differences $N_I(v_1) \backslash N_I(w_1), \ldots, N_I(v_t) \backslash N_I(w_t)$ equals $I$. Then $(S, T_1, T_2)$ is called an $(s, t)$-partition of $C$.*

*Note:* The ordering of the vertices in $T_1$ and $T_2$ are important. So, we always consider them as *ordered* sets.

**Lemma 2.** *Let $G = (C \cup I, E)$ be a connected split graph with $|C| = q$. Let $s$ and $t$ be any non-negative integers such that $s + 2t = q$. Then $G$ has a star partition of size $s + t$ if and only if $C$ has an $(s, t)$-partition.*

**Lemma 3.** *Let $G = (C \cup I, E)$ be a connected split graph with $|C| = q$ and $|I| = p$ and let $s, t$ be any non-negative integers. Then we can decide whether $C$ has an $(s, t)$-partition in time $O(q^{2t+1}p)$.*

**Theorem 2.** *Let $G = (C \cup I, E)$ be a connected split graph with $C = \{x_1, \ldots, x_q\}$ as a maximum clique of $G$ so that $\omega(G) = |C| = q$. Then $sp(G) = \omega(G)$ if and only if for every ordered pair $(i, j)$ with $1 \leq i, j \leq q$ and $i \neq j$, either $N_I(x_j)$ has a vertex of degree one or $N_I(x_i) \cap N_I(x_j)$ has a vertex of degree two (or both).*

*Proof.* By Lemma 2, $sp(G) = \omega(G) = q$ if and only if $C$ has no $(s, t)$-partition with $t > 0$. Also $C$ has no $(s, t)$-partition with $t > 0$ means that if we form even one difference, say $N_I(x_i) \backslash N_I(x_j)$ for some $(i, j)$ with $1 \leq i, j \leq q$ and $i \neq j$, then the union of $N_I(x_i) \backslash N_I(x_j)$ and $N_I(x_k)$'s, where $1 \leq k \leq q$ and $k \notin \{i, j\}$, does not include some element of $I$. But this means that for every $(i, j)$ with $1 \leq i, j \leq q$ and $i \neq j$, some element of $I$ belongs only to $N_I(x_j)$ or only to $N_I(x_i) \cap N_I(x_j)$. In other words, this means that for every $(i, j)$ with $1 \leq i, j \leq q$ and $i \neq j$ either $N_I(x_j)$ has a vertex of degree 1 or $N_I(x_i) \cap N_I(x_j)$ has a vertex of degree 2 (or both). □

### 3.1 The Case of 2-Split Graphs

We begin with a definition that highlights a natural transition that is possible between arbitrary simple graphs and 2-split graphs.

*Note:* In a graph $G$, if $u$ and $v$ is a pair of non-adjacent vertices, then we call $uv$ a *non-edge* of $G$.

**Definition 3.** *1. Let $G = (V, E)$ be any graph. Then the split division of $G$, denoted $G_S$, is the 2-split graph $G_S = (C \cup I, E_S)$ that has the clique part $C$ equal to $V(G)$ and the independent part $I$ equal to $E(G)$ and has vertex $e = uv$ in $I$ adjacent to (its end) vertices $u$ and $v$ in $C$.*

*2. Let $G = (C \cup I, E)$ be a 2-split graph. Then the kernel of $G$, denoted $G_K$, is the graph $G_K = (V_K, E_K)$ with vertex set $V_K$ equal to the clique part $C$ of $G$ and edge set $E_K = \{vw \mid N_G(z) = \{v, w\}$ for some $z \in I\}$.*

**Lemma 4.** *Let $G = (C \cup I, E)$ be a 2-split graph and let $G_K = (V_K, E_K)$ be its kernel. Then $G$ has $S = \{u_1, \ldots, u_s\}$, $T_1 = \{v_1, \ldots, v_t\}$ and $T_2 = \{w_1, \ldots, w_t\}$ as an $(s, t)$-partition of $C$ if and only if $G_K$ has $v_j w_j$ as a non-edge for each $1 \leq j \leq t$ and $\{w_1, \ldots, w_t\}$ as an independent set.*

## 4  Improved NP-Completeness Results

In this section, we present three NP-completeness results for STAR PARTITION on split graphs. Each of them, in a strict sense, improves the best known results.

STAR PARTITION is NP-complete for $K_{1,5}$-free split graphs [25]. This result, in conjunction with a result from [12], implies that STAR PARTITION is NP-complete for even for $K_{1,5}$-free $(1, 3)$-split graphs. In the theorem below, we extend the result to $K_{1,5}$-free 2-split graphs.

**Theorem 3.** STAR PARTITION *is NP-complete even when restricted to $K_{1,5}$-free 2-split graphs.*

*Proof.* Given a graph $G = (V, E)$ and a positive integer $k$, we can verify whether any partition of $V$ into at most $k$ sets is a star partition of $G$ in polynomial time. Hence it follows that STAR PARTITION lies in NP.

We prove the NP-hardness result by providing a reduction from the INDE-PENDENT SET problem restricted to certain special graphs. Given a graph $G$ and a positive integer $k$, the INDEPENDENT set problem asks whether $G$ has an independent set of size $k$.

Let $G$ be any graph on $2\ell$ vertices and suppose $G$ has a perfect matching. Also suppose that each vertex in $G$ has degree either two or three. A graph of this form is given in Fig. 3. Any such graph has the independence number $\alpha(G) \leq \ell$. The INDEPENDENT SET problem remains NP-complete even when restricted to graphs of this form and the corresponding parameter $k \leq \ell - 2$. This NP-completeness result on independent sets follows from a simple reduction from the MAX2SAT problem restricted to those instances in which each clause has exactly two literals, each variable occurs exactly *thrice* and each literal occurs at least *once*; MAX2SAT restricted to such instances is NP-hard [23].

We indeed prove the NP-hardness of STAR PARTITION for $K_{1,5}$-free 2-split graphs by providing a polynomial reduction from the above restriction of the INDEPENDENT SET problem.

Let $(G, k)$ be an instance of the restricted INDEPENDENT SET problem presented above. Suppose $G$ has $2\ell$ vertices. Then, since $1 \leq k \leq \ell - 2$, $k < \ell$ and $\ell \geq 3$. Our reduction transforms this $G$ into its split division $G_S = (C \cup I, E_S)$. By definition, $G_S$ is a split graph in which each vertex in $I$ has degree two. In other words, it is a 2-split graph; see Definition 3. Since each vertex in $G$ has degree at most three, it now follows that each vertex in the clique part $C$ of $G_S$ has at most three neighbours in its independent part $I$. Thus, $G_S$ is $K_{1,5}$-free too. Finally, we set the parameter equal to $k' = 2\ell - k$.

The construction of $(G_S, 2\ell - k)$ from $G$ can be carried out in polynomial time. We now show that $G$ has an independent set of size $k$ if and only if $G_S$ has a star partition of size $k' = 2\ell - k$.

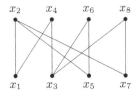

**Fig. 3.** The structure of an INDEPENDENT SET instance $G$ considered in the NP-hardness reduction.

Suppose $G$ has an independent set of size $k$, say $A$. Then $A$ has exactly one end vertex from a set of $k$ matching edges of $G$ ($G$ has a perfect matching of size $\ell(> k)$). Let $B = V(G)\backslash A$.

We now consider the bipartite graph $H = (A \cup B, E_H)$ with vertex set $A \cup B = V(G)(= C)$ and the edge set $E_H = \{ab : a \in A, b \in B$ and $ab$ is a non-edge of $G\}$. Since $G$ has maximum degree three and has $A$ as an independent set of size $k$, it follows that the bipartite graph $H$ has the degree of each vertex $v$ in $A$ at least $2\ell - k - 3$. This implies that $|N_H(v)| \geq 2\ell - (k+3)$ for any $v \in A$. But $2\ell - (k+3) \geq k$ since $k \leq \ell - 2$. Consequently, for any nonempty subset $S$ of $A$, $|N_H(S)| \geq 2\ell - (k+3) \geq k = |A| \geq |S|$. Thus, the bipartite graph $H$ satisfies Hall's condition. Hence $H$ has a matching, say $M$, saturating $A$. Now, let $A = \{w_1, \ldots, w_k\}$ and $M = \{v_1 w_1, \ldots, v_k w_k\}$. Then $G$ has $v_1 w_1, \ldots, v_k w_k$ as a matching of non-edges and $\{w_1, \ldots, w_k\}$ as an independent set; see Fig. 4.

We may now recall that $G$ is the kernel of $G_S$. Thus, if $B\backslash\{v_1, \ldots, v_k\} = \{u_1, \ldots, u_{2(\ell-k)}\}$, then, by Lemma 4, $S = \{u_1, \ldots, u_{2(\ell-k)}\}$, $T_1 = \{v_1, \ldots, v_k\}$ and $T_2 = \{w_1, \ldots, w_k\}$ form an $(2(\ell - k), k)$-partition of the clique part $C$ of the 2-split graph $G_S$. Then, by Lemma 2, $G_S$ has a star partition of size $2\ell - k$.

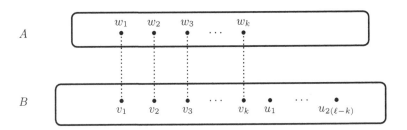

**Fig. 4.** A matching of non-edges in the graph $G$.

Conversely, suppose $G_S = (C \cup I, E_S)$ has a star partition of size $2\ell - k$. Then, since $|C| = |V(G)| = 2\ell$, by Lemmas 1 and 2, $C$ has an $(2(\ell - k), \ell)$-partition, say $S = \{u_1, \ldots, u_{2(\ell-k)}\}$, $T_1 = \{v_1, \ldots, v_k\}$ and $T_2 = \{w_1, \ldots, w_k\}$. But $G_S$ is a 2-split graph with $G$ as its kernel. So, by Lemma 4, $\{w_1, \ldots, w_k\}$ is an independent set of $G$ of size $k$. □

Consider any graph $G$. Since any star in $G$, by definition, can include at most two vertices from any clique of $G$, we have $sp(G) \geq \lceil \omega(G)/2 \rceil$. Also, the *Max2Sat* problem has polynomial time algorithms; for instance, see [22]. These facts imply that for any $K_{1,5}$-free 2-split graph $G$ arising in the NP-completeness proof, deciding whether $sp(G) = \lceil \omega(G)/2 \rceil$ has a polynomial time algorithm (i.e., we set $k = \ell = \lceil \omega(G)/2 \rceil$). But, from [25] and [12], we have that deciding whether $sp(G) = \lceil \omega(G)/2 \rceil$ is NP-complete even for $(1,3)$-split graphs. In Theorem 4 below prove that this particular case of the star partition problem remains NP-complete even for $K_{1,6}$-free 2-split graphs.

**Theorem 4.** *It is NP-complete to decide whether $sp(G) = \lceil \omega(G)/2 \rceil$ even when the instances are restricted to $K_{1,6}$-free 2-split graphs.*

The above results motivate us to resolve the computational complexity of STAR PARTITION restricted to $r$-split graphs for each fixed $r \geq 3$. We have not obtained any conclusive answers in this direction yet. But, in the following theorem, we prove that STAR PARTITION is NP-complete for $(1, r)$-split graphs for each fixed $r \geq 2$. A split graph is called a $(1, r)$ split graph if each vertex in its independent part has degree either 1 or $r$; see Definition 1.

**Theorem 5.** STAR PARTITION *is NP-complete even when restricted to $(1, r)$-split graphs for each fixed $r \geq 2$.*

## 5   Fixed Parameter (In)tractability

In this section, we study the star partition problem of split graphs in the parameterized complexity framework. We consider three natural parameterizations. Interestingly, the results obtained happen to be in three different extremes.

In our first parameterized problem, we take the *desired solution size* $k$ as the parameter and ask whether $sp(G) \leq k$. Since a split graph is disconnected if and only if it has isolated vertices, we restrict to connected split graphs.

**Theorem 6.** *Given a connected split graph $G$ and an integer $k \geq 1$ as the parameter, the problem of deciding whether $sp(G) \leq k$ is fixed parameter tractable. In fact, it has an $O((2k)^{2k+1}n)$ time algorithm.*

*Proof.* Let $G = (C \cup I, E)$ and suppose $|C| = q$ and $|I| = p$. Then $q/2 \leq sp(G) \leq q$. So, if $k < q/2$, then we can at once decide that $sp(G) \not\leq k$ and, if $k \geq q$, then we can at once decide that $sp(G) \leq k$. So, we now assume that $q/2 \leq k < q$.

Now, by Lemma 2, $G$ has a star partition of size $k$ if and only if $C$ has an $(s,t)$-partition for $(s,t) = (2k-q, q-k)$. Now $q \leq 2k$ and $t = q-k \leq k$. Also, by Lemma 3, for any non-negative integer pair $(s,t)$, we can decide whether $C$ has an $(s,t)$-partition in time $O(q^{2t+1}p)$. But, since $q \leq 2k$ and $t = q-k \leq k$, this implies that deciding whether $C$ has an $(s,t)$-partition with $(s,t) = (2k-q, q-k)$ can be decided in time $O((2k)^{2k+1}n)$.    □

For any graph $G$, $sp(G) \geq \omega(G)/2$. This motivates the following parameterized problem: Given a graph $G$ and an integer $k \geq 0$ as the parameter, what is the parameterized complexity of deciding whether $sp(G) \leq \lceil \omega(G)/2 \rceil + k$? Here we are *parameterizing above a guaranteed value*. In the theorem below, we provide some interesting answers to this question.

**Theorem 7.** *Given a graph $G$ and an integer $k \geq 0$ as the parameter, the problem of deciding whether $sp(G) \leq \lceil \omega(G)/2 \rceil + k$ is para-NP-hard even when restricted to either (1) $K_{1,6}$-free $(0, 2)$-split graphs or (2) $K_{1,5}$-free $(0, 1, 3)$-split graphs.*

*Proof.*(1) From Theorem 4, given a $K_{1,6}$-free 2-split graph $G$, deciding whether $sp(G) = \lceil \omega(G)/2 \rceil$ is NP-complete. For $k > 0$, the NP-completeness result can be obtained by adding $k$ isolated vertices to the 2-split graph constructed in the reduction of Theorem 4; this results in a $(0, 2)$-split graph. Thus the problem is para-NP-hard even when restricted to $K_{1,6}$-free $(0, 2)$-split graphs.
(2) From [25] and [12], given a $K_{1,6}$-free $(1, 3)$-split graph $G$, deciding whether $sp(G) = \lceil \omega(G)/2 \rceil$ is NP-complete. For $k > 0$, the NP-completeness result can be obtained by adding $k$ isolated vertices to the $(1, 3)$-split graph constructed in the reduction; this results in a $(0, 1, 3)$-split graph. Thus problem is para-NP-hard even when restricted to $K_{1,6}$-free $(0, 1, 3)$-split graphs. ☐

For any connected split graph $G$, $sp(G) \leq \omega(G)$. This motivates the following parameterized problem: Given a graph $G$ and an integer $k \geq 0$ as the parameter, what is the parameterized complexity of deciding whether $sp(G) \leq \omega(G) - k$?. In this parameterization, we are trying to *save $k$ stars*. We have the following theorem.

**Theorem 8.** *Given a graph $G$ and an integer $k \geq 0$ as the parameter, the problem of deciding whether $sp(G) \leq \omega(G) - k$ is $W[1]$-hard even for $(1, 2)$-split graphs. Also this problem is in $W[3]$ for connected split graphs.*

## 6   Polynomial Time Algorithmic Results

By Theorem 3, STAR PARTITION is NP-complete even for $K_{1,5}$-free 2-split graphs. Also, by Theorem 5, the problem is NP-hard for $(1, r)$-split graphs for each fixed $r \geq 2$. In this section, we present some polynomial time algorithms complementing these hardness results. In particular, we obtain a polynomial time $3/2$-approximation algorithm for *certain* 2-split graphs. This may, in some sense, be considered as an improvement on a polynomial time 2-approximation algorithm for split graphs in [25]. We also obtain a linear time exact algorithm for MIN STAR PARTITION on 1-split graphs. Both these results make use of the structural result of Lemma 4.

In the following definition, we define a class of 2-split graphs, namely $\mathcal{S}(\mathcal{G})$, corresponding to any graph class $\mathcal{G}$. In Theorem 9, we present a polynomial time $3/2$-approximation algorithm for MIN STAR PARTITION on the graph class $\mathcal{S}(\mathcal{G})$ provided the maximum independent set problem has a polynomial time exact algorithm for the graph class $\mathcal{G}$.

**Definition 4.** *Let $\mathcal{G}$ be any graph class. Then $\mathcal{S}(\mathcal{G})$ denotes the set of those 2-split graphs for which the kernel is in $\mathcal{G}$.*

**Theorem 9.** *Let $\mathcal{G}$ be any graph class for which the maximum independent set problem has a polynomial time algorithm. Then* MIN STAR PARTITION *has a polynomial time $3/2$-approximation algorithm for the graph class $\mathcal{S}(\mathcal{G})$.*

We now present a linear time exact algorithm MIN STAR PARTITION on 1-split graphs (which implies a similar algorithm for $(0,1)$-split graphs). This complements the NP-hardness result for 2-split graphs and $(1,r)$-split graphs ($r \geq 2$) presented in Theorems 3 and 5, respectively. Interestingly, we prove that for any 1-split graph $G$, $sp(G) = \max(\lceil \omega(G)/2 \rceil, \alpha(G^2))$.

If $Z$ is a star in a graph $G$, then any two vertices in $Z$ are at distance at most two assuming $|Z| \geq 2$. Thus if $G$ has two vertices that are at distance three or more, then they cannot be part of a single star of $G$. We also recall that the *square of a graph $G = (V, E)$*, denoted $G^2$, is a graph with vertex set $V$ and the edge set consisting of those unordered pairs of vertices $u, v \in V$ for which $d_G(u, v) \leq 2$. Thus two vertices in $G^2$ are non-adjacent if and only if the distance between them is three or more in $G$. Thus, we have the following lemma.

**Lemma 5.** *Let $G$ be any graph and let $I$ be any set of vertices in $G$ such that any pair of vertices in $I$ are at distance three or more in $G$. Then $sp(G) \geq |I|$. Consequently, $sp(G) \geq \alpha(G^2)$.*

*Remark:* Computing $\alpha(G^2)$ is NP-hard even for 3-split graphs. This follows from a simple reduction from the NP-complete EXACT3COVER problem [11].

**Corollary 1.** *Let $G = (C \cup I, E)$ be a split graph and let $I_1$ denote the set of all degree one vertices in the independent part $I$. Then $sp(G) \geq |N(I_1)|$.*

*Proof.* Let $N(I_1) = \{x_1, \ldots, x_k\}$. Then, for each $1 \leq i \leq k$, choose one $y_i$ from $N_I(x_i)$. Then any two vertices in $\{y_1, \ldots, y_k\}$ are at a distance of three from each other in $G$. Thus, by Lemma 5, we have that $sp(G) \geq k$. Now the corollary follows since $|N(I_1)| = k$. ☐

**Corollary 2.** *If $G = (C \cup I, E)$ is a 1-split graph, then $|N(I)| = \alpha(G^2)$.*

In the following theorem, we present a simple formula for the star partition number of a 1-split graph.

**Theorem 10.** *If $G$ is a 1-split graph, then $sp(G) = \max(\lceil \omega(G)/2 \rceil, \alpha(G^2))$. Consequently,* MIN STAR PARTITION *has a linear time exact algorithm for $(0,1)$-split graphs.*

# 7    Conclusion

We leave the computational complexity of STAR PARTITION for $r$-split graphs unresolved for each fixed $r \geq 3$; we believe it is NP-complete. It would be interesting to obtain a factor $3/2$ (or better) polynomial time approximation algorithm for MIN STAR PARTITION on at least *all* of 2-split graphs. It appears that further study of 2-split graphs can possibly help in proving matching inapproximability results. Designing better than factor 2 approximation algorithms for MIN STAR PARTITION on split graphs remains an interesting algorithmic problem. We also have the complexity status of the problem open even for $K_{1,4}$-free split graphs.

# References

1. Andreatta, G., De Francesco, C., De Giovanni, L., Serafini, P.: Star partitions on graphs. Discret. Optim. **33**, 1–18 (2019)
2. Bang-Jensen, J., Huang, J., MacGillivray, G., Yeo, A.: Domination in convex bipartite and convex-round graphs. Technical report, University of Southern Denmark (1999). https://doi.org/10.5555/870702
3. Björklund, A., Husfeldt, T., Koivisto, M.: Set partitioning via inclusion-exclusion. SIAM J. Comput. **39**, 546–563 (2009). https://doi.org/10.1137/070683933
4. Brandstädt, A., Kratsch, D.: On the restriction of some NP-complete graph problems to permutation graphs. In: Budach, L. (ed.) FCT 1985. LNCS, vol. 199, pp. 53–62. Springer, Heidelberg (1985). https://doi.org/10.1007/BFb0028791
5. Cockayne, E., Goodman, S., Hedetniemi, S.: A linear algorithm for the domination number of a tree. Inf. Process. Lett. **4**, 41–44 (1975). https://doi.org/10.1016/0020-0190(75)90011-3
6. Damaschke, P., Müller, H., Kratsch, D.: Domination in convex and chordal bipartite graphs. Inf. Process. Lett. **36**, 231–236 (1990). https://doi.org/10.1016/0020-0190(90)90147-P
7. Dor, D., Tarsi, M.: Graph decomposition is np-complete: a complete proof of Holyer's conjecture. SIAM J. Comput. **26**, 1166–1187 (1997). https://doi.org/10.1137/S0097539792229507
8. Downey, R., Fellows, M.: Fundamentals of Parameterized Complexity. Texts in Computer Science. Springer, London (2013)
9. Duginov, O.: Partitioning the vertex set of a bipartite graph into complete bipartite subgraphs. Discrete Math. Theor. Comput. Sci. **16**, 203–214 (2014). https://doi.org/10.46298/dmtcs.2090
10. Farber, M., Keil, J.: Domination in permutation graphs. J. Algorithms **6**, 309–321 (1985). https://doi.org/10.1016/0196-6774(85)90001-X
11. Garey, M., Johnson, D.: Computers and Intractability: A Guide to the Theory of NP-Completeness. W. H. Freeman, New York (1979)
12. Gonzalez, T.: Clustering to minimize the maximum intercluster distance. Theor. Comput. Sci. **38**(1985), 293–306 (1985)
13. Kelmans, A.: Optimal packing of induced stars in a graph. Discret. Math. **173**, 97–127 (1997). https://doi.org/10.1016/S0012-365X(96)00121-5
14. Kirkpatrick, D., Hell, P.: On the completeness of a generalized matching problem. In: Proceedings of the Tenth Annual ACM Symposium on Theory of Computing, pp. 240–245 (1978). https://doi.org/10.1145/800133.804353

15. Maffray, F., Preissmann, M.: On the np-completeness of the k-colorability problem for triangle-free graphs. Discret. Math. **162**, 313–317 (1996). https://doi.org/10.1016/S0012-365X(97)89267-9
16. McMorris, F., Shier, D.: Representing chordal graphs on $k_{1,n}$. Comment. Math. Univ. Carol. **24**(3), 489–494 (1983)
17. Mondal, J., Vijayakumar, S.: Star covers and star partitions of cographs and butterfly-free graphs. In: Proceedings 10th International Conference on Algorithms and Discrete Applied Mathematics, CALDAM (2024). Accepted
18. Mondal, J., Vijayakumar, S.: Star covers and star partitions of double-split graphs. J. Comb. Optim. Accepted
19. Monnot, J., Toulouse, S.: The path partition problem and related problems in bipartite graphs. Oper. Res. Lett. **35**, 677–684 (2007). https://doi.org/10.1016/j.orl.2006.12.004
20. Müller, H., Brandstädt, A.: The np-completeness of Steiner tree and dominating set for chordal bipartite graphs. Theoret. Comput. Sci. **53**, 257–265 (1987). https://doi.org/10.1016/0304-3975(87)90067-3
21. Nguyen, X.: Induced star partition of graphs with respect to structural parameters. Charles University in Prague, Technical report (2023)
22. Papadimitriou, C.: Computational Complexity. Addison-Wesley, Boston (1994)
23. Raman, V., Ravikumar, B., Srinivasa Rao, S.: A simplified NP-complete MAXSAT problem. Inf. Process. Lett. **65**, 1–6 (1998)
24. Raman, V., Saurabh, S.: Short cycles make W-hard problems hard: FPT algorithms for W-hard problems in graphs with no short cycles. Algorithmica **52**, 203–225 (2008). https://doi.org/10.1007/s00453-007-9148-9
25. Shalu, M., Vijayakumar, S., Sandhya, T., Mondal, J.: Induced star partition of graphs. Discret. Appl. Math. **319**, 81–91 (2022). https://doi.org/10.1016/j.dam.2021.04.015
26. Van Bevern, R., et al.: Partitioning perfect graphs into stars. J. Graph Theory **85**, 297–335 (2017). https://doi.org/10.1002/jgt.22062
27. Vazirani, V.V.: Approximation Algorithms. Springer, Heidelberg (2003). https://doi.org/10.1007/978-3-662-04565-7
28. West, D.: Introduction to Graph Theory, 2nd edn. Pearson, London (2018)
29. Zuckerman, D.: Linear degree extractors and the inapproximability of max clique and chromatic number. Theory Comput. **3**, 103–128 (2007). https://doi.org/10.4086/toc.2007.v003a006

# Star Covers and Star Partitions of Cographs and Butterfly-free Graphs

Joyashree Mondal and S. Vijayakumar[(✉)]

Indian Institute of Information Technology, Design and Manufacturing (IIITDM)
Kancheepuram, Chennai 600127, India
{mat18d001,vijay}@iiitdm.ac.in

**Abstract.** A graph that is isomorphic to $K_{1,r}$ for some $r \geq 0$ is called a *star*. For a graph $G = (V, E)$, any subset $S$ of its vertex set $V$ is called a *star* of $G$ if the subgraph induced by $S$ is a star. A collection $\mathcal{C} = \{V_1, \ldots, V_k\}$ of stars in $G$ is called a *star cover* of $G$ if $V_1 \cup \ldots \cup V_k = V$. A star cover $\mathcal{C}$ of $G$ is called a *star partition* of $G$ if it is also a partition of $V$. Given a graph $G$, the problem STAR COVER asks for a star cover of $G$ of minimum size. Given a graph $G$, the problem STAR PARTITION asks for a star partition of $G$ of minimum size. Both the problems are NP-hard even for bipartite graphs [24]. In this paper, we obtain exact $O(n^2)$ time algorithms for both STAR COVER and STAR PARTITION on $(C_4, P_4)$-free graphs and on $(2K_2, P_4)$-free graphs. We also prove that STAR COVER and STAR PARTITION are polynomially equivalent, up to the optimum value, for butterfly-free graphs and present an $O(n^{14})$ time $O(\log n)$-approximation algorithm for these equivalent problems on butterfly-free graphs. We also obtain $O(\log n)$-approximation algorithms for STAR COVER on *hereditary* graph classes.

**Keywords:** Star Cover · Star Partition · Cographs · Butterfly-free Graphs · Polynomial Time Algorithms · Approximation Algorithms

## 1 Introduction

A graph is called a *star* if it is isomorphic to the complete bipartite graph $K_{1,r}$ for some $r \geq 0$; please see Fig. 1. When a graph models a network, like a road or computer network, each *induced* subgraph that is a star corresponds to a sub-network that is a star network. The *center* of an induced star in such a graph potentially corresponds, in the underlying network, to either a bottleneck or a point that is desirable for locating some facilities; this is especially the case when the star has two or more *non-center* vertices. These practical considerations as well as their combinatorial appeal motivate a study of two optimization problems, namely STAR COVER and STAR PARTITION, that will be defined below.

By extension, any subset $S$ of the vertex set of a graph $G = (V, E)$ is called a *star* of $G$ if the subgraph induced by $S$ is a star. A collection $\mathcal{C} = \{V_1, \ldots, V_k\}$

---

The second author is supported by DST-SERB MATRICS: MTR/2022/000870.

S. Kalyanasundaram and A. Maheshwari (Eds.): CALDAM 2024, LNCS 14508, pp. 224–238, 2024.
https://doi.org/10.1007/978-3-031-52213-0_16

**Fig. 1.** Some examples of stars.

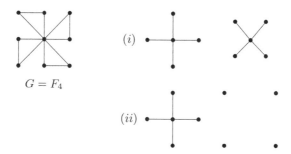

**Fig. 2.** The friendship graph $G = F_4$ along with (i) an optimal star cover and (ii) an optimal star partition of it; thus $sc(G) = 2$ and $sp(G) = 5$.

of stars in $G$ is called a *star cover* of $G$ if $V_1 \cup \ldots \cup V_k = V$. A star cover $\mathcal{C}$ of $G$ is called a *star partition* of $G$ if it is also a partition of $V$. A star cover $\mathcal{C}$ consisting of $k$ distinct stars of $G$ is called a star cover of $G$ of *size* $k$. The *size* of a star partition is defined similarly. The minimum $k$ for which $G$ admits a star cover of size $k$ is called the *star cover number* of $G$ and is denoted by $sc(G)$. The minimum $k$ for which $G$ admits a star partition of size $k$ is called the *star partition number* of $G$ and is denoted by $sp(G)$. For instance, the graph $G$ in Fig. 2 has $sc(G) = 2$ and $sp(G) = 5$.

In this paper, we study the following problems on cographs, butterfly-free graphs and hereditary graph classes.

STAR COVER
**Input:** A graph $G$.
**Goal:** A minimum star cover of $G$.

STAR PARTITION
**Input:** A graph $G$.
**Goal:** A minimum star partition of $G$.

Consider any graph $G$. Since a star partition of $G$ is also a star cover of $G$, we have $sc(G) \leq sp(G)$. Also the centers of any star cover of $G$ form a dominating set of $G$. So, we further have $\gamma(G) \leq sc(G) \leq sp(G)$, where $\gamma(G)$ is the domination number of $G$. Thus, if $G$ is triangle-free, we also have $sp(G) = sc(G) = \gamma(G)$; this easily follows from Proposition 1. Consequently, STAR COVER and STAR PARTITION are both polynomially equivalent to the minimum dominating set problem on triangle-free graphs.

In general, $sp(G) \neq sc(G)$. For instance, for the friendship graph $F_n = K_1 \oplus (nK_2)$, where $n \geq 1$, $sc(F_n) = 2$ but $sp(G) = n + 1$; see Fig. 2. In this paper, we prove that $sp(G) = sc(G)$ for any butterfly-free graph. This proof indeed implies that STAR COVER and STAR PARTITION are polynomially equivalent for butterfly-free graphs. Butterfly-free graphs includes many natural graph classes such as bipartite graphs, split graphs and cluster graphs.

In [16], stars have been considered in the context of generalizing the well-known maximum matching problem. It considers the problem of covering a maximum number of vertices of the input graph by vertex-disjoint *induced* stars of the form $K_{1,i}, 1 \leq i \leq r$, where $r \geq 1$ is any fixed positive integer. This problem is shown to have a polynomial time algorithm in [16]. This algorithm leads to a simple $\frac{r}{2}$-approximation algorithm for STAR PARTITION on $K_{1,r}$-free graphs, $r \geq 2$, that remains NP-hard even for line graphs, a subclass of claw-free graphs [24].

A summary of known results on STAR COVER and STAR PARTITION follows. Deciding whether an input graph can be covered by *or* partitioned into at most two stars has polynomial time algorithms but deciding whether an input graph can be covered by *or* partitioned into three stars is NP-complete [24]; these problems remain NP-complete even for $K_4$-free graphs. Both STAR COVER and STAR PARTITION are NP-hard for (a) chordal bipartite graphs [22], (b) $(C_4, C_6, \ldots, C_{2t})$-free bipartite graphs for every fixed $t \geq 2$ [13], (c) subcubic bipartite planar graphs [15,24], (d) $K_{1,5}$-free split graphs in which each vertex in the independent part has degree exactly two [10], (e) line graphs (and hence claw-free graphs and, more generally $K_{1,r}$-free graphs for any fixed $r \geq 3$) [11,24] and (f) co-tripartite graphs [19,24].

It is NP-hard to approximate STAR PARTITION within $n^{1/2-\epsilon}$ for all $\epsilon > 0$ [24,28]. Also both STAR COVER and STAR PARTITION do not have any polynomial time $c \log n$-approximation algorithm for some constant $c > 0$ unless P = NP [25]. Both STAR COVER and STAR PARTITION have a polynomial time (a) 2-approximation algorithm for split graphs [24]; (b) $O(\log n)$-approximation algorithms for triangle-free graphs [25]; (c) $(d + 1)$-approximation algorithm for triangle-free graphs of degree at most $d$ [25]. STAR PARTITION has a polynomial time $r/2$-approximation algorithm for $K_{1,r}$-free graphs (which implies a $3/2$-approximation algorithm for, for instance, line graphs and cobipartite graphs as they are claw-free) [16,24].

With the number of stars in a star cover/partition as the parameter, we have the following: (1) Both the problems are W[2]-complete for bipartite graphs and are fixed parameter tractable for graphs of girth at least five [23]. (2) The problems are fixed parameter tractable for split graphs [10]. (For an introduction to parameterized complexity one may refer to [12].)

The problems have exact (a) $2^n n^{O(1)}$ time and exponential space algorithms and (b) $3^n n^{O(1)}$ time and polynomial space algorithms [5].

Both STAR COVER and STAR PARTITION have exact polynomial time algorithms for bipartite permutation graphs [6,14], convex bipartite graphs [3,9], doubly-convex bipartite graphs [3], trees [7] and double-split graphs [20].

*Related Literature:* The problem of covering a maximum number of vertices of an input graph by vertex-disjoint *induced* stars of the form $K_{1,i}, 1 \leq i \leq r$, where $r \geq 1$ is any fixed positive integer, has a polynomial time algorithm [16].

The problem of partitioning (the vertex set of) an input graph into induced paths of length $t$ ($t \geq 3$ is fixed) is NP-complete even for bipartite graphs with maximum degree three [21]. This result implies that the problem of partitioning a graph into induced $K_{1,2}$'s is NP-complete, even when restricted to bipartite graphs with maximum degree three.

The problem of partitioning an input graph into equal but fixed size stars, not necessarily induced, is investigated for many natural subclasses of perfect graphs in [4]. (This problem is also referred to as STAR PARTITION in that paper.) This problem had already been shown to be NP-complete even when the star size is fixed to be three in [17]. Given a graph $G$ and a positive integer $k$, the problem of deciding whether $G$ can be partitioned into exactly $k$ (not necessarily induced) stars, each of size at least two, is investigated in [1].

*Our Results:* We obtain an $O(t(n)n^2)$ time $O(\log n)$-approximation algorithm for STAR COVER on any *hereditary* graph class $\mathcal{G}$ for which the maximum independent set problem has an $O(t(n))$ time exact algorithm. As a corollary, for STAR COVER on perfect graphs, we also obtain a polynomial time $O(\log n)$-approximation algorithm.

We prove that $sp(G) = sc(G)$ for any butterfly-free graph $G$. We also present an $O(n^{14})$ time $O(\log n)$-approximation algorithm for both STAR COVER and STAR PARTITION on butterfly-free graphs.

For both STAR COVER and STAR PARTITION, we obtain $O(n^2)$ time exact algorithms on (1) trivially perfect graphs (($C_4, P_4$)-free graphs) and on (2) complements of trivially perfect graphs (($2K_2, P_4$)-free graphs). We also obtain linear time algorithms for the problems on threshold graphs (($2K_2, C_4, P_4$)-free graphs).

## 2   Preliminaries

All graphs considered in this paper are finite. For basic graph theory terminology, we refer to [26]. Thus $K_n$, $P_n$ and $C_n$ denote the complete graph, the path and the cycle on $n$ (unlabeled) vertices. Also $K_{m,n}$ denotes the complete bipartite graph with independent bipartitions of sizes $m$ and $n$. In particular, the complete bipartite graph $K_{1,n}$ is the *star* on $n + 1$ vertices, where $n \geq 0$.

A *clique (independent set)* of a graph $G$ is any set of pairwise adjacent (non-adjacent) vertices of $G$. The size of a maximum clique (maximum independent set) in $G$ is denoted by $\omega(G)$ ($\alpha(G)$). A *k-(vertex) coloring* of a graph $G$ is a partition $(V_1, V_2, \ldots, V_k)$ of $V(G)$ such that $V_i$ is an independent set in $G$ for $1 \leq i \leq k$. The minimum $k$ for which $G$ admits a $k$-coloring of $G$ is called the chromatic number of $G$ and is denoted by $\chi(G)$.

For any vertex $v$ of a graph $G$, $N(v)$ denotes the set of all vertices adjacent to $v$ in $G$ and $N[v]$ denotes $\{v\} \cup N(v)$ (these are called the open and closed neighbourhoods of $v$, respectively). A set $D \subseteq V(G)$ is called a *dominating set* of $G$ if $\cup_{v \in D} N[v] = V(G)$. The size of a minimum dominating set of $G$ is called the domination number of $G$ and is denoted by $\gamma(G)$.

For any set $X$ of vertices in a graph $G$, $G[X]$ denotes the graph induced by $X$ in $G$. Given a graph $H$, a graph $G$ is said to be $H$-free if it has no induced subgraphs isomorphic to $H$. More generally, given a family (finite or infinite) of graphs $\mathcal{F}$, a graph $G$ is said be $\mathcal{F}$-free if $G$ does not have any induced subgraph $H$ isomorphic to some graph $H \in \mathcal{F}$.

The complement of a graph $G$ is denoted by $\overline{G}$. Let $G_1 = (V_1, E_1)$ and $G_2 = (V_2, E_2)$ be two vertex-disjoint graphs. Then the graph $G = (V_1 \cup V_2, E_1 \cup E_2)$ is called the *union* of $G_1$ and $G_2$ and is denoted by $G_1 \cup G_2$. In particular, $pK_n$ denotes the union of $p$ vertex-disjoint complete graphs, each isomorphic to $K_n$. Also the graph $G$ obtained by joining each vertex of $G_1$ to each vertex of $G_2$ in the union $G_1 \cup G_2$ is called the *join* of $G_1$ and $G_2$ and is denoted by $G_1 \oplus G_2$. Both union and join operations of graphs are associative. So, union (join) of more than two graphs are defined in the natural manner.

As already defined in the introductory section, a graph $G$ is called a *star* if $G \cong K_{1,r}$ for some integer $r \geq 0$. Thus any star $G$ has a vertex $x$ such that $[\{x\}, N(x)]$ is a partition of its vertex set into two independent sets. Such an $x$ in $G$ is called a *center* of $G$ and the vertices in $N(x)$ are called the non-center vertices of the star $G$. Thus, any star that is not isomorphic to $K_{1,1} = K_2$ has a *unique* center; see Fig. 1.

If $G = (V, E)$ is any graph, then, as already defined in the introductory section, we also call a subset $S$ of $V$ a *star* of $G$ if the induced subgraph $G[S]$ is a star.

*Notation:* Let $G$ be any graph. Then any star $S$ in $G$ can be written as a disjoint union, namely $\{x\} \cup I$, where $x$ is a *center* vertex of the star $G[S]$ and $I = S \setminus \{x\}$ is the independent set consisting of the non-center vertices of $G[S]$ (where $|I| \geq 0$).

We conclude this section with a proposition.

**Proposition 1** ([20]). *Consider any graph $G$. Then the following hold:*

1. *If a graph $G$ has a star cover of size $k$, then it also has a star cover of size $k$ such that the stars in the cover share, if anything, only the centers.*
2. *If a graph $G$ has a star cover of size $k$ such that the stars in it have their centers distinct, then it also has a star partition of size $k$.*

## 3   Hereditary Graph Classes: Star Cover

A class $\mathcal{G}$ of graphs is called a hereditary graph class if, for every graph $G$ in $\mathcal{G}$, all the induced subgraphs of $G$ are also in $\mathcal{G}$. The family of all graphs is an obvious example of a hereditary graph class. Any graph class with a forbidden induced subgraph characterization is also hereditary.

In this section, we show that STAR COVER on any hereditary graph class $\mathcal{G}$ has an $O(t(n)n^2)$ time $O(\log n)$-approximation algorithm assuming the maximum independent set problem on the class $\mathcal{G}$ has an $O(t(n))$ time algorithm.

This result, in particular, implies that STAR COVER has a polynomial time $O(\log n)$-approximation algorithm for *perfect graphs*. Since perfect graphs include bipartite graphs, we also have a matching inapproximabilty result assuming P $\neq$ NP [25]. Another interesting consequence concerns butterfly-free graphs; it appears in Sect. 4.

**Algorithm:** *Approx-Hereditary*

> **Input:** A graph $G$ from a hereditary graph class $\mathcal{G}$.
> **Output:** A star cover of $G$.

1. Set $S = \emptyset$.
2. Color all vertices of $G$ black.
3. While $G$ has a black vertex repeat the following:
   - Find a star $Z$ with a maximum number of black vertices.
   - Color the black vertices of $G$ that are in $Z$ grey.
   - Set $S = S \cup \{Z\}$.
4. Output $S$.

**Fig. 3.** An $O(\log n)$-approximation algorithm for the star cover number of a graph from a hereditary graph class.

**Theorem 1.** *Let $\mathcal{G}$ be any hereditary graph class for which the maximum independent set problem has an $O(t(n))$ time algorithm. Then STAR COVER has an $O(\log n)$-approximation algorithm running in time $O(n^2 t(n))$ for any $G$ in $\mathcal{G}$.*

A graph $G$ is called a perfect graph if each of its induced subgraph $H$ has $\chi(H) = \omega(H)$. Perfect graphs is a well-known hereditary graph class containing several natural graph classes. Many optimization problems, including the maximum independent set problem, are polynomially solvable for perfect graphs. So, we have the following interesting consequence.

**Corollary 1.** STAR COVER *has a polynomial time $O(\log n)$-approximation algorithm for perfect graphs. Moreover, for some $c > 0$, the problem has no polynomial time $c \log n$ approximation algorithm when restricted to perfect graphs assuming $P \neq NP$.*

The second half of the corollary above follows from the imapproximability result on the minimum dominating set problem restricted to bipartite graphs [25] and the equivalence of dominating set problem and STAR COVER on bipartite graphs ($sc(G) = \gamma(G)$ for any bipartite graph $G$; this easily follows from Proposition 1).

## 4    Butterfly-free Graphs

**Fig. 4.** The butterfly graph.

A graph $G$ that is isomorphic to $K_1 \oplus (2K_2)$ is called a butterfly; see Fig. 4. Butterfly-free graphs include many natural graph classes such as bipartite graphs (indeed all triangle-free graphs), split graphs (indeed all $2K_2$-free graphs) and cluster graphs.

In this section, we prove that $sp(G) = sc(G)$ for any butterfly-free graph $G$. Our proof indeed implies that STAR COVER and STAR PARTITION are polynomially equivalent for butterfly-free graphs. We also obtain $O(n^{14})$ time $O(\log)$-approximation algorithms for both the problems on butterfly-free graphs. We further note that this $O(\log n)$ approximation guarantee is essentially the best possible for butterfly-free graphs assuming P $\neq$ NP.

**Theorem 2.** *Let $G$ be any butterfly-free graph. If $G$ has a star cover of size $k$, then it also has a star partition of size at most $k$. Consequently, $sp(G) = sc(G)$. Moreover, given a star cover of $G$, a star partition of at most the same size can be computed in time $O(n^2 \log n)$.*

*Proof.* Let $\mathcal{C}$ be any star cover of the butterfly-free graph $G$. By Proposition 1, we shall assume that the stars in $\mathcal{C}$ share, if anything, only the centers. Thus, if the stars in $\mathcal{C}$ have their centers distinct, then $\mathcal{C}$ is already a star partition of $G$. So, let us assume that some stars in $\mathcal{C}$ share their centers.

Suppose $Z_1 = \{x\} \cup I_1$ and $Z_2 = \{x\} \cup I_2$ are two distinct stars in $\mathcal{C}$ with a common center, namely $x$. Then, by our assumption on $\mathcal{C}$, $I_1$ and $I_2$ are disjoint independent sets of $G$ and do not include the center of any star in $\mathcal{C}$. If $I_1 \cup I_2$ is an independent set, we replace $Z_1$ and $Z_2$ in $\mathcal{C}$ by $Z_1 \cup Z_2$ to obtain another star cover $\mathcal{C}'$. Else, since $G$ is butterfly-free, the induced subgraph $G[I_1 \cup I_2]$ is a $2K_2$-free bipartite graph. So, the vertices in $I_1$ has an ordering, say $(u_1, \ldots, u_p)$, such that $N_{I_2}(u_1) \subseteq \ldots \subseteq N_{I_2}(u_p)$ [27]; see Fig. 5. (Indeed such an ordering of vertices in $I_1$ can be computed in time $O(n \log n)$ by ordering the vertices $I_1$ according to their degrees in $G[I_1 \cup I_2]$.) So, we have that $I_3 = \{u_1, u_2, \ldots, u_{p-1}\} \cup [I_2 \backslash N_{I_2}(u_p)]$ is an independent set in $G$. In this case, we replace the stars $Z_1$ and $Z_2$ in $\mathcal{C}$ with the stars $\{x\} \cup I_3$ and $\{u_p\} \cup N_{I_2}(u_p))$ to obtain another star cover $\mathcal{C}'$ of $G$.

Also, by our assumption on $\mathcal{C}$, no star in it has $u_p$ as its center. Thus we have that the stars in $\mathcal{C}'$ have the number of *reused* centers one less when compared to those in $\mathcal{C}$. Moreover, the stars in the new star cover $\mathcal{C}'$ also share, if anything, only the centers. Thus, if we proceed similarly as long as the resulting star cover has a pair of distinct stars with a common center, we obtain, in at most $n$ steps, a star cover $\mathcal{C}^*$ of $G$ of size at most $|\mathcal{C}|$ that is also a star partition of $G$. Thus, the computation can be carried out in time $O(n^2 \log n)$.

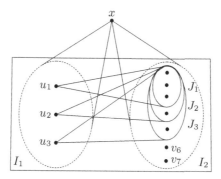

**Fig. 5.** The subgraph of a butterfly-free graph induced by a pair of its stars with a common center vertex.

The above arguments imply that $sp(G) \leq sc(G)$. Since any star partition of $G$ is also a star cover of $G$, we also have $sc(G) \leq sp(G)$. Hence we have $sp(G) = sc(G)$. $\qquad\square$

**Theorem 3.** *The problems* STAR COVER *and* STAR PARTITION *have* $O(n^{14})$ *time* $O(\log n)$*-approximation algorithms for butterfly-free graphs. Moreover, for some* $c > 0$*, neither of the problems has a polynomial time* $c \log n$ *approximation algorithm when restricted to butterfly-free graphs assuming* $P \neq NP$.

*Proof.* Butterfly-free graphs form a hereditary graph class. The maximum independent set problem has an $O(n^{12})$ time exact algorithm for this graph class [2,18]. So, the theorem follows from Theorems 1 and 2.

The second half of the theorem follows from a similar imapproximability result on the maximum dominating set problem restricted to bipartite graphs [25] and the equivalence of dominating set problem, STAR COVER and STAR PARTITION on bipartite graphs $(sp(G) = sc(G) = \gamma(G)$ for any bipartite graph $G$; this follows, for instance, from Proposition 1. $\qquad\square$

## 5   Cographs ($P_4$-free Graphs)

The class of cographs (i.e., $P_4$-free graphs) is also the class of graphs that contains $K_1$ and is closed under the union and join operations of graphs. This suggests a bottom-up construction of any cograph $G$ by starting from its vertices and by iteratively taking the union or join of induced subgraphs already built. Such a construction of a cograph is modelled by what is called its *cotree*.

In this section, we present exact $O(n^2)$ time algorithms for STAR COVER and STAR PARTITION on (1) trivially perfect graphs and (2) co-trivially perfect graphs. All these algorithms happen to be some natural greedy algorithms. We also obtain a linear time algorithm for threshold graphs.

*Canonical Cotrees:* In the bottom-up construction of a cograph, any maximal sequence of successive union (join) operations can be replaced by a single union (join) operation. In the corresponding cotree model, the child of any 0-node (1-node) will not be a 0-node (1-node). We call this special cotree of a cograph its *canonical cotree.*

## 5.1 Trivially Perfect Graphs

In this section, we present an $O(n^2)$ time exact algorithms for STAR COVER and STAR PARTITION on trivially perfect graphs. Indeed our algorithms for both the problems on this graph class are certain simple and natural greedy algorithms. But proving their correctness is fairly involved.

We note that STAR COVER and STAR PARTITION are essentially two distinct computational problems on trivially perfect graphs. For instance, for the friendship graph $F_n \cong K_1 \oplus (nK_2)$, $n \geq 2$, which is trivially perfect, $sc(G) = 2$ but $sp(G) = n + 1$.

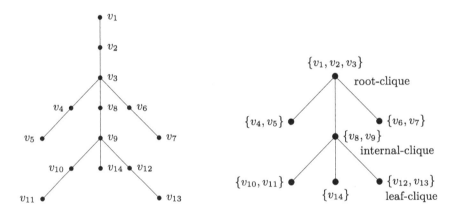

**Fig. 6.** A rooted tree underlying a trivially perfect graph $G$ and the corresponding clique-tree $T(G)$.

*Note on Canonical Cotrees of Trivially Perfect Graphs:* The *canonical cotree* of any trivially perfect graph $G$ has the additional property that every 1-node of it (that corresponds to a join), if any exists, has at most one 0-node (that corresponds to an induced disconnected subgraph). Otherwise $G$ will have a $C_4$.

It is well-known that any trivially perfect graph is a comparability graph of a partial order defined by a rooted tree $T$. But for the study of STAR COVER and STAR PARTITION, we prefer a different but related representation of $G$ that we call its *clique-tree* representation and denote it by $T(G)$. An example is provided in Fig. 6.

*Note:* The clique-tree of any connected trivially perfect graph has the property that every internal node of it, if any exists, has at least two children. Please see Fig. 6.

The construction in the following lemma is realized from the corresponding canonical cotree of the input trivially perfect graph. The canonical cotree of any cograph can be computed in linear time [8].

**Lemma 1.** *The clique-tree forest representation of any trivially perfect graph can be constructed in linear time.*

**Lemma 2.** *Let $G \not\cong K_n$ be a connected trivially perfect graph and suppose that it is given by its clique tree representation. Then* any *optimal star cover (partition) of $G$ has a star with the center* alone *from the root-clique of $G$.*

**Lemma 3.** *Let $G$ be any connected trivially perfect graph that is not a complete graph. Then $G$ has an optimal star cover (partition) containing some maximum star of $G$.*

**Algorithm** *Star Partition-Trivially Perfect*

> **Input:** A trivially perfect graph $G$.
>
> **Output:** A star partition $C$ of $G$.
>
> 1. Set $C = \emptyset$.
> 2. While $G$ is not the null graph, repeat the following:
>    − Pick a component $H$ of $G$.
>    − Find a maximum star $X$ in $H$.
>    − Set $C = C \cup \{X\}$ .
>    − set $G = G \setminus X$.
> 3. Output $C$.

**Fig. 7.** An $O(n^2)$ time algorithm for STAR PARTITION on trivially perfect graphs

The following is the main structural theorem on connected trivially perfect graphs.

**Theorem 4.** *Let $G$ be any connected trivially perfect graph. Then any maximum star of $G$ belongs to some optimal star cover (partition) of $G$.*

**Theorem 5.** STAR PARTITION *has an $O(n^2)$ time exact algorithm for trivially perfect graphs.*

We now move on to our study of STAR COVER on trivially perfect graphs. In the following theorem, we furnish the main structural result.

**Algorithm** *Star Cover-Trivially Perfect*

    **Input:** A trivially perfect graph $G$.

    **Output:** A star cover $\mathcal{C}$ of $G$.

1. Set $\mathcal{C} = \emptyset$.
2. While $G$ is not the null graph, repeat the following:
   - Pick a component $H$ of $G$.
   - Find a maximum star $X = \{x\} \cup I$ of $H$.
   - Set $\mathcal{C} = \mathcal{C} \cup \{X\}$.
   - If $H \setminus X$ is disconnected, set $G = G \setminus I$. Else set $G = G \setminus X$.
3. Output $\mathcal{C}$.

**Fig. 8.** An $O(n^2)$ time algorithm for STAR COVER on trivially perfect graphs

**Theorem 6.** *Let $G = (V, E)$ be any connected trivially perfect graph and let $X = \{v\} \cup I$ be any maximum induced star of $G$. Let $H = G \backslash I$ if $G \backslash X$ is disconnected and let $H = G \backslash X$ otherwise. Let $\mathcal{C}'$ be any optimal star cover of $H$. Then $\mathcal{C} = \{X\} \cup \mathcal{C}'$ is an optimal star cover of $G$.*

**Theorem 7.** STAR COVER *has an $O(n^2)$ time exact algorithm for trivially perfect graphs.*

### 5.2 Co-trivially Perfect Graphs ($(2K_2, P_4)$-Free Graphs)

Co-trivially perfect graphs form a subclass of butterfly-free graphs. Thus, the problems STAR COVER and STAR PARTITION are equivalent for this class of graphs by Theorem 2. So, we will design an algorithm for STAR PARTITION on co-trivially perfect graphs.

    We will consider only connected $(2K_2, P_4)$-free graphs. A connected $(2K_2, P_4)$-free graph $G \not\cong K_1$ can be obtained by taking the complement of a disconnected trivially perfect graph ($(C_4, P_4)$-free graph) $G'$. Thus, complements of *clique-tree forests* provide a reasonable representation for $(2K_2, P_4)$-free graphs.

    The complement of a clique-tree will be called a *co-clique-tree*. The complement of a clique-tree forest will be called a *co-clique-tree forest*.

    The co-clique-tree forest of an input connected co-trivially perfect graph $G \not\cong K_1$ can be computed in linear time by suitably modifying the algorithm of Lemma 1.

**Lemma 4.** *Let $G$ be a connected $(2K_2, P_4)$-free graph. Suppose that $G$ is given by its co-clique-tree forest representation. Then $G$ has a minimum star partition $\mathcal{C}$ such that the centers of stars in $\mathcal{C}$ are from the bottom most nodes of co-clique-trees constituting $G$.*

**Lemma 5.** *Let $G$ be a connected $(2K_2, P_4)$-free graph. Suppose that $G$ is given by its co-clique-tree forest representation. Let $L$ be a leaf node of smallest size in $G$. Let $G' = L \oplus (G \backslash L)$. Then $sp(G') \leq sp(G)$.*

*Notation:* If $G = (L_1, \ldots, L_p)$ is a complete multipartite graph, then we will always assume that $|L_1| \leq \ldots \leq |L_p|$.

---

**Input:** A connected $(2K_2, P_4)$-free graph $G$ in its co-clique-tree representation $F$.
**Output:** A complete multipartite graph $G' = (L_1, \ldots, L_p)$.

1. Set $H = G_1 = G_2 = G$.
2. While $H$ is not the null graph, repeat the following:
   - Find a smallest leaf node $L$ in $H$ ($L$ is possibly a root also).
   - Set $G_1 = G_2$.
   - Set $G_2 = L \oplus (G_1 \setminus L)$.
   - Set $H = H \setminus L$.
3. Set $G' = G_2(L_1, \ldots, L_p)$.
4. Find the largest $q$ such that $|L_1| + \ldots + |L_q| \leq p - q$.
5. If $|L_1| + \ldots + |L_q| = p - q$ and $L_1, \ldots, L_q$ all belong to one co-clique-tree of $G$ having more than $p - q$ vertices, do the following:
   - If any $L_j$ with $q < j \leq p$ and $|L_j| = |L_q|$ belongs to some other co-clique-tree of $G$, set $L = L_q$, $L_q = L_j$ and $L_j = L$.
6. Output $G' = G_2(L_1, \ldots, L_p)$.

---

**Fig. 9.** An algorithm that converts a $(2K_2, P_4)$-free graph into a specific complete multipartite graph.

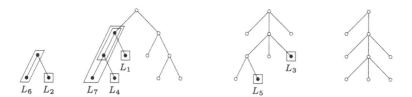

**Fig. 10.** An illustration of the execution of the algorithm in Fig. 9.

*Note:* Let us consider the algorithm in Fig. 9.

1. Step 5 of this algorithm ensures that, if $|L_1| + \ldots + |L_q| = p - q$ and a choice of such $L_1, \ldots, L_q$ is possible from more than one clique-tree of $G$, then such a choice is necessarily made.
2. If a leaf node $L$ chosen for removal has a single sibling node $L'$ in any iteration of the while loop of the algorithm, then $L'$ can be merged with its parent node after the removal of $L$ to ensure that the modified graph has the co-clique-tree forest structure; see Fig. 10.

By Lemma 5, the while loop of the algorithm in Fig. 9 has $sp(G_2) \leq sp(G_1)$ as a loop invariant. So, we have the following lemma.

**Lemma 6.** *Let $G$ be a connected $(2K_2, P_4)$-free graph in its co-clique-tree forest representation and suppose $G'$ is the complete multipartite graph output by the algorithm in Fig. 9 on giving $G$ as the input. Then $sp(G) \geq sp(G')$.*

**Lemma 7.** *Let $G = (L_1, \ldots, L_p)$ be a complete multipartite graph with $|L_1| \leq \ldots \leq |L_p|$. Let $q$ be the largest integer such that $|L_1| + \ldots + |L_q| \leq p - q$. Then $sp(G) \geq p - q$.*

**Lemma 8.** *Let $G$ be a connected $(2K_2, P_4)$-free graph in its co-clique-tree forest representation and suppose $G' = (L_1, \ldots, L_p)$ is the complete multipartite graph output by the algorithm in Fig. 9 on giving $G$ as the input and has $|L_1| \leq \cdots \leq |L_p|$. Let $q$ be the largest integer such that $|L_1| + \cdots + |L_q| \leq p - q$. Then $sp(G) \geq p - q$.*

*Suppose further that (1) $|L_1| + \cdots + |L_q| = p - q$, (2) $L_1, \ldots, L_q$ and every $L_j$ with $q < j \leq p$ and $|L_j| = |L_q|$ (if any exists) are all from the same tree $T$ of $G$ and (3) $T$ has more vertices than $p - q$. Then $sp(G) \geq p - q + 1$.*

**Lemma 9.** *Let $G$ be a connected $(2K_2, P_4)$-free graph in its co-clique-tree forest representation and suppose $G' = (L_1, \ldots, L_p)$ is the complete multipartite graph output by the algorithm in Fig. 9 on giving $G$ as the input and has $|L_1| \leq \cdots \leq |L_p|$. Let $q$ be the largest integer such that $|L_1| + \cdots + |L_q| \leq p - q$.*

*If $|L_1| + \cdots + |L_q| = p - q$, $L_1, \ldots, L_q$ and every $L_j$ with $q < j \leq p$ and $|L_j| = |L_q|$ (if any exists) are all from the same tree $T$ of $G$ and $T$ has more vertices than $p - q$, then $sp(G) \leq p - q + 1$. Else $sp(G) \leq p - q$.*

**Theorem 8.** *Let $G$ be a connected co-trivially perfect graph in its co-clique-tree forest representation and suppose $G' = (L_1, \ldots, L_p)$ is the complete multipartite graph output by the algorithm in Fig. 9 on giving $G$ as the input and has $|L_1| \leq \cdots \leq |L_p|$. Let $q$ be the largest integer such that $|L_1| + \cdots + |L_q| \leq p - q$.*

*If $|L_1| + \cdots + |L_q| = p - q$, $L_1, \ldots, L_q$ and every $L_j$ with $q < j \leq p$ and $|L_j| = |L_q|$ (if any exists) are all from the same tree $T$ of $G$ and $T$ has more vertices than $p - q$, then $sp(G) = p - q + 1$. Else $sp(G) = p - q$.*

*Moreover, an optimal star partition of $G$ can be computed in time $O(n^2)$.*

**Corollary 2.** *. Let $G = (L_1, \ldots, L_p)$ be a connected complete multipartite graph with $|L_1| \leq \cdots \leq |L_p|$. If $q$ is the largest integer such that $|L_1| + \cdots + |L_q| \leq p - q$, then $sp(G) = p - q$.*

### 5.3 Threshold Graphs ($(C_4, 2K_2, P_4)$-free Graphs)

A graph is called a threshold graph if it is $(C_4, 2K_2, P_4)$-free. In this section, we present linear time algorithm for STAR PARTITION on threshold graphs. This suffices since STAR COVER and STAR PARTITION are equivalent for threshold graphs by Theorem 2.

It is well-known that any threshold graph $G$, can be represented by string of its vertices, along with a label ($u$ or $j$, meaning union or join) attached to each vertex. For instance, the sequence $\epsilon u_1 u_2 u_3 j_1 u_4 u_5 j_2$ corresponds to a graph that is obtained by starting with $\epsilon$ and simply adding the $u_i$'s and joining the $j_i$'s. The following theorem is fairly obvious.

**Theorem 9.** *Let $G$ be a connected threshold graph. Then $sp(G) = \lceil \omega(G)/2 \rceil$. Indeed $G$ can be partitioned into a clique and at most one star. Moreover, an optimal star partition of $G$ can be computed in linear time.*

## 6    Conclusion

We leave the computational complexity of STAR COVER and STAR PARTITION for cographs open. There is a nice duality in the objectives of our greedy algorithms for trivially perfect graphs and co-trivially perfect graphs. Further understanding in this direction may possibly lead to a polynomial time algorithm to each of these problems *provided* it is not NP-hard for cographs. We remark that both STAR COVER and STAR PARTITION are NP-hard for $P_5$-free graphs as they include split graphs and the problems are NP-hard for split graphs.

## References

1. Andreatta, G., De Francesco, C., De Giovanni, L., Serafini, P.: Star partitions on graphs. Discrete. Optim. **33**, 1–18 (2019)
2. Balas, E., Yu, C.S.: On graphs with polynomially solvable maximal-weight clique problem. Networks **19**, 247–253 (1989)
3. Bang-Jensen, J., Huang, J., MacGillivray, G., Yeo, A.: Domination in convex bipartite and convex-round graphs. Technical report, University of Southern Denmark (1999)
4. van Bevern, R., et al.: Partitioning perfect graphs into stars. J. Graph Theory **85**, 297–335 (2017)
5. Björklund, A., Husfeldt, T., Koivisto, M.: Set partitioning via inclusion-exclusion. SIAM J. Comput. **39**, 543–563 (2009)
6. Brandstädt, A., Kratsch, D.: On the restriction of some NP-complete graph problems to permutation graphs. In: Budach, L. (ed.) FCT 1985. LNCS, vol. 199, pp. 53–62. Springer, Heidelberg (1985). https://doi.org/10.1007/BFb0028791
7. Cockayne, E.J., Goodman, S., Hedetniemi, S.T.: A linear algorithm for the domination number of a tree. Inform. Process. Lett. **4**, 41–44 (1975)
8. Corneil, D.G., Perl, Y., Stewart, L.K.: A linear recognition algorithm for cographs. SIAM J. Comput. **14**(4), 926–934 (1985)
9. Damaschke, P., Müller, H., Kratch, D.: Domination in convex and chordal bipartite graphs. Inf. Process. Lett. **36**, 231–236 (1990)
10. Divya, D., Vijayakumar, S.: On star partition of split graphs. In: Proceedings of the 10th International Conference on Algorithms and Discrete Applied Mathematics, CALDAM (2024). Accepted
11. Dor, D., Tarsi, M.: Graph decomposition is NP-complete: a complete proof of Holyer's conjecture. SIAM J. Comput. **26**(4), 1166–1187 (1997)

12. Downey, R.G., Fellows, M.R.: Fundamentals of Parameterized Complexity. Undergraduate Texts in Computer Science. Springer, London (2013). https://doi.org/10.1007/978-1-4471-5559-1
13. Duginov, O.: Partitioning the vertex set of a bipartite graph into complete bipartite subgraphs. Discrete Math. Theor. Comput. Sci. **16**(3), 203–214 (2014)
14. Farber, M., Keil, J.M.: Domination in permutation graphs. J. Algorithms **6**, 309–321 (1985)
15. Garey, M.R., Johnson, D.S.: Computers and Intractability; A Guide to the Theory of NP-Completeness. W. H. Freeman & Co., New York (1990)
16. Kelmans, A.K.: Optimal packing of stars in a graph. Discrete Math. **173**, 97–127 (1997)
17. Kirkpatrick, D.G., Hell, P.: On the completeness of a generalized matching problem. In: Lipton, R., Burkhard, W., Savitch, W., Friedman, E., Aho, A., (eds.) (STOC 1978) 10th ACM Symposium on Theory of Computing, pp. 240–245 (1978)
18. Lokshantov, D., Vatshelle, M., Villanger, Y.: Independent set in $P_5$-free graphs in polynomial time. In: Proceedings of the 25th ACM-SIAM Symposium On Discrete Algorithms, SODA, pp. 570–581 (2014)
19. Maffray, F., Preissmann, M.: On the NP-completeness of the k-colorability problem for triangle-free graphs. Discret. Math. **162**, 313–317 (1996)
20. Mondal, J., Vijayakumar, S.: Star covers and star partitions of double-split graphs. J. Comb. Optim. Accepted
21. Monnot, J., Toulouse, S.: The path partition problem and related problems in bipartite graphs. Oper. Res. Lett. **35**, 677–684 (2007)
22. Müller, H., Brandstädt, A.: The NP-completeness of Steiner tree and dominating set for chordal bipartite graphs. Theoret. Comput. Sci. **53**, 257–265 (1987)
23. Raman, V., Saurabh, S.: Short cycles make W-hard problems hard: FPT algorithms for W-hard problems in graphs with no short cycles. Algorithmica **52**, 203–225 (2008)
24. Shalu, M.A., Vijayakumar, S., Sandhya, T.P., Mondal, J.: Star partition of graphs. Discret. Appl. Math. **319**, 81–91 (2022)
25. Vazirani, V.V.: Approximation Algorithms. Springer, Heidelberg (2001)
26. West, D.B.: Introduction to Graph Theory, 2nd edn. Prentice-Hall, New Jersey (2000)
27. Yannakakis, M.: Node-deletion problems on bipartite graphs. SIAM J. Comput. **10**, 310–327 (1981)
28. Zuckerman, D.: Linear degree extractors and the inapproximability of max clique and chromatic number. Theory Comput. **3**, 103–128 (2007)

# Open Packing in $H$-free Graphs and Subclasses of Split Graphs

M. A. Shalu[iD] and V. K. Kirubakaran[(✉)][iD]

Indian Institute of Information Technology, Design and Manufacturing,
Kancheepuram, Chennai, India
{shalu,mat19d002}@iiitdm.ac.in

**Abstract.** Given a graph $G(V, E)$, a vertex subset $S$ of $G$ is called an open packing in $G$ if no pair of distinct vertices in $S$ have a common neighbour in $G$. The size of a largest open packing in $G$ is called the open packing number of $G$ and is denoted by $\rho^o(G)$. It is interesting to note that the open packing number is a lower bound for the total domination number in graphs with no isolated vertices [Henning and Slater, 1999]. Given a graph $G$ and a positive integer $k$, OPEN PACKING problem tests whether $G$ has an open packing of size at least $k$.

It is known that OPEN PACKING is NP-complete on split graphs (i.e., the class of $\{2K_2, C_4, C_5\}$-free graphs) [Ramos et al. 2014]. In this work, we complete the study on the complexity of OPEN PACKING on $H$-free graphs for every graph $H$ on at least three vertices by proving that OPEN PACKING is (i) NP-complete on $K_{1,3}$-free graphs and (ii) polynomial-time solvable on $(P_4 \cup rK_1)$-free graphs for every $r \geq 1$. Further, we prove that OPEN PACKING is (i) NP-complete on $K_{1,4}$-free split graphs and (ii) polynomial-time solvable on $K_{1,3}$-free split graphs. We prove a similar dichotomy result on split graphs with degree restrictions on the vertices in the independent set of the clique-independent set partition of the split graph.

**Keywords:** Total dominating set · Open packing · Split graphs · $H$-free graphs

## 1 Introduction

*Covering* and *packing* problems are a kind of primal-dual problems that have always attracted the researchers. In this article, we study one such covering-packing dual problem: total dominating set and open packing in graphs [8]. In a graph $G$, a vertex subset $D$ of $G$ is called a *total dominating set* in $G$ if every vertex in $V(G)$ is adjacent to some vertex in $D$. In other words, $V(G) = \cup_{u \in D} N_G(u)$, where $N_G(u)$ denotes the set of vertices in $G$ that are adjacent to $u$ in $G$. Note that by the definition of total dominating set, it is evident that a graph $G$ admits a total dominating set if and only if $G$ has no isolated vertices. The cardinality of a smallest total dominating set in $G$ is called the *total domination*

© The Author(s), under exclusive license to Springer Nature Switzerland AG 2024
S. Kalyanasundaram and A. Maheshwari (Eds.): CALDAM 2024, LNCS 14508, pp. 239–251, 2024.
https://doi.org/10.1007/978-3-031-52213-0_17

$number, \gamma_t(G)$, of $G$. A vertex subset $S$ of a graph $G$ is called an *open packing* in $G$ if for every pair of distinct vertices $x, y \in S$, $N_G(x) \cap N_G(y)$ is empty. The *open packing number* of $G$ denoted by $\rho^o(G)$ is the size of a largest open packing in $G$. It is known that $\rho^o(G) \leq \gamma_t(G)$ [8]. The following are the decision and optimization versions of the problems total dominating set and open packing.

| TOTAL DOMINATING SET |
| --- |
| **Instance** : A graph $G$ and $k \leq |V(G)|$. |
| **Question** : Does $G$ has a total dominating set of size at most $k$? |

| MIN-TOTAL DOMINATING SET |
| --- |
| **Instance** : A graph $G$. |
| **Task** : Find $\gamma_t(G)$. |

| OPEN PACKING |
| --- |
| **Instance** : A graph $G$ and $k \leq |V(G)|$. |
| **Question** : Is there an open packing of size at least $k$ in $G$? |

| MAX-OPEN PACKING |
| --- |
| **Instance** : A graph $G$. |
| **Task** : Find $\rho^o(G)$. |

Total dominating set is one of the well studied problems in literature and an extensive list of results can be found in [5, 7, 10]. For our work, it is interesting to note that TOTAL DOMINATING SET is NP-complete on $K_{1,5}$-free split graphs [19] and polynomial-time solvable on $K_{1,4}$-free split graphs [15]. Also, TOTAL DOMINATING SET is NP-complete on $K_{1,3}$-free graphs [12] and an optimal total dominating set of a chordal bipartite graph can be found in polynomial-time [3]. The study on open packing of graphs was initiated by Henning and Slater [9] and Rall [13] proved that for every non-trivial tree $T$, $\gamma_t(G) = \rho^o(G)$. In a recent work [16] (yet to be published), we extended this result by proving that the total domination number and the open packing number are equal when the underlying graph is a chordal bipartite graph with no isolated vertices. It is also known that OPEN PACKING is NP-complete on split graphs (equivalently, the class of $\{2K_2, C_4, C_5\}$-free graphs) [14] and bipartite graphs (a subclass of $K_3$-free graphs) [16,17]. In this work, we complete the study of OPEN PACKING on $H$-free graphs for every graph $H$ on at least three vertices by proving the following theorems.

**Theorem 1.** *Let $H$ be a graph on three vertices. Then, an optimal open packing in $H$-free graphs can be found in polynomial-time if and only if $H \not\cong K_3$ unless $P = NP$.*

**Theorem 2.** *For $p \geq 4$, let $H$ be a graph on $p$ vertices. Then, OPEN PACKING is polynomial-time solvable on the class of $H$-free graphs if and only if $H \in \{pK_1, (K_2 \cup (p-2)K_1), (P_3 \cup (p-3)K_1), (P_4 \cup (p-4)K_1)\}$ unless $P = NP$.*

In order to prove the above theorem, we proved the following results.

(i) OPEN PACKING is NP-complete on $K_{1,3}$-free graphs.
(ii) For every $r \geq 1$ and for every connected $(P_4 \cup rK_1)$-free graph $G$, $\rho^o(G) \leq 2r + 1$. Also, we prove this bound to be tight.

We also proved the following set of dichotomy results in subclasses of split graphs.

1. OPEN PACKING is NPC on $K_{1,r}$-free split graphs for $r \geq 4$ and is polynomial-time solvable for $r \leq 3$.
2. OPEN PACKING is NPC on $I_r$-split graphs (see Sect. 2 for definition) for $r \geq 3$ and is polynomial-time solvable for $r \leq 2$.

The above list of dichotomy results gave us the class of graphs where TOTAL DOMINATING SET and OPEN PACKING differ from each other in the view of classical complexity (Table 1 emphasises this difference).

**Table 1.** Complexity difference between TOTAL DOMINATING SET and OPEN PACKING in subclasses of split graphs

| Graph Classes | TOTAL DOMINATING SET | OPEN PACKING |
|---|---|---|
| $I_2$-Split Graphs | NPC [1] | P[*][a] |
| $K_{1,4}$-free Split Graphs | P [15] | NPC[*] |

[a][*] denotes the results in this work

## 2  Preliminaries

We follow West [18] for terminology and notation. The graphs considered in this work are simple and undirected unless specified otherwise. Given a graph $G(V, E)$, let $n$ denote the number of vertices in $G$. Given a vertex $x \in V(G)$, the (open) neighbourhood of $x$ in $G$ is defined as $N_G(x) = \{y \in V(G) : xy \in E(G)\}$ and the degree of $x$ in $G$ is defined as $\deg_G(x) = |N_G(x)|$. The closed neighourhood of a vertex $x$ in $G$ is defined as $N_G[x] = \{x\} \cup N_G(x)$. A vertex $x$ in $G$ is called an isolated vertex in $G$, if $N_G(x) = \emptyset$. Given $U \subseteq V(G)$, the subgraph of $G$ induced by $U$ is denoted as $G[U]$. Given a graph $H$, $G$ is said to be $H$-free if no induced subgraph of $G$ is isomorphic to $H$. For a vertex $x \in V(G)$, let $E_G(x)$ denote the set of all edges incident on $x$, and for an edge $e \in E(G)$, let $V_G(e)$ denote the end vertices of $e$ in $G$. Note that for $u \in V(G)$ and $e \in E(G)$, the edge $e \in E_G(u)$ if and only if $u \in V_G(e)$. For a graph $G$, the line graph $L(G)$ of $G$ is a graph with vertex set as $E(G)$ and two entries $e, e' \in V(L(G))$ are adjacent in $L(G)$ if $V_G(e) \cap V_G(e') \neq \emptyset$. Given two graphs $H$ and $H'$, the graph union $H \cup H'$ is defined as $V(H \cup H') = V(H) \cup V(H')$ and $E(H \cup H') = E(H) \cup E(H')$. For $p \in \mathbb{N} \cup \{0\}$ and a graph $H$, the graph $pH$ is defined as the union of $p$ disjoint copies of $H$.

Given $C \subseteq V(G)$, if every pair of vertices in $C$ are adjacent in $G$, then $C$ is known as a *clique* in $G$. Given $I \subseteq V(G)$ if no pair of vertices in $I$ are adjacent in $G$, then $I$ is called an *independent set* in $G$. The size of a largest independent set in $G$ is called the *independence number* of $G$, and is denoted by $\alpha(G)$. Let $P_n$, $C_n$ and $K_n$ denote the path, cycle and complete graph on $n$ vertices, respectively.

A set $D \subseteq V(G)$ is called a *dominating set* in $G$, if every vertex in $V(G) \backslash D$ is adjacent to some vertex in $D$.

A graph $G(V, E)$ is said to be a *split graph*, if there exists a partition $V(G) = C \cup I$ such that $C$ is a clique and $I$ is an independent set in $G$ and is denote as $G(C \cup I, E)$. Note that for $r \geq 2$, if a split graph $G$ is $K_{1,r}$-free, then $|N_G(v) \cap I| \leq r - 1$ for every $v \in C$. In accordance with this observation the class of $I_r$-*split graphs* is defined for a fixed natural number $r$ as a split graph $G(C \cup I, E)$ such that $\deg_G(v) = r$ for every $v \in I$.

Given a graph $G$ and a positive integer $k \leq |V(G)|$, the problem INDEPENDENT SET asks whether $G$ has an independent set of size at least $k$. Given a graph $G$, the problem MAX-INDEPENDENT SET asks for the value of $\alpha(G)$. For $r \in \mathbb{N}$, given a collection of sets $X_1, X_2, \ldots, X_r$ each of cardinality $q$ for some $q \in \mathbb{N}$ and a subset $M$ of $\prod_{i=1}^{r} X_i$, the $r$-DIMENSIONAL MATCHING problem asks whether there exists $L \subseteq M$ such that (i) $|L| = q$ and (ii) for every pair of $r$-tuples $x = (x_1, x_2, \ldots, x_r)$ and $y = (y_1, y_2, \ldots, y_r)$ in $L$, $x_i \neq y_i$ for $1 \leq i \leq r$.

Also, we refer the reader to [2,5] for a brief note on parameterized algorithms, intractability and W-hierarchy.

# 3   *H*-free Graphs

We dedicate this section to prove the dichotomy result on the open packing problem stated in Theorem 2. Observation 1 helps us to prove the necessary part of Theorem 2.

**Observation 1.** *For $p \geq 4$, let $H$ be a graph on $p$ vertices such that $H \notin \{P_4 \cup (p-4)K_1, P_3 \cup (p-3)K_1, K_2 \cup (p-2)K_1, pK_1\}$. Then, $H$ contains one of $K_3, 2K_2, C_4, K_{1,3}$ or $C_5$ as an induced subgraph.*

Proof of Observation 1 is omitted in this article. The following remark about OPEN PACKING on $H$-free graphs for $H \in \{K_3, 2K_2, C_4, C_5\}$ holds by the fact that OPEN PACKING is NP-complete on split graphs (i.e., the class of $\{2K_2, C_4, C_5\}$-free graphs) [14] and bipartite graphs (a subclass of $K_3$-free graphs) [16,17].

*Remark 1.* OPEN PACKING is NP-complete on (i) $2K_2$-free graphs (ii) $C_4$-free graphs, (iii) $C_5$-free graphs and (iv) triangle-free graphs.

Next, we prove that OPEN PACKING is NP-complete on $K_{1,3}$-free graphs.

## 3.1   $K_{1,3}$-free Graphs

The following construction gives a polynomial-time reduction from INDEPENDENT SET in simple graphs to OPEN PACKING in $K_{1,3}$-free graphs where the former problem is known to be NP-complete.

**Construction 1.**

*Input:* A simple graph $G$ with $V(G) = \{u_1, u_2, \ldots, u_n\}$.

*Output:* A $K_{1,3}$-free graph $G'$.

*Guarantee:* $G$ has an independent set of size $k$ if and only if $G'$ has a open packing of size $k$.

*Procedure:*

Step 1 : Replace each edge $e = uu'$ in $G$ by a three vertex path $ueu'$ in $G'$.

Step 2 : For every pair of edges $e, e' \in E(G)$, add an edge $ee'$ in $G'$ if $e, e' \in E_G(u)$ for some $u \in V(G)$.

Step 3 : For every vertex $u_i \in V(G)$ with exactly one edge, say $e$ incident on it in $G$, introduce a vertex $v_i$ and two edges $u_i v_i, v_i e$ in $G'$.

An example of Construction 1 is given in Fig. 1. Also, note that Step 2 of Construction 1 can be viewed as an embedding (attachment) of the line graph of $G$ into the graph obtained in Step 1. Further, $V(G') = V(G) \cup E(G) \cup \{v_i : 1 \leq i \leq n$ and $\deg_G(u_i) = 1\}$ and $E(G) = \{ue : u \in V(G), e \in E(G)$ and $e \in E_G(u)\} \cup E(L(G)) \cup \{v_i u_i : 1 \leq i \leq n$ and $\deg_G(u_i) = 1\} \cup \{v_i e : 1 \leq i \leq n$ and $E_G(u_i) = \{e\}\}$. Hence, $|V(G')| \leq 2n + m$ and $|E(G')| \leq 2m + \binom{m}{2} + 2n$. Thus, the graph $G'$ can be constructed in quadratic time in the input size. Note that the graph $G'$ is $K_{1,3}$-free. Proof of Guarantee of Construction 1 is omitted in this article. The following theorem follows from Construction 1 and the fact

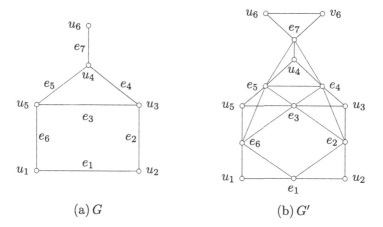

(a) $G$ (b) $G'$

**Fig. 1.** (a) a simple graph $G$, (b) $K_{1,3}$-free graph $G'$ obtained from $G$ using Construction 1.

that INDEPENDENT SET is NP-complete on simple graphs [11].

**Theorem 3.** OPEN PACKING *is NP-complete on* $K_{1,3}$-*free graphs*.

The following theorem is used to give an approximation hardness result for OPEN PACKING in $K_{1,3}$-free graphs.

**Theorem 4** ([6]). MAX-INDEPENDENT SET *cannot be approximated within a factor of* $n^{(1-\epsilon)}$ *for any* $\epsilon > 0$, *in general graphs unless* $P = NP$.

Next theorem follows from Construction 1 and Theorem 4.

**Theorem 5.** MAX-OPEN PACKING *is hard to approximate within a factor of* $N^{\frac{1}{2}-\epsilon}$ *for any* $\epsilon > 0$ *in* $K_{1,3}$-*free graphs unless* $P = NP$, *where* $N$ *denotes the number of vertices in a* $K_{1,3}$-*free graph.*

We use the following lemma by Rall [13] to prove that OPEN PACKING parameterized by solution size is in $W[1]$.

**Lemma 1** ([13]). *Given a graph* $G$, *let the neighbourhood graph* $G^{[o]}$ *of* $G$ *be a simple graph with* $V(G^{[o]}) = V(G)$ *and* $E(G^{[o]}) = \{xy : x, y \in V(G), x \neq y \text{ and } N_G(x) \cap N_G(y) \neq \emptyset\}$. *Then, a vertex subset* $S$ *is an open packing in* $G$ *if and only if* $S$ *is an independent set in* $G^{[o]}$.

**Theorem 6.** OPEN PACKING *parameterized by solution size is* $W[1]$-*complete in* $K_{1,3}$-*free graphs.*

*Proof.* Note that by Lemma 1, every instance $(G, k)$ of OPEN PACKING in $K_{1,3}$-free graphs can be reduced into an instance $(G^{[o]}, k)$ of INDEPENDENT SET in general graphs. Since INDEPENDENT SET parameterized by solution size is $W[1]$-complete [4], OPEN PACKING parameterized by solution size is $W[1]$-complete in $K_{1,3}$-free graphs by Lemma 1 and Construction 1.    □

Now, we prove the necessary part of Theorem 2.

*Necessary Part of Theorem 2.* Let $H$ be a graph on $p$ vertices for some $p \geq 4$ such that OPEN PACKING is polynomial-time solvable on $H$-free graphs. Then, we prove that $H \in \{P_4 \cup (p-4)K_1, P_3 \cup (p-3)K_1, K_2 \cup (p-2)K_1, pK_1\}$. On the contrary, assume that $H \notin \{P_4 \cup (p-4)K_1, P_3 \cup (p-3)K_1, K_2 \cup (p-2)K_1, pK_1\}$. Then, by Observation 1, $H$ contains at least one of (i) $K_3$ (ii) $2K_2$ (iii) $C_4$ (iv) $K_{1,3}$ or (v) $C_5$ as its induced subgraph. So, (i) the class of $H'$-free graphs is a subclass of $H$-free graphs for some $H' \in \{K_3, 2K_2, C_4, K_{1,3}, C_5\}$ and (ii) OPEN PACKING is NP-complete on $H'$-free graphs for every $H' \in \{K_3, 2K_2, C_4, K_{1,3}, C_5\}$ (by Remark 1 and Theorem 3). Thus by (i) and (ii), we can conclude that OPEN PACKING is NP-complete on $H$-free graphs, a contradiction.    □

In the next section, we prove the sufficiency part of Theorem 2.

### 3.2    $(P_4 \cup rK_1)$-free Graphs

In this section, we show that the open packing number is bounded above by a function of $r$ in the class of $(P_4 \cup rK_1)$-free graphs for every $r \geq 1$. Lemma 2 shows that this bound would imply that the open packing number of a $(P_4 \cup rK_1)$-free graph can be found in polynomial-time.

**Lemma 2.** *Given a graph class* $\mathcal{G}$, *if there exists* $k \in \mathbb{N}$ *such that* $\rho^o(G) \leq k$ *for every* $G \in \mathcal{G}$, *then (i)* $G$ *contains at most* $O(n^k)$ *open packings and (ii) all open packings in* $G$ *can be computed in* $O(n^{k+1})$ *time for every* $G \in \mathcal{G}$. *So,* $\rho^o(G)$ *can be computed in* $O(n^{k+1})$ *time.*

Proof of Lemma 2 is not included in this article.

Next, we prove that for $r \geq 1$, $\gamma_t(G) \leq 2r + 2$ and $\rho^o(G) \leq 2r + 1$ in a connected $(P_4 \cup rK_1)$-free graph $G$ (the bound on the open packing number of $(P_4 \cup rK_1)$-free graphs along with Lemma 2 will solve the sufficiency part of Theorem 2). Further, we show that the bound on the open packing number in these graph classes are tight (in Remark 3) and that the bound on the total domination number is tight for the class of $(P_4 \cup K_1)$-free graphs (in Remark 2). The following known lemma solves the problems for $P_4$-free graphs.

**Lemma 3 (Folklore).** *Let $G$ be a connected $P_4$-free graph. Then, $\rho^o(G) \leq \gamma_t(G) = 2$.*

The lemma below shows that the total domination number of a connected $(P_4 \cup rK_1)$-free graph is bounded above by a function of $r$ and its proof is omitted in this article.

**Lemma 4.** *For $r \geq 1$, if $G$ is a connected $(P_4 \cup rK_1)$-free graph, then $\gamma_t(G) \leq 2r + 2$.*

*Remark 2.* The bound given in Lemma 4 is tight for the case $r = 1$, and a cycle on six vertices $(C_6)$ is a connected $(P_4 \cup K_1)$-free graph that satisfies this bound. Also, the graphs obtained by replacing every vertex of $C_6$ with a clique will remain as a $(P_4 \cup K_1)$-free graph with total domination number four.

The next lemma proves a bound on the open packing number in the class of $(P_4 \cup rK_1)$-free graphs.

**Lemma 5.** *For $r \geq 1$, if $G$ is a connected $(P_4 \cup rK_1)$-free graph, then $\rho^o(G) \leq 2r + 1$.*

Proof of Lemma 5 is omitted in this article.

*Remark 3.* The bound given in Lemma 5 is tight. For example, let $G_r$ be a graph with $V(G_r) = (\cup_{i=1}^{r}\{x_i, y_i, z_i\}) \cup \{u, v\}$ and $E(G_r) = (\cup_{i=1}^{r}\{x_iy_i, y_iz_i, z_iu\}) \cup \{uv\}$. Then, $G_r$ is $(P_4 \cup rK_1)$-free and $S_r = (\cup_{i=1}^{r}\{x_i, y_i\}) \cup \{v\}$ is an open packing in $G$ of size $2r + 1$. The graph $G_3$ is given in Fig. 2. Also, note that the graph obtained by replacing the vertex $u$ in $G_r$ by a clique will remain as a $(P_4 \cup rK_1)$-free graph with open packing number $2r + 1$.

**Theorem 7.** *For $r \geq 0$, let $G$ be a connected $(P_4 \cup rK_1)$-free graph. Then, there are polynomially many open packings in $G$ and all open packings in $G$ can be found in polynomial-time. Hence, $\rho^o(G)$ can be found in polynomial-time.*

The above theorem follows from Lemmas 2, 3, 5 and the fact that for a graph $G$ with components $G_1, G_2, \ldots, G_k$, $\rho^o(G) = \sum_{i=1}^{k} \rho^o(G_i)$. Next, we explicitly state the complexity status of (i) $rK_1$-free graphs (ii) $(K_2 \cup rK_1)$-free graphs and (iii) $(P_3 \cup rK_1)$-free graphs as Corollary 1 for the sake of Theorem 2.

**Corollary 1.** OPEN PACKING *is in P in the class of (i) $rK_1$-free graphs for $r \geq 3$, (ii) $(K_2 \cup rK_1)$-free graphs for $r \geq 1$ and (iii) $(P_3 \cup rK_1)$-free graphs for $r \geq 0$.*

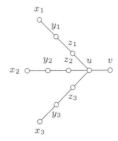

**Fig. 2.** A $(P_4 \cup 3K_1)$-free graph $G_3$ defined in Remark 3 with an open packing $S_3 = \{x_1, x_2, x_3, y_1, y_2, y_3, v\}$ of size $7 = (2(3) + 1)$.

Corollary 1 follows from the fact that for $r \geq 3$ (resp. $r \geq 1$, $r \geq 0$), the class of $rK_1$-free graphs (resp. $(K_2 \cup rK_1)$-free graphs, $(P_3 \cup rK_1)$-free graphs) is a subclass of $(P_4 \cup (r-2)K_1)$-free graphs (resp. $(P_4 \cup (r-1)K_1)$-free graphs, $(P_4 \cup rK_1)$-free graphs). Note that the sufficiency part of Theorem 2 follows from Theorem 7 and Corollary 1. Also, note that the proof of Theorem 1 follows from Corollary 1 and Remark 1.

## 4    Split Graphs

Ramos et al. [14] proved that OPEN PACKING is NP-complete on split graphs. In this section, we study the complexity of OPEN PACKING in (i) $K_{1,r}$-free split graphs for $r \geq 2$ and (ii) $I_r$-split graphs for $r \geq 1$. This study shows the complexity difference between TOTAL DOMINATING SET and OPEN PACKING in split graphs (see Table 1).

### 4.1    $K_{1,r}$-free Split Graphs

It is known that TOTAL DOMINATING SET is NP-complete on $K_{1,5}$-free split graphs [19] and is polynomial-time solvable on $K_{1,4}$-free split graphs [15]. In this section, we give a dichotomy result for OPEN PACKING in $K_{1,r}$-free split graphs by proving that the problem is (i) NP-complete for $r \geq 4$ and (ii) polynomial-time solvable for $r \leq 3$.

We begin this section with a construction to prove that OPEN PACKING is NP-complete in $K_{1,4}$-free split graphs.

**Construction 2.**

*Input:* A simple graph $G(V, E)$.

*Output:* A $K_{1,4}$-free split graph $G'(C \cup I, E)$.

*Guarantee:* $G$ has an independent set of size $k$ if and only if $G'$ has an open packing of size $k + 1$.

*Procedure:*

Step 1 : Subdivide each edge $e = uu'$ in $G$ into a three vertex path $ueu'$ in $G'$.

Step 2 : Introduce three new vertices $x$, $y$, $z$ and two edges $xy, xz$ in $G'$.

Step 3 : Make $E(G) \cup \{y, z\}$ a clique in $G'$.

An example of Construction 2 is shown in Fig. 3. The vertex set and the edge set of the graph $G'$ can be stated as $V(G') = V(G) \cup E(G) \cup \{x, y, z\}$ and $E(G') = \{ue : u \in V(G), e \in E(G) \text{ and } e \in E_G(u)\} \cup \{xy, xz, yz\} \cup \{ye : e \in E(G)\} \cup \{ze : e \in E(G)\} \cup \{ee' : e, e' \in E(G) \text{ and } e \neq e'\}$. Note that $|V(G')| = n + m + 3$ and $|E(G')| = 2m + 3 + 2m + \binom{m}{2} = O(m^2)$. Hence, the graph $G'$ can be constructed in polynomial-time from $G$. Also, by Construction 2, $V(G') = C \cup I$ is a clique-independent set partition of $G'$ with $C = E(G) \cup \{y, z\}$ and $I = V(G) \cup \{x\}$. Thus, $G'(C \cup I, E)$ is a split graph. Clearly, $G'$ is $K_{1,4}$-free. Proof of the guarantee of Construction 2 is not included in this proof.

The following theorems holds by Construction 2 and the fact that INDEPENDENT SET is NP-complete on simple graphs [11].

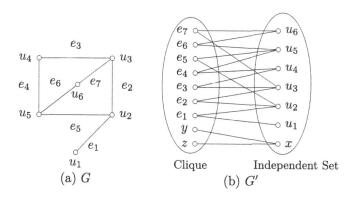

**Fig. 3.** (a) a graph $G$ and (b) the graph $G'$ constructed from $G$ using Construction 2

**Theorem 8.** OPEN PACKING *is NP-complete on $K_{1,4}$-free split graphs.*

Similar to that of Theorems 5 and 6, it can be proved that in $K_{1,4}$-free split graphs (i) OPEN PACKING parameterized by solution size is W[1]-complete and (ii) MAX-OPEN PACKING is hard to approximate within a factor of $N^{(\frac{1}{2}-\epsilon)}$ for any $\epsilon > 0$ unless $P = NP$, where $N$ denotes the number of vertices in a $K_{1,4}$-free split graph.

Theorem 8 showed the hardness of OPEN PACKING on $K_{1,4}$-free split graphs. In Theorem 9, we show that OPEN PACKING is in P in the class of $K_{1,3}$-free split graphs.

**Theorem 9.** OPEN PACKING *is polynomial-time solvable on $K_{1,3}$-free split graphs.*

We proved that for a connected $K_{1,3}$-free split graph $G(C \cup I, E)$, $\rho^o(G) \in \{1, 2, |I|\}$ and it can be found in polynomial-time whether $\rho^o(G) = 1, 2$ or $|I|$. Hence, the open packing number of a $K_{1,3}$-free graph can be found in polynomial-time (proofs of the above statements are not included in this article).

## 4.2   $I_r$-split Graphs

We observed that a minor modification in Corneil and Perl's reduction [1] will show that TOTAL DOMINATING SET is NP-complete on $I_2$-split graphs. We have also obtained a dichotomy result by proving that TOTAL DOMINATING SET in $I_r$-split graphs is (i) NP-complete for $r \geq 2$ and (ii) polynomial-time solvable for $r = 1$ (this result is not included in this article). In this section, we prove that OPEN PACKING is (i) NP-complete on $I_r$-split graphs for $r \geq 3$ and (ii) polynomial-time solvable on $I_r$-split graphs for $r \leq 2$. The following construction is used to prove that OPEN PACKING is NP-complete on $I_r$-split graphs for $r \geq 3$.

**Construction 3.**
*Input:* A collection of sets $X_1, X_2, \ldots, X_r$ such that $|X_i| = q$ for some $q \in \mathbb{N}$ for $i = 1, 2 \ldots, r$ and a non-empty set $M \subseteq \prod\limits_{i=1}^{r} X_i$.
*Output:* An $I_r$-split graph $G(C \cup I, E)$.
*Guarantee:* $(X_1, X_2, \ldots, X_r, M)$ is an yes-instance of $r$-DIMENSIONAL MATCH-ING if and only if $G$ has an open packing of size $q$.
*Procedure:*
Step 1 : For every $i \in \{1, 2, \ldots, r\}$ and for every $x \in X_i$, create a vertex $z_{(x,i)}$ in $G$. Similarly, for $w \in M$, create a vertex $y_w$ in $G$.
Step 2 : Introduce an edge between every pair of distinct vertices of $C = \cup_{i=1}^{r}\{z_{(x,i)} : x \in X_i\}$ in $G$.
Step 3 : Introduce an edge between the vertex $z_{(x,i)} \in C$ and $y_w$ in $G$ if the $i^{th}$ coordinate of $w$ is $x$.

An example of Construction 3 is given in Fig. 4. The vertex and edge sets of the graph $G$ can be stated as $V(G) = C \cup I$, where $I = \{y_w : w \in M\}$ and $E(G) = \{xx' : x, x' \in C\} \cup \{z_{(x,i)}y_w : x \in X_i, w \in M$ and the $i^{th}$ coordinate of $w$ is $x\}$. Then, $|V(G)| = (r \cdot q) + |M|$ and $|E(G)| = r \cdot |M| + \binom{r \cdot q}{2}$. Hence, the graph $G$ can be constructed in polynomial-time with respect to $r, q$ and $|M|$. Note that by the construction of $G$, $C$ is a clique and $I$ is an independent set, and hence $V(G) = C \cup I$ is a clique-independent set partition of $V(G)$. Thus, $G(C \cup I, E)$ is a split graph. Further, note that by the construction of $G$, for every $y_w \in I$, there exists a $r$-tuple $w \in W$ such that $N_G(y_w) = \{z_{(x_1,1)}, z_{(x_2,2)}, \ldots, z_{(x_r,r)}\}$ if $w = (x_1, x_2, \ldots, x_r)$. Hence, $\deg_G(y_w) = r$ for every $y_w \in I$. Thus, $G$ is an $I_r$-split graph. Proof of Guarantee of Construction 3 is not included in this article.

**Theorem 10.** *For $r \geq 3$, OPEN PACKING is NP-complete on $I_r$-split graphs.*

*Proof.* Given an $I_r$-split graph $G$ and a vertex subset $S$ of $G$, it can be tested in polynomial-time whether $N_G(u) \cap N_G(v)$ is empty for every distinct $u, v \in S$. Hence, OPEN PACKING is in the class NP on $I_r$-split graphs. Also, Construction 3 and the fact that $r$-DIMENSIONAL MATCHING is NP-complete for $r \geq 3$ [11] implies that OPEN PACKING is NP-complete on $I_r$-split graphs for $r \geq 3$.

$$X_1 = \{x_1, u_1, v_1\} \; X_2 = \{x_2, u_2, v_2\} \; X_3 = \{x_3, u_3, v_3\} \; X_4 = \{x_4, u_4, v_4\}$$

$$M = \{w_1, w_2, w_3, w_4, w_5\}$$

where $w_1 = (x_1, u_2, v_3, v_4) \quad w_2 = (u_1, u_2, u_3, u_4) \quad w_3 = (v_1, v_2, v_3, x_4)$
$$w_4 = (v_1, v_2, x_3, x_4) \quad w_5 = (x_1, x_2, v_3, v_4)$$

(a) 4-DIMENSIONAL MATCHING

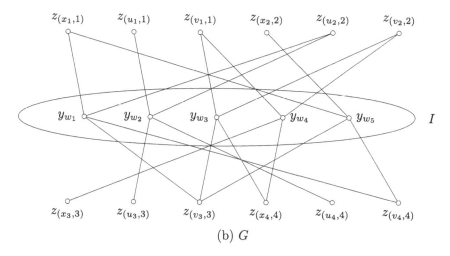

(b) $G$

**Fig. 4.** (a) An instance $(X_1, X_2, X_3, X_4, M)$ of a 4-DIMENSIONAL MATCHING and (b) An $I_4$-split graph $G(C \cup I, E)$ corresponding to the instance $(X_1, X_2, X_3, X_4, M)$ using Construction 3 where $C = \{z_{(x_1,1)}, z_{(u_1,1)}, z_{(v_1,1)}, z_{(x_2,2)}, z_{(u_2,2)}, z_{(v_2,2)}, z_{(x_3,3)}, z_{(u_3,3)}, z_{(v_3,3)}, z_{(x_4,4)}, z_{(u_4,4)}, z_{(v_4,4)}\}$ is a clique in $G$ and $I = \{y_{w_1}, y_{w_2}, y_{w_3}, y_{w_4}, y_{w_5}\}$ is an independent set in $G$. The edges between the vertices of $C$ aren't given in the figure for clarity.

The following theorem shows that OPEN PACKING is polynomial-time solvable on a superclass of $I_1$-split graphs and $I_2$-split graphs. Thus, completing our dichotomy result for OPEN PACKING in $I_r$-split graphs.

**Theorem 11.** *A maximum open packing in a split graph $G(C \cup I, E)$ with $1 \leq \deg_G(v) \leq 2$ for every $v \in I$ can be found in $O(n^3)$ time.*

Proof of Theorem 11 is omitted.

## 5   Conclusion

In this article, we completed the study on the complexity of OPEN PACKING in $H$-free graphs for every graph $H$ on at least three vertices by proving that OPEN PACKING is (i) NP-complete on $K_{1,3}$-free graphs and (ii) polynomial-time solvable on $(P_4 \cup rK_1)$-free graphs for every $r \geq 1$. Further, we proved that OPEN PACKING is (i) NP-complete on $K_{1,4}$-free split graphs and (ii) polynomial-time solvable on $K_{1,3}$-free split graphs. We also showed that OPEN PACKING is (i) NP-complete on $I_r$-split graphs for $r \geq 3$ and (ii) polynomial-time solvable on $I_r$-split graphs for $r \leq 2$. We have also proved that MAX-OPEN PACKING is hard to approximate within a factor of $N^{(\frac{1}{2}-\epsilon)}$ for any $\epsilon > 0$ unless $P = NP$ in (i) $K_{1,4}$-free split graphs and (ii) $K_{1,3}$-free graphs, where $N$ denotes the number of vertices of a graph in these graph classes. Also, we proved that OPEN PACKING parameterized by solution size is W[1]-complete on these two graph classes.

## References

1. Corneil, D.G., Perl, Y.: Clustering and domination in perfect graphs. Discret. Appl. Math. **9**(1), 27–39 (1984). https://doi.org/10.1016/0166-218X(84)90088-X
2. Cygan, M., et al.: Parameterized Algorithms, 1st edn. Springer, Cham (2015). https://doi.org/10.1007/978-3-319-21275-3
3. Damaschke, P., Muller, H., Kratsch, D.: Domination in convex and chordal bipartite graphs. Inf. Process. Lett. **31**, 231–236 (1990). https://doi.org/10.1016/0020-0190(90)90147-P
4. Downey, R.G., Fellows, M.R.: Fixed-parameter tractability and completeness II: on completeness for W[1]. Theoret. Comput. Sci. **141**(1), 109–131 (1995). https://doi.org/10.1016/0304-3975(94)00097-3
5. Downey, R.G., Fellows, M.R.: Fundamentals of Parameterized Complexity. Springer, London (2016). https://doi.org/10.1007/978-1-4471-5559-1
6. Håstard, J.: Clique is hard to approximate within $n^{1-\epsilon}$. Acta Math. **182**, 105–142 (1999). https://doi.org/10.1007/BF02392825
7. Haynes, T.W., Hedetniemi, S., Slater, P.: Fundamentals of Domination in Graphs, 1st edn. CRC Press, Boca Raton (1998). https://doi.org/10.1201/9781482246582
8. Henning, M.A.: Packing in trees. Discret. Math. **186**(1), 145–155 (1998). https://doi.org/10.1016/S0012-365X(97)00228-8
9. Henning, M.A., Slater, P.J.: Open packing in graphs. J. Comb. Math. Comb. Comput. **29**, 3–16 (1999)
10. Henning, M.A., Yeo, A.: Total Domination in Graphs. Springer, New York (2015). https://doi.org/10.1007/978-1-4614-6525-6
11. Karp, R.M.: Reducibility among combinatorial problems. In: Miller, R.E., Thatcher, J.W., Bohlinger, J.D. (eds.) Complexity of Computer Computations, pp. 85–103. Springer, Cham (1972). https://doi.org/10.1007/978-1-4684-2001-2_9
12. McRae, A.A.: Generalizing NP-completeness proofs for bipartite graphs and chordal graphs. Ph.D. thesis, Clemson University, USA (1995)
13. Rall, D.F.: Total domination in categorical products of graphs. Discussiones Mathematicae Graph Theory **25**, 35–44 (2005). https://doi.org/10.7151/dmgt.1257
14. Ramos, I., Santos, V.F., Szwarcfiter, J.L.: Complexity aspects of the computation of the rank of a graph. Discrete Math. Theor. Comput. Sci. **16** (2014). https://doi.org/10.46298/dmtcs.2075

15. Renjith, P., Sadagopan, N.: The steiner tree in $K_{1,r}$-free split graphs-a dichotomy. Discret. Appl. Math. **280**, 246–255 (2020). https://doi.org/10.1016/j.dam.2018.05.050

16. Shalu, M.A., Kirubakaran, V.K.: Total domination number and its lower bound in some subclasses of bipartite graphs. Manusript-UnderReview (2023)

17. Shalu, M.A., Vijayakumar, S., Sandhya, T.P.: A lower bound of the cd-chromatic number and its complexity. In: Gaur, D., Narayanaswamy, N.S. (eds.) CALDAM 2017. LNCS, vol. 10156, pp. 344–355. Springer, Cham (2017). https://doi.org/10.1007/978-3-319-53007-9_30

18. West, D.B.: Introduction to Graph Theory, 2nd edn. Pearson, London (2018)

19. White, K., Farber, M., Pulleyblank, W.: Steiner trees, connected domination and strongly chordal graphs. Networks **15**(1), 109–124 (1985). https://doi.org/10.1002/net.3230150109

# Graph Theory

# Location-Domination Type Problems Under the Mycielski Construction

Silvia M. Bianchi[1], Dipayan Chakraborty[2,3(✉)] [iD], Yanina Lucarini[1],
and Annegret K. Wagler[2]

[1] Deprtamento de Matemática, Universidad Nacional de Rosario, Av. Pellegrini 250,
S2000BTP Rosario, Argentina
{sbianchi,lucarini}@fceia.unr.edu.ar

[2] Université Clermont-Auvergne, CNRS, Mines de Saint-Étienne,
Clermont-Auvergne-INP, LIMOS, 63000 Clermont-Ferrand, France
{dipayan.chakraborty,annegret.wagler}@uca.fr

[3] Department of Mathematics and Applied Mathematics, University of
Johannesburg, Auckland Park 2006, South Africa

**Abstract.** We consider the following variants of the classical minimum
dominating set problem in graphs: locating-dominating set, locating total-
dominating set and open locating-dominating set. All these problems are
known to be hard for general graphs. A typical line of attack, therefore, is
to either determine the minimum cardinalities of such sets in general or to
establish bounds on these minimum cardinalities in special graph classes.
In this paper, we study the minimum cardinalities of these variants of the
dominating set under a graph operation defined by Mycielski in [21] and
is called the Mycielski construction. We provide some general lower and
upper bounds on the minimum sizes of the studied sets under the Mycielski construction. We apply the Mycielski construction to stars, paths and
cycles in particular, and provide lower and upper bounds on the minimum
cardinalities of such sets in these graph classes. Our results either improve
or attain the general known upper bounds.

**Keywords:** Locating-dominating set · Open locating-dominating set ·
Locating total-dominating set · Mycielski construction

## 1 Introduction

For a graph modeling a facility, the placement of monitoring devices, for example, fire detectors or surveillance cameras, motivates the study of various location-
domination type problems in graphs. The problem of placing monitoring devices
so that every site of a facility is visible from a monitor leads to a domination problem. In addition, the position of a fire, a thief, or a saboteur in the facility can

This work was sponsored by a public grant overseen by the French National Research
Agency as part of the "Investissements d'Avenir" through the IMobS3 Laboratory of
Excellence (ANR-10-LABX-0016) and the IDEX-ISITE initiative CAP 20-25 (ANR-
16-IDEX-0001).

S. Kalyanasundaram and A. Maheshwari (Eds.): CALDAM 2024, LNCS 14508, pp. 255–269, 2024.
https://doi.org/10.1007/978-3-031-52213-0_18

be uniquely located by a specific subset of the monitoring devices which leads to location problems. During the last decades, several combined location-domination problems of this type have been actively studied, see, for example, the bibliography maintained by Lobstein and Jean [19]. In this work, we study three different location-domination type problems under a graph operation known as the Mycielski construction and defined by Mycielski himself in [21].

All graphs in this paper are finite, simple and connected. Given a graph $G = (V, E)$, the (open) neighborhood of a vertex $u \in V$ is the set $N(u) = N_G(u)$ of all vertices of $G$ adjacent to $u$, and $N[u] = N_G[u] = \{u\} \cup N(u)$ is the closed neighborhood of $u$. A subset $C \subseteq V$ is *dominating* (respectively, *total-dominating*) if the set $N[u] \cap C$ (respectively, $N(u) \cap C$) is non-empty for all $u \in V$. In addition, a subset $C \subseteq V$ *separates* (respectively, *total-separates*) a pair $u, v \in V$ if $N[u] \cap C \neq N[v] \cap C$ (respectively, $N(u) \cap C \neq N(v) \cap C$). In such a case, we also say that $u, v \in V$ are *separated by* $C$ (respectively, *total-separated by* $C$). A subset $C \subseteq V$ is called

- a *locating-dominating set* [25] (or $LD$-set for short) of $G$ if it is a dominating set of $G$ that separates all pairs of distinct vertices outside of $C$, that is, $N(u) \cap C \neq N(v) \cap C$, for all distinct $u, v \in V - C$;
- a *locating total-dominating set* [15] (or $LTD$-set for short) of $G$ if it is a total-dominating set of $G$ that separates all pairs of distinct vertices outside of $C$, that is, $N(u) \cap C \neq N(v) \cap C$, for all distinct $u, v \in V - C$;
- an *open locating-dominating set* [23] (or $OLD$-set for short) of $G$ if it is a total-dominating set of $G$ that total-separates all pairs of distinct vertices of the graph, that is, $N(u) \cap C \neq N(v) \cap C$, for all distinct $u, v \in V$.

Two distinct vertices $u, v$ of a graph $G = (V, E)$ are called *false twins* if $N(u) = N(v)$, see [23]. Similarly, any two vertices $u, v \in V$ with $N[u] = N[v]$ are called *true twins*. Now, for $X \in \{LD, LTD, OLD\}$, the $X$-problem on $G$ is the problem of finding an $X$-set of minimum size in $G$. The size of such a set is called the $X$-number of $G$ and is denoted by $\gamma_X(G)$. Note that a graph $G$ without isolated vertices admits an $OLD$-set if there are no false twins in $G$. On the other hand, LD-sets and LTD-sets are admitted by all graphs.

From the definitions themselves, the following relations hold for any graph $G$ admitting any two $X$-sets for $X \in \{LD, LTD, OLD\}$:

$$\gamma_{LD}(G) \leq \gamma_{LTD}(G) \leq \gamma_{OLD}(G). \tag{1}$$

It has been shown that determining $\gamma_X(G)$ is in general NP-hard for all $X \in \{LD, LTD, OLD\}$. Apart from determining $\gamma_{LD}(G)$ being NP-hard in general [23], it remains so for bipartite graphs [8] and some subclasses of chordal graphs like split graphs and interval graphs [13]. This result is also extended to planar bipartite unit disk graphs in [20] and intersection graphs in [12]. Closed formulas for the exact values of $\gamma_{LD}(G)$ have so far been found for restricted graph families, for example, for paths [25], cycles [5], stars, complete multipartite graphs, some subclasses of split graphs and thin suns [1,4]. Bounds for the $LD$-number of trees were provided in [6]. A linear-time algorithm to determine $\gamma_{LD}(G)$ for $G$ being a tree was provided by Slater in [26] and has been extended to block graphs (graphs which generalize the concept of trees in that any 2-connected subgraph in a block graph is complete) in [2]. Moreover, in connection to block graphs and hence, trees, tight upper and lower bounds for $LD$-numbers of block graphs and twin-free block graphs have been established in [7].

Determining $\gamma_{OLD}(G)$ is NP-hard not only in general [23] but also on other graph classes like perfect elimination bipartite graphs [22], interval graphs [13] and is APX-hard on chordal graphs of maximum degree 4 [22]. Closed formulas for the exact value of $\gamma_{OLD}(G)$ have so far been found only for restricted graph families such as cliques and paths [23], some subclasses of split graphs and thin suns [1]. Tight lower and upper bounds for $OLD$-numbers certain classes of graphs like trees [23], block graphs [7], lower bounds for interval graphs, permutation graphs and cographs [12] and upper bounds for cubic graphs [17] have been established. Lastly, some algorithmic aspects of the problem have been discussed in [2,22].

Concerning $LTD$-sets, it can be checked that it is as hard as the $OLD$-problem by using the same arguments as in [23]. Bounds for the $LTD$-number of trees are given in [15,16]. In addition, the $LTD$-number in special families of graphs, including cubic graphs, grid graphs, complete multipartite graphs, some subclasses of split graphs and thin suns is investigated in [1,16].

In fact, giving bounds for the $X$-numbers in special graphs is a popular way to tackle the problems. In this work, we study the behavior of the three $X$-sets of graphs under the following graph operation defined by Mycielski in [21]. Given a graph $G = (V, E)$ with $V = \{v_1, \ldots, v_n\}$, a new graph $M(G)$ is constructed as follows: for every vertex $v_i$ of $G$, add a new vertex $u_i$ and make $u_i$ adjacent to all vertices in $N_G(v_i)$. Finally add a vertex $u$ which is adjacent to all $u_i$. Let the set containing all the vertices $u_i$'s be called $U$, that is, $U = \{u_1, u_2, \ldots, u_n\}$.

Originally, Mycielski introduced this construction in the context of graph coloring and used it to generate graphs $M(G)$ whose chromatic number increases by one compared to the chromatic number of $G$. In [11], it is proved that the application of the Mycielski construction also increases the dominating number by one. In this paper, we show that the same holds for total domination and study the $X$-numbers of the graphs $M(G)$, where $G$ is a star $K_{1,n}$, a path $P_n$ and a cycle $C_n$ (see Fig. 1, Fig. 2 and Fig. 3, respectively, for examples of their illustrations). As far as previous works on such variants of the dominating sets of Mycielski constructions is concerned, we know of only one such, namely, in [24] where the authors find tight upper bounds of $ID$-numbers of $M(G)$ for $G$ being an identifiable graph (that is, a graph without true twins). The $ID$-number of an identifiable graph $G$ is the minimum cardinality of a dominating set $C$ of $G$ such that $N[u] \cap C \neq N[v] \cap C$ for all distinct pairs $u, v$ of vertices of $G$ (see [18]).

In Sect. 2, we show that the application of the Mycielski construction increases the total-dominating number by at least one and give a general lower bound on the studied $X$-numbers of the graphs $M(G)$ in terms of $\gamma_X(G)$ when $G$ is either a path or a cycle. We then combine this bound with previously

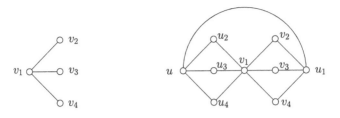

**Fig. 1.** The star $K_{1,3}$ and the resulting graph $M(K_{1,3})$

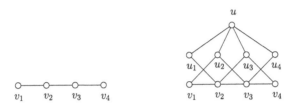

**Fig. 2.** The path $P_4$ and the resulting graph $M(P_4)$

known results on $\gamma_X(P_n)$ (respectively, on $\gamma_X(C_n)$) to obtain lower bounds on the $X$-numbers of $M(P_n)$ (respectively, of $M(C_n)$).

In Sect. 3, we give a general upper bound on the $X$-numbers of the graphs $M(G)$. We also show that this bound is attained when $G$ is a star and improve the bound for the cases when $G$ is a path or cycle.

We note that there are some particularities in applying the Mycielski construction to paths and cycles with a small number of vertices. In fact, we have $M(P_2) = C_5$. While we have $\gamma_{LD}(P_2) = 1$ and $\gamma_{LTD}(P_2) = \gamma_{OLD}(P_2) = 2$, it is easy to see that $\gamma_{LD}(C_5) = 2$, $\gamma_{LTD}(C_5) = 3$, and $\gamma_{OLD}(C_5) = 4$ hold. Moreover, $P_3$ and $C_4$ have false twins, and so do $M(P_3)$ and $M(C_4)$. Hence, there exist no $OLD$-sets of these graphs. However, we have $\gamma_X(P_3) = \gamma_X(C_4) = 2$ and $\gamma_X(M(P_3)) = \gamma_X(M(C_4)) = 4$ for $X \in \{LD, LTD\}$. Hence, in the rest of what follows, we study paths $P_n$ and cycles $C_n$ with larger values of $n$.

We close with some concluding remarks and open problems for future research.

## 2    Lower Bounds on $X$-Numbers of Graphs $M(G)$

To start with, observe the following fact that, for every graph $G$,

1. two vertices $v_i$ and $v_j$ are false twins in $G$ if and only if the vertices $v_i, v_j$ and $u_i, u_j$ are pairs of false twins in $M(G)$; and
2. $M(G)$ has no true twins.

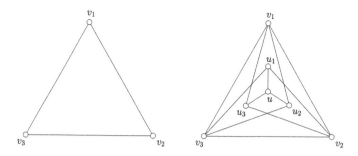

**Fig. 3.** The cycle $C_3$ and the resulting graph $M(C_3)$

In [11], it is proved that, for every graph $G$, the equality $\gamma(M(G)) = \gamma(G)+1$ holds, where $\gamma(G)$ is the dominating number of $G$. Analogously, for the total-dominating number $\gamma_t(M(G))$, we can prove:

**Lemma 1.** *For every graph $G$ without isolated vertices, we have $\gamma_t(M(G)) = \gamma_t(G) + 1$.*

*Proof (sketch).* Let $C \subseteq V$ be a total-dominating set of $G$ and let $u_i \in U$. We define $C_i = C \cup \{u_i\}$. As $G$ has no isolated vertices, every vertex in $V(M(G))$ is adjacent to a vertex in $C_i$ and so, $\gamma_t(M(G)) \leq \gamma_t(G) + 1$.

Now, let $C_M$ be a total-dominating set of $M(G)$ of cardinality $\gamma_t(M(G))$. Every vertex in $V$ is adjacent to a vertex in $C_M$. Let us define the sets $C_V = C_M \cap V$ and $C_U = C_M \cap U$. Then it can be verified that the set $C_V \cup \{v_i : u_i \in C_U\}$ is a total-dominating set of $G$ and $|C_V| \leq |C_M - \{u\}|$. Thus, if $u \in C_M$, we are done. Therefore, let us assume that $u \notin C_M$. Then, there exists $u_j \in C_M$ in order for $C_M$ to total-dominate $u$. Now, for every vertex $u_i \in C_U$, any neighbor $v_k$ ($\in V$) of $u_i$ also has a neighbor in $C_V$ (the same vertex in $C_V$ that is a neighbor of $u_k$). This implies tat $C_M - C_U$ is a total-dominating set of $V$. Since, $u_j \in C_U$, we have $|C_U| \geq 1$ and hence, the result follows.  □

This motivates us to study the parameter $\gamma_X(M(G))$ in terms of $\gamma_X(G)$. In doing so, we now establish a general lower bound on the $X$-numbers of the graphs $M(G)$, where $G$ is either a path $P_n$ or a cycle $C_n$ and $X \in \{LD, LTD, OLD\}$.

**Theorem 1.** *Let $X \in \{LD, LTD, OLD\}$. For a graph $G$ that is either a path $P_n$ or a cycle $C_n$ admitting an $X$-set, we have*

$$\gamma_X(M(G)) \geq \gamma_X(G) + 1.$$

*Proof (sketch).* As a proof sketch, we provide here the proof of the theorem only for the case that $X = LD$. The proof in the other cases when $X \in \{LTD, OLD\}$ follows with similar proof techniques. To begin with, let us assume that $G$ is any graph (not necessarily a path or a cycle) and that $C_M$ is a minimum $LD$-set of $M(G)$. Let $C_V = C_M \cap V$ and $C_U = C_M \cap U$. Then define the set $C = C_V \cup \{v_i : u_i \in C_U\}$.

*Claim. C* is an *LD*-set of *G*.

*Proof of Claim.* Firstly, since $C_M$ is a dominating set of $M(G)$, one can verify that the set $C$ is also a dominating set of $G$. We now show that $C$ is also a separating set of $G$. Let $v_i, v_j \in V$ be any pair of arbitrary vertices such that $v_i, v_j \notin C$. Then we show that $v_i, v_j$ are separated by $C$ in $G$. Let $w_k \in C_M$ with $k \notin \{i, j\}$ separate $v_i$ and $v_j$ in $M(G)$, where $w_k \in \{u_k, v_k\}$. If $w_k = v_k$, then $C$ clearly separates $v_i, v_j$. So, let $w_k = u_k$ and that $u_k$ is a neighbour of $v_i$ and not of $v_j$ in $M(G)$. Now, if $k = j$, then $v_j \in C$ and is trivially separated from every other vertex of $G$ by the definitions of $LD$–sets. So, let $k \neq j$. Then again, $v_k$ is a neighbour of $v_i$ and not of $v_j$ in $G$ and thus $C$ separates $v_i, v_j$. This establishes the claim.                                                                          □

Thus we have,

$$\gamma_{LD}(M(G)) \geq |C_M - \{u\}| \geq |C| \geq \gamma_{LD}(G). \tag{2}$$

So, if $u \in C_M$, then $\gamma_{LD}(M(G)) > |C_M|$ and hence, the statement of the theorem holds. So, let us assume that $u \notin C_M$, in which case, we have $\gamma_{LD}(M(G)) \geq \gamma_{LD}(G)$. Toward contradiction, let us assume that $\gamma_{LD}(M(G)) = \gamma_{LD}(G)$. Then, by the assumed equalities in (2), $C$ must be a minimum $LD$-set of $G$. This in turn implies that for each $i$, we have $|\{u_i, v_i\} \cap C_M| \leq 1$. In other words, if $u_i \in C_M$, then $v_i \notin C_M$ and vice-versa.

For the rest of the proof sketch, let us assume that $G$ is either a path $P_n$ or a cycle $C_n$. First of all, we observe that if any three consecutive vertices $v_i, v_{i+1}, v_{i+2} \in C$, then $C$ cannot be a minimum $LD$-set of $G$, as one can discard $v_{i+1}$ from $C$ and the latter still remains an $LD$-set. Similarly, for some $i$, if $v_i, v_{i+1}, v_{i+3}, v_{i+4} \in C$, then again $C$ cannot be a minimum $LD$-set of $G$, as one can discard $v_{i+1}, v_{i+3}$ and include $v_{i+2}$ in $C$ and the latter still remains an $LD$-set of $G$. With those observations, let us first assume that some vertex $u_i \in C_M$ in order to dominate $u$. If one of its neighbours in $G$, say $v_{i+1}$, without loss of generality, belongs to $C_M$, then we must also have $v_{i+2} \in C_M$ in order for $C_M$ to dominate $u_{i+1}$ (note that $v_i \notin C_M$). Thus, $v_i, v_{i+1}, v_{i+2} \in C$, a contradiction to the minimality of $C$ by our earlier observation. Hence, let us assume that for all $u_i \in C_M$, none of its neighbours in $G$, that is, $v_{i-1}$ and $v_{i+1}$, belong to $C_M$. So, fix one such $u_i \in C_M$. Then $v_i \notin C_M$. Therefore, without loss of generality, let $u_{i+1} \in C_M$ in order for the latter to dominate $v_i$. If any of $u_{i-1}, v_{i-1} \in C_M$, then again we would have three consecutive vertices of $G$ in $C$, a contradiction. So, let us assume that $u_{i-1}, v_{i-1} \notin C_M$. In order for $v_{i-1}, v_{i+1}$ to be separated, we must have either $w_{i-2} \in C_M$ or $w_{i+2} \in C_M$, where $w_{i-2} \in \{u_{i-2}, v_{i-2}\}$ and $w_{i+2} \in \{u_{i+2}, v_{i+2}\}$. However, we cannot have $w_{i+2} \in C_M$, as otherwise, we would have $v_i, v_{i+1}, v_{i+2} \in C$, the same contradiction as before. Hence, $w_{i-2} \in C_M$. If $w_{i-2} = v_{i-2}$, then $u_{i-2} \notin C_M$. This implies that $v_{i-3} \in C_M$ for $C_M$ to dominate $u_{i-2}$. This implies that $v_{i-3}, v_{i-2}, v_i, v_{i+1} \in C$, a contradiction by our earlier observation. Moreover, if $w_{i-2} = u_{i-2}$, then $v_{i-2} \notin C$ and hence, $w_{i-3} \in C_M$ for $v_{i-2}$ to be dominated by $C_M$, where $w_{i-3} \in \{u_{i-3}, v_{i-3}\}$. Here again, we have $v_{i-3}, v_{i-2}, v_i, v_{i+1} \in C$, the same contradiction.

This proves that our assumption of $\gamma_{LD}(M(G)) = \gamma_{LD}(G)$ is wrong which, in turn, proves the theorem for the case that $X = LD$. The other cases when $X \in \{LTD, OLD\}$ follow with similar proof techniques.    □

Note that this lower bound in Theorem 1 is tight:

- for $X \in \{LD, LTD\}$, we have $\gamma_X(C_3) = 2$ as $\{v_1, v_2\}$ is a minimum $X$-set, and $\gamma_X(M(C_3)) = 3$ as $\{v_1, v_2, u\}$ is a minimum $X$-set (see Fig. 3 for $C_3$ and $M(C_3)$),
- for $X = OLD$, no tight examples are yet known in this case.

We deduce lower bounds for $\gamma_X(M(P_n))$ and $\gamma_X(M(C_n))$ from the respective values of $\gamma_X(P_n)$ and $\gamma_X(C_n)$. Theorem 1 together with the results from [5] on $\gamma_{LD}(P_n)$ and $\gamma_{LD}(C_n)$ yield:

**Corollary 1.** *If $G$ equals $P_n$ or $C_n$ for $n \geq 3$, we have as lower bound:*

$$\gamma_{LD}(M(G)) \geq \left\lceil \frac{2n}{5} \right\rceil + 1.$$

The exact $OLD$-numbers of path and cycles are studied in [23] and [3], respectively. However, the latter result for cycles of even order needed to be corrected and as such, we state and prove the result in its entirety as follows.

**Theorem 2.** *For any cycle $C_n$ on $n$ vertices such that $n \geq 3$ and $n \neq 4$, we have*

$$\gamma_{OLD}(C_n) = \begin{cases} \left\lceil \frac{2n}{3} \right\rceil, & \text{for odd } n, \\ 2 \left\lceil \frac{n}{3} \right\rceil, & \text{for even } n. \end{cases}$$

*Proof.* We prove the theorem by first showing that $\left\lceil \frac{2n}{3} \right\rceil$ for odd $n$ and $2 \left\lceil \frac{n}{3} \right\rceil$ for even $n$ is a lower bound on $\gamma_{OLD}(C_n)$ and then providing an $OLD$-set of $C_n$ of exactly the same cardinality as the lower bound. We start with establishing the lower bound first.

Seo and Slater showed in [23] that if $G$ is a regular graph on $n$ vertices, of regular-degree $r$ and with no open twins, then we have $\gamma_{OLD}(G) \geq \frac{2}{1+r}n$. Using this result in [23] for the cycle $C_n$, therefore, we have $\gamma_{OLD}(C_n) \geq \frac{2}{3}n$, that is, $\gamma_{OLD}(C_n) \geq \left\lceil \frac{2}{3}n \right\rceil$. Now, for $n \neq 6k + 4$ for any non-negative integer $k$, we have

$$\left\lceil \frac{2n}{3} \right\rceil = \begin{cases} \left\lceil \frac{2n}{3} \right\rceil, & \text{for odd } n, \\ 2 \left\lceil \frac{n}{3} \right\rceil, & \text{for even } n. \end{cases}$$

Thus, the only case left to prove is the following claim.

*Claim.* For $n = 6k + 4$ with $k \geq 1$, we have $\gamma_{OLD}(C_n) \geq 2 \left\lceil \frac{n}{3} \right\rceil = 4k + 4$.

*Proof (of Claim).* The proof of the last claim is by induction on $k$ with the base case being for $k = 1$, that is, when $C_n$ is a cycle on $n = 10$ vertices. We fisrt show the result for $n = 10$.

*Subclaim.* $\gamma_{OLD}(C_{10}) \geq 8$.

*Proof (of Subclaim).* Let $V(C_{10}) = \{v_1, v_2, \ldots, v_{10}\}$ and $S$ be a minimum open $OLD$-set of $C_{10}$. Then, $|S| < 10$ by a charaterization result by Foucaud et al. [10] on the extremal graphs $G$ for which $\gamma_{OLD}(G) = |V(G)|$. Hence, there exists a vertex $v_1$ (without loss of generality) such that $v_1 \notin S$. We consider the induced 5-paths $P_1 : v_1v_2v_3v_4v_5$ and $P_2 : v_1v_{10}v_9v_8v_7$. Then, from the path $P_1$, the vertex $v_3 \in S$ for the latter to total-dominate $v_2$ and the vertex $v_5 \in S$ for the latter to separate the pair $v_2, v_4$. By the same argument, from path $P_2$, the vertices $v_9, v_7 \in S$. Moreover, at least one vertex from each of the pairs $(v_2, v_4), (v_4, v_6), (v_6, v_8), (v_8, v_{10}), (v_{10}, v_2)$ must belong to $S$ for the latter to total-dominate $v_3, v_5, v_7, v_9, v_1$, respectively. Hence, the result follows from counting. □

Thus, the result holds for the base case of the induction hypothesis. We, therefore, assume $k \geq 2$ and that $\gamma_{OLD}(C_m) \geq 4q + 4$ for all cycles $C_m$ with $|V(C_m)| = 6q + 4$ and $q \in \{1, 2, \ldots, k-1\}$. Toward contradiction, let us assume that $\gamma_{OLD}(C_n) < 4k + 4$. Moreover, let $V(C_n) = \{v_1, v_2, \ldots, v_n\}$. Then again, by the charaterization result in [10], we have $\gamma_{OLD}(C_n) < n$. This implies that, for any minimum $OLD$-set $S$ of $C_n$, there exists a pair $(v_{n-6}, v_{n-5})$ (by a possible renaming of vertices) such that $v_{n-6} \in S$ and $v_{n-5} \notin S$. Let $C'_{n-6}$ be the cycle on $n - 6$ vertices formed by adding the edge $v_1v_{n-6}$ in the graph $C_n - \{v_{n-5}, v_{n-4}, \ldots, v_n\}$. Note that $|V(C'_{n-6})| = 6(k-1)+4$ and hence, the induction hypothesis applies to it to give

$$\gamma_{OLD}(C'_{n-6}) \geq 4(k-1) + 4 = 4k. \tag{3}$$

Now, let $S' = S - \{v_{n-5}, v_{n-4}, \ldots, v_n\}$.

*Subclaim. $S'$ is an $OLD$-set of $C'_{n-6}$.*

*Proof (of Subclaim).* To show that $S'$, first of all, is a total-dominating set of $C'_{n-6}$, we notice that the vertices $v_1, v_5 \notin S$. Therefore, all vertices in the set $\{v_2, v_3, \ldots, v_{n-6}\}$ remain total-dominated by $S'$. Moreover, $S'$ total-dominates $v_1$ by virtue of $v_{n-6} \in S'$. This proves that $S'$ is a total-dominating set of $C'_{n-6}$.

We now show that $S'$ is also a total-separating set of $C'_{n-6}$. To that end, since $v_{n-5} \notin S$, if now the vertex $v_n \notin S$ as well, then $S'$ clearly total-separates every pair of vertices in $C'_{n-6}$ and hence, is an $OLD$-set. If however, $v_n \in S$ and total-separates a pair of vertices in $C'_{n-6}$, the pair can either be $(v_1, v_2)$ or $(v_1, v_3)$. Since $n \geq 16$, we have $3 < n - 6$ and hence, $v_{n-6} \in S'$ total-separates the pairs $(v_1, v_2)$ and $(v_1, v_3)$ in $C'_{n-6}$. Therefore, $S'$ is an $OLD$-set of $C'_{n-6}$. □

*Subclaim. $|S \cap \{v_{n-5}, v_{n-4}, \ldots, v_n\}| \geq 4$.*

*Proof (of Subclaim).* Since $v_{n-6} \notin S$, it implies that $v_{n-3} \in S$ in order for the latter to total-dominate the vertex $v_{n-4}$. Moreover, we also have $v_{n-1} \in S$ in order for $S$ to total-separate the pair $(v_{n-2}, v_{n-4})$. Furthermore, we must have at least one vertex each from the pairs $(v_{n-4}, v_{n-2})$ and $(v_{n-2}, v_n)$ belonging to $S$ in order for the latter to total-dominate the vertices $v_{n-3}$ and $v_{n-1}$, respectively. Finally, at least one vertex from the pair $(v_{n-4}, v_n)$ must also belong to $S$ in order for $S$ to total-separate the pair $(v_{n-3}, v_{n-1})$. This proves that the result holds.                                                                                  □

Recall that $|S| = \gamma_{OLD}(C_n) < 4k + 4$, by assumption. Thus, we have

$$\gamma_{OLD}(C'_{n-6}) \leq |S'| = |S| - |S \cap \{v_{n-5}, v_{n-4}, \ldots, v_n\}| < 4k + 4 - 4 = 4k,$$

a contradiction to the Inequality (3). This proves the claim and establishes the lower bound on $\gamma_{OLD}(C_n)$.                                                              □

The theorem is, therefore, proved by providing an $OLD$-set $S$ of $C_n$ of the exact same cardinality as the lower bound, that is,

$$|S| = \begin{cases} \left\lceil \frac{2n}{3} \right\rceil, & \text{for odd } n, \\ 2 \left\lceil \frac{n}{3} \right\rceil, & \text{for even } n. \end{cases} \tag{4}$$

Let $= V(C_n) = \{v_1, v_2, \ldots, v_n\}$ and that $n = 6k + r$, where $r \in \{0, 1, 2, 3, 4, 5\}$. For $k = 0$, that is, $C_n$ being either a 3-cycle or a 5-cycle, it can be checked that the sets $\{v_1, v_2\}$ and $\{v_1, v_2, v_3, v_4\}$ are the respective $OLD$-sets. Thus, the result holds in this case. For the rest of this proof, therefore, we assume that $n \geq 6$, that is, $k \geq 1$. We now construct a vertex subset $S$ of $C_n$ by including in $S$ the vertices

1. $v_{6i-4}, v_{6i-3}, v_{6i-1}, v_{6i}$ for all $i \in \{1, 2, \ldots, k\}$ if $r = 0, 3$. In this case, we have $|S| = 4k$ for $r = 0$ and $|S| = 4k + 2$ for $r = 3$.
2. $v_{6i-4}, v_{6i-3}, v_{6i-2}, v_{6i-1}$ for all $i \in \{1, 2, \ldots, k\}$ if $r \neq 0$; with
   (a) the vertices $v_{6k}, v_{6k+1}, \ldots, v_{6k+r-1}$ if $r = 1, 2, 4$. In this case, we have $|S| = 4k + r$; and
   (b) the vertices $v_{6k+1}, v_{6k+2}, v_{6k+3}, v_{6k+4}$ if $r = 5$. In this case, we have $|S| = 4k + 4$.

It can be checked that the constructed set $S$ is, indeed, an $OLD$-set of $C_n$ and of the cardinality as in Eq. (4). This proves the result.                              □

Combining Theorem 1 with results on $\gamma_{OLD}(P_n)$ in [23] and on $\gamma_{OLD}(C_n)$ in Theorem 2, we deduce:

**Corollary 2.** *Consider $P_n$ with $n = 6k + r$ for $k \geq 1$ and $r \in \{0, \ldots, 5\}$, then we have:*

$$\gamma_{OLD}(M(P_n)) \geq \begin{cases} 4k + r + 1 & \text{if } r \in \{0, \ldots, 4\}, \\ 4k + 5 & \text{if } r = 5; \end{cases}$$

*and for $n \geq 3$ and $n \neq 4$, we have*

$$\gamma_{OLD}(M(C_n)) \geq \begin{cases} \left\lceil \frac{2n}{3} \right\rceil + 1, & \text{for odd } n, \\ 2 \left\lceil \frac{n}{3} \right\rceil + 1, & \text{for even } n. \end{cases}$$

Theorem 1 together with the results from [15] on $\gamma_{LTD}(P_n)$ and from [16] on $\gamma_{LTD}(C_n)$ imply:

**Corollary 3.** *If $G$ equals $P_n$ or $C_n$ for $n \geq 3$, we have as lower bound:*

$$\gamma_{LTD}(M(G)) \geq \left\lfloor \frac{n}{2} \right\rfloor - \left\lfloor \frac{n}{4} \right\rfloor + \left\lceil \frac{n}{4} \right\rceil + 1.$$

# 3    Upper Bounds on $X$-Numbers of Graphs $M(G)$

We first establish a general upper bound on $X$-numbers of graphs $M(G)$ in terms of $\gamma_X(G)$.

**Theorem 3.** *Let* $X \in \{LD, LTD, OLD\}$. *For a graph* $G$ *admitting an* $X$-*set, we have*

$$\gamma_X(M(G)) \le 2\gamma_X(G).$$

*Proof (sketch).* Let $C$ be a minimum $X$-set of $G$. Then, we construct a new set $C' = C \cup \{u_i : v_i \in C\}$. It can be checked that if $C$ is a dominating (respectively, total-dominating) set of $G$, then so is it of $M(G)$. For any $x \in V(M(G))$, let $N_M(x)$ (respectively, $N_M[x]$) denote the neighborhood (respectively, closed neighborhood) of $x$ in $M(G)$. If $C$ is a total-separating set of $G$, then for any $x \in V(M(G))$, we have

- $C' \cap N_M(u) = \{u_i : v_i \in C\}$
- $C' \cap N_M(u_j) = (C \cup \{u_i : v_i \in C\}) \cap N_M(u_j) = C \cap N(v_j)$
- $C' \cap N_M(v_j) = (C \cup \{u_i : v_i \in C\}) \cap N_M(v_i) = \{v_k, u_k : v_k \in N(v_j) \cap C\}$

As is evident, the set $C' \cap N_M(x)$ is unique for each $x \in V(M(G))$. Thus, $C'$ is also a total-separating set of $M(G)$. Moreover, $|C'| = 2|C|$. This proves the result.    □

Based on results from [4,14,16] on $X$-numbers of stars and the relation (1), we can show that the bound given in Theorem 3 is tight for stars (see Fig. 4 for illustration):

**Theorem 4.** *For stars* $K_{1,n}$ *with* $n \ge 3$, *we have* $\gamma_X(K_{1,n}) = n$ *and*

$$\gamma_X(M(K_{1,n})) = 2n$$

*whenever* $X \in \{LD, LTD\}$.

Note that stars $K_{1,n}$ have false twins and, therefore, so does $M(K_{1,n})$. Hence, $M(K_{1,n})$ does not admit an $OLD$-set. Figure 5 provides an example for $\gamma_X(M(G)) = 2\gamma_X(G)$ when $X = OLD$. For $OLD$-sets, we can further prove the following.

**Theorem 5.** *Let* $G$ *be a graph without isolated vertices and false twins. Then* $\gamma_{OLD}(M(G)) \le \gamma_{OLD}(G) + 2$.

*Proof (sketch).* Let $C \subset V$ be an $OLD$-set of $G$ and let $u_i \in U$. We define $C_i = C \cup \{u, u_i\}$. As $G$ has no isolated vertices, every vertex in $V(M(G))$ is adjacent to a vertex in $C_i$. This implies that $C_i$ is a total-dominating set of $M(G)$. Moreover, by the fact that $C$ is a total-separating set of $G$, it can be checked that each of the following sets is unique.

$$C_i \cap N_M(u) = \{u_i\};$$
$$C_i \cap N_M(v_j) = (C \cap N(v_j)) \cup \{u_i\} \quad \text{for } v_j \in N(u_i);$$
$$C_i \cap N_M(v_j) = C \cap N(v_j) \qquad\qquad \text{for } v_j \notin N(u_i); \text{ and}$$
$$C_i \cap N_M(u_j) = (C \cap N(v_j)) \cup \{u\} \quad \text{for } u_j \in U.$$

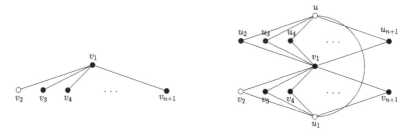

**Fig. 4.** $K_{1,n}$ and $M(K_{1,n})$ (black vertices form a minimum $X$-set when $X \in \{LD, LTD\}$)

This proves that $C_i$ is an $OLD$-set of $M(G)$ and since, $|C_i| = |C| + 2$, the result follows. □

The bound given in Theorem 5 is tight, as $M(P_2)$ and $M(C_3)$ (in Fig. 5) show. In addition, it enables us to prove the following for $\gamma_{OLD}(M(P_n))$ and $\gamma_{OLD}(M(C_n))$.

**Theorem 6.** *For all $n \geq 2$ and $n \neq 3$, we have*

$$\gamma_{OLD}(M(P_n)) = \gamma_{OLD}(P_n) + 2$$

*and for all $n \geq 3$, we have*

$$\gamma_{OLD}(C_n) + 1 \leq \gamma_{OLD}(M(C_n)) \leq \gamma_{OLD}(C_n) + 2.$$

*Proof (sketch).* The result for cycles follows directly from Theorems 1 and 5. For paths, again using Theorem 5, we only need to show that $\gamma_{OLD}(P_n) \geq \gamma_{OLD}(P_n) + 2$ for all $n \geq 2$ and $n \neq 3$. As far as small paths a concerned, it can be checked that $\gamma_{OLD}(P_2) = 2$, $\gamma_{OLD}(P_4) = \gamma_{OLD}(P_5) = 4$; and $\gamma_{OLD}(M(P_2)) = 4$, $\gamma_{OLD}(M(P_4)) = \gamma_{OLD}(M(P_5)) = 6$. Thus, the result holds for these small paths. Therefore, we assume that $n = 6k + r$ with $k \geq 1$, where $r \in \{0, 1, \ldots, 5\}$. If $V(P) = \{v_1, v_2, \ldots, v_n\}$, the proof follows by partitioning the vertex set of

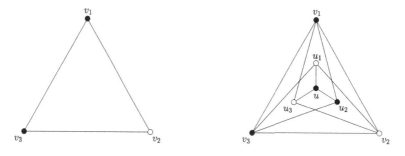

**Fig. 5.** $C_3$ and $M(C_3)$ (black vertices form a minimum $OLD$-set)

$M(P_n)$ into $\lceil \frac{n}{6} \rceil$ parts, the first $\lfloor \frac{n}{6} \rfloor$ of which are given by $B_i = \{v_j, u_j : 6i - 5 \leq j \leq 6i\}$ for all $1 \leq i \leq \lfloor \frac{n}{6} \rfloor$; and the last (if exists, that is, if $r \neq 0$) part $B_l = \{v_j, u_j : 6k + 1 \leq j \leq r\}$. Further analysis of any block $B_i$ for $1 \leq i \leq \lfloor \frac{n}{6} \rfloor$ shows that any $OLD$-set $C$ of $M(P_n)$ must contain at least 4 vertices from $B_i$. Moreover, we would have $|C \cap U| \geq 2$. This gives the total count for the cardinality of $C$ to be $4k + 2$ and thus proves the theorem for $r = 0$. Moreover, each other case for $r \in \{1, 2, 3, 4, 5\}$ is dealt with separately where it can be shown that, for $r \in \{1, 2, 3, 4\}$, exactly $r$ vertices and, for $r = 5$, exactly 4 vertices need to be included in $C$. This proves the theorem by comparison to the results for $\gamma_{OLD}(P_n)$ in [23].    □

Concerning $LD$-numbers, we note that $\gamma_{LD}(M(P_2)) = 2$ and $\gamma_{LD}(M(P_3)) = \gamma_{LD}(M(P_4)) = \gamma_{LD}(M(P_5)) = 4$ holds. We can improve the general upper bounds for $\gamma_{LD}(M(P_n))$ and $\gamma_{LD}(M(C_n))$ as follows:

**Theorem 7.** *Consider $P_n$ with $n = 3k + r$ for $k \geq 2$, $r \in \{0, 1, 2\}$ and $C_n$ with $n \geq 3$, then we have:*

$$\gamma_{LD}(M(P_n)) \leq \begin{cases} 2k + 1 \ if\, r = 0 \\ 2k + 2 \ if\, r \in \{1, 2\} \end{cases}$$

*and*

$$\gamma_{LD}(M(C_n)) \leq \begin{cases} n - \lfloor \frac{n}{3} \rfloor + 1 & if\, n \ is\ odd \\ n - 2\lfloor \frac{n}{6} \rfloor + 1 & if\, n \ is\ even \end{cases}$$

*Proof (sketch).* We provide the proof sktech for paths to illustrate the proof technique. The proof for cycles follows with similar techniques. Let $n \geq 6$, $n = 3k + r$ with $k \geq 2$ and $r \in \{0, 1, 2\}$. Then, according to three possible values of $r$, we define the following sets.

- If $r = 0$, we define $C = \{v_2\} \cup \{v_i, v_{i+1} : i = 3\ell + 1, \ell \in \{1, \ldots, k - 1\}\} \cup \{u_{3k}, u\}$. In this case, we have $|C| = 2k + 1$.
- If $r = 1$, we define $C_1 = (C - \{u_{3k}\}) \cup \{v_{3k+1}, u_{3k+1}\}$. Here we have, $|C_1| = 2k + 2$
- If $r = 2$, we define $C_2 = (C - \{u_{6k}\}) \cup \{v_{6k+1}, v_{6k+2}\}$. In this case, we have $|C_2| = 2k + 2$

Further analysis of the above sets $C$, $C_1$ and $C_2$ shows that in each of the above three cases, the sets are $LD$-sets of $M(P_n)$. The result then follows by the cardinalities of the sets in the above three cases.    □

We observe that the upper bounds are tight for $M(P_n)$ with $6 \leq n \leq 8$ and for $M(C_n)$ with $n \in \{3, 6, 7\}$, but are not tight for $M(C_4)$ and $M(C_5)$, for example. There are no examples yet known where the upper bounds are not tight for $M(P_n)$.

The next theorem provides an upper bound for the $LTD$-numbers of $M(P_n)$ and $M(C_n)$. However, before coming to it, as far as small graphs of these graph classes are concerned, we note that $\gamma_{LTD}(M(P_2)) = 3$, $\gamma_{LTD}(M(P_3)) = 4$ and $\gamma_{LTD}(M(P_4)) = \gamma_{LTD}(M(P_5)) = 5$. The next result improves the general upper bounds for $\gamma_{LTD}(M(P_n))$ and $\gamma_{LTD}(M(C_n))$ as follows.

**Theorem 8.** *Consider $P_n$ with $n = 6k + r$ for $k \geq 1$, $r \in \{0, \ldots, 5\}$ and $C_n$ with $n \geq 3$, then we have:*

$$\gamma_{LTD}(M(P_n)) \leq \begin{cases} 4k + 2 & if\, r = 0 \\ 4k + r + 1 & if\, r \in \{1, 2, 3\} \\ 4k + r & if\, r \in \{4, 5\} \end{cases}$$

*and*

$$\gamma_{LTD}(M(C_n)) \leq \begin{cases} n - \lfloor \frac{n}{3} \rfloor + 2 & if\, n\ is\ odd \\ n - 2 \lfloor \frac{n}{6} \rfloor + 2 & if\, n\ is\ even \end{cases}$$

*Proof (sketch).* The upper bound on the $LTD$-number of $M(P_n)$ follows by the fact that $\gamma_{LTD}(M(P_n)) \leq \gamma_{OLD}(M(P_n)) = \gamma_{OLD}(P_n) + 2$ (by Theorem 6) and by the known exact values of $\gamma_{OLD}(P_n)$ from [23].

For the upper bound on the $LTD$-number of $M(C_n)$, we consider the following two graphs $G$ and $G'$. Let $G = (V, E)$ be the graph such that $V = \{v_1, v_2, \ldots, v_n\}$ and $E = \{\{v_i, v_{i+2}\}, \{v_i, v_{i+4}\}, \{v_i, v_{i+3}\} : i \in \{1, \ldots, n\}\}$ (where the sum of the indices is taken modulo $n$). Renaming the vertices of $V$ in such a way that $w_i = v_{1+2i}$ for $i \in \{0, \ldots, n-1\}$, we consider the second graph $G'$ with vertex set $\{w_i : i \in \{0, \ldots, n-1\}\}$ and edge set $\{\{w_i, w_{i+1}\}, \{w_i, w_{i+2}\} : i \in \{0, \ldots, n-1\}\}$. We then look at the graph $G'$ with vertex set $V(C_n)$, and denote by $C_{G'}$ its minimum vertex cover. Then, it is easy to check that $C_{G'} \cup \{u_i, u\}$ for some $i \in \{1, \ldots, n\}$ is an $LTD$-set of $M(C_n)$. The theorem for $C_n$ therefore follows by the use of another result proven separately that the size of a minimum vertex cover of $G$ is

- $n - \lfloor \frac{n}{3} \rfloor$ when $n$ odd and not a multiple of 3,
- $n - \lfloor \frac{n}{4} \rfloor$ when $n$ is odd and multiple of 3,
- $n - 2 \lfloor \frac{n}{6} \rfloor$ when $n$ is even.

$\square$

We observe that for $\gamma_{LTD}(M(C_n))$, there are values of $n$ where the upper bound is attained (for example, $n \in \{6, 9\}$), but also where this is not the case (for example, $n \in \{3, 4, 5, 7, 8\}$). There are no examples yet known where the upper bounds are not tight for $M(P_n)$.

## 4   Concluding Remarks

To summarize, we studied three location-domination type problems under the Mycielski construction. In Sect. 2, we showed that $\gamma_X(G) + 1$ is a general lower bound of $\gamma_X(M(G))$ for all paths and cycles and all $X \in \{LD, LTD, OLD\}$. Using results on $\gamma_X(P_n)$ (respectively, on $\gamma_X(C_n)$) from [3, 5, 9, 15, 23], this allowed us to deduce the appropriate lower bounds on the $X$-numbers of $M(P_n)$ (respectively, of $M(C_n)$) for $X \in \{LD, LTD, OLD\}$. As a related extension of one of the main focuses of this paper, namely, the $OLD$-numbers of $M(C_n)$, we also establish the exact $OLD$-numbers for cycles.

In Sect. 3, we firstly provided two general upper bounds on $X$-numbers of the graphs $M(G)$. We showed that the upper bound of $2\gamma_X(G)$ is attained when $G$ is a star for $X \in \{LD, LTD\}$. For $OLD$-numbers of $M(G)$, we could further establish the general upper bound of $\gamma_{OLD}(G) + 2$. We showed that this bound is attained for $\gamma_{OLD}(M(P_n))$ and, combining our results on the lower and upper bound of the $OLD$-numbers, we obtained a Vizing-type result for $M(C_n)$, namely, $\gamma_{OLD}(C_n) + 1 \leq \gamma_{OLD}(M(C_n)) \leq \gamma_{OLD}(C_n) + 2$. For the other $X$-problems with $X \in \{LD, LTD\}$, we could improve the general upper bounds for the $X$-numbers of both $M(P_n)$ and $M(C_n)$.

For the studied $X$-numbers, there are examples where the upper bounds are attained (and, therefore, cannot be improved any further). On the other hand, there are also examples where the upper bounds are not tight. Therefore, our future research includes finding these exact values. In view of the fact that lower bounds were obtained by considering the domination aspect only, we expect that the true values are closer to the upper bounds. This applies particularly to $\gamma_{LD}(M(P_n))$ and to $\gamma_{LTD}(M(P_n))$ where no examples are yet known where the upper bound is not tight.

Moreover, it would be interesting to study similar questions for other locating-dominating type problems, for example, differentiating total-dominating sets (defined as total-dominating sets that separate all vertices of the graph).

# References

1. Argiroffo, G., Bianchi, S., Lucarini, Y., Wagler, A.: Polyhedra associated with locating-dominating, open locating-dominating and locating total-dominating sets in graphs. Discrete Appl. Math. **322**, 465–480 (2022)
2. Argiroffo, G., Bianchi, S., Lucarini, Y., Wagler, A.: Linear-time algorithms for three domination-based separation problems in block graphs. Discrete Appl. Math. **281**, 6–41 (2020)
3. Argiroffo, G., Bianchi, S., Lucarini, Y., Wagler, A.: The identifying code, the locating-dominating, the open locating-dominating and the locating total-dominating problems under some graph operations. Electron. Notes Theor. Comput. Sci. **346**, 135–145 (2019)
4. Argiroffo, G., Bianchi, S., Wagler, A.: A polyhedral approach to locating-dominating sets in graphs. Electron. Notes Discrete Math. **50**, 89–94 (2015)
5. Bertrand, N., Charon, I., Hudry, O., Lobstein, A.: Identifying and locating dominating codes on chains and cycles. Eur. J. Comb. **25**, 969–987 (2004)
6. Blidia, N., Chellali, M., Maffray, F., Moncel, J., Semri, A.: Locating-domination and identifying codes in trees. Australas. J. Comb. **39**, 219–232 (2007)
7. Chakraborty, D., Foucaud, F., Parreau, A., Wagler, A.K.: On three domination-based identification problems in block graphs. In: Bagchi, A., Muthu, R. (eds.) CALDAM 2023. LNCS, vol. 13947, pp. 271–283. Springer, Cham (2023). https://doi.org/10.1007/978-3-031-25211-2_21
8. Charon, I., Hudry, O., Lobstein, A.: Minimizing the size of an identifying or locating-dominating code in a graph is NP-hard. Theoret. Comput. Sci. **290**, 2109–2120 (2003)

9. Chellali, M., Rad, N.J.: Locating-total domination critical graphs. Australas. J. Comb. **160**, 227–234 (2009)
10. Foucaud, F., Ghareghani, N., Roshany-Tabrizi, A., Sharifani, P.: Characterizing extremal graphs for open neighbourhood location-domination. Discrete Appl. Math. **302**, 76–79 (2021)
11. Fisher, D.C., McKenna, P.A., Boyer, E.D.: Hamiltonicity, diameter, domination packing and biclique partitions of Mycielski's graphs. Discrete Appl. Math. **84**, 93–105 (1998)
12. Foucaud, F., Mertzios, G.B., Naserasr, R., Parreau, A., Valicov, P.: Algorithms and complexity for metric dimension and location-domination on interval and permutation graphs. In: Mayr, E.W. (ed.) WG 2015. LNCS, vol. 9224, pp. 456–471. Springer, Heidelberg (2016). https://doi.org/10.1007/978-3-662-53174-7_32
13. Foucaud, F., Mertzios, G.B., Naserasr, R., Parreau, A., Valicov, P.: Identification, location-domination and metric dimension on interval and permutation graphs. II. Algorithms Complex. Algorithmica **78**, 914–944 (2017)
14. Gravier, S., Moncel, J.: On graphs having a $V \setminus \{x\}$-set as an identifying code. Discrete Math. **307**, 432–434 (2007)
15. Haynes, T.W., Henning, M.A., Howard, J.: Locating and total-dominating sets in trees. Discrete Appl. Math. **154**, 1293–1300 (2006)
16. Henning, M.A., Rad, N.J.: Locating-total domination in graphs. Discrete Appl. Math. **160**, 1986–1993 (2012)
17. Henning, M.A., Yeo, A.: Distinguishing-transversal in hypergraphs and identifying open codes in cubic graphs. Graphs Comb. **30**, 909–932 (2014)
18. Karpovsky, M.G., Chakrabarty, K., Levitin, L.B.: On a new class of codes for identifying vertices in graphs. IEEE Trans. Inf. Theory **44**, 599–611 (1998)
19. Lobstein, A., Jean, D.: Watching systems, identifying, locating-dominating and discriminating codes in graphs. https://dragazo.github.io/bibdom/main.pdf
20. Mueller, T., Sereni, J.S.: Identifying and locating-dominating codes in (random) geometric networks. Comb. Probab. Comput. **18**(6), 925–952 (2009)
21. Mycielski, J.: Sur le coloriage des graphes. Colloq. Math. **3**, 161–162 (1955)
22. Pandey, A.: Open neighborhood locating-dominating set in graphs: complexity and algorithms. In: 2015 International Conference on Information Technology (ICIT) IEEE (2015)
23. Seo, S.J., Slater, P.J.: Open neighborhood locating dominating sets. Austral. J. Comb. **46**, 109–119 (2010)
24. Shaminejad, A., Vatandoost, E., Mirasheh, K.: The identifying code number and Mycielski's construction of graphs. Trans. Comb. **11**, 309–316 (2022)
25. Slater, P.J.: Dominating and reference sets in a graph. J. Math. Phys. Sci. **22**(4), 445–455 (1988)
26. Slater, P.J.: Dominating and location in acyclic graphs. Networks **17**, 55–64 (1987)

# On Total Chromatic Number of Complete Multipartite Graphs

Aseem Dalal[✉] and B. S. Panda

Department of Mathematics, Indian Institute of Technology Delhi, Hauz Khas, New Delhi 110016, India
{aseem.dalal,bspanda}@maths.iitd.ac.in

**Abstract.** This paper makes progress towards settling the long-standing conjecture that the total chromatic number $\chi''(K)$ of the complete $p$-partite graph $K = K(r_1,\ldots,r_p)$ is $\Delta(K) + 1$ if and only if $K \neq K_{r,r}$ and if $K$ has an even number of vertices then $def(K) = \Sigma_{v \in V(K)}(\Delta(K) - d_K(v))$ is at least the number of parts of odd size. The problem was settled for complete 3-partite graphs by Chew and Yap in 1992, and for complete 4-partite graphs by Dong and Yap in 2000; the difficulty rises manifold with the increase in the number of parts. In 2014, Dalal and Rodger (Graphs and Combinatorics (2015), 1–15) introduced an approach using amalgamations to attack the conjecture and demonstrated its power by settling the problem for complete 5-partite graphs. Their approach required coloring of all the vertices in each part with the same color. However, the applicability of their approach is restricted because, for each $k \in \mathbb{N}$, there are complete $2k$-partite graphs $K$ for which any total coloring of $K$ in which all the vertices in each part are colored the same would require at least $\Delta(K) + 2$ colors, although $\chi''(K) = \Delta(K) + 1$. In this paper, we overcome this difficulty by providing a technique that allows the vertices in the same part to have different colors by adapting a result of Bahmanian and Rodger (J. Graph Theory (2012), 297–317) on graph amalgamations. Using our technique, we solve the classification problem for all complete 6-partite graphs.

## 1  Introduction

A total coloring of a graph $G = (V, E)$ is coloring of elements of $V(G) \cup E(G)$ so that adjacent vertices receive different colors; adjacent edges receive different colors; and if edge $e$ is incident to vertex $v$ then $e$ and $v$ receive different colors. The total chromatic number $\chi''(G)$ is the least number of colors needed to totally color $G$. The complete $p$-partite graph $K = K[V_1,\ldots,V_p]$ is the simple graph with vertex set $V(K) = \cup_{i=1}^p V_i$ (each set $V_i$ is called a part), where $V_i \cap V_j = \emptyset$ for $i \neq j$, in which two vertices are joined if and only if they belong to different parts of $K$. If the names of the vertex sets are unimportant then $K$ is simply referred to as $K(r_1,\ldots,r_p)$, where $|V_i| = r_i$ for $1 \leq i \leq p$.

The graph $K$ is of sufficient complexity that settling the values of its graph parameters is often a challenge. Finding the chromatic index $\chi'(K)$ is a typical

S. Kalyanasundaram and A. Maheshwari (Eds.): CALDAM 2024, LNCS 14508, pp. 270–285, 2024.
https://doi.org/10.1007/978-3-031-52213-0_19

example. Of course, the classic result of Vizing [15] shows that $\chi'(G)$ is $\Delta(G)$ or $\Delta(G) + 1$, thereby giving rise to the classification of whether a graph is Class 1 or Class 2, respectively. It was finally shown in 1992 that $K$ is a Class 2 graph if and only if it is overfull [13]. Similar to Vizing's result, it is conjectured that the value of $\chi''(G)$ for any simple graph $G$ is either $\Delta(G) + 1$ or $\Delta(G) + 2$ (see [2,15]), and $G$ is said to be of Type 1 or Type 2 respectively based on this value. Bermond settled the type of $K$ when it is regular [3]. Yap [16] proved that $\chi''(K) \leq \Delta + 2$, and Chew and Yap [7] showed that if $K$ has an odd number of vertices then it is of Type 1. In 1992, Chew and Yap [7] proved the following result.

**Theorem 1** [7]. *Suppose that either $r_1 < r_2 \leq r_3 \leq \ldots \leq r_p$ or $p = 3$. Then $K$ is of Type 1.*

In 2000, Dong and Yap [11] extended the above result by proving the following.

**Theorem 2** [11]. *Suppose that $r_1 \leq r_2 \ldots \leq r_p$ and that $|V(K)| = 2n$. If $r_2 \leq r_3 - 2$ then $K$ is of Type 1. Also if $K$ is not regular with $p = 4$ then it is of Type 1.*

The proof techniques in all these papers are very similar, though they get more complicated as more difficult cases are attacked. They build upon the idea of coloring the vertices so that all vertices in one part, say $V_\beta$, receive the same color, while all other vertices receive different colors (a so-called $\beta$-biased total coloring). Such total colorings were characterized in [14] by Hoffman and Rodger, thereby producing the following theorem. It is most easily stated in terms of the deficiency, which is the measure of how far a graph $G$ is from being regular, and is defined by $def(G) = \Sigma_{v \in V(G)}(\Delta(G) - d_G(v))$.

**Theorem 3** [14]. *Suppose that $r_1 \leq r_2 \leq \ldots \leq r_p$ and that $|V(K)| = 2n$. If*

$$def(K) \geq \begin{cases} 2n - r_1 & \text{if } p = 2 \text{ or} \\ & \text{if } p \text{ is even, } r_1 \text{is odd, and } r_1 = r_{p-1}, \\ 2n - r_p & \text{otherwise,} \end{cases}$$

*then $K$ is Type 1.*

Hoffman and Rodger [14] made the following conjecture for necessary and sufficient conditions for $K$ to be of Type 2.

**Conjecture 4** [14]. *A complete multipartite graph $K = K(r_1, \ldots, r_p)$ is of Type 2 if and only if*

1. *$p = 2$ and $K$ is regular, or*
2. *$|V(K)|$ is even and $def(K)$ is less than the number of parts of odd size.*

Dalal and Rodger [8] introduced, in 2014, a novel approach using amalgamations to attack the problem. They exemplified the power of the approach by settling the classification problem for all complete 5-partite graphs, thereby extending the result for $p = 4$ in Theorem 2. More precisely, they proved:

**Theorem 5** [8]. $K = K(r_1, \ldots, r_5)$ *is Type 2 if and only if* $|V(K)| \equiv 0 \mod 2$ *and* $def(K)$ *is less than the number of parts in $K$ of odd size.*

Dalal, Panda and Rodger [9] made use of the amalgamation technique, blended with the traditional approach, to improve upon Theorem 2. More specifically, they proved the following.

**Theorem 6** [9]. *Suppose that* $r_1 \leq r_2 \ldots \leq r_p$ *and that* $|V(K)| = 2n$. *If* $r_2 < r_3$ *then $K$ is of Type 1.*

Recently, using amalgamations, Dalal et al. [10] proved the following result which provides that if the sizes of the third and fourth parts are not equal, then $K$ is of Type 1.

**Theorem 7** [10]. *Let* $K(r_1, r_2, \ldots, r_p)$ *be a complete multipartite graph such that* $r_1 \leq r_2 \leq \ldots \leq r_p$ *and* $|V(K)| = 2n$. *If* $r_3 < r_4$ *then $K$ is of Type 1.*

Thus, the amalgamation approach introduced by Dalal and Rodger [8] has been quite useful. To obtain a total coloring of $K$, their approach requires that all the vertices in each part are colored the same. However, for each $k \in \mathbb{N}$, there are complete $2k$-partite graphs $K$ such that any total coloring of $K$ in which all the vertices in each part are colored the same would require at least $\Delta(K) + 2$ colors although $\chi''(K) = \Delta(K) + 1$. Therefore, this puts limitation on the applicability of the approach by Dalal and Rodger. To illustrate this, we take an example of a complete 6-partite graph $K[V_1, \ldots, V_6] = K(r, r, r, r, r, r + 2)$, $r$ is odd. Any total coloring of $K$ in which all the vertices of $V_6$ are colored the same would leave at least one vertex $v \in V_1 \cup \ldots \cup V_5$ unsaturated. However, all the vertices in $V_1 \cup \ldots \cup V_5$ are of the maximum degree, and thus such a total coloring would require at least $\Delta(K) + 2$ colors. However, $K(r, r, r, r, r, r + 2)$ is of Type 1 (as proved by Theorem 8 of this paper).

In this paper, we overcome this difficulty by providing a generalized amalgamation approach which allows the vertices in the same part to have different colors. Our approach thus has a wider applicability and we demonstrate this by solving the classification problem for all complete 6-partite graphs. More precisely, we prove the following.

**Theorem 8.** $K = K(r_1, \ldots, r_6)$ *is Type 2 if and only if* $|V(K)| \equiv 0 \mod 2$ *and* $def(K)$ *is less than the number of parts in $K$ of odd size.*

We first introduce some notations. $H$ is an *amalgamation* of $G$ if there exists a function $\psi$ called an amalgamation function from $V(G)$ onto $V(H)$ and a bijection $\phi' : E(G) \to E(H)$ such that $e$ joins $u$ and $v$ in $E(G)$ if and only if $\phi'(e)$ joins $\psi(u)$ and $\psi(v)$ in $E(H)$. Note that $\phi'$ is completely determined by $\psi$. Figure 1 gives an illustration of an amalgamation of $K(2, 3, 3)$ to $K'$. Associated with $\psi$ is the *number function* $\eta : V(H) \to \mathbb{N}$ defined by $\eta(v) = |\psi^{-1}(v)|$, for each $v \in V(H)$. $G$ is a *detachment* of $H$ if there exists an amalgamation function $\psi$ of $G$ onto $H$ such that $|\psi^{-1}(\{u\})| = \eta(u)$ for every $u \in V(H)$. Some authors refer to detachments as *disentanglements*. The subgraph of $G$ induced by the edges colored $j$ is denoted by $G(j)$. For a graph $G$, $m_G(u,v)$ denotes the number of edges joining vertices $u$ and $v$ in $G$, and $l_G(u)$ denotes the number of loops incident to vertex $u$. If $x, y$ are real numbers, then $\lfloor x \rfloor$ and $\lceil x \rceil$ denote the integers such that $x - 1 \leq \lfloor x \rfloor \leq x \leq \lceil x \rceil \leq x + 1$, and $x \approx y$ means $\lfloor y \rfloor \leq x \leq \lceil y \rceil$. We denote the set of the first $k$ natural numbers by $N_k$.

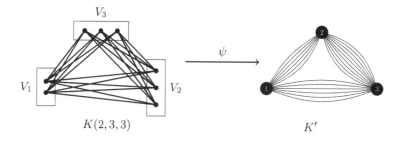

**Fig. 1.** Amalgamation of $K(2,3,3)$ into $K'$ where $V(K') = \{1,2,3\}$.

## 2   Generalized Approach

Our technique makes use of the following theorem of Bahamanian and Rodger [1] which generalizes several existing amalgamation results in various ways.

**Theorem 9** [1]. *Let $H$ be a $k$-edge-colored graph and $\eta$ be a function from $V(H)$ into $\mathbb{N}$ such that for each $w \in V(H)$, $\eta(w) = 1$ implies $l_H = 0$. Then there exists a loopless $\eta$-detachment $G$ of $H$ with amalgamation function $\psi : V(G) \longrightarrow V(H)$, $\eta$ being the number function associated with $\psi$ such that $G$ satisfies the following conditions:*

1. *$d_G(u) \approx d_H(w)/\eta(w)$ for each $w \in V(H)$ and each $u \in \psi^{-1}(w)$;*
2. *$d_{G(j)}(u) \approx d_{H(j)}(w)/\eta(w)$ for each $w \in V(H)$, each $u \in \psi^{-1}(w)$ and each $j \in \mathbb{N}_k$;*
3. *$m_G(u,v) \approx m_H(w,z)/(\eta(w)\eta(z))$ for every pair of distinct vertices $w, z \in V(H)$, each $u \in \psi^{-1}(w)$ and $v \in \psi^{-1}(z)$; and*
4. *$m_{G(j)}(u,v) \approx m_{H(j)}(w,z)/(\eta(w)\eta(z))$ for every pair of distinct vertices $w, z \in V(H)$, each $u \in \psi^{-1}(w)$, $v \in \psi^{-1}(z)$ and each $j \in \mathbb{N}_k$.*

Let $K'$ be an amalgamation of complete multipartite graph $K$, where $|V(K')| = k'$, $\psi$ and $\eta$ be the *amalgamation function* and the associated *number function*, respectively. For simplicity, we assume that $V(K') = \mathbb{N}_{k'}$. We would find $\Delta(K) + 1$ subgraphs of $K'$, each with degree sequence majorized by $(\eta(1)), \ldots, \eta(k'))$, so that their union forms a graph in which for $1 \leq i < j \leq \eta(k')$ vertices $i$ and $j$ are joined by exactly $\eta(i)\eta(j)$ or $0$ edges as the case may be, and on disentangling $K'$ to get $K$ (using Theorem 9), the vertices of $K$ are colored using only the $\Delta(K) + 1$ colors while ensuring that the properties of total coloring are satisfied. It is convenient to consider each subgraph as a color class. The technique follows four general steps:

1. $k'$ or $k'-1$ color classes of $K'$ are defined in which each vertex $i \in \{1, \ldots, k'\}$ has degree as close to $\eta(i)$ as possible except for one which has degree 0.
2. Many color classes of $K'$ with degree sequence $(\eta(1)), \ldots, \eta(k'))$ saturating all the vertices are found.
3. The remaining edges of $K'$ are partitioned into few remaining color classes.
4. Using Theorem 9, $K'$ is disentangled to get $K$. Most vertices of $K$ are then colored using colors in step (1) and left ones are colored using color(s) in step (3).

This may be seen in contrast with the technique of Dalal and Rodger, though not stated in the above form in [8]. For a complete multipartite $K = K(V_1, \ldots, V_p)$, their approach would always have a fixed *amalgamation function* $\psi : V(K) \to V(K')$ such that $\psi(V_i) = \{i\}$ and $\eta(i) = r_i$ for $1 \leq i \leq p$. In Step (1), $p$ color classes would be defined. In Step (4), only these $p$ color classes would be used to color the vertices of $K$ by coloring all the vertices in each of $p$ parts by one of the $p$ colors.

Thus, it is apparent that our technique is much more general. The limitation of the approach of Dalal and Rodger [8] and the power of our technique is apparent when we deal with two subcases in the proof of Theorem 8.

## 3    Lemmas and Proofs

The following lemma of Hilton [12] characterizes the graphs of order $2n$ and having maximum degree $2n - 1$ according to their total chromatic number.

**Lemma 1**  [12]. *Let $n \geq 1$, let $J$ be a subgraph of $K_{2n}$, let $e = |E(J)|$ and let $j$ be the maximum size (i.e., number of edges) of a matching in $J$. Then $\chi''(K_{2n} - E(J)) = 2n + 1$ if and only if $e + j \leq n - 1$.*

To apply the above lemma for complete multipartite graphs $K(r_1, r_2, \ldots, r_p)$ of order $2n$ with maximum degree $2n - 1$ (i.e. $r_1 = 1$), take $J = K^c$, then $e = |E(J)| = \frac{def(K)}{2}$ and by some simple calculations we get $j = n - \frac{o(K)}{2}$ where $o(K)$ is the number of parts of odd size in $K$. So for the complete multipartite graph of order $2n$ and maximum degree $2n - 1$, the above lemma gives the following result:

**Lemma 2.** *Let $K(r_1, r_2, \ldots, r_p)$ be a complete multipartite graph of order $2n$ and maximum degree $2n - 1$ (i.e., $r_1 = 1$). Then $K$ is Type 2 if and only if $def(K)$ is less than the number of parts of odd size in $K$.*

Similarly, the following lemma of Chen and Fu  [5] characterizes the graphs of order $2n$ and maximum degree $2n - 2$, and therefore also applies to complete multipartite graphs $K(r_1, r_2, \ldots, r_p)$ of order $2n$ with $r_1 = 2$.

**Lemma 3**  [5]. *Let $G$ be a graph of order $2n$ and $\Delta(G) = 2n - 2$. Then $G$ is of Type 2 if and only if $G^c$ is a disjoint union of an edge and a star having $2n - 3$ edges.*

A well-known reason for a graph to be of Type 2 was established in a more general setting in [6], but the following suffices for our purposes.

**Lemma 4** [6]. *If $K$ has an even order and if $def(K)$ is less than the number of parts of odd size then $K$ is Type 2.*

We also make use of the following result by Bryant et al. [4] which gives necessary and sufficient conditions for a decomposition of complete multigraphs into cycles of varying lengths.

**Lemma 5** [4]. *Let $\lambda$, $n$ and $m$ be integers with $n, m \geq 3$ and $\lambda \geq 1$. There exists a decomposition of $\lambda K_n$ into $m$-cycles if and only if (i) $m \leq n$; (ii) $\lambda(n-1)$ is even; and (iii) $m$ divides $\lambda\binom{n}{2}$. There exists a decomposition of $\lambda K_n$ into $m$-cycles and a perfect matching if and only if (i) $m \leq n$; (ii) $\lambda(n-1)$ is odd; and (iii) $m$ divides $\lambda\binom{n}{2} - \frac{n}{2}$.*

For any set $S$, let $r * S$ be the multi-set consisting of $r$ copies of each element in $S$, and for any graph $G$, let $r * G$ denote the multigraph formed by replacing each edge in $G$ with $r$ edges. A sequence $s = (s_1, \ldots, s_n)$ is said to majorize a sequence $t = (t_1, \ldots, t_n)$ if $s_i \geq t_i$ for $1 \leq i \leq n$. In any total coloring, the set of edges colored $C_i$ is known as the $i^{th}$ color class, denoted by $C_i$; it will cause no confusion to also refer to the subgraph induced by the edges colored $i$ as $C_i$. For any pair of disjoint subsets $A$ and $B$ of the vertex set of a graph $G$, let $G[A, B]$ denote the bipartite subgraph of $G$ induced by the edge set $\{\{a, b\} | a \in A, b \in B\}$. If the vertex set of $G$ is a subset of the vertex set of $H$, then let $G + H$ be the multigraph with vertex set $V(H)$ formed from $H$ by adding the edges in $G$ to $H$ (so, for example, if an edge appears once in $G$ and in $H$ then it appears twice in $G + H$). For any vertex $x$ and $y$, $E_{xy}$ or $\{x, y\}$ denotes the edge between $x$ and $y$. An edge-coloring of a multigraph is *equitable* if, among the edges incident to each vertex, the number of edges of each color class differs by at most one from the number of edges of each other class. Equitable edge-coloring has been used in scheduling and timetabling problems.

**Lemma 6.** *For all $r \geq 3$, there exists an equitable 10-edge-coloring of $2(r-1) * K_6 + 5 * K[\{1, 2, 3, 4\}, \{5, 6\}] + 10 * E_{56}$, where $V(K_6) = \mathbb{N}_6$, such that the degree sequence of each color class is $(r, r, r, r, r+2, r+2)$.*

*Proof.* Let $R_1, \ldots, R_5$ be a 1-factorization of $K_6$. The required color classes are as follows:

$C_{2i} = (r-1) * R_i + E_{56} + [\{1, 5\}, \{2, 5\}, \{3, 6\}, \{4, 6\}]$ for $1 \leq i \leq 5$.
$C_{2i-1} = (r-1) * R_i + E_{56} + [\{3, 5\}, \{4, 5\}, \{1, 6\}, \{2, 6\}]$ for $1 \leq i \leq 5$.
$\cup_{i=1}^{10} C_i = 2(r-1) * K_6 + 5 * K[\{1, 2, 3, 4\}, \{5, 6\}] + 10 * E_{56}$.

**Lemma 7.** *For all $r \geq 3$, there exists an equitable 20-edge-coloring of $4(r-1) * K_6 + 5 * K[\{1, 2, 3, 4\}, \{5\}] + 15 * K[\{1, 2, 3, 4\}, \{6\}] + 20 * E_{56}$, where $V(K_6) = \mathbb{N}_6$, such that the degree sequence of each color class is $(r, r, r, r, r+1, r+3)$.*

*Proof.* Let $R_1, \ldots, R_5$ be a 1-factorization of $K_6$. The required color classes are as follows:

$C_{4i-3} = (r-1) * R_i + E_{56} + [\{1,5\}, \{2,6\}, \{3,6\}, \{4,6\}]$ for $1 \le i \le 5$.
$C_{4i-2} = (r-1) * R_i + E_{56} + [\{2,5\}, \{1,6\}, \{3,6\}, \{4,6\}]$ for $1 \le i \le 5$.
$C_{4i-1} = (r-1) * R_i + E_{56} + [\{3,5\}, \{2,6\}, \{1,6\}, \{4,6\}]$ for $1 \le i \le 5$.
$C_{4i} = (r-1) * R_i + E_{56} + [\{4,5\}, \{2,6\}, \{3,6\}, \{1,6\}]$ for $1 \le i \le 5$.
$\cup_{i=1}^{20} C_i = 4(r-1) * K_6 + 5 * K[\{1,2,3,4\}, \{5\}] + 15 * K[\{1,2,3,4\}, \{6\}] + 20 * E_{56}$.

**Lemma 8.** *For odd $r \ge 3$, there exists an equitable 5-edge-coloring of $(r-1) * K_6 + 4 * K[\{1,2,3,4,5\}, \{6\}] + K[\{1,2,3,4,5\}, \{7\}]$, where $V(K_6) = \mathbb{N}_6$, such that the degree sequence of each color class is $(r,r,r,r,r+3,1)$.*

*Proof.* Let $R_1, \ldots, R_5$ be a 1-factorization of $K_6$. The required color classes are as follows:

$C_i = (r-1) * R_i + (K[\{1,2,3,4,5\}, \{6\}] \setminus \{i,6\}) \cup \{i,7\}$ for $1 \le i \le 5$.
$\cup_{i=1}^{5} C_i = (r-1) * K_6 + 4 * K[\{1,2,3,4,5\}, \{6\}] + K[\{1,2,3,4,5\}, \{7\}]$.

**Lemma 9.** *For odd $r \ge 3$, there exists an equitable 10-edge-coloring of $2(r-1) * K_6 + 4 * K[\{1,2,3,4,5\}, \{6\}] + 2 * K[\{1,2,3,4,5\}, \{7\}] + K_5$, $(V(K_6) = \mathbb{N}_6$, $V(K_5) = \mathbb{N}_5)$, such that degree sequence of each color class is $(r,r,r,r,r+2,1)$.*

*Proof.* Let $R_1, \ldots, R_5$ be 1-factorization of $K_6$. The required color classes are as follows:

$C_1 = (r-1) * R_1 + K[\{1,2,3,4,5\}, \{6\}] \setminus [\{1,6\}, \{2,6\}, \{3,6\}] + \{1,2\} + \{3,7\}$.
$C_2 = (r-1) * R_2 + K[\{1,2,3,4,5\}, \{6\}] \setminus [\{2,6\}, \{3,6\}, \{4,6\}] + \{2,3\} + \{4,7\}$.
$C_3 = (r-1) * R_3 + K[\{1,2,3,4,5\}, \{6\}] \setminus [\{3,6\}, \{4,6\}, \{5,6\}] + \{3,4\} + \{5,7\}$.
$C_4 = (r-1) * R_4 + K[\{1,2,3,4,5\}, \{6\}] \setminus [\{4,6\}, \{5,6\}, \{1,6\}] + \{4,5\} + \{1,7\}$.
$C_5 = (r-1) * R_5 + K[\{1,2,3,4,5\}, \{6\}] \setminus [\{5,6\}, \{1,6\}, \{2,6\}] + \{5,1\} + \{2,7\}$.
$C_6 = (r-1) * R_6 + K[\{1,2,3,4,5\}, \{6\}] \setminus [\{1,6\}, \{3,6\}, \{2,6\}] + \{1,3\} + \{2,7\}$.
$C_7 = (r-1) * R_7 + K[\{1,2,3,4,5\}, \{6\}] \setminus [\{3,6\}, \{5,6\}, \{4,6\}] + \{3,5\} + \{4,7\}$.
$C_8 = (r-1) * R_8 + K[\{1,2,3,4,5\}, \{6\}] \setminus [\{5,6\}, \{2,6\}, \{3,6\}] + \{5,2\} + \{3,7\}$.
$C_9 = (r-1) * R_9 + K[\{1,2,3,4,5\}, \{6\}] \setminus [\{2,6\}, \{4,6\}, \{1,6\}] + \{2,4\} + \{1,7\}$.
$C_{10} = (r-1) * R_{10} + K[\{1,2,3,4,5\}, \{6\}] \setminus [\{4,6\}, \{1,6\}, \{5,6\}] + \{4,1\} + \{5,7\}$.
$\cup_{i=1}^{10} C_i = 2(r-1) * K_6 + 4 * K[\{1,2,3,4,5\}, \{6\}] + 2 * K[\{1,2,3,4,5\}, \{7\}] + K_5$.

**Lemma 10.** *For even $r \ge 2$, there exists an equitable 5-edge-coloring of $(r-1) * K_6 + 5 * K[\{1,2,3,4,5\}, \{6\}]$, (where $V(K_6) = \mathbb{N}_6$) such that the degree sequence of each color class is $(r,r,r,r,r,r+4)$).*

*Proof.* Let $R_0, \ldots, R_4$ be a 1-factorization of $K_6$. We partition $(r-1) * K_6 + 5 * K[\{1,2,3,4,5\}, \{6\}]$, (where $V(K_6) = \mathbb{N}_6$) into (5)-color classes $C_0, \ldots, C_4$ such that the degree sequence of each color class is $(r,r,r,r,r,r+4)$ as follows:
For $0 \le i \le 4$, $C_i = (r-1) * R_i + K[\{1,2,3,4,5\}, \{6\}]$
and $\cup_{i=0}^{4} C_i = K_6 + 5 * K[\{1,2,3,4,5\}, \{6\}]$. □

We now provide proof of the Theorem 8.

*Proof (Sketch of Proof of Theorem 8).* If $def(K)$ is less than the number of parts of odd size in $K$, then by Lemma 4, $K$ is of Type 2. So, we assume the $def(K)$ to be greater or equal to the number of parts of odd size. If $r_1 = 1$ or $r_1 = 2$, then $K$ is of Type 1 by Lemma 2 and Lemma 3. Therefore, we can assume $r_1 = r \geq 3$. By using Theorem 1, Theorem 3, Theorem 6, and Theorem 7, only the following five cases are required to be settled:

1. $K(r, r, r, r, r+2, r+2)$
2. $K(r, r, r, r, r+1, r+3)$
3. $K(r, r, r, r, r, r+4)$
4. $K(r, r, r, r, r, r+2)$
5. $K(r, r, r, r, r+1, r+1)$

It is useful to define notation which describes the uncolored edges of the amalgamated graph $K'$. So if $G$ is the subgraph of $K'$ induced by the edges colored so far, then $\phi(G, K')$ is defined as follows: for each $\{i, j\} \subseteq \{1, \ldots, k'\}$ such that $i$ and $j$ are joined by an edge in $K'$, let $\phi(G, K')(\{i, j\}) = \eta(i)\eta(j) - \epsilon_{i,j}$, where $\epsilon_{i,j}$ is the number of edges joining $i$ and $j$ in $G$.

**Case 1 :** $K(r, r, r, r, r+2, r+2)$.

Let $K'$ be an amalgamation of $K$ with *amalgamation function* $\psi : V(K) \to V(K')$ such that $\psi(V_i) = \{i\}$ and $\eta(i) = r_i$ for $1 \leq i \leq 6$. Let $V(K_6) = \{1, \ldots, 6\}$ and let $C_i$ be the set of edges colored $c_i$ for $1 \leq i \leq \Delta(K) + 1 = 5r + 5$, defined as follows:

Sub – Case 1, $r$ is odd

(1) Let $\{W_1, \ldots, W_6\}$ be a decomposition of $2 * K_6$ into 6 cycles of length 5 with vertex $i$ missing in $W_i$ for $1 \leq i \leq 6$ (existence by Lemma 5).

$C_1 = \frac{r-1}{2} * W_1 + E_{56} + [\{2, 5\}, \{3, 6\}, \{4, 6\}].$
$C_2 = \frac{r-1}{2} * W_2 + E_{56} + [\{3, 5\}, \{4, 6\}, \{1, 6\}].$
$C_3 = \frac{r-1}{2} * W_3 + E_{56} + [\{2, 5\}, \{4, 5\}, \{1, 6\}].$
$C_4 = \frac{r-1}{2} * W_4 + E_{56} + [\{1, 5\}, \{3, 5\}, \{2, 6\}].$
$C_5 = \frac{r-1}{2} * W_5 + [\{2, 6\}, \{3, 6\}] + \{1, 4\}.$
$C_6 = \frac{r-1}{2} * W_6 + [\{1, 5\}, \{4, 5\}] + \{2, 3\}.$
$\bigcup_{i=1}^6 C_i = (r-1) * K_6 + 4 * E_{56} + 2 * K[\{1, 2, 3, 4\}, \{5, 6\}] + [\{2, 3\}, \{1, 4\}].$

(2) Use Lemma  6 to produce (10)-color classes $M_1, M_2, \ldots, M_{10}$ in $K'$ such that $\cup_{i=1}^{10} M_i$ induces $2(r-1) * K_6 + 5 * K[\{1, 2, 3, 4\}, \{5, 6\}] + 10 * E_{56}$. Take $\frac{r-1}{2}$ copies of each of $M_1, \ldots, M_{10}$ to form $(5)(r-1)$ new color classes. So in steps (1-2), $5r + 1$ color classes have been defined, and their union $G$ is $(r^2 - r) * K_6 + \frac{5r-1}{2} * K[\{1, 2, 3, 4\}, \{5, 6\}] + (5r-1) * E_{56} + [\{2, 3\}, \{1, 4\}]$ such that $\phi = \phi(G, K')$ satisfies: $\phi(\{1, 2\}) = r = \phi(\{3, 4\}) = \phi(\{1, 3\}) = \phi(\{2, 4\})$, $\phi(\{2, 3\}) = r - 1 = \phi(\{1, 4\})$; for $1 \leq i \leq 4$ and $5 \leq j \leq 6$, $\phi(\{i, j\}) = \frac{r+1}{2}$, and $\phi(\{5, 6\}) = 5$.

(3) Let $M_0', M_1'$ and $M_2'$ be a decomposition of $K_4$ into 1-factors such that $M_0' = [\{2,3\},\{1,4\}]$. We use the 4 colors to color the remaining edges as follows:

$$C_{5r+2} = \lfloor \tfrac{r+1}{4} \rfloor * K[\{1,2,3,4\},\{5\}] + \lceil \tfrac{r+1}{4} \rceil * K[\{1,2,3,4\},\{6\}] + \bigcup_{i=1}^{\frac{r-1}{2}} M_i' \text{ mod 3}.$$

$$C_{5r+3} = \lfloor \tfrac{r+1}{4} \rfloor * K[\{1,2,3,4\},\{6\}] + \lceil \tfrac{r+1}{4} \rceil * K[\{1,2,3,4\},\{5\}] + \bigcup_{i=\frac{r+1}{2}}^{r-1} M_i' \text{ mod 3}.$$

$$C_{5r+4} = 2 * E_{56} + \cup_{i=r}^{2r-1} M_i \text{ mod 3}; \quad C_{5r+5} = 3 * E_{56} + \cup_{i=2r}^{3r-1} M_i' \text{ mod 3}.$$

Using Theorem 9, we disentangle $K'$ to get $K$ with all its edges properly colored using $\Delta(K)+1$ colors $c_1,\ldots,c_{\Delta+1}$. We now require to color the vertices of $K$ using only the colors $c_1,\ldots,c_{\Delta+1}$ to get a total coloring with $\Delta(K)+1$ colors. It may be observed that for $1 \leq i \leq 6$ the color $c_i$ is absent from the vertex $\{i\}$ in $K'$. Therefore, for $1 \leq i \leq 6$, the color $c_i$ is absent from all the vertices in $\psi^{-1}(\{i\}) = V_i \in K$, and we color them with $c_i$.

Sub $-$ Case 2, $r$ is even : Let $K'$ be an amalgamation of $K$ with *amalgamation function* $\psi : V(K) \to V(K')$ such that $\psi(V_i) = \{i\}$ and $\eta(i) = r_i$ for $1 \leq i \leq 6$. Let $V(K_6) = \{1,\ldots,6\}$ and let $C_i$ be the set of edges colored $c_i$ for $1 \leq i \leq \Delta(K)+1 = 5r+5$, defined as follows:

(1) Let $\{W_1,\ldots,W_6\}$ be a decomposition of $2*K_6$ into 6 cycles of length 5 with vertex $i$ missing in $W_i$ for $1 \leq i \leq 6$.

$$C_1 = \tfrac{r}{2} * W_1 + 2 * E_{56}; \; C_2 = \tfrac{r}{2} * W_2 + 2 * E_{56}; \; C_3 = \tfrac{r}{2} * W_3 + 2 * E_{56};$$
$$C_4 = \tfrac{r}{2} * W_4 + 2 * E_{56}; \; C_5 = \tfrac{r}{2} * W_5; \; C_6 = \tfrac{r}{2} * W_6. \cup_{i=1}^{6} C_i = r * K_6 + 8 * E_{56}.$$

(2) Use Lemma 6 to produce (10)-color classes $M_1, M_2, \ldots, M_{10}$ in $K'$ such that $\cup_{i=1}^{10} M_i$ induces $2(r-1) * K_6 + 5 * K[\{1,2,3,4\},\{5,6\}] + 10 * E_{56}$. Take $\frac{r-2}{2}$ copies of each of $M_1,\ldots,M_{10}$ to form $5(r-2)$ new color classes. We take another 5 colors to color the edges as follows:

$$C_{5r-3} = (r-1) * R_1 + E_{56} + [\{1,5\},\{2,5\},\{3,6\},\{4,6\}].$$
$$C_{5r-2} = (r-1) * R_2 + E_{56} + [\{3,5\},\{4,5\},\{1,6\},\{2,6\}].$$
$$C_{5r-1} = (r-1) * R_3 + E_{56} + [\{1,5\},\{2,5\},\{3,6\},\{4,6\}].$$
$$C_{5r} = (r-1) * R_4 + E_{56} + [\{3,5\},\{4,5\},\{1,6\},\{2,6\}].$$
$$C_{5r+1} = (r-1) * R_5 + E_{56} + [\{1,5\},\{2,5\},\{3,6\},\{4,6\}].$$

where $\{R_1,\ldots,R_5\}$ is a 1-factor decomposition of $K_6$.
So in steps (1-2), $5r+1$ color classes have been defined, and their union $G$ is $(r^2-r+1)*K_6 + \frac{5r-6}{2}*K[\{1,2,3,4\},\{5,6\}] + [\{1,5\},\{2,5\},\{3,6\},\{4,6\}] + (5r+3)*E_{56}$ such that $\phi = \phi(G,K')$ satisfies: for $1 \leq i \neq j \leq 4$, $\phi(\{i,j\}) = r-1$; for $1 \leq i \leq 2$, $\phi(\{i,5\}) = \tfrac{r}{2}+1$, $\phi(\{i,6\}) = \tfrac{r}{2}+2$; for $3 \leq i \leq 4$, $\phi(\{i,5\}) = \tfrac{r}{2}+2$; $\phi(\{i,6\}) = \tfrac{r}{2}+1$, and $\phi(\{5,6\}) = 0$.

(3) Let $M_0', M_1'$ and $M_2'$ be a decomposition of $K_4$ into 1-factors. We use the 4 colors to color the remaining edges as follows:

$$C_{5r+2} = \tfrac{r}{2} * K[\{1,2,3,4\},\{5,6\}].$$
$$C_{5r+3} = K[\{1,2,3,4\},\{5,6\}] + [\{3,5\},\{4,5\},\{1,6\},\{2,6\}] + \cup_{i=1}^{r-3} M_i' \text{ mod 3}.$$
$$C_{5r+4} = \cup_{i=r-2}^{2r-3} M_i' \text{ mod 3}; \quad C_{5r+5} = \cup_{i=2r-2}^{3r-3} M_i' \text{ mod 3}.$$

Using Theorem 9, we disentangle $K'$ to get $K$ with all its edges properly colored using $\Delta(K)+1$ colors $c_1, \ldots, c_{\Delta+1}$. We now require to color the vertices of $K$ using only the colors $c_1, \ldots, c_{\Delta+1}$ to get a total coloring with $\Delta(K)+1$ colors. It may be observed that for $1 \leq i \leq 6$ the color $c_i$ is absent from the vertex $\{i\}$ in $K'$. Therefore, for $1 \leq i \leq 6$, the color $c_i$ is absent from all the vertices in $\psi^{-1}(\{i\}) = V_i \in K$, and we color them with $c_i$.

**Case 2:** $K(r, r, r, r, r+1, r+3)$.
Let $K'$ be an amalgamation of $K$ with *amalgamation function* $\psi : V(K) \to V(K')$ such that $\psi(V_i) = \{i\}$ and $\eta(i) = r_i$ for $1 \leq i \leq 6$. Let $V(K_6) = \{1, \ldots, 6\}$ and let $C_i$ be the set of edges colored $c_i$ for $1 \leq i \leq \Delta(K) + 1 = 5r + 5$, defined as follows:

Sub $-$ Case 1, $r$ is odd

(1) Let $\{W_1, \ldots, W_6\}$ be a decomposition of $2 * K_6$ into 6 cycles of length 5 with vertex $i$ missing in $W_i$ for $1 \leq i \leq 6$.
$C_1 = \frac{r-1}{2} * W_1 + E_{56} + [\{2,5\}, \{3,6\}, \{4,6\}]$; $C_2 = \frac{r-1}{2} * W_2 + E_{56} + [\{1,5\}, \{3,6\}, \{4,6\}]$.
$C_3 = \frac{r-1}{2} * W_3 + E_{56} + [\{4,5\}, \{1,6\}, \{2,6\}]$; $C_4 = \frac{r-1}{2} * W_4 + E_{56} + [\{3,5\}, \{1,6\}, \{2,6\}]$.
$C_5 = \frac{r-1}{2} * W_5 + K[\{1,2,3,4\}, \{6\}]$; $C_6 = \frac{r-1}{2} * W_5 + [\{1,5\}, \{2,5\}] + \{3,4\}$.
$\bigcup_{i=1}^{6} C_i = (r-1) * K_6 + 4 * E_{56} + 3 * K[\{1,2,3,4\}, \{6\}] + K[\{1,2,3,4\}, \{5\}] + [\{1,5\}, \{2,5\}] + \{3,4\}$.

$r \equiv 1 \pmod 4$

(2) Use Lemma 7 to produce (20)-color classes $M_1, M_2, \ldots, M_{20}$ in $K'$ such that $\bigcup_{i=1}^{20} M_i$ induces $4(r-1) * K_6 + 5 * K[\{1,2,3,4\}, \{5\}] + 15 * K[\{1,2,3,4\}, \{6\}] + 20 * E_{56}$. Take $\frac{r-1}{4}$ copies of each of $M_1, \ldots, M_{20}$ to form $5(r-1)$ new color classes. So, in steps (1-2), $5r + 1$ color classes have been defined, and their union $G$ is $\bigcup_{i=1}^{5r+1} C_i = (r^2 - r) * K_6 + \frac{15r-3}{4} * K[\{1,2,3,4\}, \{6\}] + (5r-1) * E_{56} + \frac{5r-1}{4} * K[\{1,2,3,4\}, \{5\}] + [\{1,5\}, \{2,5\}] + \{3,4\}$ such that $\phi = \phi(G, K')$ satisfies:
$\phi(\{1,2\}) = r = \phi(\{2,3\}) = \phi(\{1,4\}) = \phi(\{1,3\}) = \phi(\{2,4\})$, and $\phi(\{3,4\}) = r - 1$; for $1 \leq i \leq 2$, $\phi(\{i,5\}) = \frac{3r-3}{4}$ and for $3 \leq i \leq 4$, $\phi(\{i,5\}) = \frac{3r+1}{4}$; and for $1 \leq i \leq 4$, $\phi(\{i,6\}) = \frac{r+3}{4}$; $\phi(\{5,6\}) = 4$.

(3) Let $M_0', M_1'$ and $M_2'$ be a decomposition of $K_4$ into 1-factors such that $M_0' = [\{3,4\}, \{1,2\}]$. We use the 4 colors to color the remaining edges as follows:

$C_{5r+2} = \frac{r+3}{4} * K[\{1,2,3,4\}, \{6\}] + \frac{r-1}{4} * K[\{1,2,3,4\}, \{5\}] + \bigcup_{i=1}^{\frac{r-3}{2}} M_{i \bmod 3}' + \{1,2\} + [\{3,5\}, \{4,5\}].$

$C_{5r+3} = \frac{r-1}{4} * K[\{1,2,3,4\}, \{5\}] + \bigcup_{i=\frac{r-3}{2}+1}^{\frac{5r-5}{4}} M_{i \bmod 3}' + 2 * E_{56}.$

$C_{5r+4} = \frac{r-1}{4} * K[\{1,2,3,4\}, \{5\}] + \bigcup_{i=\frac{5r-5}{4}+1}^{2r-1} M_{i \bmod 3}' + 2 * E_{56}.$

$C_{5r+5} = \bigcup_{i=2r}^{3r-1} M_{i \bmod 3}'.$

## $r \equiv 3 \pmod 4$

(2) Use Lemma 7 to produce (20)-color classes $M_1, M_2, \ldots, M_{20}$ in $K'$ such that $\bigcup_{i=1}^{20} M_i$ induces $4(r-1)*K_6 + 5*K[\{1,2,3,4\}, \{5\}] + 15*K[\{1,2,3,4\}, \{6\}] + 20 * E_{56}$. Take $\frac{r-3}{4}$ copies of each of $M_1, \ldots, M_{20}$ to form $5(r-3)$ new color classes. We color some more edges using 10 colors as follows:

$C_{5r-8} = (r-1) * R_1 + E_{56} + [\{1,5\}, \{2,6\}, \{3,6\}, \{4,6\}].$
$C_{5r-7} = (r-1) * R_2 + E_{56} + [\{2,5\}, \{1,6\}, \{3,6\}, \{4,6\}].$
$C_{5r-6} = (r-1) * R_3 + E_{56} + [\{3,5\}, \{2,6\}, \{1,6\}, \{4,6\}].$
$C_{5r-5} = (r-1) * R_4 + E_{56} + [\{4,5\}, \{2,6\}, \{3,6\}, \{1,6\}].$
$C_{5r-4} = (r-1) * R_5 + E_{56} + [\{1,5\}, \{2,6\}, \{3,6\}, \{4,6\}].$
$C_{5r-3} = (r-1) * R_1 + E_{56} + [\{2,5\}, \{1,6\}, \{3,6\}, \{4,6\}].$
$C_{5r-2} = (r-1) * R_2 + E_{56} + [\{3,5\}, \{2,6\}, \{1,6\}, \{4,6\}].$
$C_{5r-1} = (r-1) * R_3 + E_{56} + [\{4,5\}, \{2,6\}, \{3,6\}, \{1,6\}].$
$C_{5r} = (r-1) * R_4 + E_{56} + [\{3,5\}, \{2,6\}, \{1,6\}, \{4,6\}].$
$C_{5r+1} = (r-1) * R_5 + E_{56} + [\{4,5\}, \{2,6\}, \{3,6\}, \{1,6\}].$

where $\{R_1, \ldots, R_5\}$ is a 1-factor decomposition of $K_6$. So, in steps (1-2), $5r+1$ color classes have been defined, and their union $G$ is $\bigcup_{i=1}^{5r+1} C_i = (r^2 - r)*K_6 + \frac{15r-5}{4}*K[\{1,2,3,4\}, \{6\}] + (5r-1)*E_{56} + \frac{5r+1}{4}*K[\{1,2,3,4\}, \{5\}] + \{3,4\} + \{1,6\} + \{2,6\}$ such that $\phi = \phi(G, K')$ satisfies: $\phi(\{1,2\}) = r = \phi(\{2,3\}) = \phi(\{1,4\}) = \phi(\{1,3\}) = \phi(\{2,4\})$, and $\phi(\{3,4\}) = r - 1$; for $1 \leq i \leq 2$, $\phi(\{i,6\}) = \frac{r+1}{4}$, and for $3 \leq i \leq 4$, $\phi(\{i,6\}) = \frac{r+5}{4}$; for $1 \leq i \leq 4$, $\phi(\{i,5\}) = \frac{3r-1}{4}$; $\phi(\{5,6\}) = 4$.

(3) Let $M_0', M_1'$ and $M_2'$ be a decomposition of $K_4$ into 1-factors such that $M_0' = [\{1,2\}, \{3,4\}]$. We use the 4 colors to color the remaining edges as follows:

$C_{5r+2} = \frac{r+1}{4} * K[\{1,2,3,4\}, \{6\}] + \frac{r+1}{4} * K[\{1,2,3,4\}, \{5\}] + \bigcup_{i=1}^{\frac{r-3}{2}} M_{i \bmod 3}' + \{1,2\} + [\{3,6\}, \{4,6\}].$

$C_{5r+3} = \frac{r+1}{4} * K[\{1,2,3,4\}, \{5\}] + \bigcup_{i=\frac{r-3}{2}+1}^{\frac{5r-7}{4}} M_{i \bmod 3}'.$

$C_{5r+4} = \frac{r-3}{4} * K[\{1,2,3,4\}, \{5\}] + \bigcup_{i=\frac{5r-7}{4}+1}^{2r-1} M_{i \bmod 3}' + 4 * E_{56}.$

$C_{5r+5} = \bigcup_{i=2r}^{3r-1} M_{i \bmod 3}'.$

Using Theorem 9, we disentangle $K'$ to get $K$ with all its edges properly colored using $\Delta(K) + 1$ colors $c_1, \ldots, c_{\Delta+1}$. We now require to color the vertices of $K$ using only the colors $c_1, \ldots, c_{\Delta+1}$ to get a total coloring with $\Delta(K) + 1$ colors. It may be observed that for $1 \leq i \leq 6$ the color $c_i$ is absent from the vertex $\{i\}$ in $K'$. Therefore, for $1 \leq i \leq 6$, the color $c_i$ is absent from all the vertices in $\psi^{-1}(\{i\}) = V_i \in K$, and we color them with $c_i$.

<u>Sub − Case 2, $r$ is even</u>: Proof omitted due to space constraints.

**Case 3 :** $K(r, r, r, r, r, r + 4)$.

<u>Sub − Case 1, $r$ is odd</u>
This subcase requires special attention as any total coloring of $K$ in which all the vertices of $V_6$ are colored the same would leave at least one vertex $v \in V_1 \cup \ldots \cup V_5$ unsaturated. However, all the vertices in $V_1 \cup \ldots \cup V_5$ are of the maximum degree, and thus such a total coloring would require at least $\Delta(K) + 2$ colors. Thus, the approach by Dalal and Rodger [8] would not yield a $\Delta(K) + 1$ total coloring. We solve this subcase using our generalized approach as follows:
    Let $K'$ be an amalgamation of $K$ (where $V(K') = \{1, 2, \ldots, 7\}$) with amalgamation function $\psi : V(K) \to V(K')$ as defined below:
$\psi(V_i) = \{i\}$ for $1 \leq i \leq 5$, $\psi(V_6 \setminus \{a\}) = \{6\}$, and $\psi(\{a\}) = 7$, where $a$ is an arbitrary vertex in $V_6$. The associated *number function* $\eta : V(K') \to \mathbb{N}$ thus is defined as: $\eta(\{i\}) = r$ for $1 \leq i \leq 5$, $\eta(\{6\}) = r + 3$ and $\eta(\{7\}) = 1$. Let $V(K_6) = \{1, \ldots, 6\}$ and let $C_i$ be the set of edges colored $c_i$ for $1 \leq i \leq \Delta(K) + 1 = 5r + 5$, defined as follows: (note that each color class is majorized by $(\eta(1), \ldots, \eta(6), \eta(7)) = (r, r, r, r, r, r + 3, 1)$).

(1) Let $\{W_1, \ldots, W_6\}$ be a decomposition of $2 * K_6$ into 6 cycles of length 5 with vertex $i$ missing in $W_i$ for $1 \leq i \leq 6$.

$C_1 = \frac{r-1}{2} * W_1 + K[\{2, 3, 4, 5\}, \{6\}]$
$C_2 = \frac{r-1}{2} * W_2 + K[\{1, 3, 4, 5\}, \{6\}] \setminus \{5, 6\} + \{5, 7\}$
$C_3 = \frac{r-1}{2} * W_3 + K[\{1, 2, 4, 5\}, \{6\}] \setminus \{4, 6\} + \{4, 7\}$
$C_4 = \frac{r-1}{2} * W_4 + K[\{1, 2, 3, 5\}, \{6\}] \setminus \{3, 6\} + \{3, 7\}$
$C_5 = \frac{r-1}{2} * W_5 + K[\{1, 2, 3, 4\}, \{6\}] \setminus \{2, 6\} + \{2, 7\}$.
$C_6 = \frac{r-1}{2} * W_6 + \{1, 7\} + [\{2, 4\}, \{3, 5\}]$.
$\bigcup_{i=1}^{6} C_i = (r - 1) * K_6 + 3 * K[\{1, 2, 3, 4, 5\}, \{6\}] + K[\{1, 2, 3, 4, 5\}, \{7\}]$
$+ [\{2, 4\}, \{3, 5\}] + \{1, 6\}$

(2) Use Lemma 8 to produce (5)-color classes $M_1, M_2, \ldots, M_5$ in $K'$ such that $\cup_{i=1}^{5} M_i$ induces $(r-1) * K_6 + 4 * K[\{1, 2, 3, 4, 5\}, \{6\}] + K[\{1, 2, 3, 4, 5\}, \{7\}]$. Take $(r − 1)$ copies of each of $M_1, \ldots, M_5$ to form $(5r − 5)$ new color classes. So, in steps (1-2), $5r + 1$ color classes have been defined, and their union $G$ is $\bigcup_{i=1}^{5r+1} C_i = (r^2 − r) * K_6 + (4r − 1) * K[\{1, 2, 3, 4, 5\}, \{6\}] + r * K[\{1, 2, 3, 4, 5\}, \{7\}] + \{1, 6\} + [\{2, 4\}, \{3, 5\}]$ such that $\phi = \phi(G, K')$ satisfies: For $1 \leq i, j \leq 5$, $i \neq j$, $\phi(\{i, j\}) = r$ except $\phi(\{2, 4\}) = r − 1 = \phi(\{3, 5\})$, For $2 \leq i \leq 5$, $\phi(\{i, 6\}) = 1$ and $\phi(\{1, 6\}) = 0$; and for $1 \leq i \leq 5$, $\phi(\{i, 7\}) = 0$.

(3) Let $C_1$ and $C_2$ be a decomposition of $K_5$ into cycles of length 5 where $C_1 = [\{2,4\}, \{3,5\}, \{1,4\}, \{2,3\}, \{1,5\}]$ and $C_2 = [\{3,4\}, \{1,3\}, \{2,5\}, \{4,5\}, \{1,2\}]$. We define the 4 color class as follows:

$$C_{5r+2} = \{5,6\} + \{1,4\} + \{2,3\} + \tfrac{r-1}{2} * C_1$$
$$C_{5r+3} = \{2,6\} + \{1,5\} + \{3,4\} + \tfrac{r-1}{2} * C_2$$
$$C_{5r+4} = \{4,6\} + \{1,3\} + \{2,5\} + \tfrac{r-1}{2} * C_1$$
$$C_{5r+5} = \{3,6\} + \{4,5\} + \{1,2\} + \tfrac{r-1}{2} * C_2$$

Using Theorem 9, we disentangle $K'$ to get $K$ with all its edges properly colored using $\Delta(K) + 1$ colors $c_1, \ldots, c_{\Delta+1}$. It may be observed that for $1 \leq i \leq 5$ the color $c_i$ is absent from the vertex $\{i\}$ in $K'$. Therefore, for $1 \leq i \leq 5$, the color $c_i$ is absent from all the vertices in $\psi^{-1}(\{i\}) = V_i \in K$, and we color them with $c_i$. In $K'$, vertex $\{6\}$ is unsaturated by color class $C_6$ and thus color $c_6$ is absent from all the vertices in $\psi^{-1}(\{6\}) = V_6 \setminus \{a\}$. We color all the vertices in $V_6 \setminus \{a\}$ with $c_6$. Also, in $K'$ the vertex $\{7\}$ is unsaturated by the color class $C_{5r+2}$ and thus color $c_{5r+2}$ is absent from all the vertices in $\psi^{-1}(\{7\}) = \{a\}$. We color vertex $a \in V_6$ with $c_{5r+2}$.

$Sub - Case\ 2,\ r$ is even: Let $K'$ be an amalgamation of $K$ with *amalgamation function* $\psi : V(K) \to V(K')$ such that $\psi(V_i) = \{i\}$ and $\eta(i) = r_i$ for $1 \leq i \leq 6$. Let $V(K_6) = \{1, \ldots, 6\}$ and let $C_i$ be the set of edges colored $c_i$ for $1 \leq i \leq \Delta(K) + 1 = 5r + 5$, defined as follows:

(1) Let $\{W_1, \ldots, W_6\}$ be a decomposition of $2 * K_6$ into 6 cycles of length 5 with vertex $i$ missing in $W_i$ for $1 \leq i \leq 6$.

$$C_1 = \tfrac{r-2}{2} * W_1 + K[\{1,2,3,4,5\}, \{6\}] \setminus \{1,6\} + [\{2,6\}, \{3,6\}] + \{4,5\}$$
$$C_2 = \tfrac{r-2}{2} * W_2 + K[\{1,2,3,4,5\}, \{6\}] \setminus \{2,6\} + [\{3,6\}, \{4,6\}] + \{1,5\}$$
$$C_3 = \tfrac{r-2}{2} * W_3 + K[\{1,2,3,4,5\}, \{6\}] \setminus \{3,6\} + [\{4,6\}, \{5,6\}] + \{2,1\}$$
$$C_4 = \tfrac{r-2}{2} * W_4 + K[\{1,2,3,4,5\}, \{6\}] \setminus \{4,6\} + [\{5,6\}, \{1,6\}] + \{3,2\}$$
$$C_5 = \tfrac{r-2}{2} * W_5 + K[\{1,2,3,4,5\}, \{6\}] \setminus \{5,6\} + [\{1,6\}, \{2,6\}] + \{3,4\}.$$
$$C_6 = \tfrac{r-2}{2} * W_5 + [\{1,4\}, \{4,2\}, \{2,5\}, \{5,3\}, \{3,1\}].$$
$$\textstyle\bigcup_{i=1}^{6} C_i = (r-2) * K_6 + 6 * K[\{1,2,3,4,5\}, \{6\}] + K_5$$

where the vertex set of $K_5$ is $\mathbb{N}_5$.

(2) Use Lemma 10 to produce (5)-color classes $M_1, M_2, \ldots, M_5$ in $K'$ such that $\cup_{i=1}^{5} M_i$ induces $(r-1) * K_6 + 5 * K[\{1,2,3,4,5\}, \{6\}]$. Take $(r-1)$ copies of each of $M_1, \ldots, M_5$ to form $(5r-5)$ new color classes. So, in steps (1-2), $5r + 1$ color classes have been defined, and their union $G$ is $\bigcup_{i=1}^{5r+1} C_i = (r^2 - r - 1) * K_6 + (5r+1) * K[\{1,2,3,4,5\}, \{6\}] + K_5$ such that $\phi = \phi(G, K')$ satisfies $\phi(\{i,j\}) = r$ for $1 \leq i \neq j \leq 5$, and $\phi(\{i,6\}) = 0$.

(3) Let $C_1$ and $C_2$ be a decomposition of $K_5$ into cycles of length 5. We use the 4 colors to color the remaining edges as follows:

$$C_{5r+2} = C_{5r+4} = \tfrac{r}{2} * C_1; C_{5r+3} = C_{5r+5} = \tfrac{r}{2} * C_2.$$

Using Theorem 9, we disentangle $K'$ to get $K$ with all its edges properly colored using $\Delta(K) + 1$ colors $c_1, \ldots, c_{\Delta+1}$. We now require to color the vertices of $K$ using only the colors $c_1, \ldots, c_{\Delta+1}$ to get a total coloring with $\Delta(K) + 1$ colors. It may be observed that for $1 \leq i \leq 6$ the color $c_i$ is absent from the vertex $\{i\}$ in $K'$. Therefore, for $1 \leq i \leq 6$, the color $c_i$ is absent from all the vertices in $\psi^{-1}(\{i\}) = V_i \in K$, and we color them with $c_i$.

**Case 4 :** $K(r, r, r, r, r, r + 2)$.

The approach by Dalal and Rodger [8] would not yield a $\Delta(K) + 1$ total coloring. We solve this subcase using our generalized approach as follows:

Let $K'$ be an amalgamation of $K$ (where $V(K') = \{1, 2, \ldots, 7\}$) with amalgamation function $\psi : V(K) \rightarrow V(K')$ defined as follows: $\psi(V_i) = \{i\}$ for $1 \leq i \leq 5$, $\psi(V_6 \setminus \{a\}) = \{6\}$, and $\psi(\{a\}) = 7$, where $a$ is an arbitrary vertex in $V_6$. The associated *number function* $\eta : V(K') \rightarrow \mathbb{N}$ thus is defined as: $\eta(\{i\}) = r$ for $1 \leq i \leq 5$, $\eta(\{6\}) = r + 1$ and $\eta(\{7\}) = 1$. Let $V(K_6) = \{1, \ldots, 6\}$ and let $C_i$ be the set of edges colored $c_i$ for $1 \leq i \leq \Delta(K) + 1 = 5r + 5$, defined as follows: (note that each color class is majorized by $(\eta(1), \ldots, \eta(6), \eta(7)) = (r, r, r, r, r, r + 1, 1)$)

Sub $-$ Case 1, $r$ is odd

(1) Let $\{W_1, \ldots, W_6\}$ be a decomposition of $2 * K_6$ into 6 cycles of length 5 with vertex $i$ missing in $W_i$ for $1 \leq i \leq 6$. Let $\{M_1, \ldots, M_5\}$ be a decomposition of $K_5$ into 5 near 1-factors with vertex $i$ missing in $M_i$ for $1 \leq i \leq 5$. Let $C_1 = [\{1, 2\}, \{2, 3\}, \{3, 4\}, \{4, 5\}, \{5, 1\}]$, and $C_2 = [\{1, 3\}, \{3, 5\}, \{5, 2\}, \{2, 4\}, \{4, 1\}]$ be cyclic decomposition of $K_5$ into two 5-cycles.

$C_1 = \frac{r-1}{2} * W_1 + [\{2, 6\}, \{3, 6\}] + \{4, 5\}$.
$C_2 = \frac{r-1}{2} * W_2 + [\{3, 7\}, \{4, 6\}] + \{5, 1\}$
$C_3 = \frac{r-1}{2} * W_3 + [\{4, 7\}, \{5, 6\}] + \{1, 2\}$
$C_4 = \frac{r-1}{2} * W_4 + [\{5, 7\}, \{1, 6\}] + \{2, 3\}$
$C_5 = \frac{r-1}{2} * W_5 + [\{2, 7\}, \{1, 6\}] + \{3, 4\}$
$C_6 = \frac{r-1}{2} * W_6 + [\{1, 7\}] + [\{2, 4\}, \{3, 5\}]$.
$\bigcup_{i=1}^{6} C_i = (r - 1) * K_6 + K[\{1, 2, 3, 4, 5\}, \{6\}] + K[\{1, 2, 3, 4, 5\}, \{7\}] + C_1 + [\{1, 6\}] + [\{2, 4\}, \{3, 5\}]$

(2) Use Lemma 9 to produce (10)-color classes $M_1, M_2, \ldots, M_{10}$ in $K'$ such that $\cup_{i=1}^{10} M_i$ induces $2(r - 1) * K_6 + 4 * K[\{1, 2, 3, 4, 5\}, \{6\}] + 2 * K[\{1, 2, 3, 4, 5\}, \{7\}] + K_5$. Take $(\frac{r-1}{2})$ copies of each of $M_1, \ldots, M_{10}$ to form $(5r - 5)$ new color classes. So, in steps (1-2), $5r + 1$ color classes have been defined, and their union $G$ is $\bigcup_{i=1}^{5r+1} C_i = (r^2 - r) * K_6 + (2r - 1) * K[\{1, 2, 3, 4, 5\}, \{6\}] + r * K[\{1, 2, 3, 4, 5\}, \{7\}] + C_1 + [\{1, 6\}] + [\{2, 4\}, \{3, 5\}] + (\frac{r-1}{2}) * K_5$ such that $\phi = \phi(G, K')$ satisfies: for $1 \leq i, j \leq 5$, $i \neq j$, $\phi(\{i, j\}) = \frac{r-1}{2}$ except $\phi(\{1, 3\}) = \frac{r+1}{2} = \phi(\{1, 4\}) = \phi(\{2, 5\})$; for $2 \leq i \leq 5$, $\phi(\{i, 6\}) = 1$, $\phi(\{1, 6\}) = 0$; and for $1 \leq i \leq 5$, $\phi(\{1, 7\}) = 0$.

(3) Let $C_1$ and $C_2$ be a decomposition of $K_5$ into cycles of length 5. We use the 2 colors to color the remaining edges as follows

$$C_{5r+2} = [\{2,6\},\{3,6\},\{5,6\}] + \{1,4\} + \tfrac{r-1}{2} * C_1$$
$$C_{5r+3} = [\{4,6\}] + \{2,5\} + \{1,3\} + \tfrac{r-1}{2} * C_2$$

Using Theorem 9, we disentangle $K'$ to get $K$ with all its edges properly colored using $\Delta(K)+1$ colors $c_1, \ldots, c_{\Delta+1}$. It may be observed that for $1 \le i \le 5$ the color $c_i$ is absent from the vertex $\{i\}$ in $K'$. Therefore, for $1 \le i \le 5$, the color $c_i$ is absent from all the vertices in $\psi^{-1}(\{i\}) = V_i \in K$, and we color them with $c_i$. In $K'$, vertex $\{6\}$ is unsaturated by color class $C_6$ and thus color $c_6$ is absent from all the vertices in $\psi^{-1}(\{6\}) = V_6 \setminus \{a\}$. We color all the vertices in $V_6 \setminus \{a\}$ with $c_6$. Also, in $K'$ the vertex $\{7\}$ is unsaturated by the color class $C_{5r+2}$ and thus color $c_{5r+2}$ is absent from all the vertices in $\psi^{-1}(\{7\}) = \{a\}$. We color vertex $a \in V_6$ with $c_{5r+2}$.

$\underline{\text{Sub} - \text{Case 2, } r \text{ is even.}}$: We omit the proof due to space constraints.

**Case 5 :** $K(r,r,r,r,r+1,r+1)$
We omit the proof due to space constraints.

**Acknowledgement.** The authors thank Prof. C. A. Rodger for introducing them to this problem.

# References

1. Bahmanian, M.A., Rodger, C.A.: Multiply balanced edge colorings of multigraphs. J. Graph Theory **70**(3), 297–317 (2012)
2. Behzad, M.: Graphs and their chromatic numbers, Doctoral Thesis, Michigan State University (1965)
3. Bermond, J.-C.: Nombre chromatique total du graphe $r$-parti complete (French). J. London Math. Soc. **9**, 279–285 (1974)
4. Bryant, D., Horsley, D., Maenhaut, B., Smith, B.R.: Cycle decompositions of complete multigraphs. J. Combin. Designs **19**, 42–69 (2011)
5. Chen, B.-L., Hung-Lin, F.: Total colorings of graphs of order 2n having maximum degree 2n–2. Graphs Comb. **8**, 119–123 (1992)
6. Chetwynd, A.G., Hilton, A.J.W.: Some refinements of the total chromatic number conjecture. Congressue Numerantium **66**, 195–215 (1988)
7. Chew, K.H., Yap, H.P.: Total chromatic number of complete $r$-partite graphs. J. Graph Theory **16**, 629–634 (1992)
8. Dalal, A., Rodger, C.A.: The total chromatic number of complete multipartite graphs with low deficiency. Graphs Comb. **31**(6), 2159–2173 (2015)
9. Dalal, A., Panda, B.S., Rodger, C.A.: Total-colorings of complete multipartite graphs using amalgamations. Discret. Math. **339**, 1587–1592 (2016)
10. Dalal, A., Panda, B.S., Rodger, C.A.: Total-colorings of complete multipartite graphs using amalgamations II, preprint
11. Dong, L., Yap, H.P.: The total chromatic number of unbalanced complete $r$-partite graphs of even order. Bull. Inst. Combin. Appl. **28**, 107–117 (2000)

12. Hilton, A.J.W.: A total-chromatic number analogue of Planthold's theorem. Discret. Math. **79**, 169–175 (1989/90)
13. Hoffman, D.G., Rodger, C.A.: The chromatic index of complete multipartite graphs. J. Graph Theory **16**, 159–163 (1992)
14. D. G. Hoffman and C. A. Rodger, The total chromatic number of complete multipartite graphs. Festschrift for C. St. J. A. Nash-Williams. Congr. Numer. **113**, 205–220 (1996)
15. Vizing, V.G.: Some unsolved problems in graph theory, (Russian) Uspehi Mat. Nauk **23**, 117–134 (1968)
16. Yap, H.P.: Total colourings of graphs. Bull. London Math. Soc. **21**, 159–163 (1989)

# The Weak-Toll Function of a Graph: Axiomatic Characterizations and First-Order Non-definability

Lekshmi Kamal K. Sheela[ID], Manoj Changat[(✉)][ID], and Jeny Jacob[ID]

Department of Futures Studies, University of Kerala, Trivandrum 695581, India
lekshmisanthoshgr@gmail.com, mchangat@keralauniversity.ac.in,
jenyjacobktr@gmail.com

**Abstract.** Toll walks on connected graphs are introduced to characterize dominating pairs of vertices in interval graphs. A weak-toll walk is an immediate generalization of a toll walk in a graph. The set of all vertices lying on weak-toll walks between two given vertices gives rise to the notion of the weak-toll function, denoted $W_T$, of a connected graph. In this paper, we characterize the weak-toll function of trees, chordal graphs, and unit interval graphs. This, in turn, provides an additional characterization of trees and unit interval graphs using a set of first-order axioms defined on an arbitrary function, known as the transit function, which is defined for every pair of elements in a non-empty finite set. Furthermore, we prove that an axiomatic characterization of the function $W_T$ of an arbitrary connected graph is impossible using a set of first-order axioms.

**Keywords:** transit function · weak-toll walk · chordal graph · unit interval graph · first-order non-definability

## 1 Introduction

To characterize dominating pairs of vertices in interval graphs, Alcon [1] introduced toll walks. A toll walk $W$ from a vertex $u$ to a different vertex $v$ of a graph $G$ is a special walk that contains exactly one neighbor of $u$, the second vertex of $W$, and exactly one neighbor of $v$, the for-last vertex of $W$. The toll interval $T(u, v)$ consists of the set of all vertices that belong to any toll walk between $u$ and $v$. This gives rise to the toll walk function $T : V(G) \times V(G) \to 2^{V(G)}$ of a graph $G$. In [11] and [13], the toll walk function $T$ of a connected graph is studied from an axiomatic point of view. More accurately, several well-known first-order betweenness axioms and new axioms are framed for the toll walk function and showed that the toll walk function of special classes of graphs, like interval graphs, chordal graphs, asteroidal triple-free graphs, are characterized by identifying the forbidden induced subgraphs of these graphs.

© The Author(s), under exclusive license to Springer Nature Switzerland AG 2024
S. Kalyanasundaram and A. Maheshwari (Eds.): CALDAM 2024, LNCS 14508, pp. 286–301, 2024.
https://doi.org/10.1007/978-3-031-52213-0_20

Weak-toll walks are the generalization of toll walks in a graph which was introduced in [4]. The original motivation for the concept of the weak-toll walk as a relaxation of the toll walk was to use it as a tool to characterize unit interval graphs as the so-called convex geometries of the convexity associated with the weak-toll walks. In this paper, we provide another characterization of unit interval graphs using weak-toll walks.

In this paper, we study the weak-toll function, a generalization of the toll walk function, and characterize the weak-toll function of unit interval graphs and trees using a set of first-order axioms framed on an arbitrary function, known as the transit function. Interestingly, we prove that there is no first-order axiomatic characterization for the weak-toll function of an arbitrary finite connected graph. In Sect. 2, we define the notion of a transit function and fix all the necessary terminologies, in Sect. 3, we frame the axioms and characterize the weak-toll function of trees, chordal graphs, and unit interval graphs, and in Sect. 4, we prove that the weak-toll function is not first-order axiomatizable using the standard tool of EF games.

## 2 Preliminaries

Let $G$ be a graph with the vertex set $V(G)$ and the edge set $E(G)$. We consider only simple connected finite graphs, that is, graphs without multiple edges and loops. The *open neighborhood* $N(v)$ of $v \in V(G)$ is the set $\{u \in V(G) : uv \in E(G)\}$ and the *closed neighborhood* $N[v]$ is $N(v) \cup \{v\}$. A *walk* $W_k$ in a graph $G$ is a sequence of $k$ vertices $w_1, \ldots, w_k$ where $w_i w_{i+1} \in E(G)$ for every $i \in \{1, \ldots, k-1\}$. We simply write $W_k = w_1 \cdots w_k$. Notice that some vertices of $W_k$ can repeat in $W_k$. If all the vertices of a walk differ, then we say that $W_k$ is a *path* $P_k$ of $G$. A path $P_k = v_1 \cdots v_k$ will also be denoted as the $v_1, v_k$-path, and we say that $P_k$ starts in $v_1$ and ends in $v_k$ and $u \xrightarrow{P} x$ denotes the subpath of a path $P$ with end vertices $u$ and $x$. The distance $d(u, v)$ between $u, v \in V(G)$ is the minimum number of edges on a $u, v$-path or infinite if such a path does not exist. Any $u, v$-path of length $d(u, v)$ is called a $u, v$-*geodesic*.

Let $G$ and $G_1, G_2, \ldots, G_k$ be connected graphs. We say that the graph $G$ is $(G_1 G_2 \cdots G_k)$-free graph if $G$ has no induced subgraphs isomorphic to $G_1, G_2, \ldots, G_k$. A connected acyclic graph is a tree. A graph class that can be defined as some induced cycle-free graphs is *chordal graphs* which are $C_k$-free graphs for every $k \geq 4$. A set of three vertices in a graph $G$ such that each pair is joined by a path that avoids the neighborhood of the third vertex is known as an *asteroidal triple* in $G$. Graph $G$ is called the *AT-free graph* if $G$ does not have an asteroidal triple. *Interval graphs* are the intersection graphs of intervals on a line. Lekkerkerker and Boland in [6] proved that a graph $G$ is an interval graph if and only if $G$ is a chordal AT-free graph. A graph $G$ is a *unit interval* graph if it can be represented by intervals on a line in which all intervals have the same length. In [12] Roberts proved that the unit interval graphs are claw-free interval graphs so that unit interval graphs are (claw, net, $S_3$, $C_{n+4}$)-free graphs (refer Fig. 1 for these graphs).

## 2.1    Transit Functions

A *transit function* on a non-empty set $V$ is a function $R : V \times V \longrightarrow 2^V$ such that for every $u, v \in V$ the following three conditions hold:

(t1)  $u \in R(u, v)$;
(t2)  $R(u, v) = R(v, u)$;
(t3)  $R(u, u) = \{u\}$.

The *underlying graph* $G_R$ of a transit function $R$ is a graph with vertex set $V$, where distinct vertices $u$ and $v$ are adjacent if and only if $R(u, v) = \{u, v\}$. The argument $x \in R(u, v)$ can be interpreted as $x$ is between $u$ and $v$. Thus, the axioms on $R$ are sometimes called "betweenness axioms". The well-studied transit functions in graphs are the interval function $I_G$ and the induced path function $J_G$. The *interval function* $I_G$ ( *induced path function* $J_G$) of a graph $G$ is a function that returns, for each pair of vertices $u, v \in G$, the set $I_G(u, v)$ ( $J_G(u, v)$) of all vertices lying on all the $u, v$ -shortest path (induced path) in $G$. It is clear that $I(u, v) \subseteq J(u, v), \forall u, v \in G$.

The axiomatic approach to the function $I_G$ of a graph $G$ has garnered attention through a series of characterizations of $I_G$, employing a set of first-order axioms framed on an arbitrary transit function by Nebesk'y (for e.g., [9]), which culminated with the elegant characterization due to Mulder and Nebesk'y [8]. Subsequently, this work has inspired numerous authors to explore the axiomatic approach with regard to other transit functions on graphs; see, for example, the case of the induced path function $J_G$ in [3] and the toll walk function in [11,13].

A *toll walk* between two different vertices $w_1$ and $w_k$ of a finite connected graph $G$ is a sequence of vertices $w_1, \ldots, w_k$ that satisfy the following conditions:

- $w_i w_{i+1} \in E(G)$ for every $i \in \{1, \ldots, k-1\}$,
- $w_1 w_i \in E(G)$ if and only if $i = 2$, $w_k w_i \in E(G)$ if and only if $i = k-1$.

That is, a toll walk $W$ from $u$ to $v$ is a walk in which $u$ is adjacent only to the second vertex of $W$, and $v$ is adjacent only to the for-last vertex of $W$. Note that if $uv \in E(G)$, then the only toll walk between $u$ and $v$ is $uv$. Additionally, we define a toll walk that starts and ends at the same vertex $w$ as $w$ itself.

The function $T : V \times V \to 2^V$ defined as

$$T_G(u, v) = \{x \in V(G) : x \text{ lies on a toll walk between } u \text{ and } v\}$$

is the *toll walk function* on $G$.

Mitre C. Dourado [4] introduced the concept of weak-toll walk. A *weak-toll walk* between $u$ and $v$ in $G$ is a sequence of vertices of the form $W : u = w_0, w_1, \ldots, w_{k-1}, w_k = v$, where the following conditions are satisfied:

- $w_i w_{i+1} \in E(G)$ for every $i \in \{1, \ldots, k-1\}$,
- $w_0 w_i \in E(G)$ implies $w_i = w_1$, $w_k w_i \in E(G)$ implies $w_i = w_{k-1}$

That is, a weak-toll walk is any walk $W : u, w_1, \ldots, w_{k-1}, v$ between $u$ and $v$ such that $u$ is adjacent only to the vertex $w_1$, which can appear more than once in the walk, and $v$ is adjacent only to the vertex $w_{k-1}$, which can appear more than once in the walk. Note that if $uv \in E(G)$ then $W : u, v$ is the only weak-toll walk between $u$ and $v$, and W: u is the only weak-toll walk that begins and ends at $u$. We define

$$W_{T_G}(u, v) = \{x \in V(G) : x \text{ lies on a weak-toll walk between } u \text{ and } v\}.$$

to be the weak-toll interval between $u$ and $v$ in $G$. We observe that the weak-toll function $W_T$ is also a well-defined transit function since $W_T$ fulfills all three transit axioms, and from the definition of the weak-toll function $W_T$ on $G$, it is clear that $G$ and $G_{W_T}$ are isomorphic. Furthermore, we have $I(u, v) \subseteq J(u, v) \subseteq T(u, v) \subseteq W_T(u, v)$.

## 3    Weak-Toll Function of Trees, Chordal and Unit Interval Graphs

In this section, we introduce some axioms for the weak-toll function. Using these axioms together with already known axioms, we characterize the weak-toll function of trees and unit interval graphs. The following lemma from Liliana Alcon et al. [2] gives a characterization of the vertices that belong to a toll walk on a graph.

**Lemma 1** [2]. *A vertex $v$ is in some toll walk between two different non-adjacent vertices $x$ and $y$ if and only if $N[x] - \{v\}$ does not separate $v$ from $y$ and $N[y] - \{v\}$ does not separate $v$ from $x$.*

The following analogous lemma provides a characterization of vertices within a weak-toll function.

**Lemma 2.** *A vertex $x$ is in some weak-toll walk between two different non-adjacent vertices $u$ and $v$ if and only if there is a $x, v$-path that includes at most one neighbor of $u$ and a $u, x$-path that includes at most one neighbor of $v$.*

*Proof.* Suppose $x \in W_T(u, v)$, then from the definition of the weak-toll function, there is at least one $x, v$-path that contains at most one neighbor of $u$ and at least one $u, x$-path that contains at most one neighbor of $v$. Conversely assume that there is a $x, v$-path, say $P$ that contains at most one neighbor of $u$ say $u'$ and a $u, x$-path say $Q$ that contains at most one neighbor of $v$ say $v'$. we may choose both $P$ and $Q$ as induced paths. Then $uu' \xrightarrow{P} x \xrightarrow{P} v$ is a $u, v$-weak-toll walk containing $x$.     □

Although toll drives are weak toll drives, the converse does not need to hold, and the following proposition characterizes the graphs for $T = W_T$.

**Proposition 1.** *Let $G$ be a graph. Then $T(u, v) = W_T(u, v)$ for all $u, v$ if and only if $G$ is a claw-free graph.*

*Proof.* Suppose $G$ contains a claw graph as an induced subgraph with $x$ being the central vertex and $u, v, y$ being the pendent vertices. Then $T(u, v) = \{u, x, v\}$ and $W_T(u, v) = \{u, x, y, v\}$. That is, $T(u, v) \neq W_T(u, v)$ on a claw graph. Conversely, let $T(u, v) \neq W_T(u, v)$ which means that $W_T(u, v) \not\subseteq T(u, v)$ for some $u, v \in V$. Assume $x \in W_T(u, v)$ and $x \notin T(u, v)$, which implies that either the $u, x$-path contains a neighbor of $v$ or the $x, v$-path contains a neighbor of $u$ and $T(u, x) \neq \{u, x\}$, $T(x, v) \neq \{x, v\}$. Let us assume that the $x, v$-path, denoted $P$, contains a neighbor of $u$, called $u'$. Then $uu' \xrightarrow{P} x \xrightarrow{P} u' \xrightarrow{P} v$ is a $u, v$- weak-toll walk, say $W$ containing $x$. Now, consider the vertices $x'$ and $v'$, which represent the neighbors of $u'$ in the $u', x$-subpath and the $u', v$-subpath of $P$, respectively. We claim that the set of vertices $u, u', x', v'$ induces a claw graph. It is obvious that $u$ cannot be adjacent to both $x'$ and $v'$ since $W$ is a weak-toll walk. If $x'v' \in E(G)$, then $uu' \xrightarrow{P} x \xrightarrow{P} x'v' \xrightarrow{P} v$ is a toll walk that contains $x$ implies $x \in T(u, v)$. So $x'v' \notin E(G)$ and, therefore, $u, u', x', v'$ induces a claw graph.     □

Consider the following axioms that are needed for the characterization of the weak-toll function of trees and unit interval graphs. Of these, axioms (b1), (J0) and (J2) are considered in [3], axioms (tr) and (JC) are taken from [13], while (TW4), (TW5), (TW6) and (bt1) are new axioms.

**Axiom (TW4).** If there exist elements $u, v, x, y$, such that $x \in R(u, v)$, $R(x, y) = \{x, y\}$ and $R(y, v) \neq \{y, v\}$ and $R(u, y) \neq \{u, y\}$, then $y \in R(u, v)$.

**Axiom (TW5).** If there exist elements $u, v, x$ such that $x \in R(u, v)$, $x \neq v$, $R(u, x) = \{u, x\}$ then there exist $v_1 \in R(x, v) \cap R(u, v), v_1 \neq x$ with $R(x, v_1) = \{x, v_1\}$ and $R(u, v_1) \neq \{u, v_1\}$.

**Axiom (TW6).** If there exist elements $u, v, x$ such that $x \in R(u, v)$, $x \neq v$ then there exist $v_1 \in R(x, v) \cap R(u, v), v_1 \neq x$ with $R(x, v_1) = \{x, v_1\}$.

**Axiom (b1).** If there exist elements $u, v, x \in V$ such that $x \in R(u, v), x \neq v$, then $v \notin R(x, u)$.

**Axiom (b2).** If there exist elements $u, v, x \in V$ such that $x \in R(u, v)$, then $R(u, x) \subseteq R(u, v)$.

**Axiom (bt1).** If there exist elements $u, v, x$ such that $x \in R(u, v)$, $R(u, x) = \{u, x\}$, $u \neq x$ then $u \notin R(x, v)$.

**Axiom (tr).** If there exist elements $u, v, x \in V$ such that $R(u, x) = \{u, x\}$, $R(x, v) = \{x, v\}$, $u \neq v$ then $x \in R(u, v)$.

**Axiom (J2).** If there exist elements $u, v, x \in V$ such that $R(u, x) = \{u, x\}$, $R(x, v) = \{x, v\}, u \neq v$ and $R(u, v) \neq \{u, v\}$, then $x \in R(u, v)$.

**Axiom (JC).** If there exist different elements $u, x, y, v \in V$ such that $x \in R(u, y)$ and $y \in R(x, v)$, $R(x, y) = \{x, y\}$ then $x \in R(u, v)$.

**Axiom (J0).** If there exist different elements $u, x, y, v \in V$ such that $x \in R(u, y)$ and $y \in R(x, v)$, then $x \in R(u, v)$.

It easily follows that, for a transit function $R$, the axiom (b1) implies the axiom (bt1), the axiom (J0) implies the axiom (JC) and the axiom (tr) implies the axiom (J2). Other axioms are independent, which we will demonstrate with examples later in this section. We have the following propositions.

**Proposition 2.** *The weak-toll function satisfies the axiom (TW4), (TW5) and (TW6) on every connected graph.*

*Proof.* In the case of axiom (TW4), suppose $x \in W_T(u,v)$. So let $W_1$ be $u, x$-walk containing at most one neighbor of $v$ and $W_2$ be $x, v$-walk containing at most one neighbor of $u$, so that $u \xrightarrow{W_1} x \xrightarrow{W_2} v$ is a $u, v$-weak-toll walk containing $x$. Since $W_T(x,y) = \{x,y\}$ and $W_T(u,y) \neq \{u,y\}$ and $W_T(y,v) \neq \{y,v\}$ it follows that $u \xrightarrow{W_1} xyx \xrightarrow{W_2} v$ is a $u, v$-weak-toll walk containing $y$ and hence $y \in W_T(u,v)$.

In the case of axiom (TW5), suppose $x \in W_T(u,v)$ and $W_T(u,x) = \{u,x\}$. Then there exists an $x, v$- induced path say $P$ without neighbor of $u$ with the exception of $x$. For the neighbor $v_1$ of $x$ on $P$ it follows that $v_1 \in W_T(x,v) \cap W_T(u,v), v_1 \neq x$ with $W_T(x,v_1) = \{x,v_1\}$ and $W_T(u,v_1) \neq \{u,v_1\}$.

In the case of (TW6), Suppose $x \in W_T(u,v)$. Let $W$ be the $u, v$-weak-toll walk containing $x$. Then $W$ contains a $x, v$-induced path $P$ that contains at most one neighbor of $u$ and $u, x$-induced path $Q$ that contains at most one neighbor of $v$. For the neighbor $v_1$ of $x$ in $P$, it follows that $u \xrightarrow{Q} xv_1 \xrightarrow{P} v$ is a $u, v$-weak-toll walk containing $v_1$ and $xv_1 \xrightarrow{P} v$ is a $x, v$- weak-toll walk containing $v_1$ so that $v_1 \in W_T(x,v) \cap W_T(u,v)$. Also $v_1 \neq x$ with $W_T(x,v_1) = \{x,v_1\}$. □

**Proposition 3.** *The weak-toll function satisfies the axioms (bt1) on chordal graphs.*

*Proof.* Suppose that $W_T$ does not satisfy the axiom (bt1) on the chordal graphs. That is, $x \in W_T(u,v)$, $W_T(u,x) = \{u,x\}$, $u \neq x$ and $u \in W_T(x,v)$. Since $x \in W_T(u,v)$ and $W_T(u,x) = \{u,x\}$ there exists a $x, v$-induced path say $P$ without a neighbor of $u$ and since $u \in W_T(x,v)$ and $W_T(u,x) = \{u,x\}$, there exists a $u, v$-induced path say $Q$ without a neighbor of $x$. Then $ux \xrightarrow{P} v \xrightarrow{Q} u$ forms a cycle of length at least 5. Since $G$ is chordal, there are chords from the path $P$ to $Q$. Let $u'$ be the neighbor of $u$ on the path $Q$ and $x'$ be the neighbor of $x$ on the path $P$. Clearly, $x$ is not adjacent to $u'$ and $u$ is not adjacent to $x'$, so the vertices $u, x, x', u'$ induce a cycle of length four if $u'x' \in E(G)$. Otherwise, the vertices $u, x, x', u'$ together with some other vertices on paths $P$ and $Q$ will induce a cycle of length at least five, the final contradiction. □

The following Theorem is a characterization of chordal graphs using the axiom (JC) on the weak toll function.

**Theorem 1.** *The weak-toll function $W_T$ of a graph $G$ satisfies the axiom (JC) if and only if $G$ is a chordal graph.*

*Proof.* If the graph $G$ contains a cycle of length greater than four with its consecutive vertices $y, x, u, v$. Clearly $x \in W_T(u, y)$ and $y \in W_T(x, v)$ but $x \notin W_T(u, v)$ since $uv$ is an edge in $G$. That is, if $W_T$ satisfies the axiom (JC), then $G$ is chordal. Conversely, suppose that $W_T$ does not satisfy the axiom (JC). That is, $x \in W_T(u, y)$, $y \in W_T(x, v)$ and $W_T(x, y) = \{x, y\}$, but $x \notin W_T(u, v)$. Since $x \in W_T(u, y)$ and $W_T(x, y) = \{x, y\}$, there is a $u, x$ path without a neighbor of $y$, say $P$ and since $y \in W_T(x, v)$ and $W_T(x, y) = \{x, y\}$, there is a $y, v$ path without a neighbor of $x$ say $Q$. Also, $x \notin W_T(u, v)$ implies that all $u, x$-paths contain more than one neighbor of $v$ or $x, v$-paths contain more than one neighbor of $u$. We may assume that the $x, v$-path contains more than one neighbor of $u$. Suppose $ux \in E(G)$, then the $x, v$-path contains at least one neighbor of $u$. Let $u'$ be the neighbor of $u$ on the path $Q$ and close to $y$. Then $uxy \xrightarrow{Q} u'u$ induces a cycle of length at least four. Now assume $ux \notin E(G)$, then all the $x, v$-path contain more than one neighbor of $u$. Let $u'$ be one neighbor of $u$ on the path $Q$ and close to $y$. Then $u \xrightarrow{P} xy \xrightarrow{Q} u'u$ forms a cycle of length at least five. There may be chords from the vertices of the path $P$ to the vertices of the $y, u'$-subpath of $Q$. Let $ab$ be a chord such that $a$ is close to $x$ in $P$. Then the sequence of vertices $a \xrightarrow{P} xy \xrightarrow{Q} ba$ induces a cycle of length at least four, a contradiction, and completes the proof. $\qquad\square$

The following theorem and lemma stated in [13] is needed to characterize the weak-toll function of trees in terms of an arbitrary transit function $R$. In this section, we consider an arbitrary transit function $R$ on a finite non-empty set $V$ and so in all the results where transit function $R$ is stated, by a transit function $R$ on $V$, we mean an arbitrary transit function $R$ on a finite non-empty set $V$.

**Proposition 4** [13]. *Let $R$ be any transit function on $V$. If $R$ satisfies $(JC)$ and $(J2)$, then the underlying graph $G_R$ of $R$ is $C_n$-free for $n \geq 4$.*

**Lemma 3** [13]. *Let $R$ be a transit function on $V$ satisfying the axioms (J2) and (JC). If $P_n$, $n \geq 2$, is an induced $u, v$-path in $G_R$, then $V(P_n) \subseteq R(u, v)$.*

We have axiom (tr) implies axiom (J2). If we replace axiom (J2) by axiom (tr) in Proposition 4, then $G_R$ is $C_n$-free for $n \geq 3$. For, if $G_R$ contains a triangle with vertices $u, x, v$, then $R(u, x) = \{u, x\}$ and $R(x, v) = \{x, v\}$, then $x \in R(u, v)$ according to the axiom (tr), a contradiction since $R(u, v) = \{u, v\}$. So, we have the following proposition and theorem.

**Proposition 5.** *Let $R$ be any transit function on $V$. If $R$ satisfies $(JC)$ and $(tr)$ then the underlying graph $G_R$ of $R$ is $C_n$-free for $n \geq 3$.*

**Theorem 2.** *The weak-toll function $W_T$ of a graph $G$ satisfies the axiom $(JC)$ and $(tr)$ if and only if $G$ is a tree.*

**Theorem 3.** *If $R$ is a transit function on $V$ that satisfies the axioms $(bt1)$, $(JC)$, $(tr)$, $(TW4)$ and $(TW6)$ then $R = W_T$ on $G_R$ and hence $G_R$ is connected.*

*Proof.* Since $R$ satisfies the axioms (JC) and (tr), by Proposition 5, we conclude that $G_R$ is a tree. Let $u \neq v$ be two vertices of $G_R$. Assume that $x \in R(u,v)$ and $x \neq v$. To show that $x \in W_T(u,v)$ on $G_R$. Clearly $x \in W_T(u,v)$ whenever $x = u$. So, let $x \notin \{u,v\}$. If $R(u,x) = \{u,x\}$ and $R(x,v) = \{x,v\}$, then $uxv$ is a weak-toll walk in $G_R$ and $x \in W_T(u,v)$. Suppose next that $R(x,v) \neq \{x,v\}$. We will construct a $x,v$-path $Q$ in $G_R$. For this, let $x = v_0$. By axiom (TW6), there exists a neighbor of $v_0$, say $v_1$ with $v_0 \neq v_1$ such that $v_1 \in R(v_0,v) \cap R(u,v)$, $R(v_0,v_1) = \{v_0,v_1\}$. By (bt1) $v_0 \notin R(v_1,v)$. If $v_1 \neq v$ and $v_1 \in R(u,v)$ then we can continue with the same procedure to get $v_2 \in R(u,v) \cap R(v_1,v)$, where $R(v_1,v_2) = \{v_1,v_2\}$, $v_1 \neq v_2$ and by (bt1) $v_1 \notin R(v_2,v)$. Similarly, we obtain the vertex $v_3$ with $v_3 \in R(u,v) \cap R(v_2,v)$, $R(v_2,v_3) = \{v_2,v_3\}$, $v_2 \neq v_3$ and by (bt1) $v_2 \notin R(v_3,v)$. So, by repeating this step, we obtain a sequence of vertices $v_0, v_1, \ldots, v_q$, $q \geq 2$, such that

1. $R(v_i, v_{i+1}) = \{v_i, v_{i+1}\}$ and $v_i \neq v_{i+1}, i \in \{0, 1, \ldots, q-1\}$,
2. $v_i \in R(u,v)$ $i \in \{0, 1, \ldots, q\}$,
3. $v_{i+1} \in R(v_i, v)$ $i \in \{0, 1, \ldots, q-1\}$,
4. $v_{i-1} \notin R(v_i, v)$ $i \in \{1, \ldots, q\}$.

Using conditions 3 and 4, it is clear that $v_{i-1} \neq v_{i+1}$. If $v_i = v_j$, $j \geq i+3$, $i \in \{0, 1, \ldots, q-3\}$, then the vertices $v_i, v_{i+1}, \ldots v_j$ induces a cycle $C_n, n \geq 3$, contradicting the assumption that $G_R$ is $C_n$-free, $n \geq 3$. Therefore $v_i \neq v_j$, $j \geq i+3$ and $i \in \{0, 1, \ldots, q-3\}$, which implies that all the $v_i$'s are distinct and this sequence needs to stop. Hence, we may assume that $v_q = v$.

Consider the case $R(u,x) = \{u,x\}$. First we claim that $u$ is not adjacent to $v_i$, $i \in \{1, \ldots, q\}$. Otherwise if possible, let $u$ is adjacent to $v_r$, then the vertices $u, x, v_1, \ldots v_r$ induce a cycle $C_n, n \geq 3$, not possible. Then $uxv_1 \ldots v_{q-1}v$ is a $u, v$-weak-toll walk and $x \in W_T(u,v)$. Now consider the case $R(u,x) \neq \{u,x\}$. If $u$ is adjacent to $v_m$, for some $m \in \{1, \ldots, q\}$, then $u$ cannot be adjacent to the vertices $v_{m+1}, \ldots v_q$, otherwise $G_R$ contains an induced cycle of length of at least three. Then $uv_m v_{m-1} \ldots v_1 x v_1 \ldots v_{q-1}v$ is a weak-toll $u, v$-walk and $x \in W_T(u,v)$. If $u$ is not adjacent to $v_i, i = 1 \ldots q-1$. we can symmetrically build a sequence $u_0, u_1, \ldots, u_r$, where $u_0 = x$, $u_r = u$ and $u_0 u_1 \ldots u_r$ is a $x, u$-path in $G_R$. Clearly, $u u_{r-1} u_{r-2} \ldots u_1 x v_1 \ldots v_{q-1}v$ is a $u, v$-weak-toll walk and in all the cases we have $x \in W_T(u,v)$ and hence $G_R$ is connected.

Now suppose that $x \in W_T(u,v)$ and $x \notin \{u,v\}$. We have to show that $x \in R(u,v)$. Let $W$ be the $u, v$-weak-toll walk containing $x$ with a minimum number of vertices. By Lemma 2, $u, x$-subpath say $Q$ of $W$ contain at most one neighbor of $v$ and $x, v$-subpath say $R$ of $W$ contain at most one neighbor of $u$. Let $P$ be the $u, v$-induced path with the maximum number of vertices in $W$. If $x$ belongs to $P$, then $x \in R(u,v)$ by Lemma 3 (since (tr) implies (J2)). Suppose $x$ not belongs to $P$, let $x'$ be the common vertex of the paths $P$ and $Q$ and close to $x$. By the choice of the induced path $P$ and weak-toll walk $W$, the $x', x$-subpath of $Q$ does not contain a neighbor of $u$ and $v$. So, $x \in R(u,v)$ by continuous application of the axiom (TW4).  $\square$

Now, using Theorems 2, and 3 and Propositions 2 and 3, we have the following theorem characterizing the weak-toll function of trees.

**Theorem 4.** *A transit function $R$ on $V$ satisfies the axioms (bt1), (tr), (JC), (TW4) and (TW6) if and only if $G_R$ is a tree and $R = W_T$ on $G_R$.*

The toll-walk function $T$ need not satisfy the axiom (b1) for arbitrary connected graphs. In [11], the graphs in which $T$ satisfies the axiom (b1) are characterized as the following theorem.

**Theorem 5** [11] *The toll walk function $T$ of a graph $G$ satisfies axiom (b1) if and only if $G$ is $(HC_5DAT)$-free graph.*

The next theorem gives the forbidden subgraph characterization of $AT$-free graphs due to Köhler from [5].

**Theorem 6** [5] *A graph $G$ is $(C_kT_2X_2X_3X_{30}\ldots X_{41}XF_2^{n+1}XF_3^nXF_4^n)$-free for $k \geq 6$ and $n \geq 1$ if and only if $G$ is $AT$-free graph.*

The weak-toll function also need not satisfy axiom (b1) for arbitrary graphs. The following theorem provides a characterization of the graphs in which the weak-toll function satisfies the axiom (b1).

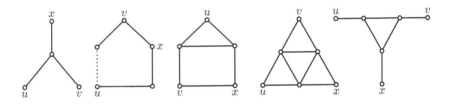

**Fig. 1.** Graphs claw, hole-$H$, house, $S_3$ and net (from left to right).

**Theorem 7.** *The weak-toll function $W_T$ of a graph $G$ satisfies the axiom (b1) if and only if $G$ is (claw, hole, house, net, $S_3$)-free graph.*

*Proof.* Clearly, if $G$ contains claw, hole, house, $S_3$ or net as an induced subgraph with vertices as shown in Figure 1, the weak-toll function $W_T$ does not satisfy the axiom (b1). That is, if $W_T$ satisfies axiom (b1), then $G$ is (claw, hole, house, net, $S_3$)-free. By Proposition 1, the weak-toll function $W_T$ coincides with toll function $T$ since $G$ is claw-free. In Theorem 5, it is proven that $T$ satisfies the axiom (b1) if and only if $G$ is $(HC_5DAT)$-free graph. Using these we can conclude that if $W_T$ satisfies axiom (b1) if and only if $G$ is a (claw, house, hole, domino, AT)-free graph. Except for the net graph and $S_3$, all the forbidden induced subgraphs of the $AT$-free graph contain claw or $C_5$ or house as induced subgraphs. The domino contains a claw as an induced subgraph. That is, $W_T$ satisfies axiom (b1), if and only if $G$ is (claw, hole, house, net, $S_3$)-free.   $\square$

The following theorem, which characterizes the weak-toll function of unit interval graphs, is established by combining Theorems 1 and 7. Since by Theorem 7,

$W_T$ satisfy the axiom (b1) if and only if $G$ is (claw, hole, house, net, $S_3$)- free and by Theorem 1, $W_T$ satisfy the axiom (JC) if and only if $G$ is $C_{n+4}$-free. By combining these two, we obtain that $W_T$ satisfies the axioms (b1) and (JC) if and only if $G$ is (claw, net, $S_3$, $C_{n+4}$)-free.

**Theorem 8.** *The weak-toll function $W_T$ of a graph $G$ satisfies the axioms (b1) and (JC) if and only if $G$ is a unit interval graph.*

Now, we can replace the axiom (JC) with the axiom (J0) in Theorem 8 to characterize the weak-toll function of unit interval graphs.

**Theorem 9.** *The weak-toll walk function $W_T$ of a graph $G$ satisfies the axioms (b1) and (J0) if and only if $G$ is a unit interval graph.*

*Proof.* By Theorem 7, it is clear that the weak-toll function $W_T$ satisfies the axiom (b1) if and only if $G$ is (claw, hole, house, net, $S_3$)-free graph. So, it suffices to show that $W_T$ satisfies the axiom (J0) if and only if $G$ is chordal. If possible, assume that $G$ contains an induced cycle of length at least four with vertices as; $u$, $v$, and $y$ as consecutive vertices, and $x$ as different from $u$, $v$, and $y$. Then $x \in W_T(u,y)$ and $y \in W_T(x,v)$ but $x \notin W_T(u,v)$. That is, $W_T$ does not satisfy the axiom (J0). Conversely, suppose that $W_T$ does not satisfy the axiom (J0). That is, $x \in W_T(u,y)$ and $y \in W_T(x,v)$ but $x \notin W_T(u,v)$. Since $G$ is claw-free, we can assume that the vertices $u, x, y$ and $v$ lies in a path. Since $x \in W_T(u,y)$ and $G$ is claw-free, there is a $u,x$-path without neighbor of $y$ say $P$ and a $x,y$-path without neighbor of $u$ say $Q$. Also, since $y \in W_T(x,v)$, there is a $x,y$-path without neighbor of $v$ say $R$ and a $y,v$-path without neighbor of $x$ say $S$. Also, $x \notin W_T(u,v)$ implies that all the $u,x$-paths contain more than one neighbor of $v$ or $x,v$-paths contain more than one neighbor of $u$. We may assume that the $x,v$-path contains more than one neighbor of $u$ say $u'$ and $u''$ in which $u'$ is close to the vertex $y$. Then the sequence of vertices $u \xrightarrow{P} x \xrightarrow{Q} y \xrightarrow{S} u'u$ forms a cycle of length at least four. There may be chords from the vertices of path $P$ to the vertices of path $S$. Let $ab$ be a chord such that $a$ is close to $x$ on the path $P$. Therefore, $a \xrightarrow{P} x \xrightarrow{Q} y \xrightarrow{S} ba$ induces a cycle of length at least four. That is, $W_T$ satisfies axioms (J0) and (b1) in $G$ if and only if $G$ is a unit interval graph. $\qquad\square$

Now we have the following theorem.

**Theorem 10.** *A transit function $R$ on $V$ satisfies the axioms (b1), (b2) (J2), (J0), (TW4), (TW5) and (TW6) if and only if $G_R$ is a connected unit interval graph and $R = W_T$ on $G_R$.*

*Proof.* Let $u$ and $v$ be two distinct vertices of $G_R$. First, assume that $x \in R(u,v)$ and $x \neq v$. We have to show that $x \in W_T(u,v)$ on $G_R$. Clearly $x \in W_T(u,v)$ whenever $x = u$. So, assume that $x \notin \{u,v\}$. If $R(u,x) = \{u,x\}$ and $R(x,v) = \{x,v\}$, then $uxv$ is a weak-toll walk of $G_R$ and $x \in W_T(u,v)$. Suppose next that $R(u,x) = \{u,x\}$ and $R(x,v) \neq \{x,v\}$. We will construct a $x,v$-path $Q$ in $G_R$ without a neighbor of $u$ (except possibly $x$). For this, let $x = v_0$. By axiom

(TW5) there exists a neighbor of $v_0$, say $v_1$ with $R(u,v_1) \neq \{u,v_1\}$ such that $v_1 \in R(v_0,v) \cap R(u,v)$. By (b1) $v_0 \notin R(v_1,v)$ and by (b2) $R(v_1,v) \subseteq R(v_0,v)$ so that $R(v_1,v) \subset R(v_0,v)$ . If $v_1 \neq v$ and $v_1 \in R(u,v)$ then we can continue with the same procedure to get $v_2 \in R(u,v) \cap R(v_1,v)$, where $R(u,v_2) \neq \{u,v_2\}$, $R(v_1,v_2) = \{v_1,v_2\}$ and by (b1) and (b2), $R(v_2,v) \subset R(v_1,v)$. So, by repeating this step, we obtain a sequence of vertices $v_0, v_1, \ldots, v_q$, $q \geq 2$, such that

1. $R(v_i, v_{i+1}) = \{v_i, v_{i+1}\}, i \in \{0, 1, \ldots, q-1\}$,
2. $R(v_{i+1}, v) \subset R(v_i, v), i \in \{0, 1, \ldots, q-1\}$,
3. $R(u, v_i) \neq \{u, v_i\}, i \in \{1, \ldots, q\}$.

Clearly, this sequence needs to stop, since $V$ is finite and by the second condition. Hence, we may assume that $v_q = v$. Then we have a weak-toll $u,v$-walk $uxv_1 \ldots v_{q-1}v$ and $x \in W_T(u,v)$.

If $R(u,x) \neq \{u,x\}$, by continuous application of axioms (TW6), (b1) and (b2), we obtain a sequence of vertices $v_0, v_1, \ldots, v_q$, $q \geq 2$, satisfying the first two conditions. If $u$ is adjacent to $v_m$, for some $m \in \{1, \ldots, q\}$. Clearly $v_m \in R(u,v)$. Then we apply the axiom (TW5) to obtain $v_{m+1}, \ldots v_q$ such that $v_{m+1}, \ldots v_q$ are not adjacent to $u$ and satisfy the first two conditions listed above. Then $uv_m v_{m-1} \ldots v_1 x v_1 \ldots v_{q-1}v$ is a weak-toll $u,v$-walk and $x \in W_T(u,v)$. If $u$ is not adjacent to $v_i, i = 1 \ldots q-1$. we can symmetrically build a sequence $u_0, u_1, \ldots, u_r$, where $u_0 = x$, $u_r = u$ and $u_0 u_1 \ldots u_r$ is a $x,u$-path in $G_R$. Clearly, $uu_{r-1}u_{r-2} \ldots u_1 x v_1 \ldots v_{q-1}v$ is a $u,v$-weak-toll walk and $x \in W_T(u,v)$ and hence $G_R$ is connected. Now suppose that $x \in W_T(u,v)$, then similar to Theorem 3, we can prove that $x \in R(u,v)$. By Theorem 9, $W_T$ satisfies the axioms (J0) and (b1) which implies that $G_R$ is a unit interval graph. Also $W_T$ satisfies axioms (b1), (J0), (TW4), (TW5) and (TW6) by Theorem 9 and Proposition 2. Since unit interval graphs doesn't contain the claw, as an induced subgraph by Proposition 1, $W_T = T$ on unit interval graphs. In [11], it is proved that $T$ satisfies axiom (b1) then $T$ satisfies axiom (b2). Therefore, in unit interval graphs, $W_T$ satisfies the axiom (b2) since it satisfies axiom (b1). Also, $W_T$ satisfies axiom (J2) on every connected graph.    □

The following examples establish the independence of the axioms used in this section. In the examples (Examples 1 to 6), we define a transit function $R$ on the set $V = \{u, x, y, v\}$ and in the Example 7, 8 we define $R$ on $V = \{u, x, y, w, v\}$. Also $R(a,a) = \{a\}, \forall a \in V$.

*Example 1 ((J2), (b2), (J0), (TW4), (TW5) and (TW6), but not (b1)).*
Define $R$ as : $R(u,v) = R(x,v) = R(u,x) = V$, $R(a,b) = \{a,b\}$ for all other $a, b \in V$. Then $R$ satisfies the axioms (J2), (b2), (J0), (TW4), (TW5), and (TW6). Furthermore, $x \in R(u,v), x \neq v$ and $v \in R(u,x)$ and $R$ do not satisfy axiom (b1).

*Example 2 ((b1), (J2), (b2), (TW4), (TW5), (TW6), but not (J0), (JC)).*
Define $R$ as : $R(u,y) = \{u, x, v, y\}$, $R(x,v) = \{x, y, u, v\}$, $R(a,b) = \{a,b\}$ for all other $a, b \in V$. Then $R$ satisfies axioms (b1), (b2), (J2), (TW4), (TW5) and (TW6). Furthermore, $x \in R(u,y)$, $y \in R(x,v)$, and $x \notin R(u,v)$ and hence $R$ does not satisfy the axioms (J0) and (JC).

*Example 3 ((b1), (b2), (J0), (TW4), (TW5), (TW6) but not (J2), (tr)).*
Define $R$ as : $R(u,v) = \{u,x,v\}$, $R(a,b) = \{a,b\}$ for all other $a,b \in V$. Then $R$ satisfies axioms (b1), (b2), (J0), (TW4), (TW5) and (TW6). In addition $R(u,y) = \{u,y\}$, $R(y,v) = \{y,v\}$, $R(u,v) \neq \{u,v\}$ but $y \notin R(u,v)$ so that $R$ does not satisfy the axiom (J2) and (tr).

*Example 4 ((b1), (b2), (J2), (J0), (TW5) and (TW6), but not (TW4)).*
Define $R$ as : $R(u,v) = \{u,y,v\}$, $R(u,x) = \{u,y,x\}$, $R(x,v) = \{x,y,v\}$, $R(a,b) = \{a,b\}$ for all other $a,b \in V$. Then $R$ satisfies the axioms (b1), (b2), (J2), (J0), (TW5) and (TW6). Furthermore, $y \in R(u,v)$, $R(x,y) = \{x,y\}$, $R(u,x) \neq \{u,x\}$, $R(v,x) \neq \{v,x\}$, and $x \notin R(u,v)$ and $R$ do not satisfy the axiom (TW4).

*Example 5 ((b1), (b2), (J2), (J0), (TW4) and (TW6), but not (TW5)).*
Define $R$ as: $R(u,v) = \{u,x,y,v\}$, $R(x,v) = \{x,y,v\}$, $R(a,b) = \{a,b\}$ for all the other $a,b \in V$. Then, $R$ satisfies axioms (b1), (b2), (J2), (J0), (TW4), and (TW6). In addition to $x \in R(u,v)$, $R(u,x) = \{u,x\}$, there does not exist $v_1$ such that $v_1 \in R(x,v) \cap R(u,v), v_1 \neq x$ with $R(x,v_1) = \{x,v_1\}$, $R(u,v_1) \neq \{u,v_1\}$ so $R$ does not satisfy axiom (TW5).

*Example 6 ((JC), (TW4), (tr) and (TW6), but not (bt1).).*
Define $R$ as $R(u,v) = R(x,v) = V$, $R(u,y) = \{u,x,y\}$, $R(a,b) = \{a,b\}$ for all other $a,b \in V$. Then $R$ satisfies axioms (JC), (TW4), (tr) and (TW6). But $x \in R(u,v)$, $R(u,x) = \{u,x\}$, and $u \in R(v,x)$ and $R$ do not satisfy axiom (bt1).

*Example 7 ((b1), (b2), (J2), (J0), (TW4) and (TW5), but not (TW6)).*
Define $R$ as: $R(u,v) = \{u,x,w,v\}$, $R(x,v) = \{x,y,u,v\}$, $R(y,v) = \{y,u,v\}$, $R(u,x) = \{u,w,x\}$, $R(a,b) = \{a,b\}$ for all other $a,b \in V$. We can verify that $R$ satisfies the axioms (b1), (b2), (J2), (J0), (TW4), and (TW5). Furthermore, $x \in R(u,v)$, $R(u,x) \neq \{u,x\}$, but there does not exist $v_1$ such that $v_1 \in R(x,v) \cap R(u,v), v_1 \neq x$ with $R(x,v_1) = \{x,v_1\}$, $R(v_1,v) \subseteq R(x,v)$ so $R$ does not satisfy the axiom (TW6).

*Example 8 ((b1), (J2), (J0), (TW4), (TW5) and (TW6), but not (b2).).*
Define $R$ as : $R(u,v) = \{u,x,y,v\}$, $R(x,v) = \{x,y,w,v\}$, $R(u,y) = \{u,x,y\}$, $R(w,v) = \{w,y,v\}$, $R(a,b) = \{a,b\}$ for all other $a,b \in V$. Then $R$ satisfies the axioms (b1), (J2), (J0), (TW4), (TW5), and (TW6). Furthermore, $x \in R(u,v)$, $w \in R(x,v)$, and $w \notin R(u,v)$ so $R$ do not satisfy axiom (b2).

Note that for a transit function $R$, axioms (tr) and (TW6) implies axiom (TW5). For, take $x \in R(u,v)$, $R(u,x) = \{u,x\}$ which implies $v_1 \in R(u,v) \cap R(x,v)$ with $R(x,v_1) = \{x,v_1\}$ by (TW6). Now $R(u,x) = \{u,x\}$ and $R(x,v_1) = \{x,v_1\} \implies x \in R(u,v_1)$ by (tr) and hence axiom (TW5).

## 4  Non-definability of Weak-Toll Function

It is proved by Nebeský in [10], respectively, Changat et al. in [13] that a first-order axiomatic characterization of the induced path function $J$, respectively, the toll-walk function $T$, of an arbitrary connected graph $G$ is not possible.

Here, we show that, like the function $J$ and $T$, it is impossible to characterize the weak-toll function $W_T$ of a connected graph using a set of first-order axioms. The idea of proof of the impossibility of such a characterization is the following. First, we construct two non-isomorphic graphs $H_d$ and $H'_d$ and a first-order axiom which may not be satisfied by the weak-toll function $W_T$ of an arbitrary connected graph. The following axiom is defined for an arbitrary transit function $R$ in a finite non-empty set $V$ and is called a *scant property* following Nebeský [10].

Axiom (SP): If $R(x,y) \neq \{x,y\}$, then $R(x,y) = V$, for any $x,y \in V$.

First, we define certain concepts and terminology of first-order logic [7]. The tuple $\mathbf{X} = (X, \mathcal{S})$ is called a *structure* when $X$ is a nonempty set called *universe*, and $\mathcal{S}$ is a finite set of function symbols, relation symbols, and constant symbols called *signature*. Here, we assume that the signature contains only relation symbols. The *quantifier rank* of a formula $\phi$ is its depth of quantifier nesting and is denoted by $qr(\phi)$. Let $\mathbf{A}$ and $\mathbf{B}$ be two structures with the same signatures. A map $q$ is said to be a *partial isomorphism* from $\mathbf{A}$ to $\mathbf{B}$ if and only if $dom(q) \subset A$, $rg(q) \subset B$, $q$ is injective and for any $n$-ary relation $R$ in the signature and $a_0$, ..., $a_{l-1} \in dom(q)$, $R^{\mathcal{A}}(a_0, \dots, a_{l-1})$ if and only if $R^{\mathcal{B}}(q(a_0), \dots, q(a_{l-1}))$.

Let $r$ be a positive integer. The *$r$-move Ehrenfeucht-Fraisse game* on $\mathbf{A}$ and $\mathbf{B}$ is played between 2 players called the *Spoiler* and the *Duplicator*, according to the following rules.

Each run of the game has $r$ moves. In each move, Spoiler plays first and picks an element from the universe $A$ of the structure $\mathbf{A}$ or from the universe $B$ of the structure $\mathbf{B}$; Duplicator then responds by picking an element from the universe of the other structure. Let $a_i \in A$ and $b_i \in B$ be the two elements picked by the Spoiler and Duplicator in their $i$th move, $1 \leq i \leq r$. The Duplicator wins the run $(a_1, b_1), \dots, (a_r, b_r)$ if the mapping $a_i \to b_i$, where $1 \leq i \leq r$ is a partial isomorphism from the structure $\mathbf{A}$ to $\mathbf{B}$. Otherwise, Spoiler wins the run $(a_1, b_1), \dots, (a_r, b_r)$.

*Duplicator wins the $r$-move EF-game on $\mathbf{A}$ and $\mathbf{B}$* if Duplicator can win every run of the game, regardless of how Spoiler plays. The following theorem is our main tool in proving the inexpressibility results.

**Theorem 11** [7]. *The following statements are equivalent for two structures $\mathbf{A}$ and $\mathbf{B}$ in a relational vocabulary.*

1. *$\mathbf{A}$ and $\mathbf{B}$ satisfy the same sentence $\sigma$ with $qr(\sigma) \leq n$.*
2. *The Duplicator has an $n$-round winning strategy in the EF game on $\mathbf{A}$ and $\mathbf{B}$.*

By a *ternary structure* we mean an ordered pair $(X, D)$ where $X$ is a finite nonempty set and $D$ is a ternary relation on $X$. By the *underlying graph* of a ternary structure $(X, D)$ we mean the graph $G$ with the properties that $X$ is its vertex set and distinct vertices $u$ and $v$ of $G$ are adjacent if and only if

$$\{x \in X : D(u,x,v)\} \cup \{x \in X : D(v,x,u)\} = \{u,v\}.$$

We call a ternary structure $(X, D)$, 'the $W'$- *structure* of a graph $G$', if $X$ is the vertex set of $G$ and $D$ is the ternary relation corresponding to $W_T$ (that is, $(x, y, z) \in D$ if and only if $y$ lies in some $x, z$- weak-toll walk). Obviously, if $(X, D)$ is a $W'$-structure, then it is the $W'$-structure of the underlying graph of $(X, D)$. Let $F : X \times X \to 2^X$ be defined as $F(x, y) = \{u \in X : D(x, u, y)\}$. So, for any ternary structure $(X, D)$, we can associate the function $F$ corresponding to $D$ and vice versa. We say that $(X, D)$ is *scant* if the function $F$ corresponding to the ternary relation $D$, satisfies the axiom (SP) together with the axioms (t1), (t2) and (t3); in other words, $F$ is a transit function satisfying the axiom (SP).

Next, we present two graphs $H_d$ and $H'_d$ such that the $W'$-structure of one of them is scant and the other is not. Moreover, the proof will settle, once we prove that Duplicator wins the EF game on $H_d$ and $H'_d$. For $d \geq 2$ let $H_d$ be a graph with vertices

$$V(H_d) = \{u_1, u_2, \dots, u_{4d}, v_1, v_2, \dots, v_{4d}, x_1\},$$

and edges (indices are via modulo $4d$)

$$E(H_d) = \{u_i u_{i+1}, v_i v_{i+1}, u_i v_i, v_1 x_1, v_{2d+1} x_1 : i \in \{1, \dots, 4d\}\}.$$

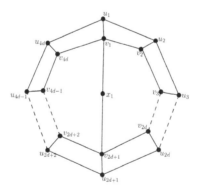

**Fig. 2.** Graph $H_d$.

For $d \geq 2$ let $H'_d$ be a graph with vertices

$$V(H'_d) = \{u'_1, u'_2, \dots, u'_{4d}, v'_1, v'_2, \dots, v'_{4d}, x'_1\},$$

and edges (indices are via modulo $2d$)

$$E(H'_d) = \{u'_1 u'_{2d}, u'_i u'_{i+1}, u'_{2d+1} u'_{4d}, u'_{2d+i} u'_{2d+i+1}, v'_1 v'_{2d}, v'_i v'_{i+1}, v'_{2d+1} v'_{4d},$$

$$v'_{2d+i} v'_{2d+i+1}, u'_j v'_j, v'_1 x'_1, v'_{2d+1} x'_1 : i \in \{1, \dots, 2d-1\}, j \in \{1, \dots, 4d\}\}.$$

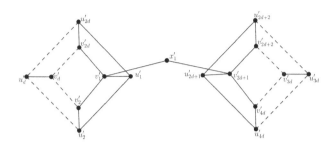

**Fig. 3.** Graph $H'_d$.

Graphs $H_d$ and $H'_d$ are shown in Figs. 2 and 3, respectively. The first-order non-definability of the toll-walk function is obtained in [13] using the graphs $H_d$ and $H'_d$, and the scant property (SP). In [13] it is proved that the toll walk function of $H_d$ satisfies the axiom (SP), but the toll walk function of $H'_d$ does not satisfy the axiom (SP). That is, for the toll walk function $T$ of $H_d$, $T(u,v) = V(H_d)$ for all $u, v \in V(H_d)$ and $uv \notin E(H_d)$. Since all toll-walks are weak-toll walks, we get $W_T(u,v) = V(H_d)$ for all $u, v \in V(H_d)$ and $uv \notin E(H_d)$. Also, note that $W_T(v'_2, x'_1) = \{v'_2, x'_1, v'_1, u'_1, v'_{2d}, v'_{2d-1}, \ldots, v'_4, u'_{2d}, u'_{2d-1}, \ldots, u'_4\} \neq V(H'_d)$. From these observations, we obtain the following lemma.

**Lemma 4.** *Let $d \geq 2$.*

*i. The $W'$-structure of $H_d$ is scant.*
*ii. The $W'$-structure of $H'_d$ is not scant.*

Again, since the graphs used for proving the non-definability of the toll function in [13] are similar to $H_d$ and $H'_d$, we obtain the following lemma.

**Lemma 5.** *Let $n \geq 1$ and $d > 2^{n+1}$. If $(X_1, D_1)$ and $(X_2, D_2)$ are scant ternary structures such that the underlying graph of $(X_1, D_1)$ is $H_d$ and the underlying graph of $(X_2, D_2)$ is $H'_d$, then $(X_1, D_1)$ and $(X_2, D_2)$ satisfy the same sentence $\psi$ with $qr(\psi) \leq n$.*

From Lemma 4 and Lemma 5, we can conclude the following result.

**Theorem 12.** *There exists no sentence $\sigma$ of the first-order logic of vocabulary $\{D\}$ such that a connected ternary structure is a $W'$-structure if and only if it satisfies $\sigma$.*

In conclusion, we observe the following. In [13], it is proved that the function $T$ of an arbitrary bipartite graph cannot also be first-order definable. Therefore, using the same structures (odd and even cycles as in the case of $T$ in [13]), we can prove that the function $W_T$ of a bipartite graph is also not first-order definable. We further observe that Ptolemaic graphs are a subclass of chordal graphs, a superclass of unit interval graphs, and chordal graphs are perfect. Thus, the following problem seems interesting.

**Problem:** Is there a first-order axiomatic characterization of the weak-toll function $W_T$ of Ptolemaic graphs and perfect graphs?

**Acknowledgments.** L.K.K.S acknowledges the financial support of CSIR, Government of India, for providing CSIR Senior Research Fellowship (CSIR-SRF) (No 09/102(0260)/2019-EMR-I). J.J acknowledges the financial support of the University of Kerala, India, for providing University JRF (No: 445/2020/UOK, 391/2021/UOK, 3093/2022/ UOK, 4202/2023/UOK).

# References

1. Alcon, L.: A note on path domination. Discuss. Math. Graph Theory **36**, 1021–1034 (2016). https://doi.org/10.7151/dmgt.1917
2. Alcon, L., et al.: Toll convexity. Euro. J. Combin. **46**, 161–175 (2015). https://doi.org/10.1016/j.ejc.2015.01.002
3. Changat, M., Mathew, J., Mulder, H.M.: The induced path function, monotonicity and betweenness. Discrete Appl. Math. **158**(5), 426–433 (2010). https://doi.org/10.1016/j.dam.2009.10.004
4. Dourado, M.C., Gutierrez, M., Protti, F., Tondato, S.: Weakly toll convexity and proper interval graphs (2022). arXiv:2203.17056. https://doi.org/10.48550/arXiv.2203.17056
5. Köhler, E.: Graphs without asteroidal triples, Ph.D. Thesis, Technische Universität Berlin, Cuvillier Verlag, Göttingen (1999)
6. Lekkerkerker, C.G., Boland, J.C.: Representation of a finite graph by a set of intervals on the real line. Fundamenta Math. **51**, 45–64 (1962)
7. Libkin, L.: Elements of Finite Model Theory. Springer, Heidelberg (2013). https://doi.org/10.1007/978-3-662-07003-1
8. Mulder, H.M., Nebeský, L.: Axiomatic characterization of the interval function of a graph. Europ. J. Combin. **30**, 1172–1185 (2009). https://doi.org/10.1016/j.dam.2018.07.018
9. Nebeský, L.: Characterizing the interval function of a connected graph. Math. Bohem. **123**(2), 137–144 (1998). https://doi.org/10.21136/MB.1998.126307
10. Nebeský, L.: The induced paths in a connected graph and a ternary relation determined by them. Math. Bohemica **127**(3) 397–408 (2002). https://doi.org/10.21136/MB.2002.134072
11. Sheela, L.K.K., Changat, M., Peterin, I.: Axiomatic characterization of the toll walk function of some graph classes. In: Bagchi, A., Muthu, R. (eds.) CALDAM 2023. LNCS, vol. 13947, pp. 427–446. Springer, Cham (2023). https://doi.org/10.1007/978-3-031-25211-2_33
12. Roberts, F.S.: Indifference graphs. In: Harary, F. (ed.) Proof Techniques in Graph Theory, pp. 139–146. Academic Press, New York (1969)
13. Changat, M., Jacob, J., Sheela, L.K.K., Peterin, I.: The toll walk transit function of a graph: axiomatic characterizations and first-order non-definability (2023). arXiv:2310.20237v1. https://doi.org/10.48550/arXiv.2310.20237

# Total Coloring of Some Graph Operations

T. Kavaskar and Sreelakshmi Sukumaran[✉]

Central University of Tamil Nadu, Thiruvarur 610 005, India
lakshmisk16g@gmail.com

**Abstract.** The total chromatic number $\chi_T(G)$ of $G$ is the least positive integer $k$ for which $G$ admits a $k$-total coloring. Clearly, $\chi_T(G) \geq \Delta(G) + 1$. A long standing Total Coloring Conjecture (TCC) asserts that every graph $G$ has $\chi_T(G) \leq \Delta(G) + 2$. If $\chi_T(G) = \Delta(G) + 1$, then $G$ is a type-1 graph and if $\chi_T(G) = \Delta(G) + 2$, then $G$ is a type-2 graph. Weak TCC states that any simple graph $G$ has $\chi_T(G) \leq \Delta(G) + 3$. In this paper, we give an upper bound for the total chromatic number of the join $G \vee H$ of graphs $G$ and $H$. Also, we verify that if $G$ satisfies TCC, then $G \vee G$ satisfies TCC and the join of two type-1 graphs having the same order satisfies TCC. We show that $G \vee H$ satisfies weak TCC under certain constrains. Moreover, we show that the join of any two graphs $G$ and $H$ of same order satisfies weak TCC if both $G$ and $H$ are satisfying TCC. Also, we prove that if $G$ and $H$ are any two $k$-regular graphs with same odd order, then $G \vee H$ is not type-1. In addition, we verify that the join of any two cycles satisfies TCC. We give an upper bound for the total chromatic number of generalized join of graphs and as a result we obtain an upper bound for the total chromatic number of the lexicographic product $G \circ H$ of $G$ and $H$ in terms of the maximum degrees of $G$ and $H$ if $H$ satisfies TCC. Also, we show that the lexicographic product of a graph with compliment of complete graphs satisfies weak TCC. In particular, when the graph is Type-1 then this lexicographic product will satisfy TCC.

**Keywords:** Total coloring conjecture · Join of graphs

## 1 Introduction

All graphs considered here are finite, simple and undirected. Let $G = (V(G), E(G))$ be a graph with vertex set $V(G)$ and edge set $E(G)$. For $v \in V(G)$, let $N_G(v)$ denote the open neighborhood of $v$ and $N_G[v]$ denote the closed neighborhood of $v$. $\Delta(G)$ denote the maximum degree of $G$. For any $A$ which is a subset of $V(G)$, $\langle A \rangle$ denotes the subgraph induced by $A$. Graph coloring is the process of assigning colors to the elements of a graph. Graph coloring has various practical applications also. There are different kinds of graph colorings like, vertex coloring, edge coloring, total coloring etc. If the coloring is for the vertices only, then it is said to be vertex coloring or simply coloring. If the coloring is for edges only, then it is said to be the edge coloring. Total coloring is the coloring

S. Kalyanasundaram and A. Maheshwari (Eds.): CALDAM 2024, LNCS 14508, pp. 302–312, 2024.
https://doi.org/10.1007/978-3-031-52213-0_21

in which we assign colors to both the vertices and edges of a graph and it is said to be proper if no two adjacent or incident elements are receiving the same color. A $k$-total coloring is the total coloring in which we are using $k$ colors. Total chromatic number of a graph $G$ denoted as $\chi_T(G)$ is the minimum number of colors required for coloring the vertices and edges of the graph properly (Similarly we can define chromatic number ($\chi(G)$) and chromatic index number ($\chi'(G)$) corresponding to the vertex and edge coloring, respectively). Graphs with $\chi'(G) = \Delta(G)$ are called *class-1* and graphs with $\chi'(G) = \Delta(G) + 1$ are called *class-2*. Also, graphs with $\chi_T(G) = \Delta(G) + 1$ are called *type-1* and graphs with $\chi_T(G) = \Delta(G) + 2$ are called *type-2*.

In the year 1953 Behzad [1] conjectured that $\Delta(G) + 2$ is an upper bound for $\chi_T(G)$. It is known as the Total Coloring Conjecture (TCC), which is one among the classic open problems in graph theory. TCC is studied widely by various mathematicians. During $1980's$, Sánchez-Arroyo [15] proved that deciding whether a graph is type-1 or not is $NP$-complete and also the total coloring of a complete bipartite graph is $NP$-hard. Moreover, McDiarmid and Sánchez-Arroyo [11] proved that determining the total chromatic number is NP-hard even for $r$-regular bipartite graphs, for each fixed $r \geq 3$. It can be easily seen that TCC is true for the complete graphs [2], cycles and bipartite graphs.

In case of planar graphs, so many results related to TCC have been done and are mainly based on the maximum degree and the girth constraints. For planar graphs with maximum degree at most 5, TCC was verified by A. V. Kostochka [7] during the late 90's. Yap [21] verified it for planar graphs with maximum degree at least 8 and Kowalik et. al. [8] proved that for any planar graph with maximum degree at least 9 is type-1. For the planar graphs with maximum degree 6 and 7, TCC was verified under certain conditions. For the non-planar case, TCC is verified for so many classes of graphs. TCC for the cartesian product of two graphs is verified for many cases [5], and still there are cases for which it is not verified. But regarding the other graph products only a few results are proved on TCC yet [3,20]. Geetha et al. [4] have produced an excellent survey on total coloring, which is a valuable source of information in the state of art.

Even though many well-known researchers from different parts of the world have studied TCC for over 60 years, it remains open till now. So it make sense for the current researchers to go for some relaxed version of TCC which is known as the Weak TCC. Before defining the weak TCC we define some more weaker version of TCC called the $k$-TCC which was introduced by Manu Basavaraju et al. in [10].

**$k$-Total coloring Conjecture ($k$-TCC).**

*For any graph $G$, $\chi_T(G) \leq \Delta(G) + k$, for some fixed positive integer $k \geq 2$.*

The 2-TCC is nothing but the original TCC and 3-TCC is known as the weak TCC. Molloy and Reed [13] showed a probabilistic approach to prove that for a sufficiently large $\Delta$, $\chi_T(G) \leq \Delta(G) + C$, where $C = 10^{26}$.

Let $G$ be a graph with $n$ vertices and $H_1, H_2, \ldots, H_n$ be a collection of graphs. The $G$-*generalized join* of $H_1, H_2, \ldots, H_n$, denoted by $G[H_1, H_2, \ldots, H_n]$, is the

graph $G'$ with vertex set $V(G') = \bigcup\limits_{i=1}^{n} V(H_i)$ and edge set $E(G') = \left( \bigcup\limits_{i=1}^{n} E(H_i) \right) \cup$
$\left( \bigcup\limits_{ij \in E(G)} \{xy | x \in V(H_i), y \in V(H_j)\} \right)$.

If $H_i \cong H$ for $1 \le i \le n$, then $G[H, H, \dots, H]$ is the standard *lexicographic product of $G$ and $H$* and it is denoted as $G \circ H$. If $G = K_2$, then $K_2[H_1, H_2]$ is the well known *join* of graphs $H_1$ and $H_2$ and it is denoted by $H_1 \vee H_2$.

A complement reducible graph (also called a *co-graph*) is defined recursively as follows:

i) A graph on a single vertex is a complement reducible graph.
ii) If $G_1, G_2, \dots, G_k$ are complement reducible graphs, then so is their union $G_1 \cup G_2 \cup \dots \cup G_k$.
iii) If $G$ is a complement reducible graph, then so is its complement.

The co-graphs have arisen in many disparate areas of mathematics and have been independently rediscovered by various researchers. The verification of TCC for the join of two graphs will automatically shows that the co-graphs also satisfies TCC. But verifying TCC even for the join of some simple classes of graphs will pause many difficulties, which can be seen from some proofs that we have given in this paper. In our journey to verify TCC for co-graphs, we find some results that contribute more power to the validity of TCC in general but, TCC for the join of two arbitrary graphs remains still open.

Some works that have been done regarding the verification of TCC for the join of certain classes of graphs are as follows : Seoud et al. [16,17] calculated the total chromatic number of the join of two paths. Guanggrong Li and Limin Zhang [9] proved that the join of a complete in-equipartite graph and a path is type-1. In their proof the difficulty in proving TCC for the join of such graphs (that is $K_{n_1, n_2}$ for $n_1 \ne n_2$ and $P_m$) is easily visible as there arises various sub cases for a single proof (see [9]). Further Wang et al. [20] proved the equality of the vertex distinguishing total chromatic number and the total chromatic number of the join of a path with itself and a cycle with itself.

In [19], R. Vignesh et al. proved the validity of TCC for the join of a graph satisfying TCC with itself. But we found that the existence of a proper edge coloring that is just mentioned in the proof without any proper explanation is not always mandatory. Hence in order to overcome that here we give a rigorous proof using the coloring technique explained in the Lemma given in the second section.

Even though we do not have a proof for the existence of TCC, we have seen that it is proved for a vast range of graphs [12,14]. Here we are going to see the same for some graph operations namely the join of graphs and the lexicographic product of graphs.

The paper is organized as follows.

In the second section, we obtain a bound for the total chromatic number of the join two graphs and we verify TCC for $G \vee G$, when $G$ satisfies TCC.

As a result, we prove that $\bigvee_{i=1}^{2^m} G_i$ satisfies TCC if $G_i \cong G$ for $1 \leq i \leq 2^m$ and $G$ satisfies TCC. Also, we verify weak TCC for the join of two graphs under certain constraints and we prove that the join of two type-1 graphs with same order satisfies TCC. Moreover, we prove $G \vee H$ is not a type-1 graph if $G$ and $H$ are regular graphs with same odd number of vertices. In addition, we prove that $C_n \vee C_m$ satisfies TCC, for any positive integers $m$ and $n$.

In the third section, we produce an upper bound for the total chromatic number of the generalized join of graphs and hence we obtain an upper bound for the total chromatic number of the lexicographic product $G \circ H$ if $H$ satisfies TCC. And in particular we verify weak TCC for the lexicographic product of a graph with the compliment of a complete graph.

## 2  TCC for Join of Graphs

In this section, we first recall the Konig's Theorem.

**Theorem 1 (Konig [6]).** *For any bipartite graph, $\chi'(G) = \Delta(G)$.*

The following result gives a bound for the total chromatic number of the join of two graphs.

**Theorem 2.** *Let $G$ and $H$ be graphs with $m$ and $n$ vertices, respectively. If $\chi'(G) \leq \chi_T(H)$, then*

$$max\{\Delta(H) + m, \Delta(G) + n\} + 1 \leq \chi_T(G \vee H) \leq max\{m, n\} + \chi_T(H) + \chi(G).$$

*In general, $max\{\Delta(H) + m, \Delta(G) + n\} + 1 \leq \chi_T(G \vee H) \leq max\{m, n\} + max\{\chi'(G), \chi'(H)\} + \chi(H) + \chi(G)$.*

*Proof.* Let $r = max\{m, n\}$, $s = \chi(G)$ and $t = \chi_T(H)$. The lower bound is clear from the definition of join of graphs. For proving the upper bound, we construct a total coloring of $G \vee H$ using $r + s + t$ colors.

First, we color the vertices and edges of $G$ and $H$. Let $c$ be a total coloring of $H$ using $t$ colors, say $1, 2, \ldots, t$. It is given that $\chi'(G) \leq \chi_T(H)$. Hence we can color the edges of $G$ with some or all colors from $1, 2, \ldots, t$. Then color the vertices of $G$ with new $s$ colors, say $t + 1, t + 2, \ldots, t + s$. Thus, we colored the vertices and edges of $G$ and $H$ using $t + s$ colors properly.

Next, we color the uncolored edges of $G \vee H$ and they are precisely the edges between $G$ and $H$ and hence the subgraph induced by these edges form a bipartite graph with maximum degree $r$. Hence by Theorem 1, it can be colored properly using new $r$ colors, say $t + s + 1, t + s + 2, \ldots, t + s + r$. So we get a total coloring of $G \vee H$ using $r + s + t$ colors and therefore the result follows. The proof of second part is similar to that of the first one.

As an immediate consequence of Theorem 2, we have the following Corollary.

**Corollary 1.** *If $G$ is a bipartite graph and $H$ is a graph satisfying TCC and both having the same maximum degree, then $\chi_T(G \vee H) \leq \begin{cases} k + 4 \text{ if } H \text{ is type} - 2; \\ k + 3 \text{ if } H \text{ is type} - 1, \end{cases}$ where $k = \Delta(G \vee H)$.*

*Proof.* Let the maximum degree of both $G$ and $H$ be $\Delta$. Then, by Theorem 1, $\Delta = \chi'(G) < \Delta + 1 \leq \chi_T(H)$ and $\chi(G) = 2$. By Theorem 2, we have $\chi_T(G \vee H) \leq max\{m, n\} + \chi_T(H) + 2$.

First, if $G$ is type-1, then $\chi_T(G \vee H) \leq max\{m, n\} + \Delta(G) + 1 + 2$ and hence $\chi_T(G \vee H) \leq \Delta(G \vee H) + 3$.

Next, if $G$ is type-1, then $\chi_T(G \vee H) \leq max\{m, n\} + \Delta(G) + 2 + 2$ and thus $\chi_T(G \vee H) \leq \Delta(G \vee H) + 4$.

One can ask the following question.

**Problem 1.** When does the join two graphs satisfy $k$-TCC, for some $k \geq 2$?

The following results will give some partial answers to this. For proving these partial answers, we need the following Lemma. For a matching $M$ of $G$ and $v \in V(G)$, we say $v$ is $M$-*saturated* if $v$ is incident with some edge in $M$. Otherwise, $v$ is called $M$-*unsaturated*.

**Lemma 1.** *The edge set of $K_{n,n}$ can be partitioned into $n + 1$ matchings such that each vertex of $K_{n,n}$ is saturated by $n$ matchings among them.*

*Proof.* Let $X = \{u_1, u_2, \ldots, u_n\}$ and $Y = \{v_1, v_2, \ldots, v_n\}$ be the partition of $K_{n,n}$. Let $M_0 = \{u_i v_i : 1 \leq i \leq n\}$ and $R_0 = K_{n,n} - M_0$.

First, we successively define $R_j$'s and $M_j$'s as follows, for $1 \leq j \leq n - 2$,

$$R'_j = R_{j-1} - \{u_j, v_j\}, M_j = A_j \cup B_j, \text{ where}$$
$$A_j = \{u_i v_{i+j+1(mod\ n)} : 1 \leq i \leq j - 1 \text{ or } i = n\},$$
$$B_j = \{u_i v_{i+j(mod\ n)} : j + 1 \leq i \leq n - 1\} \text{ and}$$
$$R_j = R_{j-1} - M_j.$$

Next, we define

$$R'_{n-1} = R_{n-2} - \{u_{n-1}, v_{n-1}\},$$
$$M_{n-1} = \{u_i v_{2i+1(mod\ n)} : 1 \leq i \leq n \text{ and } i \neq n - 1\},$$
$$R_{n-1} = R_{n-2} - M_{n-1} \text{ and}$$
$$R'_n = R_{n-1} - \{u_n, v_n\},$$
$$M_n = \{u_i v_{2i(mod\ n)} : 1 \leq i \leq n - 1\}.$$

Clearly, $M_j$ is a matching in $R'_j$, for $1 \leq j \leq n$ and there are $n + 1$ matchings including $M_0$. Also note that each vertex $u_j$ (as well as $v_j$) in $K_{n,n}$ is $M_i$-saturated for all $i \in \{1, 2, \ldots, n\} \backslash \{j\}$, $|M_0| = n$, $|M_j| = n - 1$ for $1 \leq j \leq n$ and $E(R'_n) \backslash M_n = \emptyset$. Hence $\sum_{j=0}^{n} |M_j| = |E(K_{n,n})|$.

Finally, we have to prove $\{M_j\}_{j=0}^{n}$ are disjoint.

Clearly, $M_0 \cap M_j = \emptyset$, for $j \in \{1, 2, \ldots, n\}$. First, if there exist $j_1, j_2 \in \{1, 2, \ldots, n - 2\}$ and there exist $i, k \in \{1, 2, \ldots, n\}$ such that $j_1 \neq j_2$ and $u_i v_k \in$

$M_{j_1} \cap M_{j_2}$. Then, $i \neq j_1, j_2$ and $u_i v_k \in (A_{j_1} \cap A_{j_2}) \cup (A_{j_1} \cap B_{j_2}) \cup (B_{j_1} \cap A_{j_2}) \cup (B_{j_1} \cap B_{j_2})$.

When $u_i v_k \in A_{j_1} \cap B_{j_2}$, we have $i + j_1 + 1 \pmod n = k = i + j_2 \pmod n$, by the definition of $A_{j_1}$ and $B_{j_2}$. Hence $j_1 = j_2 - 1$ (as $|j_1 - j_2| < n$). Also, $1 \leq i \leq j_1 - 1$ or $i = n$ and $j_2 + 1 \leq i \leq n - 1$. In both cases, it is not possible.

When $u_i v_k \in A_{j_1} \cap A_{j_2}$, we have $i + j_2 + 1 \pmod n = k = i + j_1 + 1 \pmod n$. That means, $j_2 = j_1$ (as $j_1$ and $j_2$ are less than $n$), which is a contradiction.

When $u_i v_k \in B_{j_1} \cap B_{j_2}$, we have $j_1 = j_2$, which is not possible.

When $u_i v_k \in B_{j_1} \cap A_{j_2}$, then $j_2 = j_1 - 1$ and $j_1 + 1 \leq i \leq n - 1$ and $1 \leq i \leq j_2 - 1$, which is a contradiction.

So, for any two distinct $j_1, j_2 \in \{1, 2, \ldots, n - 2\}$, $M_{j_1} \cap M_{j_2} = \emptyset$. Similarly, we can show that $M_j$'s, $M_{n-1}$, and $M_n$ are disjoint for $j \in \{0, 1, 2, \ldots, n - 2\}$. Therefore $\{M_j\}_{j=1}^n$ are disjoint. Hence the result the follows.

Now, using Lemma 1, we prove that TCC is true for the join of a graph satisfying TCC with itself.

**Theorem 3.** *If $G$ is a graph satisfying TCC, then $G \vee G$ satisfies TCC.*

*Proof.* Let $V(G) = \{u_1, u_2, \ldots, u_n\}$ and $\Delta(G) = k$. Then $\Delta(G \vee G) = n + k$ and the graph $G \vee G$ can be considered as the union of three induced sub-graphs, that is two copies of $G$, say $G_1$ with vertex set $\{u_1, u_2, \ldots, u_n\}$, $G_2$ with vertex set $\{v_1, v_2, \ldots, v_n\}$ (i.e., $v_i \in V(G_2)$ is the corresponding vertex of $u_i \in V(G_1)$) and the edges between $G_1$ and $G_2$ (the induced subgraph of these edges is $K_{n,n}$). In order to verify TCC for $G \vee G$, we need to show that there is a total $(n + k + 2)$-coloring of $G \vee G$.

Let $c$ be a total coloring of $G$ with $k + 2$ colors, say $1, 2, \ldots, k + 2$. First we color the vertices and edges of $G_1$ totally and then color the edges of $G_2$ using $c$. Next, we color the edges between $G_1$ and $G_2$ by using Lemma 1 and finally, we assign colors to the vertices of $G_2$.

As the edges of $G_1$ are colored properly under $c$, we color the edges of $G_2$ also using $c$ as follows. For $v_i v_j \in E(G_2)$, $c(v_i v_j) = c(u_i u_j)$. For $i \in \{1, 2, \ldots, n\}$, we define $c_i \in \{1, 2, \ldots, k + 2\}$ such that $c_i$ is not represented at $u_i$ in $G_1$, that is, $c_i$ is not assigned to any of the elements in $\{u_i x : x \in N_{G_1}(u_i)\} \cup \{u_i\}$. Such a color $c_i$ will always exist as $c$ is a $(k + 2)$-total coloring of $G$ and $|\{c(u_i x) : x \in N_{G_1}(u_i)\} \cup \{c(u_i)\}| \leq k + 1$. By Lemma 1, we assign the total coloring $c'$ to the vertices and edges of $G \vee G$ using $c$ as follows. For $1 \leq i \neq j \leq n$,

$$
c'(x) = \begin{cases} c(x) \text{ if } x = u_i, \ x = u_i u_j \in E(G_1) \text{ or } x = v_i v_j \in E(G_2) \ ; \\ k + 2 + j \text{ if } x = v_j \text{ or } x \in M_j \ ; \\ c_i \text{ if } x = u_i v_i \in M_0. \end{cases}
$$

Then $c'$ colors the vertices and edges of $G \vee G$ using $n + k + 2$ colors.

Finally, we need to verify that $c'$ is proper. Note that for $x \in V(G_1) \cup E(G_1) \cup M_0 \cup E(G_2)$, $c'(x) \in \{1, 2, \ldots, k + 2\}$ and for $i, j \in \{1, 2, \ldots, n\}$ with $i \neq j$, $c'(u_i v_j)$, $c'(v_j) \in \{k + 3, k + 4, \ldots, n + k + 2\}$. Since $c' = c$ on $V(G_1) \cup$

$E(G_1) \cup E(G_2)$, $M_i$'s are matchings such that $u_i's$ and $v_i's$ are $M_i$-unsaturated and by the definition of $c_i$, we have $c'$ is proper. Hence the results follows.

By Theorem 3 and applying induction on $t$, we have the following corollary.

**Corollary 2.** *If a graph $G$ satisfies TCC, then $\bigvee\limits_{i=1}^{m} G_i$ satisfies TCC, where $G_i \cong G$ for $1 \leq i \leq n$ and $m = 2^t$ for any positive integer $t$.*

For two distinct graphs we cannot adopt the same method of proof since the missing colors in the corresponding vertices need not be same as in the above case. So, next we prove the validity of weaker version of TCC for the join of two graphs under certain restrictions.

**Theorem 4.** *If $G$ and $H$ are two graphs with $m$ and $n$ vertices respectively. Also, $\Delta(G) \geq \Delta(H)$, $m \leq n$ and $G$ satisfies TCC, then $\chi_T(G \vee H) \leq \Delta(G \vee H) + 3$.*

In the other case, that is for $m \geq n$, adding isolated vertices in $H$ and taking the join will results in a new graph whose maximum degree is different from that of our original $G \vee H$. Hence this method is not valid in that case. We now prove the following result on regular graphs with odd number of vertices.

**Theorem 5.** *If $G$ and $H$ are two $k$-regular graphs with same odd order $n$, then $G \vee H$ is not type-1.*

The equality of the number of vertices in both graphs plays a crucial role in the proof and hence in the cases of unequal number of vertices we cannot use this pattern. The following corollaries are the immediate consequences of Theorem 5 and Theorem 3.

**Corollary 3.** *For an odd ordered regular $G$ graph satisfying TCC, the join $G \vee G$ is type-2.*

**Corollary 4.** *For an odd positive integer $m \geq 3$, $C_m \vee C_m$ is a type-2 graph.*

The following result gives the validity of TCC for the join of two cycles.

**Proposition 1.** *For $m, n \geq 3$, the join of two cycles $C_m \vee C_n$ satisfies TCC .*

*Proof.* Let $G = C_m \vee C_n$. Clearly, $\Delta(G) = max\{m, n\} + 2$. Let $V(C_m) = \{u_1, u_2, \ldots, u_m\}$ and $V(C_n) = \{v_1, v_2, \ldots, v_n\}$. Without loss of generality, let us assume $m \geq n$. For $m = n = 3$, by Theorem 3 the result follows. So let us assume that, $m > 3$ and $n \geq 3$. We have to show that there exists a total coloring of $G$ using $\Delta(G) + 2$ colors, where $\Delta(G) = m + 2$.

*Case 1. (m and n are even)*
The following is a total coloring of $G$ using $m + 4$ colors.

$$c(u_i) = \begin{cases} 1 \text{ for } i \equiv 1 \text{ mod } 2 \text{ ;} \\ 2 \text{ for } i \equiv 0 \text{ mod } 2. \end{cases} \qquad c(v_i) = \begin{cases} 3 \text{ for } i \equiv 1 \text{ mod } 2 \text{ ;} \\ 4 \text{ for } i \equiv 0 \text{ mod } 2. \end{cases}$$

$$c(u_i u_{i+1}) = \begin{cases} 3 \text{ for } i \equiv 1 \text{ mod } 2 \text{ ;} \\ 4 \text{ for } i \equiv 0 \text{ mod } 2. \end{cases} \qquad c(v_i v_{i+1}) = \begin{cases} 1 \text{ for } i \equiv 1 \text{ mod } 2 \text{ ;} \\ 2 \text{ for } i \equiv 0 \text{ mod } 2. \end{cases}$$

Clearly, the subgraph induced by the uncolored edges forms a bipartite graph of maximum degree $m$ and hence using Theorem 1 we can properly color those edges using $m$ new colors and hence the result follows.

*Case 2. (m and n are odd)*
Consider the following coloring of $G$,

$$c(u_i) = \begin{cases} 1 \text{ for } i \equiv 1 \text{ mod } 2, \ i \neq m \text{ ;} \\ 2 \text{ for } i \equiv 0 \text{ mod } 2 \text{ ;} \\ 3 \text{ for } i = m. \end{cases} \qquad c(v_i) = \begin{cases} 4 \text{ for } i \equiv 1 \text{ mod } 2, \ i \neq n \text{ ;} \\ 5 \text{ for } i \equiv 0 \text{ mod } 2 \text{ ;} \\ 6 \text{ for } i = n. \end{cases}$$

$$c(u_i u_{i+1}) = \begin{cases} 5 \text{ for } i \equiv 1 \text{ mod } 2, \ i \neq m \text{ ;} \\ 4 \text{ for } i \equiv 0 \text{ mod } 2 \text{ ;} \\ 2 \text{ for } i = m, i+1 = 1. \end{cases} \qquad c(v_i v_{i+1}) = \begin{cases} 1 \text{ for } i \equiv 1 \text{ mod } 2, \ i \neq n; \\ 2 \text{ for } i \equiv 0 \text{ mod } 2 \text{ ;} \\ 5 \text{ for } i = n, i+1 = 1. \end{cases}$$

Next, we color some of the edges in between $C_m$ and $C_n$.
For $1 \leq i \leq n$, $c(u_i v_i) = 3$ and for $0 \leq k \leq n-1$,
$$c(u_{m-k} v_{n-k}) = \begin{cases} 6 \text{ for } 1 \leq k \leq n-1; \\ 1 \text{ for } k = 0. \end{cases}$$
The subgraph induced by the remaining uncolored edges forms a bipartite graph with maximum degree $m-2$ and hence the result follows from Theorem 1.

*Case 3. (m is even and n is odd.)*
We color the vertices an edges of $C_m$ and $C_n$ as follows:

$$c(u_i) = \begin{cases} 1 \text{ for } i \equiv 1 \text{ mod } 2 \text{ ;} \\ 2 \text{ for } i \equiv 0 \text{ mod } 2. \end{cases} \qquad c(v_i) = \begin{cases} 3 \text{ for } i \equiv 1 \text{ mod } 2, \ i \neq n \text{ ;} \\ 4 \text{ for } i \equiv 0 \text{ mod } 2 \text{ ;} \\ 5 \text{ for } i = n. \end{cases}$$

$$c(u_i u_{i+1}) = \begin{cases} 3 \text{ for } i \equiv 1 \text{ mod } 2 \text{ ;} \\ 4 \text{ for } i \equiv 0 \text{ mod } 2. \end{cases} \qquad c(v_i v_{i+1}) = \begin{cases} 1 \text{ for } i \equiv 1 \text{ mod } 2, \ i \neq n; \\ 2 \text{ for } i \equiv 0 \text{ mod } 2 \text{ ;} \\ 4 \text{ for } i = n \text{ and } i+1 = 1. \end{cases}$$

Next, we color some edges between $C_m$ and $C_n$. For $1 \leq i \leq n-1$, color $c(u_i v_i) = 5$ and also for $i = n$, color $c(u_{i+1} v_i) = 1$. Then the subgraph induced by the remaining uncolored edges form a bipartite graph of maximum degree $m-1$ and by Theorem 1, the result follows.

*Case 4. (m is odd and n is even.)*
First, we color the vertices and edges of both $C_m$ and $C_n$ using,

$$c(u_i) = \begin{cases} 1 \text{ for } i \equiv 1 \bmod 2 \text{ and } i \neq m \; ; \\ 2 \text{ for } i \equiv 0 \bmod 2 \; ; \\ 3 \text{ for } i = m. \end{cases} \qquad c(v_i) = \begin{cases} 4 \text{ for } i \equiv 1 \bmod 2 \; ; \\ 5 \text{ for } i \equiv 0 \bmod 2. \end{cases}$$

$$c(u_i u_{i+1}) = \begin{cases} 4 \text{ for } i \equiv 1 \bmod 2 \text{ and } i \neq m \; ; \\ 5 \text{ for } i \equiv 0 \bmod 2 \; ; \\ 2 \text{ for } i = m \text{ and } i + 1 = 1. \end{cases} \qquad c(v_i v_{i+1}) = \begin{cases} 1 \text{ for } i \equiv 1 \bmod 2 \; ; \\ 2 \text{ for } i \equiv 0 \bmod 2. \end{cases}$$

For $1 \leq i \leq n$, color $c(u_i v_i) = 3$. As $m > n$, the subgraph induced by the remaining uncolored edges form a bipartite graph of maximum degree $m - 1$. Hence the result follows.

As an immediate consequence of Proposition 1, we have the following corollary.

**Corollary 5** ([16]). *For any positive integers $m$ and $n$, $P_m \vee P_n$ satisfies TCC.*

## 3    Total Coloring of the Generalized Join of Graphs

In this section, we give an upper bound for the total chromatic number of $G[H_1, H_2, \ldots, H_n]$. Let $G$ be a class-1 graph. Then $E(G) = \bigcup_{i=1}^{k} M_i$, where $M_i$'s are disjoint matchings. Let $r$ be the least number in $\{1, 2, \ldots, k\}$ such that every vertex of $G$ is saturated by at least one of the matchings $M_{i_1}, M_{i_2}, \ldots, M_{i_r}$. Without loss of generality, we relabel $M_{i_j}$ by $M_j$ for $1 \leq j \leq r$.

**Theorem 6.** *Let $G$ be the above mentioned graph with $n$ vertices and $\{H_1, H_2, \ldots, H_n\}$ be a set of graphs with $H_i \vee H_j$ satisfying TCC, for each $i, j \in \{1, 2, \ldots, n\}$, then $\chi_T(G[H_1, H_2, \ldots, H_n]) \leq \sum_{i=1}^{r} s_i + \sum_{j=r+1}^{k} t_j$, where $s_j = max\{\Delta(H_x \vee H_y) + 2 \mid xy \in M_j\}$ for $1 \leq j \leq r$ and $t_j = max\{max\{|V(H_x)|, |V(H_y)| \mid xy \in M_j\}\}$ for $r + 1 \leq j \leq k$.*

The following corollary is a consequence of Theorems 3 and 6.

**Corollary 6.** *If $H$ is any graph satisfying TCC with $m$ vertices, then*

$$\chi_T(G \circ H) \leq \begin{cases} \Delta(G \circ H) + \Delta(H)(\Delta(G) - 1) + 2\Delta(G) & \text{if } G \text{ is class} - 1 \\ \Delta(G \circ H) + \Delta(G)\Delta(H) + 2(\Delta(G) + 1) + m & \text{if } G \text{ is class} - 2 \end{cases}$$

*Proof.* Clearly, $G \circ H \cong G[H_1, H_2, \ldots, H_n]$, where $H_i \cong H$ for $1 \leq i \leq n$ and $\Delta(G \circ H) = \Delta(H) + \Delta(G)m$. By Theorem 3, $H \vee H$ satisfies TCC. Then by Theorem 6, $s_j = \Delta(H) + m + 2$, for $1 \leq j \leq r$ and $t_j = m$, for $r + 1 \leq j \leq \chi'(G)$. By Theorem 6, $\chi_T(G \circ H) \leq \Delta(H)r + 2r + m\chi'(G) \leq (\Delta(H) + m + 2)\chi'(G)$.

If $G$ is a class-1 graph, then $\chi'(G) = \Delta(G)$. So, $\chi_T(G \circ H) \leq (\Delta(H) + m + 2)\Delta(G) \leq \Delta(G \circ H) + \Delta(H)(\Delta(G) - 1) + 2\Delta(G)$.

If $G$ is class-2, then $\chi'(G) = \Delta(G) + 1$. So, $\chi_T(G \circ H) \leq (\Delta(H) + m + 2)(\Delta(G) + 1) \leq \Delta(G \circ H) + \Delta(G)\Delta(H) + 2(\Delta(G) + 1) + m$. Hence the result follows.

As the bound above is a weaker one, we replace our $H$ with the compliment of complete graph and obtain the following result.

**Theorem 7.** *If $G$ satisfies TCC with $m$ vertices, then $G[K_n^c]$ satisfies weak TCC. In particular, if $G$ is type-1, then $G[K_n^c]$ satisfies TCC.*

# 4  Concluding Remarks and Open Problems

In this paper, one of our aims was to prove the validity of TCC for co-graphs by showing that TCC is valid for the join of any two graphs. But we could find some partial answers only and the TCC for the join of any two arbitrary graphs is still open. Also, we obtained a bound for the total chromatic number of $G$-generalized join of graphs and as a consequence we obtain an upper bound for the total chromatic number of the lexicographic product $G \circ H$.

**Acknowledgments.** The authors would like to express our deep gratitude to the reviewers for their valuable comments for improving the manuscript. For the second author, this research is supported by the CSIR-UGC, India, Junior Research Fellowship (UGC-Ref.No.: 201610138968).

# References

1. Behzad, M.: Graphs and their chromatic numbers. Ph.D. thesis, Michigan State University (1965)
2. Behzad, M., Chartrand, G., Cooper, J.K.: The color numbers of complete graphs. J. Londan Math. Soc. **42**, 226–228 (1967)
3. Geetha, J., Somasundaram, K.: Total colorings of product graphs. Graphs Combin. **34**(2), 339–347 (2018)
4. Geetha, J., Narayanan, N., Somasundaram, K.: Total colorings-a survey. AKCE Int. J. Graphs Combin. 1–13 (2023)
5. Kemnitz, A., Marangio, M.: Total colorings of cartesian product of graphs. Congr. Numer. **165**, 99–109 (2003)
6. König, D.: Über graphen und ihre anwendung auf determinantentheorie und mengenlehre. Math. Ann. **77**(4), 453–465 (1916). (German)
7. Kostochka, A.V.: The total chromatic number of any multigraph with maximum degree five is at most seven. Discrete Math. **162**(1–3), 199–214 (1996)
8. Kowalik, L., Sereni, J.S., Skrekovski, R.: Total coloring of plane graphs with maximum degree nine. SIAM J. Discrete Math. **22**(4), 1462–1479 (2008)
9. Li, G., Zhang, L.: Total chromatic number of one kind of join graphs. Discret. Math. **306**(16), 1895–1905 (2003)
10. Basavaraju, M., Chandran, L.S., Francis, M.C., Naskar, A.: Weakening total coloring conjecture: weak TCC and Hadwiger's conjecture on total graphs. arXiv:2107.09994v3 [math.CO] (2022)
11. McDiarmid, C.J.H., Sánchez-Arroyo A.: Total coloring regular bipartite graphs is NP-hard. Discret. Math. **124**, 155–162 (1994)
12. Chen, M., Guo, X., Li, H., Zhang, L.: Total chromatic number of generalized Mycielski graphs. Discret. Math. **334**, 48–51 (2014)

13. Molloy, M., Reed, B.: A bound on total chromatic number. Combinatorica **18**, 241–280 (1998)
14. Mycielski, J.: Sur le coloriage des graphs. Colloq. Math. **3**, 161–162 (1955)
15. Sánchez-Arroyo, A.: Determining the total coloring number is NP-hard. Discrete Math. **78**, 315–319 (1989)
16. Seoud, M.A.: Total chromatic numbers. Appl. Math. Lett. **5**(6), 37–39 (1992)
17. Seoud, M.A., Maqsoud, A., Wilson, R.J., Williams, J.: Total colorings of cartesian products. Int. J. Math. Educ. Sci. Technol. **28**(4), 481–487 (1997)
18. Prajnanaswaroopa, S., Geetha, J., Somasundaram, K., Hung-Lin, F., Narayanan, N.: On total coloring of some classes of regular graphs. Taiwan. J. Math. **26**(4), 667–683 (2002)
19. Vignesh, R., Geetha, J., Somasundaram, K.: total coloring conjecture for certain classes of graphs. Algorithms **11**(10), 161 (2018)
20. Wang, Z., Yan, L., Zhang, Z.: Vertex distinguishing equitable total chromatic number of join graph. Acta Math. Appl. Sinica **23**(3), 433–438 (2007)
21. Yap, H.P.: Total Colorings of Graphs. Lecture Notes in Mathematics. Springer, Berlin (1996). https://doi.org/10.1007/BFb0092895

# Star Colouring of Regular Graphs Meets Weaving and Line Graphs

M. A. Shalu[1] and Cyriac Antony[2]($\boxtimes$)

[1] Indian Institute of Information Technology, Design & Manufacturing (IIITDM)
Kancheepuram, Chennai, India
shalu@iiitdm.ac.in
[2] IIT Madras, Chennai, India
ma23r004@smail.iitm.ac.in

**Abstract.** For $q \in \mathbb{N}$, a $q$-star colouring of a graph $G$ is a proper $q$-colouring $f$ of $G$ such that there is no path $u, v, w, x$ in $G$ with $f(u) = f(w)$ and $f(v) = f(x)$ (the violating path need not be induced). For $p \geq 2$, Shalu and Antony (Discrete Math., 2022) proved that at least $p + 2$ colours are required to star colour a $2p$-regular graph $G$, and characterised the class $\mathcal{G}$ of graphs $G$ for which $p + 2$ colours suffices in terms of graph orientations. In the second author's thesis (2023), we provided a characterisation of the class $\mathcal{G}$ in terms of locally constrained graph homomorphisms. In this paper, we characterise $\mathcal{G}$ in terms of weaving patterns of edge decompositions. We also show that the study of class $\mathcal{G}$ is tied to the theory of line graphs and line digraphs of complete graphs. We prove that if a $K_{1,p+1}$-free $2p$-regular graph $G$ with $p \geq 2$ is $(p + 2)$-star colourable, then $-2$ and $p-2$ are eigenvalues of the adjacency matrix of $G$.

**Keywords:** Star coloring · Regular graphs · Cyclic plain weaving · Edge decomposition · Line digraph · Graph homomorphism · Graph orientation

## 1 Introduction

Star colouring is a variant of graph colouring used in the estimation of sparse Hessian matrices [9]. Nešetřil and Mendez [12] related the minimum number of colours required to star colour a graph to the chromatic numbers of its minors. Star colouring is studied for various graph classes, and is extensively studied for planar graphs [6, Section 14] and line graphs [11]. Speaking of line graphs, the class of line graphs has a rich theory as evidenced by the vast literature devoted to it. We refer the reader to the recent book "Line graphs and line digraphs" [5] by Beineke and Bagga. This book also devotes several chapters to line digraphs. The line digraph operation on a digraph is a natural adaptation of the line graph operation to digraphs. We do not use the notion of line digraph of a digraph in this paper. Rather, we use a closely related operation called line

© The Author(s), under exclusive license to Springer Nature Switzerland AG 2024
S. Kalyanasundaram and A. Maheshwari (Eds.): CALDAM 2024, LNCS 14508, pp. 313–327, 2024.
https://doi.org/10.1007/978-3-031-52213-0_22

digraph operation on an undirected graph discussed by the same authors in [4], which turns out to be important for star colouring of regular graphs.

For $d \geq 2$, at least $(d + 4)/2$ colours are required to star colour a $d$-regular graph, and $(d + 4)/2$ colours suffices only if $d$ is even [14]. The class $\mathcal{G}$ of $d$-regular graphs that admit a $((d+4)/2)$-star colouring (i.e., $2p$-regular $(p+2)$-star colourable graphs with $p \geq 2$) can be characterised in terms of graph orientations [14] and graph homomorphisms [3]. We emphasise that given a positive integer $p \geq 2$ and a $2p$-regular graph $G$, it is NP-complete to check whether $G \in \mathcal{G}$, even when $p = 2$ [3, Corollary 5.1]. In this paper, we show that the study of the graph class $\mathcal{G}$ is tied to the theory of line graphs, and even more so to the theory of line digraphs. We obtain a necessary condition on $K_{1,p+1}$-free graphs $G \in \mathcal{G}$ in terms of eigenvalues of $G$. More importantly, we characterise class $\mathcal{G}$ in terms of 'plain weavings' of edge decompositions (motivation and examples appear at the end of this section, and complete definitions appear in Sect. 2). The main results of this paper are the following (see Sect. 2 for definitions).

- *Theorem* 4: For every (undirected) graph $H$, the underlying undirected graph of the line digraph of $H$ admits a locally bijective homomomorphism to the line graph of $H$.
- *Theorem* 5: If a $K_{1,p+1}$-free $2p$-regular graph $G$ with $p \geq 2$ is $(p + 2)$-star colourable, then $-2$ and $p - 2$ are eigenvalues of $G$.
- *Theorem* 6: Let $G$ be a $2p$-regular graph with $p \geq 2$. Then, $G$ admits a $(p + 2)$-star colouring if and only if $G$ admits an orientation $\vec{G}$ and an edge decomposition $S = \{H_0, H_1, \ldots, H_{p+1}\}$ such that the following hold:
  (i)   each $H_i$ is $p$-regular ($i \in \mathbb{Z}_{p+2}$);
  (ii)  $S$ admits a plain weaving $\psi$ consistent with $\vec{G}$; and
  (iii) for distinct $i, j \in \mathbb{Z}_{p+2}$ and distinct $u, v \in V(H_i) \cap V(H_j)$,
        $uv \notin E(G)$ and $N_G(u) \cap N_G(v) = N_{\vec{G}}^+(u) \cap N_{\vec{G}}^+(v)$.

We also prove a result similar to Theorem 6 that characterises certain types of graph orientations called colourful Eulerian orientations.

Weaving patterns are of paramount importance not only in industries such as textiles [1], but also in digital fabrication [13], computer graphics [10], and topological graph theory [2]. A plain weave is the simplest type of weaving pattern. In a plain weave, warp and weft (threads) form a criss-cross pattern. Akleman, Chen and Gross [2] introduced a notion called cyclic plain-weaving for (topological) graphs on surfaces. A cyclic plain-weaving on a sphere is essentially equivalent to the notion of an alternating projection of a link in knot theory [2]. We define a weave of an arbitrary edge decomposition $S$ of a graph and in particular a plain weave of $S$ in Sect. 2. We exhibit an example that visualises the notion here.

Figure 1 exhibits an edge decomposition of a graph $G$. Observe that edge decompositions of a graph $G$ into $q$ or fewer subgraphs correspond to $q$-edge labellings of $G$ (i.e., functions $h \colon E(G) \to \mathbb{Z}_q$). We often represent edge decompositions in diagrams by the corresponding edge labellings. Figure 2 displays a 'cyclic plain weaving' of the edge decomposition shown in Fig. 1 (we remark that such simple visual representations are not possible if a vertex of $G$ lies in three or more members of $S$).

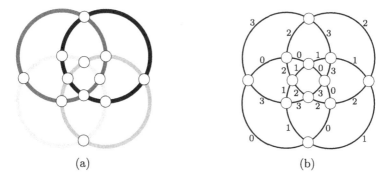

**Fig. 1.** Two visual representations of an edge decomposition $S$ of a graph $G$.

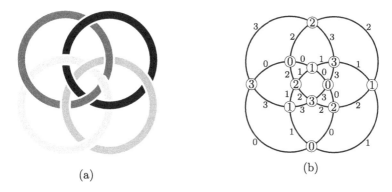

**Fig. 2.** Two visual representations of a (cyclic) plain weaving $\psi$ of the edge decomposition $S = \{H_0, H_1, H_2, H_3\}$ in Fig. 1. In (a), the 'crossing points' are the vertices of the graph, and the members of the edge decomposition are distinguished by colours. In (b), a label $i$ on a vertex $v$ denotes that $H_i$ is on top at the corresponding 'crossing point'.

This paper is organised as follows. The necessary definitions and preliminaries appear in Sect. 2. The results appear in Sect. 3, and we conclude with Sect. 4.

## 2    Definitions and Preliminaries

All graphs in this paper are finite, simple and undirected unless otherwise specified. We follow West [15] for graph theory terminology and notation. In particular, we write an edge of (an undirected) graph as $\{u, v\}$ or simply $uv$, and write an arc as $(u, v)$.

For a positive integer $q$, a (proper) $q$-colouring of a graph $G$ is a function $f \colon V(G) \to \mathbb{Z}_q$ such that $f(u) \neq f(v)$ for every edge $uv$ of $G$. For $q \in \mathbb{N}$, a $q$-star colouring of a graph $G$ is a proper $q$-colouring $f$ of $G$ such that there is no path $u, v, w, x$ in $G$ with $f(u) = f(w)$ and $f(v) = f(x)$ (the violating path need not be induced). In other words, a proper $q$-colouring of $G$ is a $q$-star colouring of

$G$ if every bicoloured subgraph of $G$ is a disjoint union of stars; hence the name star colouring.

An orientation $\vec{G}$ of a graph $G$ is the directed graph obtained by assigning a direction on each edge of $G$; that is, if $uv$ is an edge in $G$, then either $(u,v)$ or $(v,u)$ is an arc in $\vec{G}$. For a vertex $v$ of a graph $G$, the neighbourhood of $v$ in $G$, denoted by $N_G(v)$, is the set of neighbours of $v$ (i.e., vertices adjacent to $v$) in $G$. A vertex $w$ is an *out-neighbour* (resp. in-neighbour) of a vertex $v$ in an orientation $\vec{G}$ if $(v,w)$ (resp. $(w,v)$) is an arc in $\vec{G}$. For a vertex $v$ of an orientation $\vec{G}$, the out-neighbourhood of $v$ in $\vec{G}$, denoted by $N_{\vec{G}}^+(v)$, is the set of out-neighbours of $v$ in $\vec{G}$. An orientation $\vec{G}$ of a graph $G$ is *Eulerian* if for every vertex $v$ of $\vec{G}$, the number of in-neighbours of $v$ equals the number of out-neighbours of $v$.

A *homomorphism* from a graph $G$ to a graph $H$ is a function $\psi\colon V(G) \to V(H)$ such that $\psi(u)\psi(v)$ is an edge in $H$ whenever $uv$ is an edge in $G$. If $\psi$ is a homomorphism from $G$ to $H$ and $\psi(v) = w$, then we say that $v$ *is a copy of* $w$ *in* $G$ (under $\psi$).

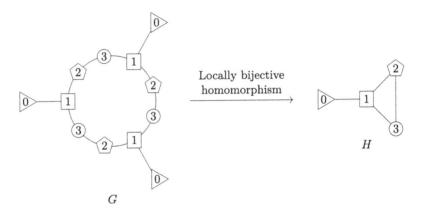

**Fig. 3.** A locally bijective homomorphism from a graph $G$ to a graph $H$. The vertices in $H$ are labelled distinct and are drawn by distinct shapes. For each vertex $w$ of $H$, each copy of $w$ in $G$ has the same label and shape as $w$.

A *Locally Bijective Homomorphism* (in short, *LBH*) from $G$ to $H$ is a function $\psi\colon V(G) \to V(H)$ such that for every vertex $v$ of $G$, the restriction of $\psi$ to the neighbourhood $N_G(v)$ is a bijection from $N_G(v)$ onto $N_H(\psi(v))$ [8] (observe that such a function $\psi$ is always a homomorphism from $G$ to $H$). In other words, a homomorphism $\psi$ from $G$ to $H$ is locally bijective if for each vertex $w$ of $H$ and each neighbour $x$ of $w$ in $H$, each copy of $w$ in $G$ has exactly one copy of $x$ in $G$ as its neighbour (see Fig. 3 for an example). An *out-neighbourhood bijective homomorphism* from $\vec{G}$ to $\vec{H}$ is a function $\psi\colon V(\vec{G}) \to V(\vec{H})$ such that for every vertex $v$ of $\vec{G}$, the restriction of $\psi$ to the out-neighbourhood $N_{\vec{G}}^+(v)$ is a bijection from $N_{\vec{G}}^+(v)$ to $N_{\vec{H}}^+(\psi(v))$ [3].

For brevity, throughout this paper, we call an eigenvalue of the adjacency matrix of a graph $G$ as an *eigenvalue of* $G$. The same convention is adopted for characteristic polynomials.

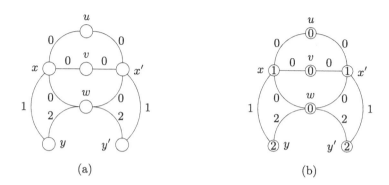

(a)                                              (b)

**Fig. 4.** An example of a plain weave. (a) an edge decomposition $S = \{H_0, H_1, H_2\}$ of a graph; (b) a weaving of $S$ which is a plain weave (a label $i$ on a vertex $v$ denotes that $(\psi(v))(1) = H_i$; note that specifying $(\psi(v))(1)$ for each vertex $v$ uniquely identifies the weaving in this example since each vertex $v$ belongs to one or two members of $S$).

An edge decomposition of $G$ is a set of subgraphs $H_0, H_1, \ldots, H_{q-1}$ of $G$ (where $q \in \mathbb{N}$) such that $\{E(H_i) : i \in \mathbb{Z}_q\}$ is a partition of $E(G)$.

Next, we define plain weavings of edge decompositions. Let $G$ be a graph without any isolated vertex, and let $S$ be an edge decomposition of $G$ such that no member of $S$ contains an isolated vertex. For each vertex $v$ of $G$, let $S_v$ denote the set of members of $S$ that include $v$ (i.e., $S_v = \{H \in S : v \in V(H)\}$). For instance, for the edge decomposition $S = \{H_0, H_1, H_2\}$ in Fig. 4a, the sets are $S_u = S_v = \{H_0\}$, $S_w = \{H_0, H_2\}$, $S_x = S_{x'} = \{H_0, H_1\}$ and $S_y = S_{y'} = \{H_1, H_2\}$.

A *weaving* of the edge decomposition $S$ of $G$ is a function $\psi$ that maps each vertex $v$ of $G$ to an ordering of members of $S_v$; formally, $\psi(v)$ is a bijection from $\{1, 2, \ldots, |S_v|\}$ onto $S_v$. A good way to visualise the notion is to imagine members of $S$ being laid out as 'threads' on a surface with layers, where for each vertex $v$, the set $S_v$ is the set of threads that are present at $v$ and $\psi$ determines which member of $S_v$ is at the top layer at $v$, which member is at the layer below it, and so on. For example, the weaving in Fig. 4b maps each vertex $z$ of the graph to a function $\psi_z$ (i.e., $\psi(z) = \psi_z$), where $\psi_u$ and $\psi_v$ are $(1 \rightarrow H_0)$; $\psi_w$ is $\begin{pmatrix} 1 \rightarrow H_0 \\ 2 \rightarrow H_2 \end{pmatrix}$; $\psi_x$ and $\psi_{x'}$ are $\begin{pmatrix} 1 \rightarrow H_1 \\ 2 \rightarrow H_0 \end{pmatrix}$; $\psi_y$ and $\psi_{y'}$ are $\begin{pmatrix} 1 \rightarrow H_2 \\ 2 \rightarrow H_1 \end{pmatrix}$. For instance, $\psi_y : \{1, 2\} \rightarrow \{H_1, H_2\}$ is defined as $\psi_y(1) = H_2$ and $\psi_y(2) = H_1$.

A weaving $\psi$ of $S$ is a *plain weave* if for each $H \in S$ and for each $uv \in E(H)$, either $(\psi(u))(1) = H$ or $(\psi(v))(1) = H$ (see Fig. 4 for an example). The definition ensures that for each $H \in S$ and for each path $P$ in $H$, the list of vertices visited while traversing the path $P$ follow an under-over-under-over-... pattern (that is, $H$ is at the top layer precisely at alternate vertices). This gives a criss-cross pattern, especially when each vertex is shared by exactly two members of $S$ (the pattern is more evident when $G$ is a 4-regular graph and members of $S$ are 2-regular subgraphs as in Fig. 2).

For integers $p \geq 2$ and $q$, an Eulerian orientation $\vec{G}$ of a $2p$-regular graph $G$ is called a *q-colourful Eulerian orientation (in short, q-CEO)* of $G$ if there exists a $q$-colouring $f$ of $\vec{G}$ such that the following hold for every vertex $v$ of $\vec{G}$ [14]:

- no out-neighbour of $v$ has the same colour as an in-neighbour of $v$;
- out-neighbours of $v$ have pairwise distinct colours; and
- all in-neighbours of $v$ have the same colour.

See Fig. 5a on page 9 for an example of a $q$-CEO. If a graph $G$ contains diamond (i.e., $K_4 - e$) or circular ladder graph $CL_{2k+1}$ as a subgraph, then $G$ does not admit a $q$-CEO for any $q \in \mathbb{N}$ [3].

Let $S$ be an edge decomposition of a graph $G$. The following condition in the definition of a plain weave $\psi$ of $S$ allows us to connect the notion to orientations of $G$: "for each $H \in S$ and for each $uv \in E(H)$, either $(\psi(u))(1) = H$ or $(\psi(v))(1) = H$". Observe that an easy way to model this condition is to ask for a fixed orientation $\vec{G}$ of $G$, and stipulate that for each edge $uv$ of $H$, the value $(\psi(u))(1) = H$ if and only if $(u, v)$ is an arc in $\vec{G}$. Thus, we have the following.

**Theorem 1.** *An edge decomposition $S$ of a graph $G$ admits a plain weaving if and only if there exists an orientation $\vec{G}$ of $G$ that satisfies the following for every vertex $v$ of $G$: (i) there exists $H \in S_v$ such that all edges incident to $v$ in $H$ are out-edges of $v$ in $\vec{G}$, and (ii) for every other member $H'$ of $S_v$, all edges incident to $v$ in $H'$ are in-edges of $v$ in $\vec{G}$.* □

Let $S$ be an edge decomposition of a graph $G$, and let $\vec{G}$ be an orientation of $G$. We say that a plain weaving $\psi$ of $S$ is *consistent* with $\vec{G}$ if for each $H \in S$ and for each $uv \in E(H)$, we have $(\psi(u))(1) = H$ if and only if $(u, v)$ is an arc in $\vec{G}$.

**Observation 1.** *Let $G$ be a $2p$-regular graph for some positive integer $p$. Then, plain weavings of edge decompositions of $G$ into $p$-regular subgraphs are in one-to-one correspondence with Eulerian orientations of $G$.* □

Next, we define the line digraph operation. The *line digraph* of (an undirected) graph $H$ has vertex set $\bigcup_{\{u,v\} \in E(H)} \{(u, v), (v, u)\}$, and there is an arc in it from $(u, v)$ to $(v, w)$ for distinct $u, v, w \in V(H)$ with $uv, vw \in E(H)$ [4]. We denote the line digraph of a graph $H$ by $\vec{L}(H)$, and the underlying undirected graph by $L^*(H)$.

## 3    Results

First, we point out that the graphs $G_{2p}$ and $\overrightarrow{G_{2p}}$ (defined in [14] and [3], respectively) are isomorphic to $L^*(K_{p+2})$ and $\vec{L}(K_{p+2})$, respectively (where $p \geq 2$). To this end, let us recall the definitions of $G_{2p}$ and $\overrightarrow{G_{2p}}$. The oriented graph $\overrightarrow{G_{2p}}$ has vertex set $\{(i,j) \colon i,j \in \mathbb{Z}_{p+2}, \ i \neq j\}$, and edge set $\{((i,j),(k,i)) \colon i,j,k \in \mathbb{Z}_{p+2}\}$. The graph $G_{2p}$ is the underlying undirected graph of $\overrightarrow{G_{2p}}$. Consider $\mathbb{Z}_q$ as the vertex set of $K_q$ for $q \in \mathbb{N}$. Observe that, by definition, the line digraph $\vec{L}(K_{p+2})$ is the graph with vertex set $\{(i,j) \colon i,j \in \mathbb{Z}_{p+2}, \ i \neq j\}$ and edge set $\{((i,j),(j,k)) \colon i,j,k \in \mathbb{Z}_{p+2}\}$. Hence, the function $\psi$ from $\{(i,j) \colon i,j \in \mathbb{Z}_{p+2}, \ i \neq j\}$ to itself defined as $\psi((i,j)) = (j,i)$ is an isomorphism from $\overrightarrow{G_{2p}}$ to $\vec{L}(K_{p+2})$. Therefore, $\overrightarrow{G_{2p}}$ is isomorphic to $\vec{L}(K_{p+2})$ (as digraphs), and thus the underlying undirected graphs $G_{2p}$ and $L^*(K_{p+2})$ are isomorphic to each other. As a result, we can rephrase Theorem 5.9 and "part I $\Longleftrightarrow$ IV" of Theorem 5.1 in [3] as follows.

**Theorem 2.** *Let $G$ be a $K_{1,p+1}$-free $2p$-regular graph with $p \geq 2$. Then, $G$ admits a $(p+2)$-star colouring if and only if $G$ admits a locally bijective homomorphism to $L^*(K_{p+2})$.* □

**Theorem 3.** *Let $G$ be a $2p$-regular graph with $p \geq 2$. Then, $G$ admits a $(p+2)$-star colouring if and only if $G$ has an orientation $\vec{G}$ that admits an outneighbourhood bijective homomorphism to $\vec{L}(K_{p+2})$.* □

Next, we show that the underlying undirected graph of the line digraph of $H$ always admits an LBH to the line graph of $H$.

**Theorem 4.** *For every graph $H$, there is an LBH from $L^*(H)$ to $L(H)$.*

*Proof.* Let $H$ be a graph. Note that the vertex set of the graph $L^*(H)$ is $\cup_{uv \in E(H)}\{(u,v),(v,u)\}$. Define $\psi \colon V(L^*(H)) \rightarrow E(H)$ as $\psi((u,v)) = \{u,v\}$ for every $(u,v) \in V(L^*(H))$. For each edge $\{(u,v),(v,w)\}$ of $L^*(H)$ where $u,v,w \in V(H)$, we have $\{\psi((u,v)),\psi((v,w))\} = \{uv,vw\}$ is an edge in $L(H)$. Hence, $\psi$ is a homomorphism from $L^*(H)$ to $L(H)$. It remains to prove that $\psi$ is locally bijective. To prove this, it suffices to show that for an arbitrary vertex $x$ of $L(H)$, and an arbitrary copy $w$ of $x$ in $L^*(H)$ under $\psi$ (i.e., $\psi(w) = x$), the members of $N_{L^*(H)}(w)$ are precisely copies of members of $N_{L(H)}(x)$ in $L^*(H)$ in a bijective fashion. To this end, consider an arbitrary vertex $u_1v_1$ of $L(H)$, where $u_1, v_1 \in V(H)$. Let $u_1, u_2, \ldots, u_k$ be the neighbours of $v_1$ in $H$, and let $v_1, v_2, \ldots, v_\ell$ be the neighbours of $u_1$ in $H$ (where $k, \ell \in \mathbb{N}$). The neighbours of $u_1v_1$ in $L(H)$ are $u_1v_2, \ldots, u_1v_\ell$ (provided $\ell > 1$) and $v_1u_2, \ldots, v_1u_k$ (provided $k > 1$). By the definition of $\psi$, for each vertex $yz$ in $L(H)$ (where $y, z \in V(H)$), the copies of $yz$ in $L^*(H)$ (under $\psi$) are $(y,z)$ and $(z,y)$. In particular, the copies of $u_1v_1$ in $L^*(H)$ are $(u_1,v_1)$ and $(v_1,u_1)$. The neighbours of $(u_1,v_1)$ in $L^*(H)$ are $(v_1,u_2), \ldots, (v_1,u_k),(v_2,u_1), \ldots, (v_\ell,u_1)$, which are precisely copies of $v_1u_2, \ldots, v_1u_k, u_1v_2, \ldots, u_1v_\ell$ in $L^*(H)$ respectively in a bijective fashion. Similarly, the neighbours of $(v_1, u_1)$ in $L^*(H)$ are $(u_1, v_2), \ldots, (u_1, v_\ell)$, $(u_2, v_1), \ldots, (u_k, v_1)$, which are precisely copies of $u_1 v_2, \ldots, u_1 v_\ell, v_1 u_2, \ldots, v_1 u_k$ in $L^*(H)$ respectively in a bijective fashion. That is, for each copy $w$ of $u_1 v_1$ in $L^*(H)$, the members of $N_{L^*(H)}(w)$ are precisely copies of members of $N_{L(H)}(u_1 v_1)$ in $L^*(H)$ in a bijective fashion. Since $u_1 v_1 \in V(L(H))$ is arbitrary, $\psi$ in an LBH from $L^*(H)$ to $L(H)$.                                                                  □

Let $G$ be a graph that admits an LBH to $L^*(K_{p+2})$, where $p \geq 2$. Since $L^*(K_{p+2})$ admits an LBH from $L(K_{p+2})$ as well, the graph $G$ admits an LBH to $L(K_{p+2})$ [7]. As a result, the characteristic polynomial of $G$ is divisible by that of $L(K_{p+2})$ [8]. Since the characteristic polynomial of $L(K_{p+2})$ in $x$ is $(x - 2p)(x - p + 2)^{p+1}(x + 2)^{(p-1)(p+2)/2}$ [5, Table 4.1], this polynomial divides the characteristic polynomial of $G$ in $x$. Since each $K_{1,p+1}$-free $2p$-regular $(p+2)$-star colourable graph admits an LBH to $L^*(K_{p+2})$ (by Theorem 2), we have the following.

**Theorem 5.** *Let $G$ be a $K_{1,p+1}$-free $2p$-regular graph with $p \geq 2$. If $G$ is $(p+2)$-star colourable, then $p-2$ and $-2$ are eigenvalues of $G$ with multiplicities at least $p + 1$ and $(p - 1)(p + 2)/2$, respectively.*                                        □

To characterise $2p$-regular $(p + 2)$-star colourable graphs in terms of plain weavings of edge decompositions, we first characterise $2p$-regular graphs that admit a $q$-CEO in terms of plain weaving of edge decompositions in Lemma 1 below (where $p \geq 2$ and $q \in \mathbb{N}$). Recall that a plain weaving $\psi$ of $S$ is *consistent* with $\vec{G}$ if for each $H \in S$ and for each $uv \in E(H)$, we have $(\psi(u))(1) = H$ if and only if $(u, v)$ is an arc in $\vec{G}$.

**Lemma 1.** *Let $p \geq 2$ and $q \geq 2$. Let $G$ be a $2p$-regular graph, and let $\vec{G}$ be an orientation of $G$. Then, $\vec{G}$ is a $q$-CEO of $G$ if and only if $G$ admits an edge decomposition $S = \{H_0, H_1, \ldots, H_{q-1}\}$ that satisfies the following:*

*(i) each $H_i$ is $p$-regular $(i \in \mathbb{Z}_q)$;*
*(ii) $S$ admits a plain weaving consistent with $\vec{G}$; and*
*(iii) for distinct $i, j \in \mathbb{Z}_q$ and distinct $u, v \in V(H_i) \cap V(H_j)$,*
      $uv \notin E(G)$ and $N_G(u) \cap N_G(v) = N_{\vec{G}}^+(u) \cap N_{\vec{G}}^+(v)$.

*Proof Overview.* Let $\vec{G}$ be a $q$-CEO of $G$ with respect to a $q$-colouring $f$ of $G$. Then, "moving" the vertex colours to arcs along the arc directions gives a $q$-arc labelling $h$ of $\vec{G}$ (see Fig. 5), and the edge decomposition (of $G$) corresponding to $h$ satisfies Properties (i) to (iii) in the lemma statement.

Conversely, suppose that the edge decomposition (of $G$) corresponding to a $q$-arc labelling $h$ of $G$ satisfies Properties (i) to (iii) in the lemma statement. Then, the function $f : V(G) \to \mathbb{Z}_q$ defined as $f(v) = i$ for each vertex $v$ of $G$ with $(\psi(v))(1) = H_i$ is a $q$-star colouring of $G$, and $\vec{G}$ is a $q$-CEO of $G$ with $f$ as the underlying colouring (see Fig. 5).

*Proof of Lemma 1.* To prove the forward direction, suppose that $\vec{G}$ is a $q$-CEO of $G$. That is, $\vec{G}$ is an Eulerian orientation and there exists a $q$-colouring $f$ of $\vec{G}$, say $f : V(\vec{G}) \to \mathbb{Z}_q$, such that the following hold for every vertex $v$ of $\vec{G}$:

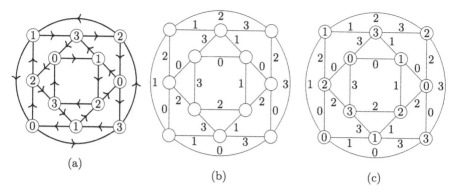

**Fig. 5. (a)** A 4-CEO of the graph $L(Q_3)$ with the underlying 4-colouring shown as a vertex labelling; **(b)** an edge decomposition of $L(Q_3)$; and **(c)** a plain weaving of the edge decomposition in (b) consistent with the orientation in (a). The edge decomposition in (b) can be obtained from (a) by "moving" the vertex labels to arcs along the arc directions (and then ignoring the arc direction). From the plain weaving in (c), one can obtain a $q$-star colouring $f$ of $L(Q_3)$ by taking the label at each vertex $v$ as the colour of $v$, and this colouring $f$ induces the orientation of $L(Q_3)$ displayed in (a).

- no out-neighbour of $v$ has the same colour as an in-neighbour of $v$,
- out-neighbours of $v$ have pairwise distinct colours, and
- all in-neighbours of $v$ have the same colour.

Let $h$ be the arc labelling of $\vec{G}$ obtained from the coloured oriented graph $(\vec{G}, f)$ by "moving" the colours from vertices to arcs along the arc directions; that is, $h: E(\vec{G}) \to \mathbb{Z}_q$ is defined as $h((u,v)) = f(u)$ for every arc $(u,v)$ of $\vec{G}$ (see Fig. 5 for an example). Clearly, $h$ induces a $q$-edge labelling $h^*$ of $G$; that is, $h^*: E(G) \to \mathbb{Z}_q$ is defined for each edge $uv$ of $G$ as $h^*(uv) = h((u,v))$ if $(u,v)$ is an arc in $\vec{G}$, and $h^*(uv) = h((v,u))$ if $(v,u)$ is an arc in $\vec{G}$. The edge labelling $h^*$ of $G$ corresponds to an edge decomposition $S$ of $G$ into $q$ subgraphs $H_0, H_1, \ldots, H_{q-1}$. Formally, each $H_i \in S$ is defined as the subgraph of $G$ induced by $E_i$, where $E_i = \{uv \in E(G): h^*(uv) = i\}$. We show that $S$ satisfies Properties (i) to (iii) in the lemma statement.

**Claim 1:** Each $H_i$ is $p$-regular ($i \in \mathbb{Z}_q$).

Let $v$ be an arbitrary vertex of $H_i$. Since $\vec{G}$ is an Eulerian orientation, $v$ has $p$ in-neighbours $w_1, w_2, \ldots, w_p$ and $p$ out-neighbours $x_1, x_2, \ldots, x_p$ in $\vec{G}$. Since $H_i$ is the subgraph of $G$ induced by $E_i$, an edge $e$ of $G$ is incident with $v$ in $H_i$ (in particular, $e \in E(H_i)$). In $\vec{G}$, the edge $e$ is oriented either towards $v$ or away from $v$.

*Case 1 (of Claim 1):* $e$ is oriented towards $v$ in $\vec{G}$.
In this case, we may assume that $e = vw_1$. Then, $i = h^*(vw_1) = h((w_1, v))$ since $vw_1 \in E(H_i)$. Therefore, $f(w_1) = h((w_1, v)) = i$ by the definition of $h$. Thus, $f(w_1) = f(w_2) = \cdots = f(w_p) = i$ because all in-neighbours of $v$ have

the same colour under $f$ in the $q$-CEO $\vec{G}$. For $1 \leq j \leq p$, we have $h^*(vw_j) = h((w_j, v)) = f(w_j) = i$, and thus $vw_j \in E(H_i)$. That is, $vw_1, vw_2, \ldots, vw_p$ are edges incident on $v$ in $H_i$. Since $f$ is a vertex colouring of $G$, we have $f(v) = \ell$ for some $\ell \in \mathbb{Z}_q \setminus \{i\}$, and $h^*(vx_j) = h((v, x_j)) = f(v) = \ell$ for $1 \leq j \leq q$. Hence, $vx_j \in E(H_\ell)$ for $1 \leq j \leq q$, and in particular $vx_j$ is not an edge in $H_i$. Thus, exactly $p$ edges are incident on $v$ in $H_i$.

*Case 2 (of Claim 1): $e$ is oriented away from $v$ in $\vec{G}$.*
In this case, we may assume that $e = vx_1$. Since $vx_1 \in E(H_i)$, we have $i = h^*(vx_1) = h((v, x_1))$, and thus $f(v) = h((v, x_1)) = i$. Therefore, $h((v, x_j)) = f(v) = i$ for $1 \leq j \leq p$. Hence, $vx_1, vx_2, \ldots, vx_p$ are edges incident on $v$ in $H_i$. For $1 \leq j \leq p$, we have $h^*(vw_j) = h((w_j, v)) = f(w_j) \neq f(v) = i$, and thus $vw_j \notin E(H_i)$. Thus, exactly $p$ edges are incident on $v$ in $H_i$.

Since $v$ is arbitrary and there are exactly $p$ edges incident on $v$ in $H_i$ in both cases, $H_i$ is $p$-regular. This proves Claim 1.

Since $G$ is $2p$-regular and each $H_i \in S$ is $p$-regular, each vertex of $G$ is in exactly two members of $S$. That is, for each vertex $v$ of $G$, the set $S_v$ of members of $S$ that include $v$ has cardinality exactly 2.

**Claim 2:** $S$ admits a plain weaving $\psi$ consistent with $\vec{G}$ (i.e., for every edge $uv \in E(H_i)$, $(\psi(v))(1) = H_i$ if and only if $(u, v)$ is an arc in $\vec{G}$).

We need to define $\psi$. To this end, we first prove that if $f(v) = i$ for some vertex $v$ of $G$, then $v \in V(H_i)$ (in other words, $H_{f(v)} \in S_v$ for each $v \in V(G)$). Let $f(v) = i$ for some $v \in V(G)$. Since $\vec{G}$ is an Eulerian orientation, $p$ edges incident on $v$ in $G$ are oriented away from $v$ by $\vec{G}$. Let $e = uv$ be one such edge in $G$. Clearly, $h^*(uv) = h((v, u)) = f(v) = i$, and thus $uv \in E_i$. Since $H_i = G[E_i]$, we have $v \in V(H_i)$. Hence, $H_{f(v)} \in S_v$ for each $v \in V(G)$.

We are ready to define $\psi$ now. Define $\psi$ as the weaving of $S$ that maps each vertex $v$ of $G$ to the function $b_v$ defined as follows: $b_v$ is the bijection from $\{1, 2\}$ to $S_v$ that maps 1 to $H_{f(v)}$ and 2 to the unique member of $S_v \setminus \{H_{f(v)}\}$ (note that $b_v$ is well-defined because $|S_v| = 2$ and $H_{f(v)} \in S_v$).

Next, we show that $\psi$ is a plain weaving of $S$. We know that each vertex of $G$ is in exactly two members of $S$. Hence, to prove that $\psi$ is a plain weaving of $S$, it suffices to show that for each $H_i \in S$ and for each $uv \in E(H_i)$, either $(\psi(u))(1) = H_i$ or $(\psi(v))(1) = H_i$. By the definition of $\psi$, we have $(\psi(u))(1) = H_{f(u)}$ and $(\psi(v))(1) = H_{f(v)}$. Thus, it suffices to show that for each $H_i \in S$ and for each $uv \in E(H_i)$, either $f(u) = i$ or $f(v) = i$. Let $uv \in E(H_i)$ for some $H_i \in S$. Since $uv \in E(H_i)$, we know that $h^*(uv) = i$. Since $uv$ is an edge in $G$, either $(u, v)$ or $(v, u)$ is an arc in $\vec{G}$. If $(u, v)$ is an arc in $\vec{G}$, then $i = h^*(uv) = h((u, v)) = f(u)$ and $f(v) \neq f(u) = i$ since $f$ is a colouring of $G$. Similarly, if $(v, u)$ is an arc in $\vec{G}$, then $i = h^*(uv) = h((v, u)) = f(v)$ and $f(u) \neq f(v) = i$. Thus, in both cases, either $f(u) = i$ or $f(v) = i$. This proves that $\psi$ is a plain weaving of $S$.

Moreover, the proof of $\psi$ being consistent with $\vec{G}$ is immediate. Recall that $\psi$ is (defined to be) consistent with $\vec{G}$ if for each $H_i \in S$ and each $uv \in E(H_i)$, we have $(\psi(u))(1) = H_i$ if and only if $(u, v)$ is an arc in $\vec{G}$. Note that $(\psi(u))(1) = H_{f(u)}$. Thus, it suffices to show that for each $H_i \in S$ and each $uv$ of $H_i$, we have $f(u) = i$ if and only if $(u, v)$ is an arc in $\vec{G}$. Let $uv \in E(H_i)$, where $i \in \mathbb{Z}_q$. That is, $h^*(uv) = i$. We know by the definition of $h^*$ that $h^*(uv) = f(u)$ if $(u, v)$ is an arc in $\vec{G}$, and $h^*(uv) = f(v) \neq f(u)$ if $(v, u)$ is an arc in $\vec{G}$. That is, $h^*(uv) = f(u)$ if and only if $(u, v)$ is an arc in $\vec{G}$. Hence, $f(u) = i$ if and only if $(u, v)$ is an arc in $\vec{G}$. This proves that $\psi$ is consistent with $\vec{G}$. This completes the proof of Claim 2.

**Claim 3:** For distinct $i, j \in \mathbb{Z}_q$ and distinct $u, v \in V(H_i) \cap V(H_j)$,
$$uv \notin E(G) \text{ and } N_G(u) \cap N_G(v) = N_{\vec{G}}^+(u) \cap N_{\vec{G}}^+(v).$$

Consider distinct $i, j \in \mathbb{Z}_q$ and distinct $u, v \in V(H_i) \cap V(H_j)$. We need to prove that $uv \notin E(G)$ and $N_G(u) \cap N_G(v) = N_{\vec{G}}^+(u) \cap N_{\vec{G}}^+(v)$. Since $H_i$ and $H_j$ are $p$-regular subgraphs of $G$ (see Claim 1) and $u \in V(H_i) \cap V(H_j)$, half of the edges of $G$ incident on $u$ are in $H_i$ and the remaining half of the edges of $G$ incident on $u$ are in $H_j$. Thus, $h^*(e) \in \{i, j\}$ for each edge $e$ incident on $u$ in $G$. Similarly, $h^*(e) \in \{i, j\}$ for each edge $e$ incident on $v$ in $G$. Since $p \geq 2$ and $\vec{G}$ is an Eulerian orientation of a $2p$-regular graph, $u$ has an in-neighbour $u' \neq v$ and an out-neighbour $u'' \neq v$ in $\vec{G}$. Similarly, $v$ has an in-neighbour $v' \neq u$ and an out-neighbour $v'' \neq u$ in $\vec{G}$. Since $uu' \in E(H_i)$ or $uu' \in E(H_j)$, we have $h^*(uu') \in \{i, j\}$. Similarly, $h^*(uu''), h^*(vv'), h^*(vv'') \in \{i, j\}$. Hence, $f(u) = h((u, u'')) = h^*(uu'') \in \{i, j\}$, $f(v) = h((v, v'')) = h^*(vv'') \in \{i, j\}$, $f(u') = h((u', u)) = h^*(uu') \in \{i, j\}$ and $f(v') = h((v', v)) = h^*(vv') \in \{i, j\}$. Thus, we have $f(u), f(v), f(u'), f(v') \in \{i, j\}$.

Next, we show that $uv \in E(G)$ leads to a contradiction. Suppose that $uv \in E(G)$. If $u' \neq v'$, then $u', u, v, v'$ is a 4-vertex path in $G$ bicoloured by $f$ (i.e., coloured using only $i$ and $j$), a contradiction since $f$ is a star colouring of $G$. If $u' = v'$, then $u', u, v$ is a triangle in $G$ bicoloured by $f$, a contradiction since $f$ is a colouring of $G$. Since we have a contradiction in both cases, $uv \notin E(G)$.

Next, we show that $N_G(u) \cap N_G(v) = N_{\vec{G}}^+(u) \cap N_{\vec{G}}^+(v)$. To produce a contradiction, assume that $N_G(u) \cap N_G(v) \neq N_{\vec{G}}^+(u) \cap N_{\vec{G}}^+(v)$. That is, $N_G(u) \cap N_G(v) \supsetneq N_{\vec{G}}^+(u) \cap N_{\vec{G}}^+(v)$, which means that there exists a vertex $w \in N_G(u) \cap N_G(v)$ such that $w \notin N_{\vec{G}}^+(u) \cap N_{\vec{G}}^+(v)$. Since $u, w, v$ is a path in $G$ and $w \notin N_{\vec{G}}^+(u) \cap N_{\vec{G}}^+(v)$, one of the edges $uw$ and $vw$ is oriented away from $w$ by $\vec{G}$. Without loss of generality, assume that $\vec{G}$ orients the edge $uw$ as $(w, u)$. Since $uw$ is an edge of $G$ incident on $u$, we have $h^*(uw) \in \{i, j\}$. Thus, $f(w) = h((w, u)) = h^*(uw) \in \{i, j\}$. Since $p \geq 2$ and $\vec{G}$ is an Eulerian orientation of a $2p$-regular graph, $u$ has at least 2 in-neighbours in $\vec{G}$. Hence, we may suppose that $u'$ and $w$ are distinct in-neighbours of $u$ (i.e., $u' \neq w$). Hence, $u', u, w, v$ is a 4-vertex path in $G$ bicoloured by $f$ (i.e., coloured using only $i$ and $j$), a contradiction since $f$ is a star colouring of $G$. Therefore, by contradiction, $N_G(u) \cap N_G(v) = N_{\vec{G}}^+(u) \cap N_{\vec{G}}^+(v)$. This completes the proof of Claim 3.

To prove the backward direction, suppose that $G$ admits an edge decomposition $S = \{H_0, H_1, \ldots, H_{q-1}\}$ that satisfies the following:

(i) each $H_i$ is $p$-regular $(i \in \mathbb{Z}_q)$;
(ii) $S$ admits a plain weaving $\psi$ consistent with $\vec{G}$ in the following sense:
    for each $H_i \in S$ and for each $uv \in E(H_i)$,
    $(\psi(u))(1) = H_i$ if and only if $(u, v)$ is an arc in $\vec{G}$; and
(iii) for distinct $i, j \in \mathbb{Z}_q$ and distinct $u, v \in V(H_i) \cap V(H_j)$,
    we have $uv \notin E(G)$ and $N_G(u) \cap N_G(v) = N_{\vec{G}}^+(u) \cap N_{\vec{G}}^+(v)$.

Consider the function $f \colon V(G) \to \mathbb{Z}_q$ defined as $f(v) = i$ for each vertex $v$ of $G$ with $(\psi(v))(1) = H_i$. First, we show that $f$ is a $q$-colouring of $G$. Consider an arbitrary edge $uv$ of $G$. Since $S$ is an edge decomposition of $G$, there exists an $H_i \in S$ such that $uv \in E(H_i)$. Since $uv \in E(H_i)$ and $\psi$ is a plain weave of the edge decomposition $S$, we have either $(\psi(u))(1) = H_i$ or $(\psi(v))(1) = H_i$. That is, either $f(u) = i$ or $f(v) = i$. In particular, $f(u) \neq f(v)$. Since $uv$ is an arbitrary edge of $G$, $f$ is a $q$-colouring of $G$.

Next, we show that $\vec{G}$ is a $q$-CEO of $G$ with $f$ as the underlying colouring. Consider an arbitrary vertex $v$ of $\vec{G}$. Since $S$ admits the plain weaving $\psi$ and each $H_i$ is $p$-regular $(i \in \mathbb{Z}_q)$, the vertex $v$ is in exactly two members of $S$, say $H_i$ and $H_j$ (where $i, j \in \mathbb{Z}_q$ and $i \neq j$). As a result, half of the edges of $G$ incident on $v$ are in $H_i$ and the remaining half of the edges of $G$ incident on $v$ are in $H_j$. Let $w_1, w_2, \dots, w_p$ be the other endpoints of the $p$ edges incident on $v$ in $G$ that are in $H_i$, and let $x_1, x_2, \dots, x_p$ be the other endpoints of the $p$ edges incident on $v$ in $G$ that are in $H_j$. Since $S_v = \{H_i, H_j\}$, we know that $(\psi(v))(1) \in \{i, j\}$. Without loss of generality, assume that $(\psi(v))(1) = j$. We know that $\psi$ consistent with $\vec{G}$ in the following sense: for each $H_i \in S$ and for each $ab \in E(H_i)$, we have $(\psi(a))(1) = H_i$ if and only if $(a, b)$ is an arc in $\vec{G}$. Thus, for $1 \leq \ell \leq p$, since $vx_\ell \in E(H_j)$ and $(\psi(v))(1) = j$, it follows that $(v, x_\ell)$ is an arc in $\vec{G}$. On the other hand, for $1 \leq \ell \leq p$, since $vw_\ell \in E(H_i)$ and $(\psi(v))(1) \neq i$, it follows that $(v, w_\ell)$ is not an arc in $\vec{G}$, and thus $(w_\ell, v)$ is an arc in $\vec{G}$. In other words, $w_1, w_2, \dots, w_p$ are in-neighbours and $x_1, x_2, \dots, x_p$ are out-neighbours of $v$ in $\vec{G}$. As a result, $\vec{G}$ is an Eulerian orientation (since $v$ is an arbitrary vertex of $\vec{G}$).

To show that $\vec{G}$ is a $q$-CEO with $f$ as the underlying colouring, it suffices to prove that $f$ satisfies the following:

- no out-neighbour of $v$ has the same colour as an in-neighbour of $v$,
- out-neighbours of $v$ have pairwise distinct colours, and
- all in-neighbours of $v$ have the same colour.

Note that for $1 \leq \ell \leq p$, we have $h^*(vw_\ell) = i$ because $vw_\ell \in E(H_i)$ and $h^*(vx_\ell) = j$ because $vx_\ell \in E(H_j)$. Thus, for $1 \leq \ell \leq p$, we have $f(w_\ell) = h((w_\ell, v)) = h^*(vw_\ell) = i$. To recap, $f(v) = j$ and $f(w_1) = f(w_2) = \cdots = f(w_p) = i$. In particular, all in-neighbours of $v$ in $\vec{G}$ have the same colour under $f$.

Clearly, $v \in V(H_i) \cap V(H_j)$ because $vw_1 \in E(H_i)$ and $vx_1 \in E(H_j)$.

Next, we show that no out-neighbour of $v$ has the same colour as an in-neighbour of $v$. To produce a contradiction, assume the contrary. That is, there exists an in-neighbour, say $w_1$, of $v$ and an out-neighbour, say $x_1$, of $v$ in $\vec{G}$ such that $f(w_1) = f(x_1)$. Thus, $f(x_1) = f(w_1) = i$. Since $\vec{G}$ is an Eulerian orientation, $x_1$ has an out-neighbour $y$ in $\vec{G}$. Clearly, $h^*(x_1y) = h((x_1, y)) = f(x_1) = i$, and thus $x_1y \in E(H_i)$. Since $x_1y \in E(H_i)$ and $vx_1 \in E(H_j)$, we have $x_1 \in V(H_i) \cap V(H_j)$. We know that $v \in V(H_i) \cap V(H_j)$. Thus, $v, x_1 \in V(H_i) \cap V(H_j)$ and $vx \in E(G)$, a contradiction to Property (iii) of $S$. Thus, by contradiction, no out-neighbour of $v$ has the same colour as an in-neighbour of $v$.

Finally, we show that out-neighbours of $v$ have pairwise distinct colours. To produce a contradiction, assume the contrary. That is, two out-neighbours of $v$, say $x_1$ and $x_2$, have the same colour under $f$. That is, $f(x_1) = f(x_2) = k$ for some $k \in \mathbb{Z}_q$. Since $\vec{G}$ is an Eulerian orientation, $x_1$ has an out-neighbour $y_1$ and $x_2$ and an out-neighbour $y_2$ in $\vec{G}$. Observe that $h^*(x_1y_1) = h((x_1, y_1)) = f(x_1) = k$; that is, $x_1y_1 \in E(H_k)$ and thus $x_1 \in V(H_i) \cap V(H_k)$. Similarly, $h^*(x_2y_2) = k$; thus, $x_2y_2 \in E(H_k)$ and consequently, $x_2 \in V(H_i) \cap V(H_k)$. Vertices $x_1$ and $x_2$ have a common neighbour $v$ in $G$ which is not a common out-neighbour of $x_1$ and $x_2$ in $\vec{G}$; that is, $N_G(x_1) \cap N_G(x_2) \supsetneq N_{\vec{G}}^+(x_1) \cap N_{\vec{G}}^+(x_2)$. Since $x_1, x_2 \in V(H_i) \cap V(H_k)$ and $N_G(x_1) \cap N_G(x_2) \neq N_{\vec{G}}^+(x_1) \cap N_{\vec{G}}^+(x_2)$, we have a contradiction to Property (iii) of $S$. Thus, by contradiction, out-neighbours of $v$ have pairwise distinct colours.

Therefore, $\vec{G}$ is indeed a $q$-CEO of $G$ with $f$ as its underlying colouring. □

Since a $2p$-regular graph $G$ with $p \geq 2$ is $(p + 2)$-star colourable if and only if $G$ admits a $(p + 2)$-CEO [14], we have the following.

**Theorem 6.** *Let $G$ be a $2p$-regular graph with $p \geq 2$. Then, $G$ admits a $(p+2)$-star colouring if and only if $G$ admits an orientation $\vec{G}$ and an edge decomposition $S = \{H_0, H_1, \ldots, H_{p+1}\}$ such that the following hold:*

*(i)   each $H_i$ is $p$-regular ($i \in \mathbb{Z}_{p+2}$);*
*(ii)  $S$ admits a plain weaving $\psi$ consistent with $\vec{G}$; and*
*(iii) for distinct $i, j \in \mathbb{Z}_{p+2}$ and distinct $u, v \in V(H_i) \cap V(H_j)$,*
       *$uv \notin E(G)$ and $N_G(u) \cap N_G(v) = N_{\vec{G}}^+(u) \cap N_{\vec{G}}^+(v)$.*                □

## 4   Conclusion

For $d \geq 2$, at least $\lceil (d+4)/2 \rceil$ colours are required to star colour a $d$-regular graph, and this bound is tight [14]. When $d$ is even, the class of $d$-regular $\lceil (d + 4)/2 \rceil$-star colourable graphs is characterised in terms of graph orientations [14], graph homomorphisms [3], and in this paper, in terms of weaving patterns of edge decompositions. This motivates the following problem.

*Problem 1:* Characterise $d$-regular $\lceil (d + 4)/2 \rceil$-star colourable graphs for odd $d$.

We remark that the tools required to answer the above problem could be useful to answer a slightly more general problem.

*Problem 2:* Characterise $d$-regular $\lceil (d + 5)/2 \rceil$-star colourable graphs.

Weaving patterns of edge decompositions merit further study in at least two directions:

(i)  Is it possible to characterise other graph-theoretic notions, such as cycle decompositions, in terms of weaving patterns?

(ii)  Characterise graphs that has an edge decomposition $S$ which admits a simple weaving pattern (such as plain weave or 2-1 twill weave) and also satisfies some basic constraints, such as distance constraints between members of $S$?

**Acknowledgement.** We thank three anonymous referees for their careful reading and valuable suggestions.

# References

1. Adanur, S.: Handbook of Weaving. CRC Press, Boca Raton (2020). https://doi.org/10.1201/9780429135828

2. Akleman, E., Chen, J., Gross, J.L.: Extended graph rotation systems as a model for cyclic weaving on orientable surfaces. Discret. Appl. Math. **193**, 61–79 (2015). https://doi.org/10.1016/j.dam.2015.04.015

3. Antony, C.: The complexity of star colouring and its relatives. Ph.D. thesis, Indian Institute of Information Technology, Design & Manufacturing, (IIITDM) Kancheepuram, Chennai, India (2023). https://doi.org/10.13140/RG.2.2.28192.66561

4. Bagga, J.S., Beineke, L.W.: A survey of line digraphs and generalizations. DML Discrete Math. Lett. **6**, 68–83 (2021). https://doi.org/10.47443/dml.2021.s109

5. Beineke, L.W., Bagga, J.S.: Line Graphs and Line Digraphs, Developments in Mathematics, vol. 68. Springer, Cham (2021). https://doi.org/10.1007/978-3-030-81386-4

6. Borodin, O.V.: Colorings of plane graphs: a survey. Discret. Math. **313**(4), 517–539 (2013). https://doi.org/10.1016/j.disc.2012.11.011

7. Fiala, J., Paulusma, D., Telle, J.A.: Locally constrained graph homomorphisms and equitable partitions. Eur. J. Comb. **29**(4), 850–880 (2008). https://doi.org/10.1016/j.ejc.2007.11.006

8. Fiala, J., Kratochvíl, J.: Locally constrained graph homomorphisms - structure, complexity, and applications. Comput. Sci. Rev. **2**(2), 97–111 (2008). https://doi.org/10.1016/j.cosrev.2008.06.001

9. Gebremedhin, A.H., Manne, F., Pothen, A.: What color is your Jacobian? Graph coloring for computing derivatives. SIAM Rev. **47**(4), 629–705 (2005). https://doi.org/10.1137/S0036144504444711

10. Hu, S.: A topological theory of weaving and its applications in computer graphics. Ph.D. thesis, USA (2013). aAI3607499

11. Lei, H., Shi, Y.: A survey on star edge-coloring of graphs. Adv. Math. **50**(1), 77–93 (2021)

12. Nešetřil, J., de Mendez, P.O.: Colorings and homomorphisms of minor closed classes. In: Aronov, B., Basu, S., Pach, J., Sharir, M. (eds.) Discrete and Computational Geometry. Algorithms and Combinatorics, vol. 25, pp. 651–664. Springer, Heidelberg (2003). https://doi.org/10.1007/978-3-642-55566-4_29

13. Ren, Y., Panetta, J., Chen, T., Isvoranu, F., Poincloux, S., Brandt, C., Martin, A., Pauly, M.: 3D weaving with curved ribbons. ACM Trans. Graph. **40**(4), 127 (2021). https://doi.org/10.1145/3450626.3459788

14. Shalu, M.A., Antony, C.: Star colouring of bounded degree graphs and regular graphs. Discret. Math. **345**(6), 112850 (2022). https://doi.org/10.1016/j.disc.2022.112850
15. West, D.B.: Introduction to Graph Theory, 2nd edn. Prentice Hall, Upper Saddle River (2001)

# Author Index

**A**
Akram, Waseem   3
Antony, Cyriac   313

**B**
Benkoczi, Robert   133
Bhattacharya, Bhaswar B.   77
Bianchi, Silvia M.   255
Bishnu, Arijit   88

**C**
Chakraborty, Dipayan   255
Changat, Manoj   148, 286

**D**
Dalal, Aseem   270
Das, Gautam Kumar   117
Das, Sandip   77
De, Koustav   14
Dey, Palash   14
Divya, D.   209

**F**
Foucaud, Florent   29
Francis, Mathew   88

**G**
Gaur, Daya   133
Gorain, Barun   44
Govindarajan, Sathish   103

**H**
Hafshejani, Sajad Fathi   133
Hellmuth, Marc   148

**I**
Islam, Sk. Samim   77

**J**
Jacob, Jeny   286

**K**
Kaur, Tanvir   44
Kavaskar, T.   302
Kirubakaran, V. K.   239

**L**
Lucarini, Yanina   255

**M**
Mahendra Kumar, R.   194
Majumder, Pritam   88
Marcille, Pierre-Marie   29
Misra, Neeldhara   14
Mittal, Harshil   14
Mittal, Rajat   59
Mondal, Joyashree   224
Mondal, Kaushik   44
Myint, Zin Mar   29

**N**
Nair, Sanjay S.   59

**P**
Panda, B. S.   270
Pandey, Arti   179
Patro, Sunayana   59
Paul, Kaustav   179

**R**
Rout, Sasmita   117

**S**
Sadagopan, N.   194
Sandeep, R. B.   29
Sarkar, Siddhartha   103

S. Kalyanasundaram and A. Maheshwari (Eds.): CALDAM 2024, LNCS 14508, pp. 329–330, 2024.
https://doi.org/10.1007/978-3-031-52213-0

Saxena, Sanjeev    3
Sen, Sagnik    29
Sen, Saumya    77
Shalu, M. A.    239, 313
Shanavas, Ameera Vaheeda    148
Sheela, Lekshmi Kamal K.    286
Srivastava, Pranjal    162
Stadler, Peter F.    148
Sukumaran, Sreelakshmi    302

**T**
Taruni, S.    29
Thakkar, Dhara    162

**V**
Vijayakumar, S.    209, 224

**W**
Wagler, Annegret K.    255

Printed in the United States
by Baker & Taylor Publisher Services